Nanostructure Science and Technology

Series editor

David J. Lockwood, FRSC
National Research Council of Canada
Ottawa, Ontario, Canada

More information about this series at http://www.springer.com/series/6331

Simona E. Hunyadi Murph
George K. Larsen · Kaitlin J. Coopersmith
Editors

Anisotropic and Shape-Selective Nanomaterials

Structure-Property Relationships

Springer

Editors
Simona E. Hunyadi Murph
Savannah River National Laboratory
Aiken, SC
USA

and

University of Georgia
Athens, GA
USA

George K. Larsen
Savannah River National Laboratory
Aiken, SC
USA

Kaitlin J. Coopersmith
Savannah River National Laboratory
Aiken, SC
USA

ISSN 1571-5744 ISSN 2197-7976 (electronic)
Nanostructure Science and Technology
ISBN 978-3-319-59661-7 ISBN 978-3-319-59662-4 (eBook)
DOI 10.1007/978-3-319-59662-4

Library of Congress Control Number: 2017943180

Printed on acid-free paper

This Springer imprint is published by Springer Nature
The registered company is Springer International Publishing AG
The registered company address is: Gewerbestrasse 11, 6330 Cham, Switzerland

Preface

This is one of the first books to provide an overview of uniquely shaped nanomaterials (anisotropic) (Part I) and their ongoing state-of-the-art applications in different fields (Part II). The reader will become familiar with a variety of top–down and bottom–up approaches to prepare nanoparticles with exquisite geometries and properties. The challenges and benefits associated with each method will allow the reader to select the best preparation approach for the targeted application based on the sample's composition, size, shape, scalability, and cost. The current characterization techniques used to evaluate nanomaterials' physico-chemical properties and how these properties could lead to diverse applications in different fields will be described in detail. This book provides an overview of the effect of the morphology and the nanometric dimension of materials on their physico-chemical properties and how this can lead to novel applications (Part II). Specifically, this book highlights current and emerging applications of nanomaterials, including sensing and imaging, catalysis, biological and environmental implications, plasmonics, energy, and national security missions. Some of the best-known, highly experienced, and well-published experts in the field will describe the most recent developments in their laboratories.

This book is designed to inform readers who are seeking a broader perspective related to shape-selective nanomaterials science and technologies. Technical readers (researchers, scientists, engineers, business developers, technology managers) will stay abreast of the latest state-of-the-art developments in nanotechnology with focus on the broad applications of these technologies at the laboratory and/or industrial scale. General readers will get the benefit of seeing the big picture of anisotropic nanostructures and a greater understanding as to why there is so much excitement in this area. Advanced readers will broaden their knowledge base and benefit from an update on the latest characterization methods and applications in the field.

It is our hope that this book will provide a detailed, interesting, and inspiring perspective of nanotechnology (particularly anisotropic nanostructures) and encourage readers to further explore this emerging area.

Aiken, USA

Simona E. Hunyadi Murph
George K. Larsen
Kaitlin J. Coopersmith

Contents

About the Editors

Prof. Dr. Simona E. Hunyadi Murph is an internationally recognized expert in the fields of nanoscience and nanotechnology. Dr. Murph is a Principal Scientist in the National and Homeland Security Directorate at Savannah River National Laboratory (SRNL) and an Adjunct Professor in the Physics and Astronomy Department at The University of Georgia (UGA), USA. She is the founder and manager of SRNL's *Group for Innovation and Advancements in NanoTechnology Sciences* (GIANTS) program that has the mission of developing young scholars' knowledge and skills in the many fields of nanoscience. Research in her group is directed toward the advancement of the field of nanotechnology by designing and controlling fabrication of colloidal materials with functional properties for sensing and imaging. These tunable nanomaterials are also being explored for environmental remediation, photothermal and magneto-thermal applications, energy conversion and storage and various catalytic processes. The remarkable advances made by Dr. Murph and her team in the field of Nanotechnology led to numerous publications, awards, grants, invention disclosures, patents, and recognitions. She holds a Doctor of Philosophy (Ph.D.) in Chemistry/Nanotechnology from the University of South Carolina, USA, an Education Specialist/Educational Leadership (EdS) degree from Augusta University, USA, and both a Master of Science (MS) in Chemistry and Bachelor of Science (BS) in Chemistry/Physics with a minor in Education from Babes-Bolyai University in Romania.

Dr. George K. Larsen is a Senior Scientist at Savannah River National Laboratory (SRNL). After receiving a B.A. and B.S. in Philosophy and Physics from Piedmont College, Dr. Larsen attended The University of Georgia, earning a Ph.D. in Physics. Dr. Larsen's research concentrates on the properties and applications of nanostructures, covering a range of topics from clean energy to the electrical properties of tilted nanorod arrays. At SRNL, Dr. Larsen's background in nanotechnology has been applied in the nuclear field through work involving the exploitation of nanostructures for remote heat generation, photocatalysis, hydrogen production, and radiation safety, among others.

Dr. Kaitlin J. Coopersmith is Senior Scientist at Savannah River National Laboratory in Aiken, SC. She received her Ph.D. in Chemistry from Syracuse University and her B.S. in Chemistry from the State University of New York at Potsdam. Her research interests include the synthesis and functionalization of metal and semiconductor nanoparticles for sensing, drug delivery, energy transfer, gas adsorption, and alternative heating mechanisms.

Part I
Introduction and Fundamentals

Chapter 1
An Introduction to Nanotechnology

Simona E. Hunyadi Murph

> *To understand the very large, we must understand the very small.*
>
> Democritus

Abstract Globalization of scientific knowledge and technological advances are sparking innovation and creativity across many fields at an unprecedented rate. Ground-breaking discoveries made in the mid-1980s, namely the development of scanning tunneling microscopy and the discovery of buckminsterfullerene, influenced scientists to envision the world at the atomic level and new paradigms emerged: nanoscience and nanotechnology. Through the manipulation of matter at the atomic level, today scientists can create novel materials with unique properties and functionalities. These new materials enable innovative technologies and applications across many fields from engineering to medicine.

Globalization of scientific knowledge and technological advances is sparking innovation and creativity across all fields at an unprecedented rate. Nearly every aspect of science and industry is driven to make advances in the (bio)medical fields, computing and electronics, environmental controls and remediation, transportation, energy production, chemical manufacturing, agriculture, and consumer products.

The technological revolution [1] that started decades ago with the introduction of electronic devices and silicon-based integrated circuitry [2] changed humanity forever. The information technology insurgency that emerged with the introduction of internet/broadband, personal computers, mobile phones, and email [3] created a global multi-dimensional world. These technologies re-defined the way we live, communicate, travel and experience the world. This burst of technological developments offered unprecedented opportunities for rapid social and economic progress in our society [4].

S.E. Hunyadi Murph (✉)
National Security Directorate, Savannah River National Laboratory, Aiken, SC, USA
e-mail: Simona.Murph@srnl.doe.gov

S.E. Hunyadi Murph
Department of Physics and Astronomy, University of Georgia, Athens, GA, USA

S.E. Hunyadi Murph et al. (eds.), *Anisotropic and Shape-Selective Nanomaterials*,
Nanostructure Science and Technology, DOI 10.1007/978-3-319-59662-4_1

Over the years, creation of smaller and smaller optical, mechanical, and electronic products and devices thrived, leading to energy efficient, lighter, and easier to operate devices. Today, the miniaturization revolution continues and scientists are manipulating the world at its smallest level: the atomic level. It was the physicist, Richard Feynman, who introduced this concept for the first time in 1959 in his famous talk "There's Plenty of Room at the Bottom". He speculated that one day, it may be possible to print all 24 volumes of Encyclopedia Brittanica on the head of a stick pin [5]. Feynman envisioned building circuits on the scale of nanometer that could be used as elements in more powerful computers and predicted the presence of nanostructures in biological systems. In 1965, he won the Nobel Prize jointly with Julian Schwinger and Sin-Itiro Tomonaga for their contributions to the development of quantum electrodynamics [6].

Two major breakthrough discoveries made in the mid-1980s opened the doors for the next generation of technological advancements in this area. It started with the development of scanning tunneling microscope (STM) in 1981, which makes it possible to image surfaces at the atomic level. This discovery, created by Gerd Binnig and Heinrich Rohrer, granted them the Nobel Prize in Physics in 1986 [7]. A few years after the development of the scanning electron microscope (SEM), in 1985, Richard Smalley and Bob Curl were able to create and isolate the first nanomaterial, named buckminsterfullerene ("bucky balls"). The buckyball molecule is a spherical fullerene molecule with the formula C_{60}. Kroto, Curl and Smalley were awarded the 1996 Nobel Prize in Chemistry for their roles in the discovery of buckminsterfullerene and the related class of molecules, the fullerenes [8].

These ground-breaking discoveries influenced scientists to envision the world at the atomic level and new paradigms emerged: nanoscience and nanotechnology. A nanometer, which is derived from the Greek prefix -nano, meaning "dwarf", is one-billionth (10^{-9}) of a meter (Fig. 1.1). Scientists associate the word "nano" to materials, devices, objects that have at least one dimension on the 1–100 nm scale. By manipulating matter at the atomic levels, scientists create new materials with unique properties and functionalities that enable new technologies and applications across many fields from engineering to medicine. A wide variety of new nanoscale building blocks have emerged in the last decades, including fullerenes, metal and semiconductor nanoparticles, and various designer molecules.

As a matter of fact, in 2013, a team of Stanford engineers built a basic computer using carbon nanotubes [9], which has the potential to launch a new generation of electronic devices that run faster, while using less energy, than those made from silicon chips. In the medical field, scientists demonstrated that gold nanoparticles can be used for the destruction of cancer cells in vitro and in vivo in mice [10]. Pilot clinical studies on human patients with refractory and/or recurrent tumors of the head and neck are underway [11]. Currently, scientists can create nanomaterials that are stronger than Kevlar, which is used in bulletproof vests, and substances 100 times stronger than steel [12]. They are able to create fluorescent materials that are one thousand times brighter than conventional dyes or dyes used to image the inside of the human body [13]. The possibilities appear endless. The possibility of finding

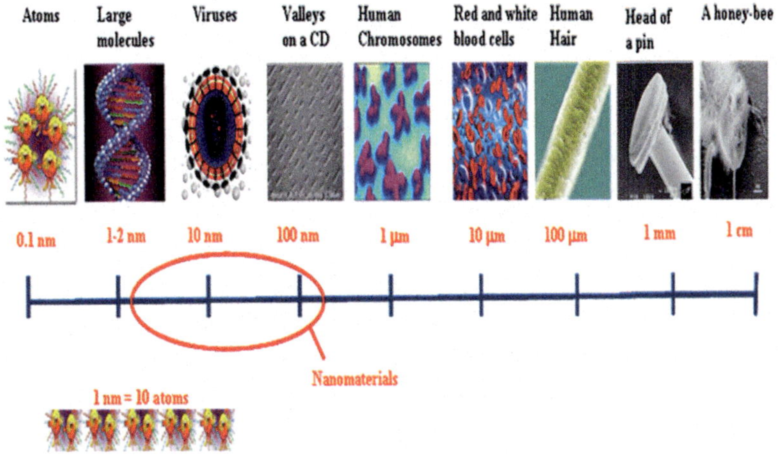

Fig. 1.1 Nano- and micro- meter scale of "things"

new and intriguing substances that can cure disease, help us explore and/or live in space, make us live longer and healthier lives, increase communications between people around the world, and discover new energy supplies is exhilarating.

References

1. Rheingold, H. 2000. *Tools for Thought: The History and Future of Mind-Expanding Technology* (p. 360). MA: MIT Press.
2. Peercy, P.S. 2000. The Drive to Miniaturization. *Nature* 406: 1023–1026.
3. MacMillan, I., et al. 2009. *Top 30 Innovations Of The Last 30 Years*. Available from: http://knowledge.wharton.upenn.edu/article/a-world-transformed-what-are-the-top-30-innovations-of-the-last-30-years/.
4. Wolf, M. 2014. Shaping Globalization. *Finance and Development* 51 (3): 22–25.
5. Hunyadi, S.E. 2007. Nanoengineered Materials: Synthesis, Design, Functionalization and Chemical Sensing Applications. In *Department of Chemistry and Biochemistry*, p. 262. Columbia, SC: University of South Carolina.
6. The Nobel Prize in Physics 1965. 2014. Available from: http://www.nobelprize.org/nobel_prizes/physics/laureates/1965/. 16 Feb 2017
7. The Nobel Prize in Physics 1986. 2014. Available from: http://www.nobelprize.org/nobel_prizes/physics/laureates/1986/.
8. Kroto, H.W., et al. 1985. C_{60}: Buckminsterfullerene. *Nature* 318: 162–163.
9. Shulaker, M.M., et al. 2013. Carbon Nanotube Computer. *Nature* 501: 526–530.
10. Park, J.H., et al. 2008. Micellar hybrid nanoparticles for simultaneous magnetofluorescent imaging and drug delivery. *Angewandte Chemie International Edition* 47: 7284–7288.
11. Echarri, M.J., A. Lopez-Martin, and R. Hitt. 2016. Targeted Therapy in Locally Advanced and Recurrent/Metastatic Head and Neck Squamous Cell Carcinoma. *Cancers* 8 (3): 27.
12. Baniasadi, M., et al. 2015. High-Performance Coils and Yarns of Polymeric Piezoelectric Nanofibers. *ACS Applied Materials & Interfaces* 7 (9): 5358–5366.
13. Mérian, J., et al. 2012. Fluorescent Nanoprobes Dedicated to in vivo Imaging: From Preclinical Validations to Clinical Translation. *Molecules* 17 (5): 5564–5591.

Chapter 2
Nanoscale Materials: Fundamentals and Emergent Properties

Simona E. Hunyadi Murph, Kaitlin J. Coopersmith and George K. Larsen

Abstract As material size decreases into the nano size regime, novel properties arise that are different from their molecular and bulk counterparts. Due to the size and shape effects in this regime, a nanoparticle's morphology has a profound effect on its properties. This chapter addresses the effect of dimensionality on the optical, electronic, chemical, and physical assets of various nanomaterials and how physical and chemical relationships can be exploited to improve their properties. Delving into the nuances of the different sizes, shapes, and compositions gives one an appreciation of the potential that nanomaterials have to improve upon today's technologies. As scientists learn to fabricate increasingly more complex nanomaterials, new opportunities develop every day. A detailed discussion on the effect of morphology and nanometric dimensions on materials' physico-chemical properties, which lead to novel applications, will be covered in Chapter 5.

Keywords Anisotropy · Dimensionality · Shape-selective nanomaterials · Polarization · Crystalline anisotropy · Optical properties · Electronic properties · Chemical · Physical properties

2.1 Introduction

The design and controlled fabrication of colloidal materials with functional properties has flourished over the last few decades. The beauty and distinctiveness of nanoscale materials is rooted in their unique properties that emerge at the 1–100 nm scale. In this transitional regime, a material's physical, chemical, and biological properties may differ in fundamental ways from the properties of both bulk matter and the constituent atoms or molecules.

S.E. Hunyadi Murph (✉) · K.J. Coopersmith · G.K. Larsen
National Security Directorate, Savannah River National Laboratory, Aiken, SC, USA
e-mail: Simona.Murph@srnl.doe.gov

S.E. Hunyadi Murph
Department of Physics and Astronomy, University of Georgia, Athens, GA, USA

S.E. Hunyadi Murph et al. (eds.), *Anisotropic and Shape-Selective Nanomaterials*, Nanostructure Science and Technology, DOI 10.1007/978-3-319-59662-4_2

For decades, a staggering amount of research has focused on the creation of new nanomaterials and the elucidation of their unique property-structure correlations. Theoretical understanding of the mechanistic principles that govern the novel properties of the nanomaterials have also been studied extensively. Highly reliable bottom-up and top-down synthetic routes that produce increasingly complex nanomaterials with highly ordered and complex geometries have been developed. Complications in scaling up and uniformity that have hampered widespread application of nanomaterials in the early developments have been addressed to some extent. High throughput fabrication procedures amenable to scaling up for industrial scale implementation have been reported for many classes of nanoparticles.

A large variety of nanomaterials of various sizes (1–100 nm), shapes (including spheres, rods, cubes, triangles, wires, ribbons, nanotubes, and polygons), and compositions (fullerenes, to metal and semiconductor nanoparticles, to various designer molecules or a combination of both) with an array of different functionalities have been reported. Thorough discussions of these synthesis procedures are covered in detail in the next chapters. In this chapter, readers will be introduced to the fundamentals of inorganic nanostructures with an emphasis on noble metals, namely gold, silver and platinum materials. A more detailed discussion on fundamentals of other types of nanomaterials (metal oxides, other metals, hybrids and quantum dots) and how their properties affect their applications will be covered in Chap. 5.

In recent years, incorporation of nanomaterials in macro-devices opened the door to potential applications in nearly every sector of science and industry, particularly in medicine, biological sciences, electronic sensors, computing and microelectronics, environmental controls and remediation, transportation, energy production, chemical manufacturing, agriculture, and consumer products. According to the USA Environmental Protection Agency, nanomaterials can be found today in more than 1300 commercial products including medical equipment, textiles, fuel additives, cosmetics, plastics and more [1]. Nanomaterials are currently produced in metric tons per year and are expected to exponentially increase as new advances emerge [2, 3]. Scientists are relentlessly pursuing different routes to create nanomaterials and understand the physical and chemical relationship between composition, size, and morphology to incorporate nanomaterials into everyday life.

2.1.1 Dimensionality and Optical Properties

The unique properties of inorganic nanoparticles are often the result of their sizes being smaller than the mean free path of their electrons (10–100 nm) or the average distance an electron travels before being scattered (scattering length) [4–6]. At this scale, electrons are confined in one, two, or three dimensions, and quantum mechanical effects dominate the characteristics of matter [7, 8]. As a result, the physical and chemical properties of materials vary with nanoparticle size and shape, as demonstrated in Table 2.1 [4–9].

Table 2.1 Size and shape effects on gold nanoparticle's optical properties

Gold nanoparticles color	Shape	Size (nm)	λ (plasmon bands) (nm)
Brown	Sphere	1.5	520
Ruby-red	Sphere	20	525
Purple	Rod	D = 20 L = 40	520 580
Blue	Rod	D = 20 L = 57	520 620
Green	Rod	D = 20 L = 75	520 720
Brown	Rod	D = 20 L = 100	520 900

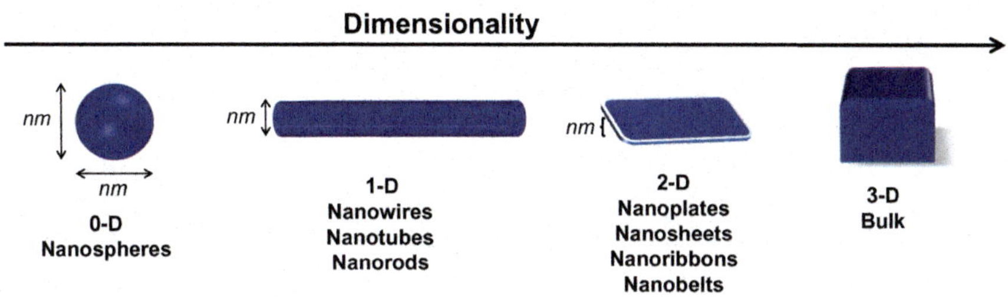

Fig. 2.1 Nanoparticle shape and dimensionality

Fundamentally, *isotropic* nanomaterials have properties and functionalities that do not depend on spatial orientation, while *anisotropic* nanomaterials have properties and functionalities that are determined by their orientation within the x, y, and z dimensions. These properties are directionally dependent on material geometry and orientation. Depending on these quantum confinement properties, nanoparticles are characterized as zero, one, two or three dimensional as demonstrated in Fig. 2.1.

All of the dimensions in zero dimensional nanostructures, or spheres, are in the nanometer size regime, so there is no electron delocalization. One dimensional structures have one dimension outside of the nanometer size regime that creates electron delocalization in one direction; the elongated axis allows for an increase in polarization and charge separation [7, 10]. This class of structure includes nanorods, nanowires, nanotubes, nanoribbons, and nanobelts [11]. Two dimensional nanoparticles are sheets or nano sized films that have two dimensions outside of the nanometer size regime. In these structures, electrons are confined in the thickness direction but not in the plane of the sheet. Three dimensional materials, or bulk materials, are not confined to the nanoscale in any direction; however, they may contain nanocrystalline structures or features that lead to interesting properties. This distinctive spatial behavior of anisotropic structures creates unexpected properties and applications that are not available in isotropic nanostructures. Ultimately, material properties can be tailored to a specific design and functionality by

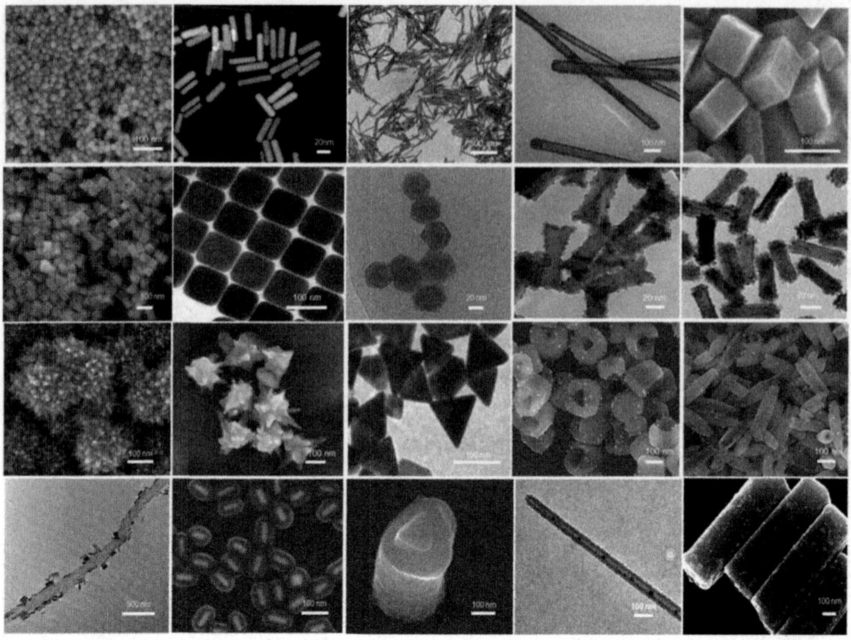

Fig. 2.2 SEM images showing a sample of the wide variety of nanoparticle shapes and sizes that can be made in our laboratory

controlling the geometry and structure of the nanometer scale structures [4–9, 12, 13]. Figure 2.2 shows a variety of different anisotropic structures that have been created in our laboratory.

Nanoparticles have high surface to volume ratios that lead to an increased fraction of atoms on the surface and fewer neighboring atoms [12]. This promotes distinctive structural and electronic changes. For example, a cube with a length of 1 m has a surface area of 6 m^2, while if this same cube was divided into a collection of smaller length cubes, each with lengths of 0.25 m, the surface area would increase to 24 m^2. Moreover, the atoms in the interior of a nanoparticle are more highly coordinated, and are therefore more stable than those at the nanoparticle surface [13, 14]. Subsequently, atoms on tips or corners of a crystal have more uncoordinated bonds and are more reactive than the edge or in-plane surface atoms. By simply decreasing the particle size, one could increase the number of surface atoms available for surface chemical reactions [12, 13]. The larger fractions of atoms at the particle surface of nanometer-sized materials promote faster reaction kinetics for various catalytic reactions. While gold is inert in its bulk form, as a nanoparticle of $\sim$3 nm, it catalyzes CO to CO_2 [6, 7, 15, 16]. Hollow nanoparticles serve as superior catalysts to their solid counterparts with the advantages of low density, increased surface area, and cost effective materials [12, 17].

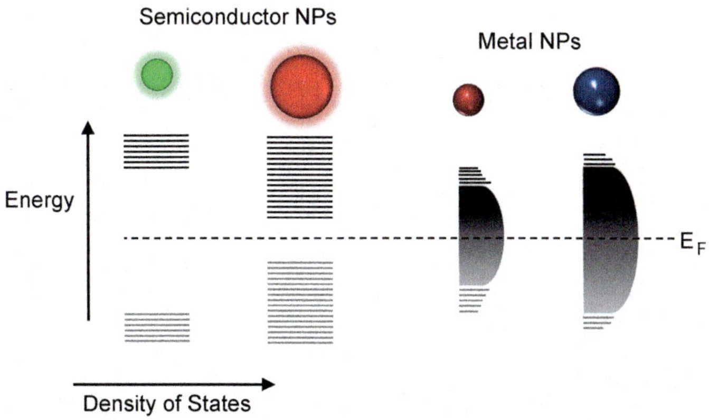

Fig. 2.3 Density of states and the location of the Fermi level (E_F) for semiconductor and metal nanocrystals

Since properties at the nanoscale are dictated by quantum mechanics, different architectures give rise to altered optical and electronic properties, which are affected by the degree of charge carrier quantum confinement, surface charge distribution and electron polarization along the various axes of the particles [18–21]. Due to the high surface to volume ratio, small surface changes, such as the binding or altering of binding events, can also lead to a significant change in the electrical and optical properties. This can lead to single molecule sensing [22].

Semiconductor nanocrystals, or quantum dots, can absorb and fluoresce in the ultraviolet to near-infrared regions of the electromagnetic spectrum. The optical properties of quantum dots are tailored by tuning the band gap, which is dependent on composition and size, as shown in Fig. 2.3. As the size increases, more atomic orbitals contribute to their density of states, leading to a smaller bandgap and a red-shift in absorption and emission. Asymmetry plays a large role in quantum dot optical properties, where photogenerated carriers can separate along longer axes, leading to a longer lifetime in the excited state [23, 24].

Unlike quantum dots, metallic nanoparticles have a continuous energy band that is responsible for their light absorption properties. Among many properties displayed by gold nanoparticles, one of the most notable is their ability to absorb and scatter light in the visible region of the electromagnetic spectrum [4–9, 12, 13, 25] as displayed in Fig. 2.4.

In metal nanoparticles, the number of the surface plasmon resonance (SPR) peaks generally increase as the symmetry of the particle decreases due to the increased polarization of the free electrons and surface charge distribution [18–20, 26]. For example, gold nanorods (AuNR) have two localized surface plasmon resonance (LSPR) bands due to the different absorption of the length and the width; [4–9, 12, 13, 25] the locations of these bands are tuned by changing the length and width of the rod [26]. Silver nanocubes can display multiple plasmon bands due to the accumulation of charges in the corners of the cube [20]. The plasmon band is

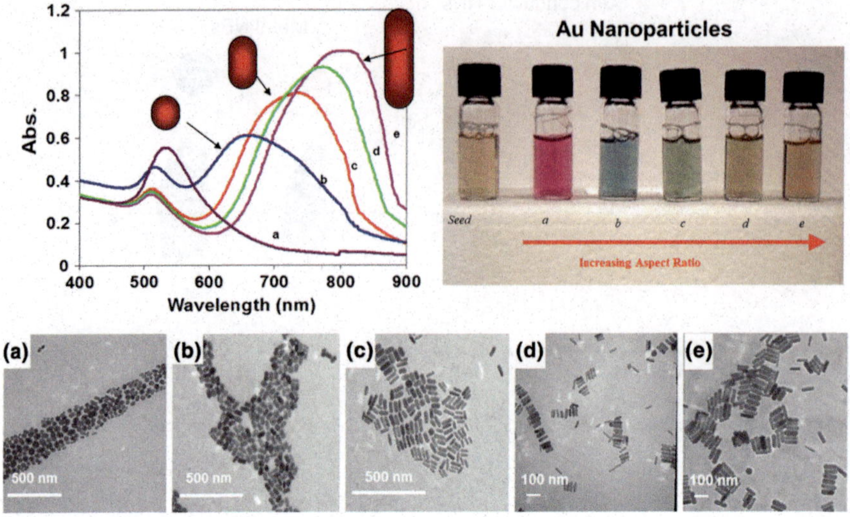

Fig. 2.4 Shape dependence of the optical properties of gold nanostructures. Transmission electron micrographs (*top*), optical spectra (*left*), and photographs of (*right*) aqueous solutions of gold nanorods of various aspect ratios. Seed sample: aspect ratio 1; sample **a**, aspect ratio ~1.4; sample **b**, aspect ratio ~2; sample **c**, aspect ratio ~3; sample **d**, aspect ratio ~3.50; sample **e**, aspect ratio ~4.4. *Scale bars* 500 nm for (**a**, **b**, **c**), and 100 nm for (**d**, **e**). Reprinted with permission from Ref. [6]. Copyright 2017 American Chemical Society

highly dependent on size, shape and refractive index, which have applications for LSPR sensing [4–9, 12, 13, 25, 27–30].

Nanoparticle shape dictates how sensitive the shift in the plasmon band is [25, 28] to changes in refractive index at the surface (Fig. 2.5), which can be exploited to detect the presence of most inorganic, organic and biological molecules and surface synthesis, catalysis and molecular binding events [6, 7, 9, 28, 31, 32]. These asymmetric properties can lead to the creation of more sensitive sensors, [32, 33] faster electronics, [34] more efficient energy storage devices [35] and better catalysts [12–14, 17, 36]. For intracellular delivery and imaging applications, [9, 27, 30] the kinetics of cellular uptake and the effect of nanoparticles on cellular functions is dependent on nanoparticle shape [37, 38].

2.1.2 *Polarization and Anisotropy*

Since most of the nanoparticles properties are highly surface dependent, morphology, composition and dimensionality have a profound effect on how nanoparticles interact with light [7, 25]. The sharp features of anisotropic shapes

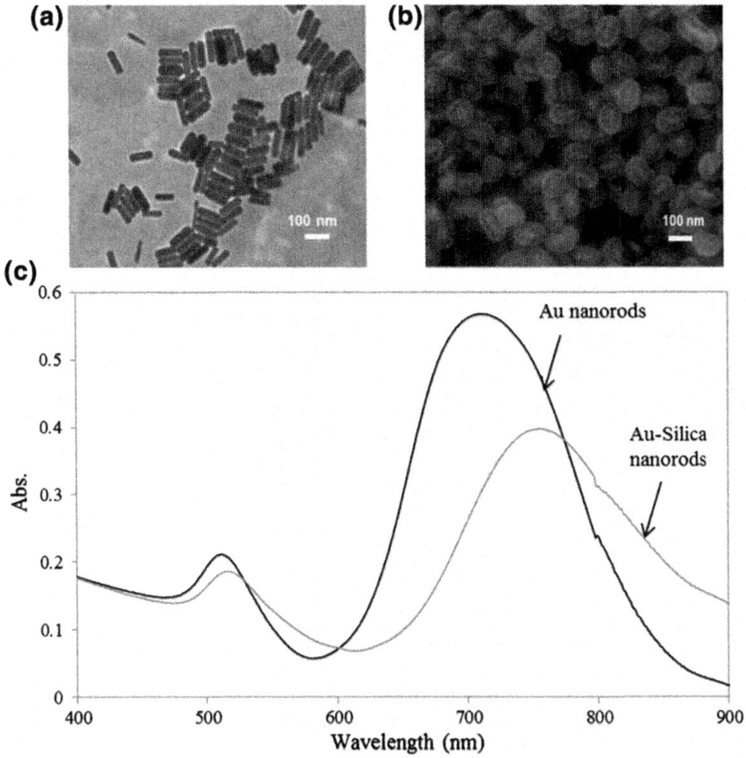

Fig. 2.5 **a** TEM image of Au nanorods, **b** SEM image of Au-silica core-shell nanorods, and **c** UV-Vis spectra of gold nanorods before and after coating with a silica shell

lead to locally enhanced electric fields that can act as "hot spots" for enhanced detection, light emitting diodes, and imaging [7, 28, 29]. Optical and magnetic properties are also controlled by the nanoparticle's anisotropy. Particles with an elongated axis, such as nanorods, have an increased charge separation that leads to an increase in catalysis and sensing efficiencies [39]. Anisotropic particles can polarize light. The degree and localization of the different planes of light is determined by the dimensions and morphologies, where the different light angles interact with the particles differently [21, 40–44]. In metal nanorods, the polarized light interacts with the longitudinal surface plasmon band, [44] whereas in spiky gold nanoshells, P-polarized light localizes on the tips of the cones and S-polarized light localizes on the regions at the bottom of the cones where the cones overlap [40]. Figure 2.6 shows finite difference time domain (FDTD) calculations for various gold nanorod shapes, where the maximum local electric field is related to the curvature [32]. The electric "hot spots" on the particles affect surface ligand bonding and optical responses [4–7, 28, 45].

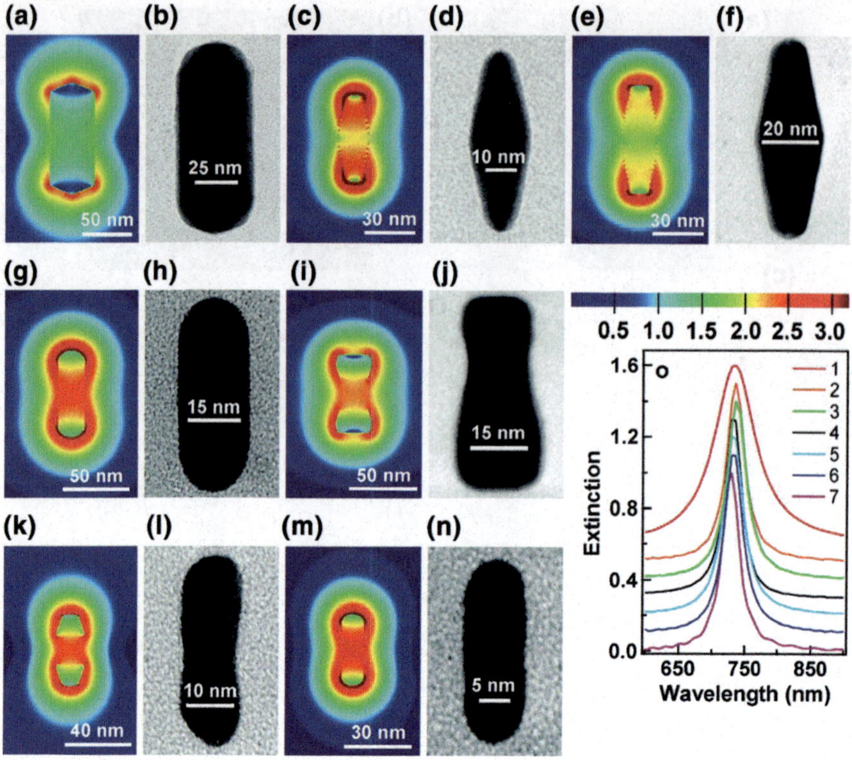

Fig. 2.6 Electric field intensity enhancement contours (*colored*) and TEM images (*gray*) of differently shaped Au nanocrystals: (**a**, **b**) Large NRs; (**c**, **d**) NBPs; (**e**, **f**) oxidized NBPs; (**g**, **h**) oxidized NRs; (**i**, **j**) dog-bone-like NRs; (**k**, **l**) peanut-like NRs; and (**m**, **n**) small NRs. (o) Calculated extinction spectra of Au nanocrystals embedded in water. *Curves 1* to *7* represent the large NRs, NBPs, oxidized NBPs, oxidized NRs, dog-bone-like NRs, peanut-like NRs, and small NRs, respectively. The field intensity enhancement is at the logarithmic scale. Reproduced with permission from Ref. [32]. Copyright © 2009 American Chemical Society

Polarization has an effect on electronic interactions; for example, the fluorophore emission of dyes attached to gold nanorods (or plasmon coupled gold nanoparticles) can become altered to match the absorption energy of the longitudinal plasmon for the anisotropic gold nanoparticle-dye conjugates, as shown in Fig. 2.7. This is attributed to the transition of the fluorophore from an excited state to a vibrational ground state close to the plasmon energy due to the enhancement from the local electric field from the plasmon resonance [42]. Polarization effects are dependent on polarization angle, so polarization studies can give information on the orientation of the nanoparticle, which is important for SERS, catalysis, biological imaging, among others.

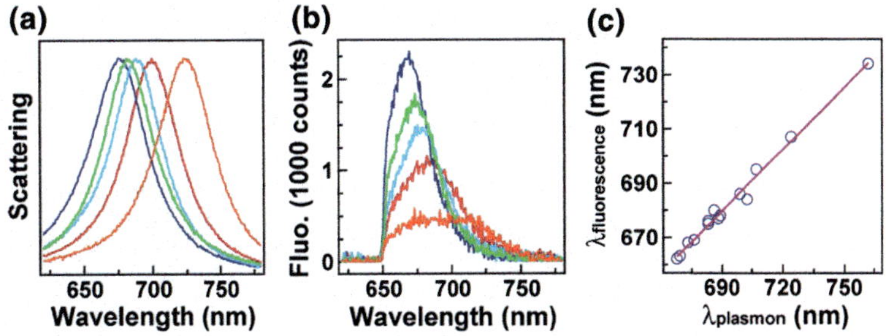

Fig. 2.7 Spectral shaping of the emission of oxazine 725. **a** Scattering spectra of the individual hybrid nanostructures containing the Au nanorods with different aspect ratios. **b** Corresponding fluorescence emission spectra. The emission spectra were recorded under the longitudinal polarization excitation. **c** Dependence of the fluorescence peak wavelength (empty circles) of oxazine 725 on the scattering peak wavelength. The line is a linear fit. Reproduced with permission from Ref. [42]. Copyright © 2009, American Chemical Society

2.1.3 *Crystalline Anisotropy*

Nanoparticle physico-chemical properties are also dependent on the interfacial atomic arrangement and coordination of the crystal. Generally, the density and symmetry of atoms in different crystallographic planes are not identical, and neither is their electronic structure, bonding, surface energy, or chemical reactivities [4, 5, 12].

The face-centered cubic crystalline lattice of single-crystal metal nanoparticles, namely gold, silver and platinum nanoparticles, typically have low-index facets: {100}, {111}, and {110} [8]. However, depending on the preparation conditions, namely surfactants, reduction agents, temperature, thermodynamic, and/or kinetic factors, etc., nanoparticles with different crystallographic facets, and subsequently, properties can be produced [4–7, 9, 12, 13, 25, 28, 30]. For example, gold nanorods of various dimensions, ~20 nm in diameter and up to 500 nm long, prepared in similar conditions through a seed mediated surfactant approach, have different crystallographic structures depending on the *presence* or *absence* of additive ions, e.g. Ag^+ (Fig. 2.8a–f). The diffraction pattern and the high resolution TEM images show that the short Au nanorods (~20 nm in diameter and up to 100 nm in length) prepared in the presence of the additive ions are single crystalline with no observable stacking faults, twins or volume dislocations [4, 5, 7]. High resolution TEM images of the short Au nanorods (Fig. 2.8d–f) show well-defined, continuous, and equally spaced fringe patterns for their atomic lattice. Surprisingly, a closer inspection of the long Au nanorods (~20 nm in diameter and up to 500 nm in length) prepared in the absence of the additive ions reveal a penta-twinned crystallographic structure with five well-defined facets typical of icosahedra structures (Fig. 2.8a–c) [20, 46].

The surfactant-directed approach is not the only procedure leading to penta-twinned crystals. Electron microscopy studies shows that silver nanowires up

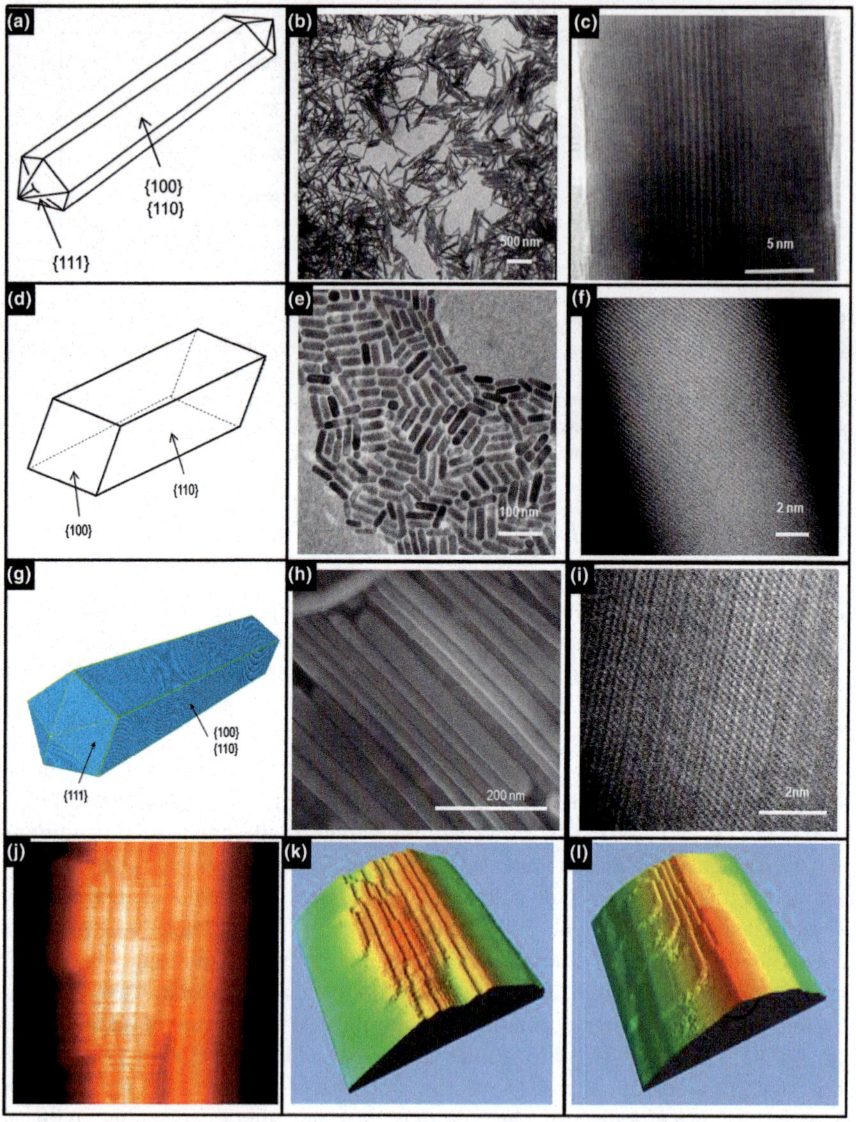
(a)
{100}
{110}
{111}
(b)
500 nm
(c)
5 nm
(d)
{110}
{100}
(e)
100 nm
(f)
2 nm
(g)
{100}
{110}
{111}
(h)
200 nm
(i)
2nm
(j)
(k)
(l)

◂**Fig. 2.8** (**a–c**) Gold nanorods prepared via a surfactant mediated approach in the absence of additive ions. **a** Cartoon of the penta twinned crystallography of Au nanorods, **b** TEM image of Au nanorods, **c** HRTEM of Au nanorods; (**d–f**) Gold nanorods prepared via a surfactant mediated approach in the presence of additive ions (Ag^+). (**a**) Cartoon of the single crystal Au nanorods, (**b**) TEM image of Au nanorods, **c** HRTEM of Au nanorods; (**g–l**) Ag nanowires prepared by an coarsening process via an oriented attachment mechanism. (**g**) Equilibrium atomic positions for a pentagonal nanowire with internal twins; (**h**) SEM image of Ag nanowires, (**i**) HRTEM image of penta-twinned Ag nanowire, (**j**) STM topography image, scale 20 · 20 nm2, of the top of the Ag nanowire, (**k**) 3 D rendering of Ag nanowires, (**l**) mage of another top section of the same wire, with the angle between the facets indicated with a black semicircle. (a) Reprinted with permission from Ref. [6]. Copyright 2017 American Chemical Society. (c) Reprinted with permission from Ref. [11]. Copyright 2017 American Chemical Society. (g) Reprinted with permission from Ref. [46]. Copyright 2017 American Chemical Society. (j-l) Reprinted with permission from Ref. [47]. Copyright 2017 with permission from Elsevier

to 10 nm long and 30 nm diameter prepared via a seedless method, in the absence of surfactants, also have penta-twinned crystallographic structures (Fig. 2.8g–l). Interestingly enough, these silver nanowires form through a coarsening process via an oriented attachment mechanism, which is completely different from the seed-mediated surfactant growth approach used to produce gold nanorods [7, 47, 48].

Ultimately, the crystallography of the final product is determined by the original "seed" nanoparticle, which can be tailored through the synthetic parameters [20, 49–51]. The crystallography of the "seed" [4–7, 9, 13, 46] (Fig. 2.9) can also determine the final nanoparticle shape, such as quantum rods versus tetrapods [49, 52] and silver spheres versus cubes or rods [20].

Crystallographic facet energies play a huge role in nanoparticle synthesis, applications, and properties [4]. In typical nanoparticle fabrication techniques, the final nanocrystal is truncated with the lowest energy facets, though this can be tuned through synthetic parameters and chemical etching [39, 53, 54]. Crystalline stability

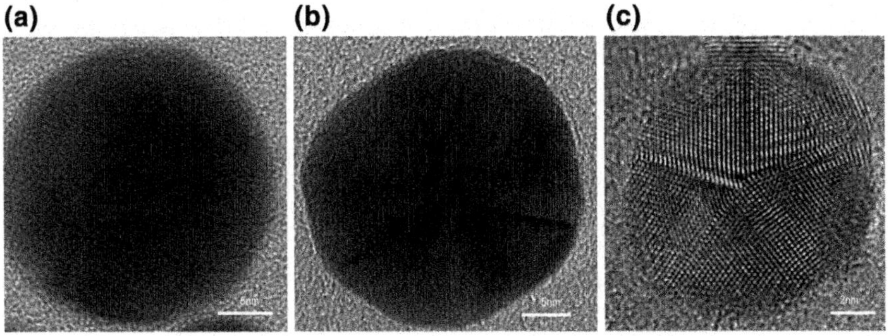

Fig. 2.9 HRTEM images of (**a**) single crystalline- Au nanospheres-"seeds", (**b**) penta-twined crystalline Au nanospheres, (**c**) penta twinned Ag nucleation centers. (**a**, **b**) Reprinted with permission from Ref. [13]. Copyright 2017 Springer. (**c**) Reprinted with permission from Ref. [46]. Copyright 2017 American Chemical Society

depends on the presence of dangling bonds, steps and terraces but this can also be tuned with facet specific capping agents. The manipulation of crystallographic stability is one aspect that drives the synthesis of anisotropic nanomaterials [50]. Multicomponent nanoparticles, such as core/shell and Janus nanoparticles, typically require a common crystallographic orientation (i.e. small lattice mismatch) to increase the binding stability between the two materials. Facet-selective etching is also used to dissolve the least stable facets, leading to the creation of nanoparticles with interesting morphologies such as hollow nanocubes [54]. Binding energy differences on certain facets is an important factor for chemical reactions and has been exploited to improve their efficiency and selectivity for applications such as catalysis, [55, 56] SERS, [57] and sensing [4, 7, 28, 45, 58]. The relative stability of crystallographic planes for surface based applications becomes increasingly important for anisotropic nanoparticles, which may contain elongated crystallographic planes and multiple different types of exposed facets.

It is important to consider how nanoparticle shape and crystalline properties relate to one another and how they impact the emergent properties of a nanoparticle. As described above, building nanoparticles from the bottom up typically depends on the manipulation of the energetics of the crystalline surfaces, and as a result, the final nanoparticle shape depends on the orientation of these crystalline planes with respect to each other. This is one reason that metallic nanocubes are fairly common the {100}, {110}, and {111} crystalline facets that form the faces, edges and corners, respectively, are typically very stable. However, for catalysis, high index facets are more reactive and desirable, and therefore different shapes are synthesized by promoting the growth of higher index facets. For example, highly catalytic Au nanoparticles enclosed by {830} and {730} facets correspondingly take the form of concave and convex nanocuboids [59]. In this case, the desire for active catalytic crystalline surfaces evolved into the generation of differently shaped nanoparticles. If one were to carve out similarly shaped Au nanoparticles from the top down, *e.g.*, using electron beam lithography, it is unlikely that these particles would have the same catalytic properties as the ones where the bottom up growth was manipulated to reveal the higher index facets. The relative surface areas of the higher energy facets determine the catalytic properties, not the overall shape of the nanoparticle. On the other hand, if one is only interested in the optical properties of the Au nanoparticles, then the bottom up and top down methods could be substituted for one other, as shape is more important than the crystallinity of the exposed facets in determining the energetics of the localized surface plasmon resonances.

2.1.4 Anisotropic Nanoparticle Structures

At the nanoscale, the shape, crystallinity, and emergent properties of a nanoparticle are all interrelated, and as demonstrated in the following sections, nanoparticles come in all shapes and sizes, where each particle can possess properties unique to their geometry [4–7, 27, 29, 30, 47]. These sections are not exhaustive, with scientists stretching their imaginations and becoming more sophisticated every day in their design and application of particles in the nanoscale regime.

2.1.4.1 Spheres

Nanospheres are the most widespread shape that is used in nanomaterial manufacturing due to the increased thermodynamic stability of spheres compared to other shapes. They are isotropic, zero-dimensional materials whose properties are highly dependent on nanoparticle diameter. An increase in size leads to more red-shifted optical properties due to the decrease in electron confinement. Metal nanoparticles do not have a bandgap so electrons are delocalized and confined in the potential well of the nanoparticle, whereas semiconductor nanoparticles, or quantum dots, have a band gap in the valence band and conduction band that is dependent on size. As the quantum dots increase in size, more atomic orbitals contribute to the conduction and valence bands, leading to a decrease in band gap size and a red shift in fluorescence emission. A variety of hollow nanospheres have also been created [60–63]. Hollow nanoparticles can be used as carriers to protect or selectively deliver a high load of toxic material in therapeutic applications. In one example, hollow mesoporous silica nanoparticles were loaded with a radionucleotide and a photosensitizer to encapsulate and deliver them to tumors [61]. Once inside of the cell, the photosensitizer reacted with the nucleotide to create reactive oxygen species to kill the cell as well as emit light as a photodynamic therapy light source [61].

Anisotropy can be achieved in isotropic nanoparticle shapes, such as spheres, through the creation of multicomponent nanoparticles. The size, shape, and composition dependent properties of nanoparticles can be combined to create nanoparticles with different chemical or physical domains [64]. These can be used to accomplish new or more complex tasks. As shown in Fig. 2.10, nanoparticles with anisotropic compositions include core/shell nanoparticles, [19, 65–67] Janus nanoparticles, [63, 68, 69] and nanoparticles with island morphologies [70].

These structures generally preserve the properties of the individual components, leading to greater tunability and functionality [4–7, 25, 28, 30, 71]. In multifunctional nanoparticles, materials with unique optical signatures can be coupled to materials with desirable physical properties in a single nanostructure. For example, magnetic-plasmonic core/shell nanoparticles have magnetic functionality along with a plasmon band characteristic of the metal [72, 73]. The addition of metallic shells leads to a plasmon band that is dependent on the strength of the coupling of the plasmon band to the core material, which is dependent on the shell thickness [73]. The development of nanomaterials with multimodal functionalities opens up new avenues for creating structures with anisotropic functionalities for applications such as nanomotors, [63, 74, 75], surface plasmon resonance sensors [76] and optical imaging.

2.1.4.2 Rods, Wires and Tubes

Nanorods, nanowires and nanotubes have an increased length in one direction that leads to electron delocalization in the lateral dimension. Elongated nanostructures have properties that are dependent on aspect ratio (r = length/width). One-dimensional nanoparticles have been synthesized from a large range of materials including: metals, (Ag, Au, Pd, Pt, Rh, Cu, etc.), semiconductors (CdSe, CdTe, etc.), carbon-based materials etc. and have been applied to surface enhanced Raman Spectroscopy (SERS), light emitting diodes, energy transfer applications, [24] drug delivery applications, catalysis, [77] among others. Nanowires (NW) and nanotubes (NT) have been used as field effect transistors to record, stimulate and inhibit neuronal signals, [78] biological and chemical sensors, [22] nanoantennas, [79] waveguides, photovoltaics, [80] lasers, [81] and electronics [34, 82]. Nanowires have potential for nanoelectronics or "lab-on-a-chip" applications due to their small sizes that allow for non-invasive probes in biological applications, the ability to transduce signals across an electrode and ability to be passivated to prevent corrosion of metal junctions that typically lead to electronic failure [78]. Nanorods have an elongated *c*-axis that causes the movement of charge carriers to be asymmetric, leading to more mobility in one dimension. In semiconductor nanorods, or quantum rods (QRs), this typically leads to a lower quantum yield due to the longer electron-hole separation that increases the recombination time and decreases recombination rate. Once the electron-hole pair is separated, there is a chance that the charge carriers can react at the interface instead of with the other charge carrier, which is beneficial for applications that require charge separation such as catalysis and solar cells. QRs and gold nanorods (AuNRs) also absorb and emit linearly polarized light [21]. The internal polarization of nanorods is not completely orthogonal to the surface normal along its entire length, leading to different surface charges that can significantly affect the properties of the nanorod [83]. In metal nanorods, a longitudinal surface plasmon peak arises that can be tuned over the visible to the near-IR region, depending on aspect ratio [26]. The elongated shape gives nanorods a higher surface area, which has a positive impact on surface based applications. For example, CeO_2-Cu_xO_y nanocubes and nanorods were evaluated for the photochemical production of H_2 from hydrazine [77]. The rods were found to have the highest photocatalytic activity due to the higher surface area of rods compared to cubes [77].

Carbon nanotubes have been gaining increasing attention due to their unique structures and high mechanical strength, which is important for electronic applications. Single walled carbon nanotubes (SWCNTs) are made up of graphene like structures in the form of a tube, whereas multi-walled carbon nanotubes (MWCNTs) are made up on one or more concentric SWCNTs, which increase the mechanical stability of the nanotubes (NTs). Carbon nanotubes can act as a metal or a semiconductor, depending on the chirality, or the orientation of the carbon's crystal lattice compared to the tube's central axis. Metal SWCNTs do not display fluorescence and they quench the fluorescence from neighboring semiconductor SWCNTs [84]. Gold nanowires have also been found to be up to 100 times mechanically stronger than bulk, with strength increasing for decreasing diameter [85].

Nanowires are typically grown orthogonally onto substrates, so nanowire arrays can be created to take advantage of their synergistic effects. For example, in semiconductor nanowires, increasing the nanowire diameter led to an increase in photocurrent density due to the increased light absorption [80]. The small diameter of these nanowires, which is much smaller than carrier diffusion lengths, can give nanowires the ability to approach the Shockley-Queisser efficiency limit for solar cells [80]. The small diameter of nanowires coupled with the longer length that allows for high mechanical stability compared to bulk, increased photocurrent and ability to be passivated with a variety of materials and molecules gives nanowires significant potential for their incorporation in electronics, sensors, waveguides, among others.

2.1.4.3 Cubes, Hexagons, Triangles

Nanocubes (NCs), nanohexagons (NHx) and nanotriangles have multiple corners that lead to accumulation of dipoles. NCs with sharp and rounded corners have been synthesized and they display multiple SPR peaks, as shown in Fig. 2.11 [18, 86, 87].

The appearance of sharp edges has been found to cause charge accumulation at the corners and lead to stronger dipoles inside of the nanoparticle that lead to interesting features in light scattering and absorption [18]. Due to their cubic shape, nanocubes (NCs) can maximize packing and assemble into lattice matched

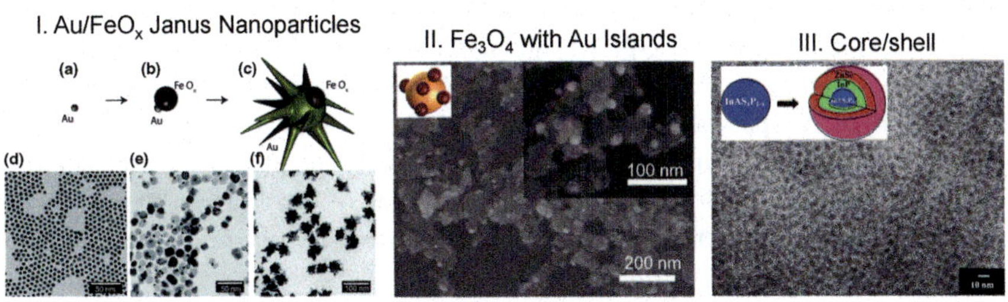

Fig. 2.10 **I**. Au/FeO$_x$ Janus nanoparticles (**a**) Au core, (**b**) Au-FeO$_x$ Janus precursor, (**c**) final Au/FeO$_x$ Janus particles, Open access from Ref. [69]. Published by the Royal Society of Chemistry (RSC); **II**. Fe_3O_4 nanoparticles with Au "island" domains, reproduced with permission from Ref. [70]. Copyright © 2016 American Chemical Society. **III.** Schematic and TEM of core/shell InAsP/InP/ZnSe nanoparticles, reproduced with permission from J. Am. Chem. Soc., 2005. 127(30): pp. 10526–10532. Copyright © 2015 American Chemical Society

superstructures for electronic and magnetic applications [88]. The different facets of nanocubes also have different activities for applications such as catalysis. Compared to other morphologies, nanocube cobalt oxides have been found to have the highest activity for the oxidation of 1,2-dichloroethane, one of the most common pollutants in waste streams [89]. Planar nanoparticles such as metallic NCs coupled to dielectric substrates display stronger substrate induced hybridization compared to spherical nanoparticles due to the larger surface area that is coupled to the dielectric material [76]. This led to the appearance of a new narrow and highly sensitive plasmon mode with powerful consequences for very sensitive surface plasmon based sensing [76]. The increased number of hot spots on the corners compared to rods and spheres generally leads to an increase in reactivity in the cubes, triangles and hexagons.

2.1.4.4 Branched and Other Shapes

Polyhedral and planar branched nanoparticles, such as stars and tetrapods [19, 33, 49, 52, 90, 91], have been created to enhance catalysis, gas sensing, and surface enhanced Raman Spectroscopy (SERS) [4, 7, 28, 29, 45, 90]. The unique structure of branched nanoparticles leads to properties that depend on the composition, length, width and number of arms. Tips, edges and vertices can act as hot spots for electric field enhancement, leading to increased sensing properties and reactivities. Branched nanoparticles tend to have absorption or emission in the red and near-IR region. The plasmon bands for star-like metal nanoparticles are the result of the hybridization of the plasmons from the core and the tips [40, 92].

Particles with a larger number of branches, such as spiky gold nanoshells, have been used in SERS sensing due to the larger number of hot spots at the junctions of the arms and at the tips of the spikes [40, 93]. As shown in Fig. 2.12, the spiky gold nanoshells showed a SERS response at lower concentrations than the smooth gold shells. These spiky structures also showed a significantly higher scattering and a surface plasmon resonance (SPR) band that was dependent on the length of the spikes, where longer spikes led to a red-shift in the band [40]. Modeling of the electric field in the spiky shells showed that the electric field is localized at the tip of the cones and the locations where spikes overlap [40]. The SPR band is believed to be a resonance that localizes mostly at the junctions of the arms [40].

CdSe/CdS nanotetrapods were found to have a higher absorption cross section, indicating that the tetrapod arms absorb more light and act as a more efficient light harvesting system compared to QRs [49]. The width of the tetrapod arms had more of an effect on quantum yield (QY) than length of the arms. For the CdSe/CdS system, the electron-hole pair is confined to the core, but quantum confinement

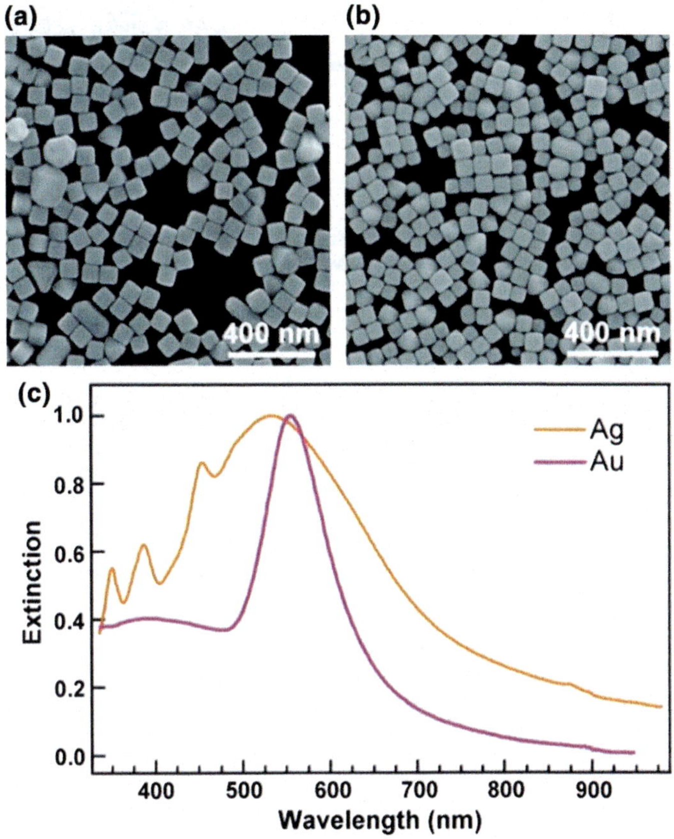

Fig. 2.11 (**a**, **b**) SEM images of the Ag and Au nanocubes, respectively. (**c**) Extinction spectra of the Ag and Au nanocubes. Reproduced with permission from Ref. [87]. Copyright © 2011 American Chemical Society

causes stronger electron localization at the branch in the narrower arms compared to the thicker arms, leading to higher QY and absorption in tetrapods with narrower arms [49]. The unique morphologies of branched nanoparticles that allow for more "hot spots" and light interactions has led to the development of a wide variety of materials and shapes that have been investigated for sensing, SERS applications, and light harvesting systems.

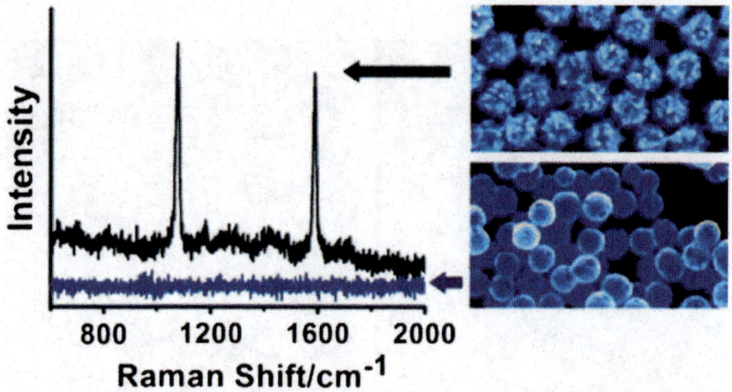

Fig. 2.12 SERS of spiky gold nanoshells (*black spectra*) compared to smooth gold nanoshells (*blue spectra*). Reproduced with permission from Ref. [40]. Copyright © 2012 American Chemical Society

2.2 Conclusions

A plethora of nanoparticle shapes have been created and employed for biological delivery and sensing, catalysis, and electronics. The chemical and physical properties are highly dependent on composition and morphology; various experimental and theoretical works have gone into exploring the nature of these properties to exploit them to their full potential.

References

1. https://www.epa.gov/chemical-research/research-nanomaterials.
2. Masciangioli, T., and W.-X. Zhang. 2003. Peer Reviewed: Environmental Technologies at the Nanoscale. *Environmental Science and Technology* 37 (5): 102A–108A.
3. Mazzola, L. 2003. Commercializing Nanotechnology. *Nature Biotechnology* 21 (10): 1137.
4. Hunyadi, S.E. 2007. Nanoengineered Materials: Synthesis, Design, Functionalization and Chemical Sensing Applications. In *Department of Chemistry and Biochemistry*, p. 262. Columbia, SC: University of South Carolina.
5. Hunyadi Murph, S.E., et al. 2012. Metallic and Hybrid Nanostructures: Fundamentals and Applications. In *Applications of Nanomaterials*, ed. J.N. Govil. USA: Studium Press LLC.
6. Murphy, C.J., et al. 2005. Anisotropic Metal Nanoparticles: Synthesis, Assembly, and Optical Applications (Feature Article; a Top Five ACS article by citations, National Chemistry Week, 2007). *Journal of Physical Chemistry B* 109: 13857–13870.
7. Murphy, C.J., et al. 2006. One-Dimensional Colloidal Gold and Silver Nanostructures. *Inorganic Chemistry* 45 (19): 7544–7554.
8. Fedlheim, D.L, and C.A. Foss. 2001. In: *Metal Nanoparticles: Synthesis, Characterization, and Applications*, Edited By D.L. Fedlheim and C.A. Foss. New York: Marcel Dekker Inc.
9. Murphy, C.J., et al. 2008. Chemical Sensing and Imaging with Metallic Nanorods. *Chemical Communications* 4 (554–557): 554.

10. Kuchibhatla, S.V.N.T., et al. 2007. One Dimensional Nanostructured Materials. *Progress in Materials Science* 52 (5): 699–913.
11. Tiwari, J.N., R.N. Tiwari, and K.S. Kim. 2012. Zero-dimensional, one-dimensional, two-dimensional and three-dimensional nanostructured materials for advanced electrochemical energy devices. *Progress in Materials Science* 57 (4): 724–803.
12. Hunyadi Murph, S.E., et al. 2011. Synthesis, Functionalization, Characterization and Application of Controlled Shape Nanoparticles in Energy Production. In *Fluorine-Related Nanoscience with Energy Applications*, ed. D.J. Nelson, and C.N. Brammer.
13. Hunyadi Murph, S.E., et al. 2011. Tuning of Size and Shape of Au-Pt Nanocatalyst for Direct Methanol Fuel Cells. *Journal of Nanoparticle Research* 13 (6347–6364): 6347.
14. Roduner, E. 2006. Size Matters: Why Nanomaterials are Different. *Chemical Society Reviews* 35 (7): 583–592.
15. Valden, M., X. Lai, and D.W. Goodman. 1998. Onset of Catalytic Activity of Gold Clusters on Titania with the Appearance of Nonmetallic Properties. *Science* 281 (5383): 1647–1650.
16. Chen, M., and D. Goodman. 2004. The Structure of Catalytically Active Gold on Titania. *Science* 306 (5694): 252–255.
17. Kim, S.-W., et al. 2002. Fabrication of Hollow Palladium Spheres and their Successful Application to the Recyclable Heterogeneous Catalyst for Suzuki Coupling Reactions. *Journal of the American Chemical Society* 124 (26): 7642–7643.
18. Wiley, B., S.H. Im, Z. Li, J. McLellan, A. Siekkinen, and Y. XIa. 2006. Maneuvering the Surface Plasmon Resonance of Silver Nanostructures through Shape-Controlled Synthesis. *The Journal of Physical Chemistry B* 110: 15666–15675.
19. Meng, M., et al. 2016. Integration of Kinetic Control and Lattice Mismatch to Synthesize Pd@AuCu Core-Shell Planar Tetrapods with Size-Dependent Optical Properties. *Nano Letters* 16: 3036–3041.
20. Wiley, B., et al. 2005. Shape-Controlled Synthesis of Metal Nanostructures: The Case of Silver. *Chemistry A European Journal* 11: 454–463.
21. Hadar, I., G.B. Hitin, A. Sitt, A. Faust, and U. Banin. 2013. Polarization Porperties of Semiconductor Nanorod Heterostructures: From Single Particles to Ensemble. *The Journal of Physical Chemistry Letters* 4: 502–507.
22. Cui, Y., et al. 2001. Nanowire Nanosensors for Highly Sensitive and Selective Detection of Biological and Chemical Species. *Science* 293 (5533): 1289–1292.
23. Giblin, J., and M. Kuno. 2010. Nanostructure Absorption: A Comparative Study of Nanowire and Colloidal Quantum Dot Absorption Cross Sections. *The Journal of Physical Chemistry Letters* 1: 3340–3348.
24. Halivni, S., A. Sitt, I. Hadar, and U. Banin. 2012. Effect of Nanoparticle Dimensionality on Fluorescence Resonance Energy Transfer in Nanoparticle-Dye Conjugated Systems. *ACS Nano* 6 (3): 2758–2765.
25. Hunyadi Murph, S.E., and C.J. Murphy. 2006. Tunable One-Dimensional Silver-Silica Nanopeapod Architectures. *The Journal of Physical Chemistry B* 110: 7226–7231.
26. Zhang, Q., L. Han, H. Jing, D.A. Blom, Y. Lin, H.L. Zin, and H. Wang. 2016. Facet Control of Gold Nanorods. *ACS Nano* 10 (2): 2960–2974.
27. Unrine, J., P. Bertsch, and S.E. Hunyadi. 2008. Bioavailability, Trophic Transfer and Toxicity of Manufactured Metal and Metal Oxide Nanoparticles in Terrestrial Environments. In *Nanoscience and Nanotechnology: Environmental and Health Impacts*, ed. V. Grassian, 343–364. New Jersey: Wiley.
28. Hunyadi, S.E., and C.J. Murphy. 2006. Bimetallic Silver-Gold Nanowires: Fabrication and Use in Surface- Enhanced Raman Scattering. *Journal of Materials Chemistry* 16 (Special Issue: Anisotropic Nanoparticles): 3929–3935.
29. Hunyadi Murph, S.E., and C.J. Murphy. 2013. Patchy Silica-Coated Silver Nanowires as SERS Substrates. *Journal of Nanoparticle Research* 15 (6): 1607.
30. Hunyadi Murph, S.E., et al. 2012. Manganese- Doped Gold Nanoparticles as Positive Contrast Agents for Magnetic Resonance Imaging (MRI). *Journal of Nanoparticle Research* 14: 658–659.

31. Chen, H., et al. 2008. Shape- and Size- Dependent Refractive Index Sensitivity of Gold Nanoparticles. *Langmuir* 24: 5233–5237.
32. Chen, H., et al. 2009. Shape-Dependent Refractive Index Sensitivities of Gold Nanocrystals with the Same Plasmon Resonance Wavelength. *The Journal of Physical Chemistry C* 113: 17691–17697.
33. Tulun, F.R., S. Öztürk, and Z.Z. Öztürk. 2016. The NO_2 Sensing Properties of the Sensors Done with Nano-Tetrapods. *Acta Physica Polonica A* 129: 797–799.
34. Gudiksen, M.S., et al. 2002. Growth of Nanowire Superlattice Structures for Nanoscale Photonics and Electronics. *Nature* 415: 617–620.
35. Langhammer, C., I. Zorić, B. Kasemo, and B.M. Clemens. 2007. Hydrogen Storage in Pd Nanodisks Characterized with a Novel Nanoplasmonic Sensing Scheme. *Nano Letters* 7: 3122–3127.
36. Subhramannia, M., and V.K. Pillai. 2008. Shape-Dependent Electrocatalytic Activity of Platinum Nanostructures. *Journal of Materials Chemistry* 18: 5858–5870.
37. Chithrani, B.D., A.A. Ghazani, and W.C.W. Chan. 2006. Determining the Size and Shape Dependence of Gold Nanoparticle Uptake into Mammalian Cells. *Nano Letters* 6: 662–668.
38. Huang, X., et al. 2010. The Effect of the Shape of Mesoporous Silica Nanoparticles on Cellular Uptate and Cell Function. *Biomaterials* 31 (3): 438–448.
39. Khon, E., K. Lambright, R. Khnayzer, P. Moroz, D. Perera, E. Butaeva, S. Lambright, F. Castellano, and M. Zamkov. 2013. Improving the Catalytic Activity of Semiconductor Nanocrystals through Selective Domain Etching. *Nano Letters* 13: 2016–2023.
40. Sanchez-Gaytan, B.L., et al. 2012. Spiky Gold Nanoshells: Synthesis and Enhanced Scattering Properties. *The Journal of Physical Chemistry C* 116: 10318–10324.
41. Sitt, A., A. Salant, G. Menagen, and U. Banin. 2011. Highly Emissive Nano Rod-in-Rod Heterostructures with Strong Linear Polarization. *Nano Letters* 11: 2054–2060.
42. Ming, T., et al. 2009. Strong Polarization Dependence of Plasmon-Enhanced Fluorescence on Single Gold Nanorods. *Nano Letters* 9: 3896–3903.
43. Hu, J., et al. 2001. Linearly Polarized Emission from Colloidal Semiconductor Quantum Rods. *Science* 292: 2060–2063.
44. Near, R.D., S.C. Hayden, R.E. Hunter, D. Thackston, and M. El-Sayed. 2013. Rapid and Efficient Prediction of Optical Extinction Coefficients for Gold Nanospheres and Gold Nanorods. *The Journal of Physical Chemistry C* 117: 23950–23955.
45. Song, C., et al. 2012. Gold-Modified Silver Nanorod Arrays: Growth Dynamics and Improved SERS Properties. *Journal of Materials Chemistry* 22: 1150–1159.
46. Hunyadi Murph, S.E., et al. 2015. A Possible Oriented Attachment Growth Mechanism for Silver Nanowire Formation. *Crystal Growth & Design* 15: 1968–1974.
47. Tao, C.G., et al. 2007. Surface Morphology and Step Fluctuations on Silver Nanowires. *Surface Science* 601: 4939–4943.
48. Lucas, M., et al. 2008. Plastic Deformation of Pentagonal Silver Nanowires: Comparison Between AFM Nanoindentation and Atomistic Simulations. *Physical Review B* 77: 2452014–2454201.
49. Talapin, D.V., et al. 2007. Seeded Growth of Highly Luminscent CdSe/CdS Nanoheterostructures with Rod and Tetrapod Morphologies. *Nano Letters* 7: 2951–2959.
50. Scarabelli, L., et al. 2015. A "Tips and Tricks" Practical Guide to the Synthesis of Gold Nanorods. *The Journal of Physical Chemistry Letters* 6: 4270–4279.
51. Sun, Y., and Y. Xia. 2002. Shape-Controlled Synthesis of Gold and Silver Nanoparticles. *Science* 298: 2176–2179.
52. Manna, L., et al. 2003. Controlled Growth of Tetrapod-Branced Inorganic Nanocrystals. *Nature Materials* 2: 382–385.
53. Lu, X., T.T. Tran, and W. Zhang. 2013. Shape-Selective Effect of Foreign Metal Ions on Growth of Noble Metal Nanocrystals with High-Index Facets. *Journal of Chemical Engineering and Process Technology* 1: 1009–1016.

54. Lyu, L., and M.H. Huang. 2011. Investigation of Relative Stability of Different Facets of Ag_2O Nanocrystals Through Face-Selective Etching. *The Journal of Physical Chemistry C* 115 (36): 17768–17773.
55. Tachikawa, T., S. Yamashita, and T. Majima. 2011. Evidence for Crystal-Face-Dependent TiO_2 Photocatalysis from Single-Molecule Imaging and Kinetic Analysis. *Journal of the American Chemical Society* 133: 7197–7204.
56. Schmidt, E., et al. 2009. Platinum Nanoparticles: The Crucial Role of Crystal Face and Colloid Stabilizer in the Diastereoselective Hydrogenation of Cinchonidine. *Chemistry European Journal* 16 (7): 2181–2192.
57. Ikeda, K., S. Suzuki, and K. Uosaki. 2011. Crystal Face Dependent Chemical Effects in Surface-Enhanced Raman Scattering at Atomically Defined Gold Facets. *Nano Letters* 11: 1716–1722.
58. Han, X., et al. 2009. Synthesis of Tin Dioxide Octahedral Nanoparticles with Exposed High-Energy 221 Facets and Enhanced Gas-Sensing Properties. *Angewandte Chemie International Edition* 121 (48): 9344–9347.
59. Zhang, Q., et al. 2015. Faceted Gold Nanorods: Nanocuboids, Convex Nanocuboids, and Concave Nanocuboids. *Nano Letters* 15 (6): 4161–4169.
60. Lal, S., et al. 2002. Light Interaction between Gold Nanoshells Plasmon Resonance and Planar Optical Waveguides. *The Journal of Physical Chemistry B* 106: 5609–5612.
61. Kamkaew, A., et al. 2016. Cerenkov Radiation Induced Photodynamic Therapy Using Chlorin e6-Loaded Hollow Mesoporous Silica Nanoparticles. *ACS Applied Materials & Interfaces* 8 (40): 26630–26637.
62. Son, S.J., X. Bai, and S.B. Lee. 2007. Inorganic Hollow Nanoparticles and Nanotubes in Nanomedicine: Part 2: Imaging, Diagnostic, and Therapeutic Applications. *Drug Discovery Today* 12: 657–663.
63. Ma, X., et al. 2015. Enzyme-Powered Hollow Mesoporous Janus Nanomotors. *Nano Letters* 15: 7043–7050.
64. Carbone, L., and P.D. Cozzoli. 2010. Colloidal Heterostructured Nanocrystals: Synthesis and Growth Mechanisms. *Nano Today* 5: 449–493.
65. Lyon, J.L., D.A. Fleming, M.B. Stone, P. Schiffer, and M.E. Williams. 2004. Synthesis of Fe oxide Core/Au Shell Nanoparticles by Iterative Hydroxylamine Seeding. *Nano Letters* 4 (4): 719–723.
66. Ji, T., et al. 2001. Preparation, Characterization, and Application of Au-shell/Polystyrene Beads and Au-shell/Magnetic Beads. *Advanced Materials* 13 (16): 1253–1256.
67. Harpeness, R., and A. Gedanken. 2004. Microwave Synthesis of Core-Shell Gold/Palladium Bimetallic Nanoparticles. *Langmuir* 20: 3431–3434.
68. Kostevsek, N., et al. 2016. The One-Step Synthesis and Surface Functionalization of Dumbbell-Like Gold-Iron Oxide Nanoparticles: A Chitosan-Based Nanotheranostic System. *Chemical Communications* 52: 378–381.
69. Reguera, J., et al. 2016. Synthesis of Janus Plasmonic-Magnetic, Star-Sphere Nanoparticles, and Their Application in SERS Detection. *Faraday Discussions* 191: 47–59.
70. Larsen, G.K., W. Farr, and S.E.H. Murph. 2016. Multifunctional Fe_2O_3-Au Nanoparticles with Different Shapes: Enhanced Catalysis, Photothermal Effects, and Magnetic Recyclability. *The Journal of Physical Chemistry C* 120 (28): 15162–15172.
71. Hunyadi, S.E., and C.J. Murphy. 2009. Synthesis and Characterization of Silver-Platinum Bimetallic Nanowires and Platinum Nanotubes. *Journal of Cluster Science* 20: 319–330.
72. Kwizera, E.A., et al. 2016. Size- and Shape-Controlled Synthesis and Properties of Magnetic-Plasmonic Core-Shell Nanoparticles. *The Journal of Physical Chemistry C* 120: 10530–10546.
73. Levin, C.S., et al. 2009. Magnetic-Plasmonic Core-Shell Nanoparticles. *ACS Nano* 3 (6): 1379–1388.
74. Liu, R., and A. Sen. 2011. Autonomous Nanomotor Based on Copper-Platinum Segmented Nanobattery. *Journal of the American Chemical Society* 133: 20064–20067.

75. Kagan, D., P. Calvo-Marzal, S. Balasubramanian, S. Sattayasamitsathit, K.M. Manesh, G. Flechsig, and J. Wang. 2009. Chemical Sensing Based on Catalytic Nanomotors: Motion-Based Detection of Trace Silver. *Journal of the American Chemical Society* 131: 12082–12083.
76. Zhang, S., et al. 2011. Substrate-Induced Fano Resonances of a Plasmonic Nanocube: A Route to Increased Sensitivity Localized Surface Plasmon Resonance Sensors Revealed. *Nano Letters* 11: 1657–1663.
77. Clavijo-Chaparro, S.L., et al. 2016. Water Splitting Behavior of Copper-Cerium Oxide Nanorods and Nanocubes Using Hydrazine as a Scavenging Agent. *Journal of Molecular Catalysis A: Chemical* 423: 143–150.
78. Patolsky, F., et al. 2006. Detection, Stimulation and Inhibition of Neuronal Signals with High-Density Nanowire Transistor Arrays. *Science* 313: 1100–1104.
79. Day, J.K., et al. 2014. Standing Wave Plasmon Modes Interact in an Antenna-Coupled Nanowire. *Nano Letters* 15: 1324–1330.
80. LaPierre, R.R., et al. 2013. III-V Nanowire Photovoltaics: Review of Design for High Efficiency. *Rapid Research Letters* 7 (10): 815–830.
81. Huang, M.H., et al. 2001. Room-Temperature Ultraviolet Nanowire Nanolasers. *Science* 292 (5523): 1897–1899.
82. Kind, M., et al. 2002. Nanowire Ultraviolet Photodetectors and Optical Switches. *Advanced Materials* 14 (2): 158–160.
83. Krishnan, R., M. Hahn, Z. Yu, J. Silcox, P. Fauchet, and T. Krauss. 2004. Polarization Surface-Charge Density of Single Semiconductor Quantum Rods. *Physical Review Letters* 92 (21): 216803.
84. Kozák, O., M. Sudolská, G. Pramanik, P. Cígler, M. Otyepka, and R. Zbořil. 2016. Photoluminescent Carbon Nanostructures. *Chemistry of Materials* 28: 4085–4128.
85. Wu, B., A. Heidelberg, and J.J. Boland. 2005. Mechanical Properties of Ultrahigh-Strength Gold Nanowires. *Nature Materials* 4: 525–529.
86. McLellan, J.M., et al. 2007. The SERS ACTIVIty of a Supported Ag Nanocube Strongly Depends on its Orientation Relative to Laser Light. *Nano Letters* 7 (4): 1013–1017.
87. Lee, Y.H., et al. 2011. Refractive Index Sensitivities of Nobel Metal Nanocrystals: The Effects of Multipolar Plasmon Resonances and the Metal Type. *Journal of Physical Chemistry C* 115 (16): 7997–8004.
88. Li, W., et al. 2013. Metal Ions to Control the Morphology of Semiconductor Nanoparticles: Copper Selenide Nanocubes. *Journal of the American Chemical Society* 135: 4664–4667.
89. González-Prior, J., et al. 2016. Oxidation of 1,2-dichloroethane Over Nanocube-Shaped Co_3O_4 Catalysts. *Applied Catalysis, B: Environmental* 199: 384–393.
90. Watson, A.M., X. Zhang., R. Alcaraz de la Osa., J.M. Sanz., F. González., F. Moreno., G. Finkelstein., J. Liu., and H.O. Everitt. 2015. Rhodium Nanoparticles for Ultraviolet Plasmonics. *Nano Letters* 15: 1095–1100.
91. Biacchi, A.J., and R.E. Schaak. 2011. The Solvent Matters: Kinetic versus Thermodynamic Shape Control in the Polyol Synthesis of Rhodium Nanoparticles. *ACS Nano* 5: 8089–8099.
92. Hao, F., et al. 2007. Plasmon Resonances of a Gold Nanostar. *Nano Letters* 7 (3): 729–732.
93. Lascola, R.J., C.S. McWhorter., and S.H. Murph. 2015. *Surface enhanced Raman scattering spectroscopic waveguide*. USPTO, Savannah River Nuclear Solutions, Llc, Aiken.

Chapter 3
Synthetic Strategies for Anisotropic and Shape-Selective Nanomaterials

Simona E. Hunyadi Murph, Kaitlin J. Coopersmith and George K. Larsen

Abstract This chapter gives an overview of the various approaches that have been taken to create anisotropic nanomaterials. The synthetic mechanisms of nanomaterials are being actively pursued due to the unique size and shape dependent properties that can be exploited for a myriad of applications. Nanomaterials have been synthesized in a gamut of shapes, sizes, and compositions. As their synthetic protocol progresses, scientists are becoming more and more creative in the fabrication of nanomaterials with very interesting architectures to tailor their properties for faster electronics, better resolution imaging, more efficient catalysts, among others.

Keywords Nanoparticle synthesis · Shape control · Growth mechanism · Chemical reduction · Seed mediated synthesis · Self-assembly · Electron beam lithography · Scanning probe lithography · Photolithography · Thin film · Vapor deposition · Templated growth

3.1 Introduction

Nanoparticles can be manufactured from the chemical approach, namely "bottom-up" (Sect. 3.1), where the synthesis begins at the atomic scale, or from the engineering approach, namely "top-down" (Sect. 3.2), where nanomaterials are "carved" out of larger materials (Fig. 3.1). Throughout history, new products were usually created by starting with large pieces of wood, stone, metal or other materials and reducing them to the right sizes and shapes.

The famous sculptor Michelangelo once stared at a large piece of stone for many months. Townspeople would ask him what he was doing and he would always

S.E. Hunyadi Murph (✉) · K.J. Coopersmith · G.K. Larsen
National Security Directorate, Savannah River National Laboratory, Aiken, SC, USA
e-mail: Simona.Murph@srnl.doe.gov

S.E. Hunyadi Murph
Department of Physics and Astronomy, University of Georgia, Athens, GA, USA

S.E. Hunyadi Murph et al. (eds.), *Anisotropic and Shape-Selective Nanomaterials*, Nanostructure Science and Technology, DOI 10.1007/978-3-319-59662-4_3

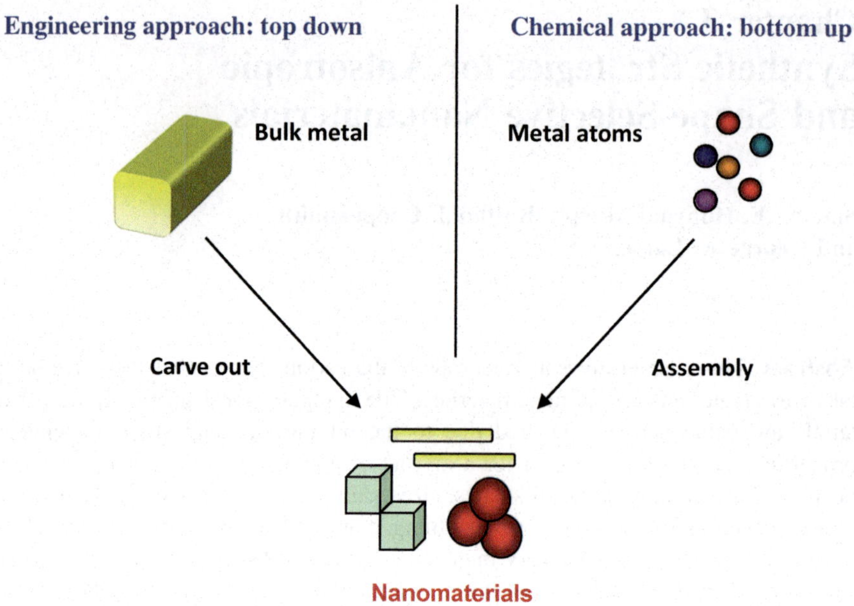

Fig. 3.1 Engineering ("*top down*") and chemical approaches ("*bottom-up*") to make nanometer scale materials

reply he was working. Three years later he unveiled the statue of David in front of the Palazzo Vecchio. A townsperson asked him how he could create something so beautiful from a large stone. He replied that David was always there, he just removed the stone that was covering him [1, 2].

Today, complex machines are used to bend, twist, cut, and meld materials into increasingly smaller sizes. As different as these might seem, both depend on "top-down" approaches to manufacturing. An alternative promising strategy towards nanotechnology on the nanometer scale is the "bottom-up" approach. Chemists assemble atoms and molecules one at a time in a predetermined and desired matter. Enormous progress has been made in the synthesis of inorganic nanospheres. Routinely, control over the diameter of nanoparticles leads to particle size distributions that are within 10% of the mean diameter, and frequently within 5%. Only since the mid-1990s and later have there been good synthetic methods to create nanoparticles of controllable sizes and shapes (other than spheres) [1, 2].

In either case, the combination of nanoscale top-down or bottom-up approaches give material and device designers a wide variety of old and new tools. The fabrication on nanomaterials is still, however, not a trivial task, especially at larger scales. Ultimately, the success of future nanotechnologies relies on the large scale production of nanoparticles of controlled size, shape, and crystal structure for successful incorporation in macrodevices [1, 2].

A wide array of nanoparticle shapes have been created from these techniques, including spheres, cubes, rods, wires and branched nanoparticles such as stars and tetrapods. This chapter gives an overview of the various strategies that have been applied to create a diverse range of nanoparticle sizes, shapes and compositions.

3.1.1 Bottom-Up Fabrication: The Chemical Approach

3.1.1.1 Overview

The most common route for the bottom-up manufacturing of metal and semiconductor nanoparticles (NPs) is through the reduction of molecular precursors (Fig. 3.2). Some popular reducing agents, such as cetyltrimethylammonium bromide (CTAB) and polyvinylpyrrolidone (PVP) also act as capping agents to stabilize the NPs. The reduction rates of the metal precursors, temperature, reductant-to-precursor ratio, ligands and strength of reductant all have an effect on the shape, size, and crystallinity, which affect the optical and physical properties [3–10]. The ligands on the surface are used to stabilize the NPs and change the surface energy at certain facets through preferential adsorption to achieve shape control [3, 4]. In NP synthesis, high energy facets grow at a faster rate until the nanoparticle is terminated with low energy facets [3]. Depending on the conditions, thermodynamic or kinetic control can be achieved at various stages of the synthesis for shape control [10]. The stabilization of crystalline facets is under thermodynamic control, which favors low surface area and termination with low energy facets [3, 5–10]. Kinetic control with slow precursor reduction can give rise to interesting morphologies and manipulating the thermodynamic-kinetic crossover can lead to heterostructured materials [5–9].

To create interesting morphologies, anisotropic growth occurs through asymmetric crystallographic direction-dependent growth. Asymmetry may exist either inherently in the crystal lattice, such as the hexagonal crystal structures, or through the breaking of cubic (e.g. face centered cubic and body centered cubic) symmetry [11]. Crystalline symmetry can be broken through the use of shape directing ions, facet specific binding coordinating compounds or by manipulating the flux of monomers that modulate the degree of supersaturation [4, 6]. The strength of the

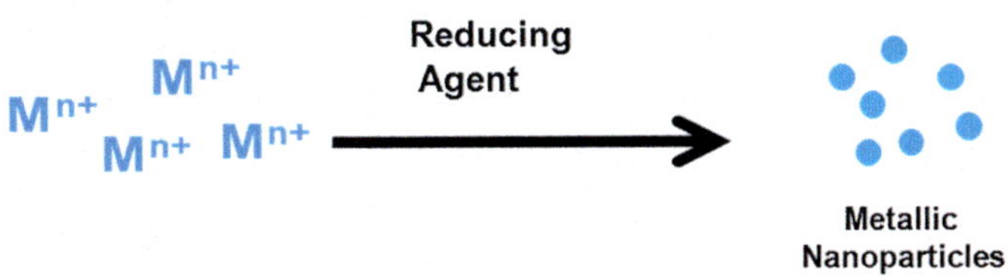

Fig. 3.2 Schematic showing the *bottom-up* manufacturing of nanoparticles through the reduction of molecular precursors

ligand-metal interaction and packing density control the rate of NP formation and the relative strength of facet stabilization controls the thermodynamically favored shape. For example, if all of the facets are equally stable, the most favored shape will be a sphere [12]. The growth and coalescence of single crystal seed NPs can lead to multi-twinned seeds, which introduce bond stretching and defects that prevent lateral growth to create nanorods and nanowires [11, 12]. Nanoparticle syntheses at low temperatures that are under kinetic control have been used to yield novel shaped nanomaterials at relatively low temperatures, including 2D triangle, rhombus and plate shapes [13]. Another route toward anisotropic growth is through polytypism, where multiple crystal structures exist in one NP. Typically, the different crystal structures in polytypic materials share a common crystal facet that promotes the growth of a material with a different crystal structure on the common facet. This has been exploited to create branched NPs such as semiconductor nano tetrapods [14]. Having high control in anisotropic NP syntheses opens the door for many new applications that require a higher surface area, specific shapes, or that have better activity with certain crystallographic facets. For example, AuNRs with the most exposed high energy facets had a higher kinetic rate constant for catalytic activity than AuNRs that contained the low energy [111] facets [4]. Shape control also allows one to tune NP optical responses, such as fluorescence or localized surface plasmon resonance (LSPR) for optical applications [4, 15]. The chemical reduction method has been used to create a variety of interesting morphologies including rods, cubes and branched nanoparticles.

Post synthetic treatment, including chemical etching, can also be applied to create interesting architectures. Chemical etchants have been employed to create interesting architectures such as CdSe/CdS dimers [16] and hollow silver-gold nanocups [7] through selective etching at certain facets and materials. The kirkendall effect, which occurs at the boundary between two metals that have different diffusion rates, has also been employed to create interesting morphologies, including dumbbell shaped bimetallic NPs.

3.1.1.2 Chemical Reduction

To create various nanoparticle shapes, such as nanorods or tetrapods, seed mediated approaches [6, 7, 10, 11] and seedless routes have emerged [17]. To create stable nanocrystals, the precipitation step must be highly controlled to prevent the formation of large, amorphous particles. Crystallization depends on the Gibbs free energy of the chemical reduction, which relies on the spontaneous formation of stable nuclei, ideally with a small size distribution and homogeneous crystal structure. The final shape typically favors a reduced total surface energy, which is the sum of the total faceted area weighted by their respective surface energies [12]. Facets are stabilized by increasing the number or strength of chemical bonds or by neutralizing the charge of the facet [12]. There are many proposed synthetic mechanisms for nanoparticle growth and the mechanism of growth that NPs undergo depend on the reaction conditions [18]. There are up to four stages of NP synthesis: (i) nucleation, (ii) growth, (iii) ripening and (iv) rapid consumption of remaining precursors [18]. These stages are dependent on concentration, the presence of surfactants, strength of reducing agent, temperature and the presence of impurities [18]. Careful selections of precursors and precursor ratios, temperature, glassware, and cleanliness can prevent issues with synthesis including the uncontrolled growth or dissolution of precursors.

During the nucleation stage, nuclei act as seeds to initiate crystal growth. Homogeneous nucleation occurs when nuclei form uniformly, leading to a small size distribution. Heterogeneous nucleation occurs due to inhomogeneities, such as impurities, bubbles, or rough container surfaces that can act as nucleation sites. Nucleation has a high energy barrier and for successful crystal growth, the nuclei have to stabilize above the critical radius (the minimum particle size needed to prevent dissolution) [12, 18]. Supersaturation is a driving force for crystal growth and leads to an increase in the nucleation rate: the higher the supersaturation, the smaller the final particle size. Supersaturation can be controlled by increasing the monomer concentration, where reducing the concentration of metal precursors decreases the chemical potential for crystallization [19]. The presence of surfactants or coordinating ligands can control crystallization by changing the surface free energy to stabilize the particles [6]. Ligands and surfactants undergo an adsorption-desorption equilibrium (i.e. "on-off" rate) that prevents aggregation and slows down growth [6, 12]. Nanocrystal growth typically requires metal reduction, so a stronger reducing agent will cause supersaturation at a lower temperature or at a faster rate.

The growth stage occurs after the nuclei form a supersaturated solution and continues until it reaches a saturated equilibrium. One proposed mechanism for the growth stage includes NP monomer diffusion and coalescence. During this mechanism, monomers coalesce into larger particles until the monomer concentration is too low or the reaction is quenched. During the growth stage, the atoms may also migrate on the surface or dissolve in the solution. NP coalescence has been measured in gold nanoparticle synthesis through small angle X-ray scattering (SAXS) by the rapid decrease in the number of particles in solution [18]. Controlling the consumption of precursors between the nucleation and growth stage is usually carried out to systematically tune NP size [6]. This is done through a hot injection method where fresh precursors are injected into the solution during NP growth. In another proposed growth mechanism, also known as the Finke-Watzky mechanism, growth occurs due to autocatalytic surface growth after a slow continuous nucleation [18]. After the nucleation stage in the Finke-Watzky mechanism, there is a fast random attachment followed by intra particle ripening until the particles surface energy are favorable against dissolution.

The ripening stage is driven by heat or energy input into the system and it causes the particle size to become more focused. There are two main proposed mechanisms: Ostwald ripening and digestive ripening [18]. In the Ostwald mechanism, smaller particles dissolve to allow for the growth of the larger particles [20]. This is based on the size dependent solubility of NPs, where the surface energy is much larger for smaller particles, leading to their dissolution. In the digestive mechanism, the larger particles are etched to allow for smaller particles to grow. The ripening stage may be quenched through temperature or precursor addition to prevent degradation or changes in the shape. The rapid consumption of precursor stage is theorized to be an autocatalytic mechanism between unreacted metal precursors at the final stage of a synthesis. This has led to an increase in NP size with a decrease in polydispersity [18]. The NP nucleation and growth mechanism depends on the reaction conditions including temperature, viscosity, precursor concentrations and pH [18].

3.1.1.3 Seed Mediated Approach

The seeded approach separates the nucleation and growth stage, which allows for more synthetic control because the energy required to grow on preformed seeds is much lower than the growth of new, independent nuclei [6]. This procedure, pioneered by Catherine J. Murphy's group, [1] is widely used to tailor the size of larger NPs and to create various NP shapes and heterostructures, such as core/shell NPs, or NPs with island morphologies. Figure 3.3 shows different seed mediated routes that have been employed.

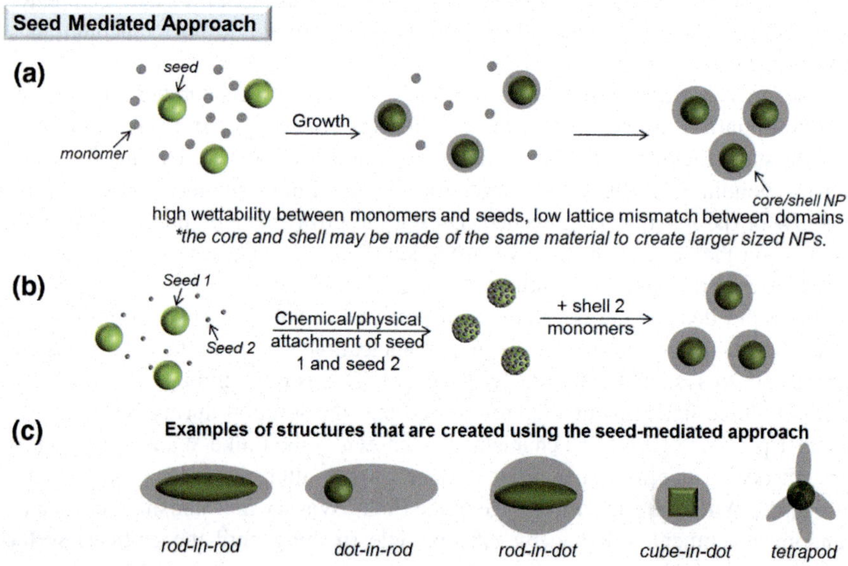

Fig. 3.3 NP seeded growth approaches; *A* seeds and monomers are combined, leading to core/shell type growth. The core and shell may be made of the same material or a different material with low lattice mismatch; *B* seeds of one type (*seed 2*, small nanocrystals or clusters) are chemically or physically attached to seeds of a different type (*seed 1*, typically nanocrystals), before being combined with the precursors to complete a shell formation which deposits onto the previously adsorbed seeds; *C* examples of some structures that have been created using the seed mediated approach

In the seeded approach, pre-synthesized seeds, which can be NPs or small metal clusters, are used as substrates for the growth of larger NPs [11]. Once the seeds are synthesized using mechanisms discussed in the beginning of this chapter, they are typically introduced into a new growth solution to create architectures such as the examples shown in Fig. 3.3c. The shape, size, composition, and crystal structure of the seed has a strong effect on the final morphology [15, 19, 21, 22]. Growth onto the seeds depends on concentration, rate of transport, and production and consumption of the precursor material; thus temperature, viscosity and rate of mixing all affect NP formation [12]. The interfacial energy, such as lattice strain, plays a role in the final morphologies since the NPs want to adopt the structure that minimizes the strain.

Gold and silver nanorods and nanowires of controllable aspect ratios were produced by a seed mediated growth approach [1, 2, 23]. In this procedure, metal salts are reduced in water, in air, at room temperature, typically with a strong reducing agent (sodium borohydride) to yield 3.5–4 nm spherical "seed" particles (Fig. 3.4). Sodium citrate or cetyltrimethylammonium bromide (CTAB) has been used as capping regents to avoid aggregation. Subsequently, growth solutions containing more metal salt, a structure-directing agent, a weak reducing agent, usually ascorbic acid (vitamin C), and "seeds" produce anisotropic nanoparticles. The presence of the structure-directing agent is crucial in obtaining nanorods as opposed to larger nanospheres; in our case, we have found that CTAB is uniquely

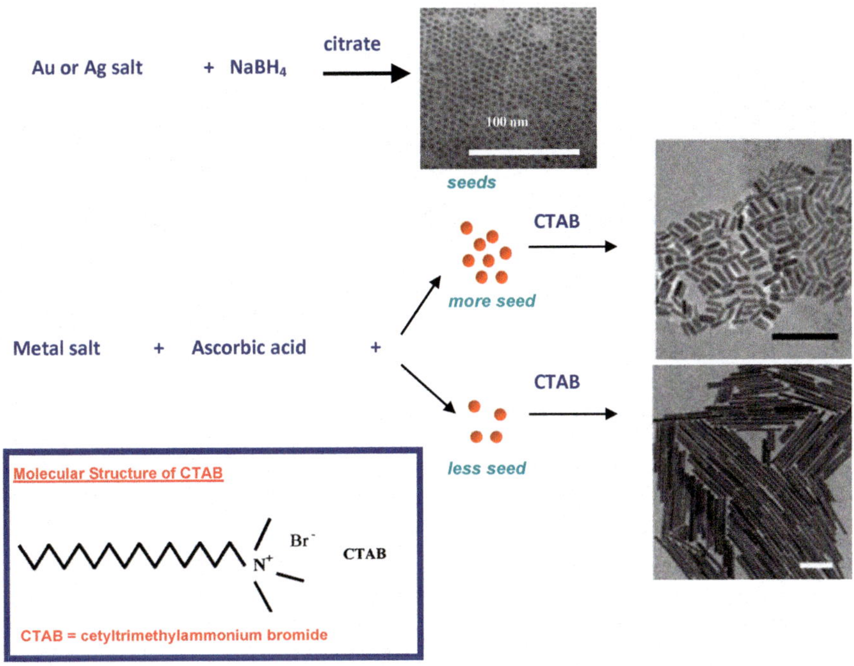

Fig. 3.4 Seed-mediated growth approach for the fabrication of gold and silver nanorods of controlled aspect ratio. Reprinted from Ref. [2]

suited to produce rods. One critical factor, that influences the growth mechanism of Au nanorods, is the presence or absence of additive metal ions. For example, the presence of ~5% Ag^+ raises the yield of gold nanorods to nearly 100%, compared to ~20–40% in the absence of Ag^+ [1, 2, 23, 24]. The highest aspect ratio gold nanorod obtainable with silver ion is about 6, compared to ~25 in the absence of silver (Fig. 3.5).

The growth mechanism of gold nanorods starts with the single crystalline seed particles. In the next step, surface binding groups are present and may preferentially bind to certain crystal faces of the seed, or leading to anisotropic growth. It was postulated that CTAB directs the growth by blocking the long axis crystal faces promoting metal growth on the short-axis faces to make nanorods. Subsequent addition of metal ions and weak reducing agent lead to metallic growth at the exposed particle faces [1, 2, 23, 24].

In gold nanorod synthesis, very small pentatwinned Au seeds can be synthesized with trisodium citrate and sodium borohydride or single crystal Au seeds can be synthesized using cetyltrimethylammonium bromide (CTAB) and sodium borohydride [5]. AuNRs that are produced from single crystal NPs have been found to be more dependent on the presence of directing ions, such as silver, than pentatwinned seeds [5]. The presence of other species also strongly influences growth.

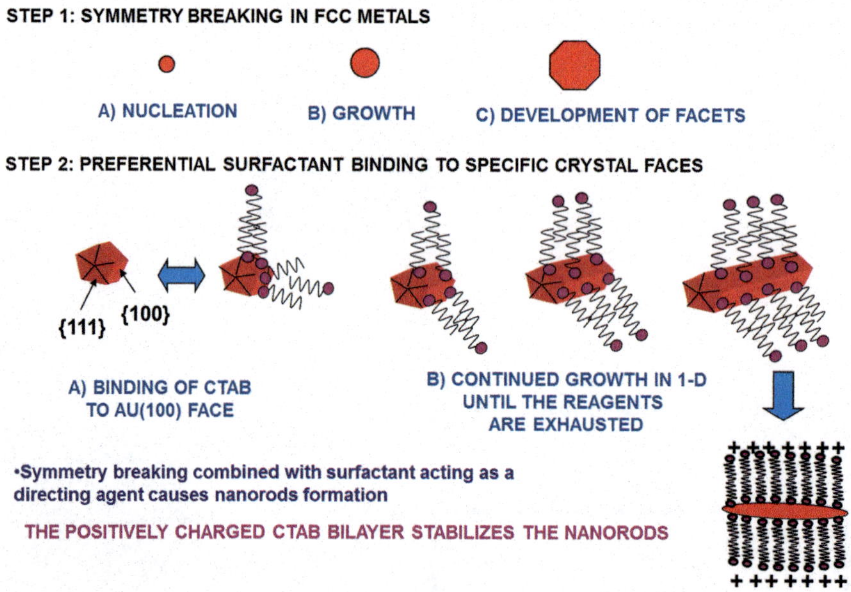

Fig. 3.5 Seed-mediated surfactant growth mechanism. The single crystalline seed particles have facets that are differentially blocked by surfactant (or an initial halide layer that then electrostatically attracts the cationic surfactant). Subsequent addition of metal ions and weak reducing agent lead to metallic growth at the exposed particle faces. In this example, the pentatetrahedral twin formation leads to Au 111 faces that are on the ends of the nanorods, leaving less stable faces of gold as the side faces, which are bound by the surfactant bilayer. Reprinted with permission from Ref. [1]. Copyright 2017 American Chemical Society

For example, AuNR synthesis includes a directing metal ion (Ag^+, Cu^{2+} or Pb^{2+}) and a cation surfactant, such as CTAB, which selectively binds to the [100] facet of AuNPs to hinder growth on that facet [4, 5, 11, 25–27]. There are several proposed mechanisms for the role of the metal ion in the selective control of the growth of high and low energy facets: (1) an underpotential deposition (UPD) mechanism, where the deposition of a monolayer of the directing metal ions (e.g. Ag^+) occurs at a lower potential than the Nernst potential, (2) a competitive surface selective binding between the metal ion and the surfactant, (3) ion guided formation of a soft template rod-shaped CTAB micelles, and (4) catalyzed oxidative etching of surface atoms [4, 5, 11, 27]. The mechanism depends on the reaction conditions and the metal ion used for shape control [11, 27]. Scanning tunneling electron microscopy (STEM) coupled with energy dispersive X-ray spectroscopy (EDS) was used to show the location of the silver in gold nanorods and "dog bone" structures. In the nanorods, the silver was deposited on the surface without any surface or facet preference and the dogbone NPs were found to have silver on the crevices, not at the locations where more Au deposited to form the dog bone shape [27]. This data suggests a mechanism that relies on non-specific silver binding such as in the UPD mechanism. AuNR facet control has also been achieved using copper ions, which was attributed to a selective competitive binding between Cu^{2+} and CTAB. As shown in Fig. 3.6, varying the copper and CTAB concentrations fine-tuned the nanorod shape.

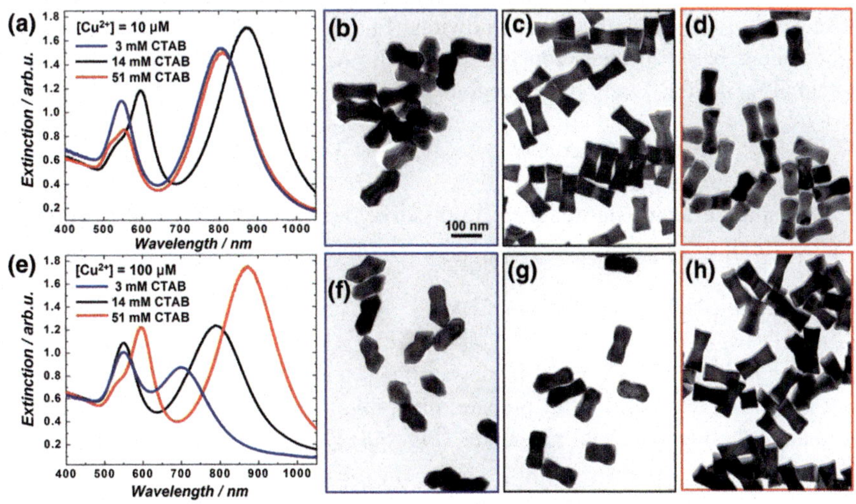

Fig. 3.6 Extinction spectra of Au NPs obtained through overgrowth of Au ETHH NPs in the presence of various concentrations of CTAB and fixed Cu^{2+} concentrations: **a** [Cu^{2+}] = 10 μM; **e** [Cu^{2+}] = 100 μM. TEM images of faceted Au nanorods obtained through overgrowth of Au ETHH NPs under various conditions: **b** [Cu^{2+}] = 10 μM, [CTAB] = 3 mM; **c** [Cu^{2+}] = 10 μM, [CTAB] = 14 mM; **d** [Cu^{2+}] = 10 μM, [CTAB] = 51 mM; **f** [Cu^{2+}] = 100 μM, [CTAB] = 3 mM; **g** [Cu^{2+}] = 100 μM, [CTAB] = 14 mM; **h** [Cu^{2+}] = 100 μM, [CTAB] = 51 mM. All TEM images share the scale bar in panel **b**. Reprinted with permission from Ref. [4] Copyright © 2016 American Chemical Society

In this work, the thermodynamic and kinetic competition between CTAB and the Cu^{2+} ions led to a unique set of NR shapes with high and low energy facets exposed [4]. The location and intensity of the plasmon bands were highly dependent on surface concavity and corner truncation, giving "fine-tuned" control of the optical parameters by altering the Cu^{2+}/CTAB ratio [4]. X-Ray photoelectron spectroscopy (XPS) was used to verify the packing density of CTAB on the various NPs. The data showed a higher CTAB concentration on the AuNRs with the exposed higher index facets and, when Cu(I) was adsorbed, it favored enclosing the NP in low index facets. The use of a shape directing metal ion has been used to create other unique architectures such as halide ions for gold nanoprisms [28], borohydride ions for Pt decahedrons [29], among others [30].

In another example, semiconductor quantum dot with CdSe/CdS core/shell morphologies were prepared using CdSe seeds either with wurtzite (w-CdSe) or zinc blende (zb-CdSe) crystal structures [21]. When *w*-CdS particles were grown onto the *w*-CdSe cores, quantum rods were produced; however, when the *w*-CdS particles were nucleated onto the *zb*-CdSe seeds, tetrapods were formed [21]. Shape control is achieved through faster growth on certain facets of the crystal [7, 11, 15].

Seeded chemical reduction can also be used to create interesting morphologies using galvanic replacement reaction (GRR). For example, gold nanocages have been prepared from silver nanocubes oxidation of the Ag atoms and reduction of the Au^{+} or Au^{3+} ions [31]. We reported earlier the synthesis of silver–gold bimetallic nanowires with tunable optical properties by a galvanic replacement reaction of gold salt with silver nanowires (Fig. 3.7) [2, 23, 32]. This reaction is spontaneous based of their reduction potentials: Ag^{+}/Ag^{0} has a standard reduction potential $E_0 = +0.8$ V and the analogous potential for $AuCl_4^{-}/Au^{0}$ is +0.99 V [32].

$$3Ag_{(s)} + HAuCl_{4(aq)} \rightarrow Au_{(s)} + 3Ag_{(aq)} + HCl_{(aq)} \quad (3.1)$$

By using an ammonia treatment to dissolve the inner silver core, according to Eq. 3.1, high aspect ratio Au nanotubes were prepared.

$$4Ag_{(s)} + 8NH_{3(aq)} + O_{2(g)} + 2H_2O_{(l)} \rightarrow 4Ag(NH_3)_2{}^{+}_{(aq)} + 4OH^{-}_{(aq)} \quad (3.2)$$

Silver nanowires were used as the sacrificial template in the galvanic reaction, according to Eqs. 3 and 4, to prepare silver–platinum bimetallic nanowires and, ultimately, hollow platinum nanotubes (Fig. 3.8) [2, 33, 34].

$$2Ag_{(s)} + [PtCl_4]^{2-}_{(aq)} \rightarrow 2AgCl_{(aq\,or\,s)} + Pt_{(s)} + 2Cl^{-}_{(aq)} \quad (3.3)$$

$$4Ag_{(s)} + [PtCl_4]^{2-}_{(aq)} \rightarrow 4AgCl_{(aq\,or\,s)} + Pt_{(s)} + 2Cl^{-}_{(aq)} \quad (3.4)$$

Figure 3.9 shows the effect of the Ag seed crystallinity on the final morphology. To produce silver cubes (AgNCs), fast nucleation and growth must occur to create

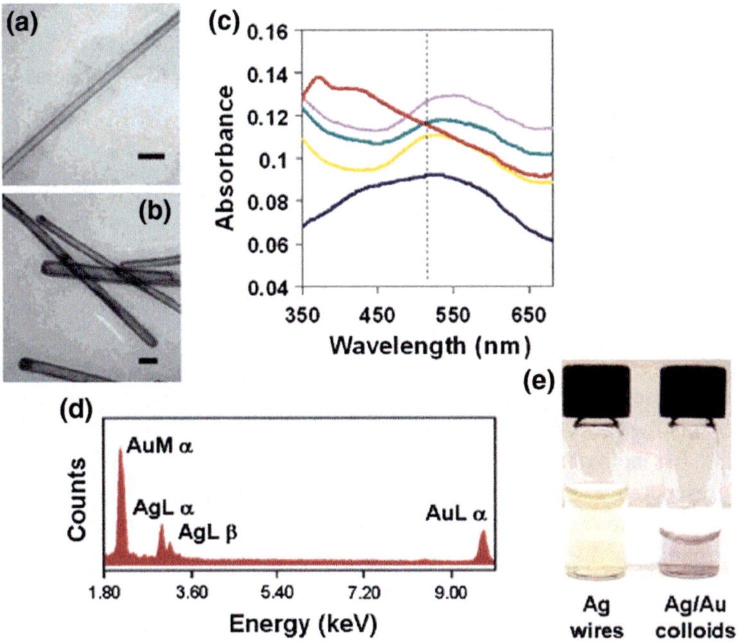

Fig. 3.7 **a** and **b** TEM micrographs of silver/gold bimetallic colloids with 19-nm wall thickness. Scale bars 100 nm. **c** Absorbance spectra of silver nanowires (*red trace*), silver/gold colloids with wall thickness 13 nm (*navy blue trace*), 19 nm (*orange trace*), 27 nm (*green trace*), and 31 nm (*purple trace*). **d** EDAX of silver/gold colloids. **e** Photograph of silver nanowires (*left*) and silver/gold colloids (*right*) with 19-nm wall thickness. Reprinted with permission from Ref. [23]. Copyright 2017 American Chemical Society

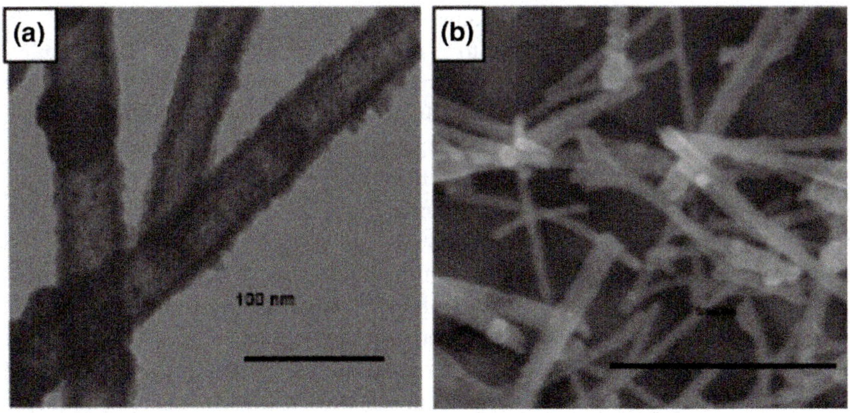

Fig. 3.8 **a** TEM, **b** SEM images of hollow Pt nanowires. *Scale bars* **a** 100 nm; and **b** 1000 nm. Reproduced with permission from Ref. [33]. Copyright 2017 American Chemical Society

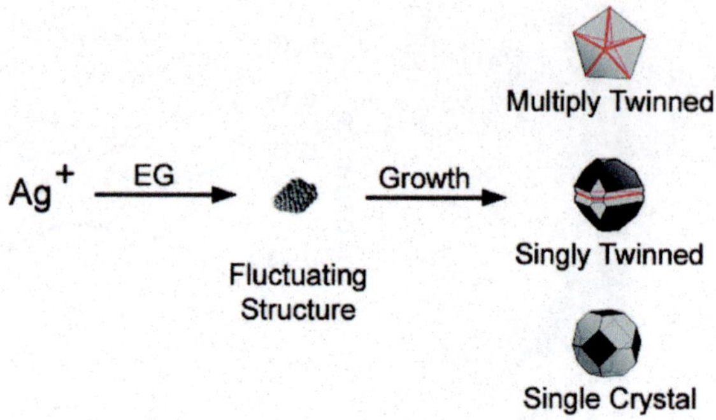

Fig. 3.9 Illustration of the reaction paths leading to well-defined silver structures as an effect of seed structures. Reprinted with permission from Ref. [35]. Copyright © 2007 American Chemical Society

single crystal seeds [19, 35]. The presence of PVP also stabilizes the {100} facet and prevents the formation of twinned seeds. If the concentration of the $AgNO_3$ is decreased, the thermodynamically stable multiple twinned seed is created due to the higher presence of the lower energy {111} facets. The defect is the highest energy site on the seed, leading to elongated growth at that site. Once the rods are grown, the PVP passivate the sides, leading to growth at the nanorod ends [19].

Seed mediated synthesis has been applied to create heterostructured materials which are typically made up two different constituents such as two metals or metallic-dielectric hybrids. Unlike alloys, the constituents remain discrete, so combining different materials allows one to couple the properties of both constituents in one particle to improve or enhance performance. The appearance of distinct metal domains rather than a continuous shell material depends on the shape and crystal structure of the seed, the wettability of the precursors material deposited onto the seed, rate of deposition and the interfacial energy between the different domains [6]. The growth of a separate domain may occur around an electron sink area, indicating the location of surface and lattice defects [36, 37]. Incompatibility in the crystal structure and lattice parameters typically leads to the creation of heterostructures with defined domains and a strong interface [6, 36–39]. By depositing one type of material onto a different type of material, the chemical, physical and electronic properties of the separate components change. For example, CeO_2 nanoparticles do not show any catalytic activity towards the decomposition of hydrazine into hydrogen; however, deposition of a metal such as Cu leads to catalytic activity [20]. These structures have also been applied to photocatalytic applications due to the charge separation at the interface [36, 37, 39], and components in electrical connectors [36, 39]. For example, we developed a seed mediated approach to precisely control the size, shape, and surface morphology of bimetallic Au–Pt nanomaterials (cubes, blocks, octahedrons, and dogbones) in a heterogenous epitaxial fashion by a wet chemical approach. Gold nanospheres, nanorods (different aspect ratios) and nanocubes were used as "seeds" for Pt growth in the presence of the mild reducing agent ascorbic acid and a cationic surfactant cethyltrimethylammonium bromide (CTAB) [33, 40] (Fig. 3.10).

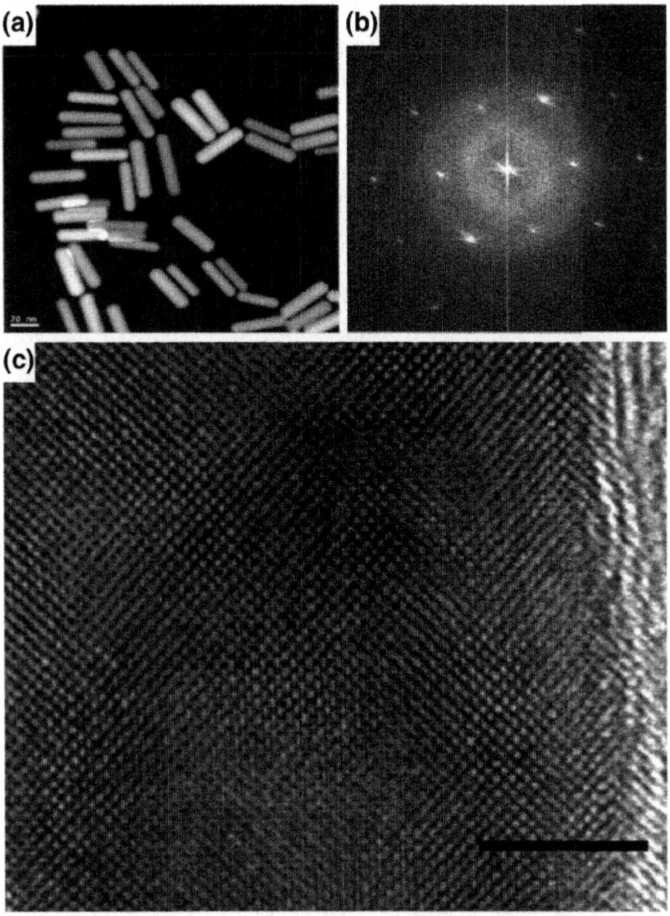

Fig. 3.10 Images of Au nanorods prepared in the presence of Ag ions. **a** SEM, **b** electron diffraction and **c** HRTEM. *Scale bars* **a** 20 nm; and **c** 5 nm. Reproduced with permission from Hunyadi Murph, S.E. et. al. [33]. Copyright 2017 American Chemical Society

Gold and platinum have a lattice mismatch of 4.08% which implies that a non-conformal growth route is favorable [41]. The high strain energy between the two lattice domains suggests that nanostructures with morphologies that differ from the initial core nanostructures would be produced. However, our studies show that the Au core nanoparticle initial shape was preserved after Pt island deposition [34]. HRTEM images and the corresponding electron diffraction pattern of individual Au–Pt nanorods show a single crystalline structure with a highly ordered continuous, parallel and uniform fringe pattern. An interplanar distance of 0.199 nm indicates a family of planes for Pt of {002}, for a *fcc* structure. [42] There is no twinning or defects present on these structures. These results demonstrate that even with a small mismatch (<5%) between lattices domains, conformal growth is possible [33, 34, 41].

The heterogeneous nucleation and growth theory postulates that the predominant growth mode of Pt onto a template is primarily determined by the lattice mismatch, as well as the differences in bond strength and electronegativity relative to the gold support. Epitaxial Pt overgrowth can occur by three possible modes: (a) layered growth (Frank-van der Merwe mode), (b) island growth (Volmer–Weber mode), and (c) intermediate or combined growth (Stranski– Krastanow mode).

HRTEM image of the Au–Pt octahedron grown from Au nanosphere seeds (Fig. 3.11) displays well-defined and continuous {111} fringes (d = 0.236 nm), which extend at a ~45° angle from Pt facet overgrowth [33]. The uninterrupted extension of the fringes across the core–shell interface suggests an epitaxial layered Pt growth mode on the spherical Au seeds. This indicates that the smaller Pt atoms (radius: 139 pm) deposit on the Au core (radius: 144 pm) by uniformly releasing

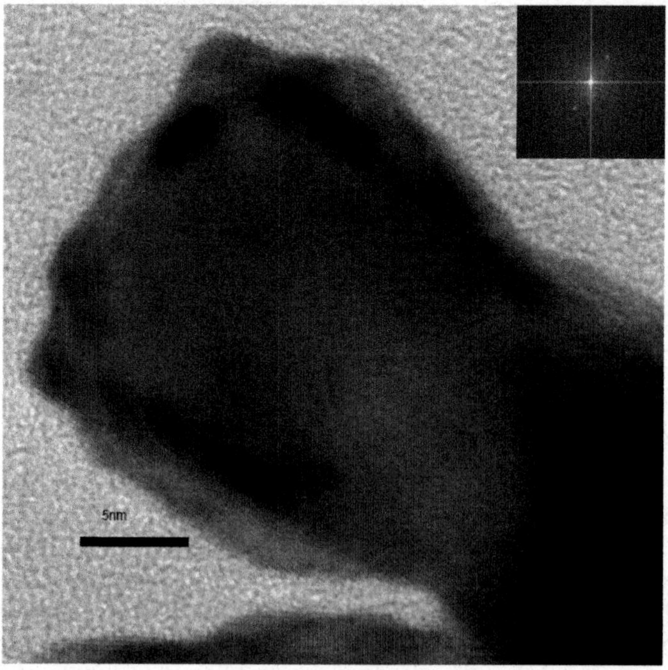

Fig. 3.11 HRTEM and electron diffraction pattern for Au–Pt nanorods. *Scale bar* 5 nm. Reproduced with permission from Ref. [33]. Copyright 2017 American Chemical Society

the lattice strain due to the minimal mismatch [33]. A HRTEM image of the Au–Pt "dogbone"-shaped nanostructure indicate a Stranski–Krastanow or a combined growth mode. Note that, although an epitaxial Pt/Au shell can be observed throughout the entire Au nanorod seeds, protrusions or nanoislands preferentially form at the endcaps and end-facet corners of the rod. These protrusions, which range in size of 3–6 nm, could also incorporate not only Pt but also excess Au since the template seeds were unpurified prior to Pt growth. The lattice arrangement of the dogbones matches that of the Au nanorod seeds because of the minimal mismatch ($\sim$4%), as evidenced by the continuous {002} fringes extending from the template [2, 33, 34] (Fig. 3.12).

Some challenges to the seeded growth approach include heterogeneous nucleation and comproportionation. Heterogeneous nucleation can occur under many different reaction conditions, especially if the chemical reaction favors monomer formation over increased growth onto the seeds [12]. If the seed concentration is too low, then the chances of seed-precursor collisions decrease, whereas too high of a concentration can lead to NP aggregation. The NP aggregates then compete with the seeds for growth to create large size distributions [12]. When metal ions are added to seed NPs of the same element, comproportionation may take effect. This is a redox reaction takes place between two atoms of the same element that have different oxidation states to form a product with same oxidation state, leading to NP dissolution [5]. Comproportionation can be avoided by adding a sufficient amount of reducing agent to reduce all of the metal ions [5].

One dimensional silver nanowires were produced by our group in an aqueous solution without any surfactant, polymer, or even externally added seed crystallites [2, 32, 43]. Recently, electron microscopy accompanied by molecular dynamics performed with the embedded atom method studies suggest that the Ag nanowires

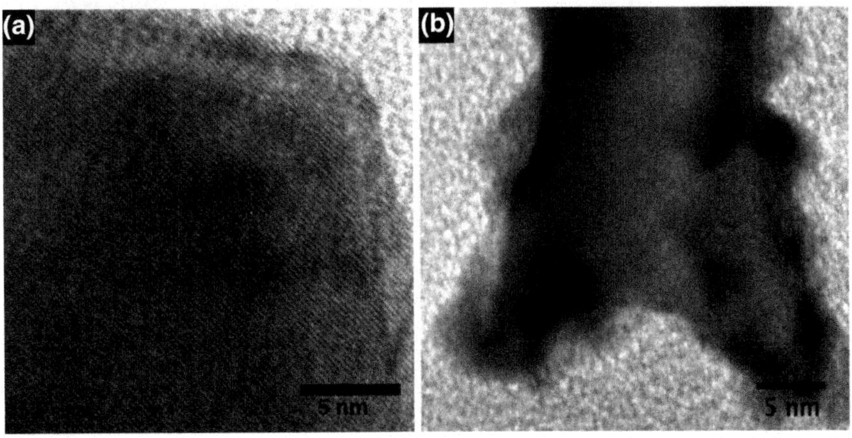

Fig. 3.12 HRTEM images of Au–Pt nanocatalysts: **a** octahedron and **b** "dogbone" shaped. Reproduced with permission from Ref. [40]. Copyright 2017 American Chemical Society

form through a seedless, surfactantless process by coarsening through an oriented attachment mechanism [1, 2, 32, 34, 41, 44]. Initially, silver nucleation centers were produced by chemical reduction of silver ions in boiling water, with sodium citrate and sodium hydroxide as additives in solution. These nucleation centers, with a twinned crystallographic orientation (Fig. 3.13), ultimately merged into fully grown silver nanowires (Fig. 3.14). This is a completely different mechanism from the seed-mediated growth approach, which has also been used to produce silver nanowires [41].

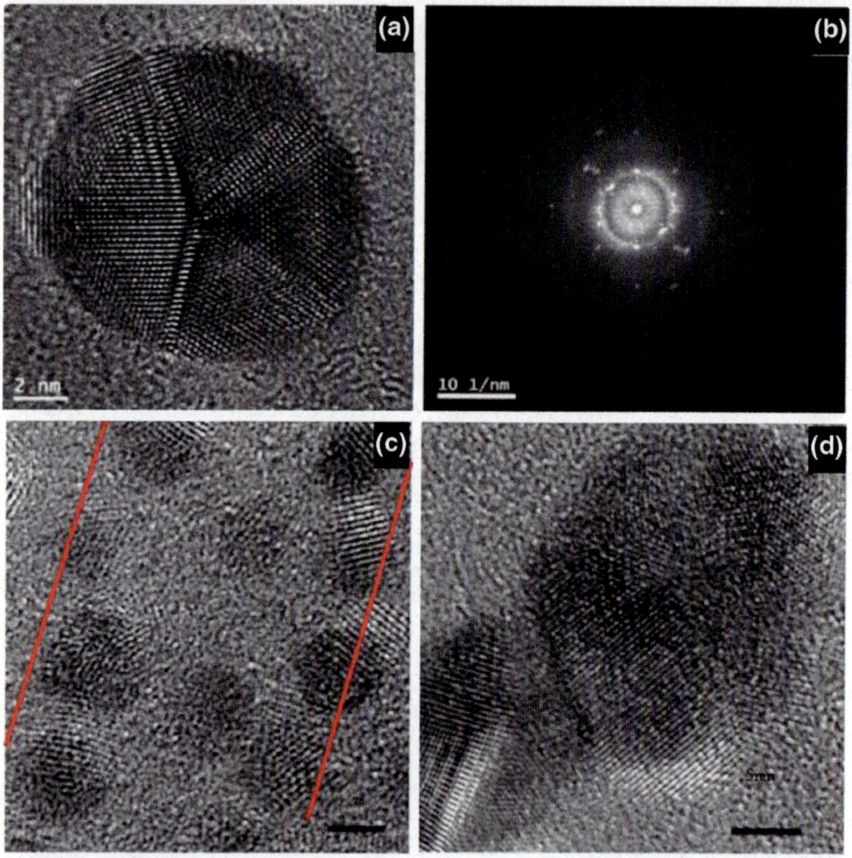

Fig. 3.13 **a** HRTEM on the 11 nm nucleation centers. A 5-fold penta-twinned crystallographic structure is observed. **b** Diffraction pattern recorded on the nucleation centers that suggest support the HRTEM regarding the penta-twinned crystallographic structure. The distance between the fringes in this case indicate (111) faces. **c** HRTEM on nucleation centers in the coarsening process via an oriented attachment mechanism; **d** an HRTEM image of two fused particles with favorable crystallographic orientations. *Scale bar* 2 nm, ac; 5 nm, c. Reproduced with permission from Ref. [43]. Copyright 2017 American Chemical Society

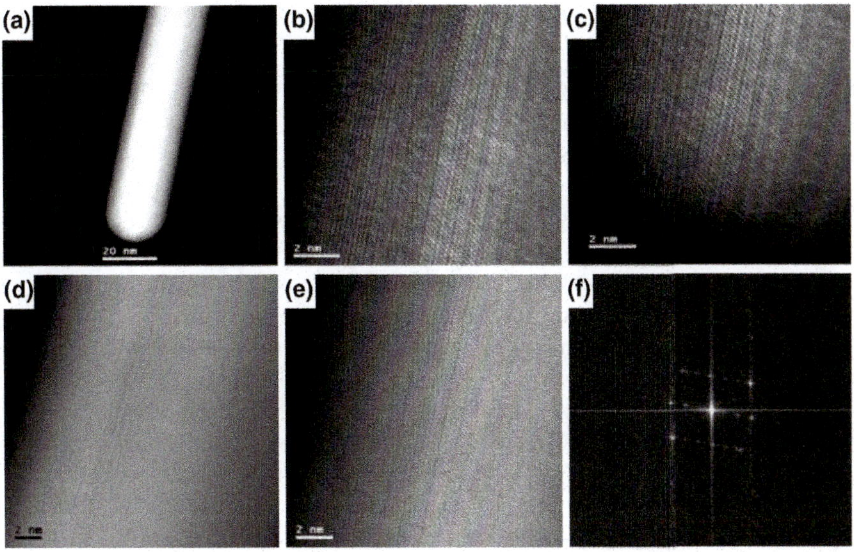

Fig. 3.14 **a–f** High resolution TEM images on silver nanowires prepared by a seedless, surfactantless method along with the **f** diffraction pattern. This suggests a staking fault of (110) or other faces of the Ag nanowires with a pentatetrahedral twin orientation. Reproduced with permission from Ref. [43]. Copyright 2017 American Chemical Society

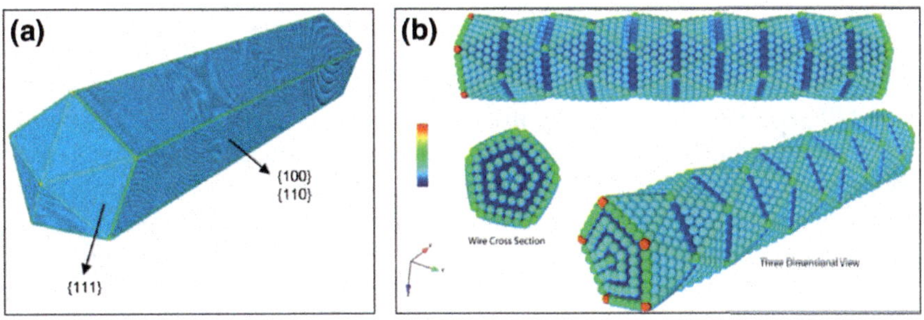

Fig. 3.15 **a** Equilibrium atomic positions for a pentagonal nanowire with internal twins. **b** Equilibrium atomic positions for a smaller pentagonal nanowire and a pentagonal wire with {111} surfaces. Reproduced with permission from Ref. [43]. Copyright 2017 American Chemical Society

Molecular statics simulations at 0 K were used to determine a few different equilibrium wire structures. Simulations on the silver nanowires reveal that the direction of evolution and growth of Ag nanowires is [100]. A combination of different faces Ag{100} or Ag{110}, or both were observed. Penta-twinned surfaces, in the form of five Ag {111} triangular faces were documented on the tips of the nanowires. Figure 3.15 shows a simulated low energy configuration, where the growth direction of the nanowire maintains low energy {111} surfaces while the sides are higher energy {100} and {110} surfaces.

3.1.2 *Solvothermal and Hydrothermal Synthesis*

NP synthesis typically requires high temperatures and solvents with high boiling points, which may be expensive and difficult to work with. For industrial applications, where efficient synthetic scale-up is required, the use of more common laboratory solvents may be more practical due to abundance or price. Unfortunately, they typically have lower boiling points that are not amenable for most synthetic protocols. To overcome these issues, the synthesis can be carried out at elevated pressures by heating them in a sealed vessel, such as an autoclave or bomb. This causes the pressure to increase above ambient pressure which increases the boiling point of the organic solvent (solvothermal) or water (hydrothermal). The increase in pressure can also change the supersaturation equilibrium and increase precursor reactivity. For example, at higher pressures, ethanol can act as a weak reducing agent [37]. Solvothermal and hydrothermal routes have been applied to synthesize an abundance of NP shapes and structures in the presence and absence of surfactants and ligands [37, 45–49]. Currently, the creation of some NPs with high crystallinity and low polydispersity, such as In_2O_3 cubes, have only been successful through solvothermal routes [48]. Solvothermal techniques allow the creation of NPs at milder conditions than typically required [49].

3.1.2.1 Microwave Irradiation

A growing trend in inorganic, organic and NP synthesis is microwave synthesis, which is typically coupled with high pressure solvothermal or hydrothermal synthesis [20, 46, 50–57]. Microwave irradiation (μWI) can directly target molecular precursors that interact with microwaves through two mechanisms: dipolar and conductive mechanisms. In the dipolar mechanism, polar molecules heat up by rotating in response to the electric component of the microwave's electromagnetic field, where in the conduction mechanism, ions move in the solvent to align to the applied electric field [50]. This leads to the conversion of electromagnetic radiation into heat. In conventional wet chemistry approaches, the vessel is heated using a hot plate, heating mantle or an oil, water or sand bath. This thermal energy is transferred to the NP precursors through the solvent which leads to temperature gradients, slower heating efficiencies and less controlled heating. By contrast, microwaves target dielectric materials directly, which results in faster kinetics and higher reproducibility thus making it more amenable to scale up. Superheating, which causes the boiling point of the solvent to increase under the influence of microwave irradiation, has also be achieved [50].

A variety of nanostructures, including metal NPs [54–56], quantum dots [46, 53] and carbon NPs [57] have been synthesized using μWI in polar and non-polar solvents [51–54]. Metal ions strongly couple to μWI to create hotspots that favor NP nucleation [54]. With meticulous selection of ligands, solvents and molecular precursors, the NPs can be selectively heated in the microwave [53, 54, 57].

The synthetic product is dependent on applied microwave power and cross-section of all of the species involved. The microwave cross-section is dependent on dielectric constant and a larger microwave cross-section, typically found in molecules with a large permanent dipole, leads to more dramatic heating in response to μWI [53]. The temperature ramping depends on the thermal conductivity of the solvent and the ability of each element to absorb μWI [53]. Compared to bulk heating methods, μWI creates NPs with larger nuclei that are crystalline and have a smaller size distribution [54]. Faster reaction rates have been measured using μWI, which is attributed to the higher local temperatures leading to increased collisions or the ability of the μWI to overcome intermediate synthetic barriers [53]. High quality NPs can be prepared using μWI within minutes [46]. An increase in QY, which is indicative of high crystallinity and low surface defects, was measured for quantum dots that were synthesized in a low boiling solvent in μWI compared to a high boiling solvent, presumably due to the increase in pressure that may cause a decrease in vacancies or defects [53]. Uniform nucleation is key for high quality NP synthesis and this can be achieved with μWI from the uniform flux of microwaves in the reaction vessel. Advances in microwave synthesis may allow for increased reproducibility between NP batches, synthetic automation and more amenable scale up due to the inside-out heating method and ability of select molecules to directly couple to μWI.

3.1.3 Self-assembly

Bottom-up fabrication also occurs via self-assembly, where individual building blocks spontaneously assemble into larger structures through molecular interactions [58–70]. These molecular interactions can be biomolecular, such as aptamer conjugation [71] and Watson-Crick base pairing of DNA [59–67], chemical such as the host-guest chemistry of curcubic[n]urils [69] and click chemistry [72], or physical such as the interfacial tension at liquid/air interfaces or through the presence of external fields. Coupling NPs can lead to interesting synergistic effects that do not occur in discrete NPs, such as the coupling of excitons or magnetic moments or energy transfer [58]. For example, phosphor-plasmonic dimers were self-assembled and displayed a plasmonically enhanced upconversion luminescence in the dimers [73]. The collective properties of various NPs can be extremely beneficial for applications such as single molecule sensing, nanoscale thermometry, imaging, electronics or catalysis. The creation of NP arrays is driven by kinetic and thermodynamic forces, where the local interactions between building blocks minimize the total energy of the system [74].

The interparticle forces can be tailored through NP shape, size and surface chemistry. A huge advantage of using NPs is the ability to couple the properties of the ligand layer with the properties of the NP, giving the ability to tune the individual components' properties. Nanoparticles prepared in solution are often capped with a variety of surfactant or capping agents used during the preparation procedures, including citrate, CTAB, PVP, etc. This provides surface passivation and helps prevent aggregation of nanoparticles. NP surface chemistry is important to provide steric or electrostatic repulsions that can be used to tune the interparticle distance and strength of assembly [74]. Whether through electrostatic or covalent bonding, nanoparticles can self-assemble to form secondary architectures with properties dictated not only by those of the individual building blocks but also by the spatial arrangement and coupling interactions between them [1, 24, 34]. Anisotropic NPs can be created, either synthetically or by tailoring the ligand layer to create Janus particles, to further modify the final assembled structure. The assembly must be controlled to prevent uncontrolled growth due to the absence of repulsive forces or the presence of non-specific interactions.

Self-assembly has been achieved in solution or using templates. A drop-casting approach followed by solvent evaporation can lead to the creation of uniquely patterned assemblies (Fig. 3.16). In this system, self-assembly is a result of the physico-chemical properties and structural affinities of the building block nanoparticles. The combination of solvent evaporation from the sample drop and interfacial hydrophilicity or hydrophobicity as well as the balance between van der Waals forces, capillary action, surface tension, and hydrophobic or electrostatic interactions, etc. is responsible for a variety of assemblies and agglomeration features (agglomeration, end-to-end, end-to-side, etc.).

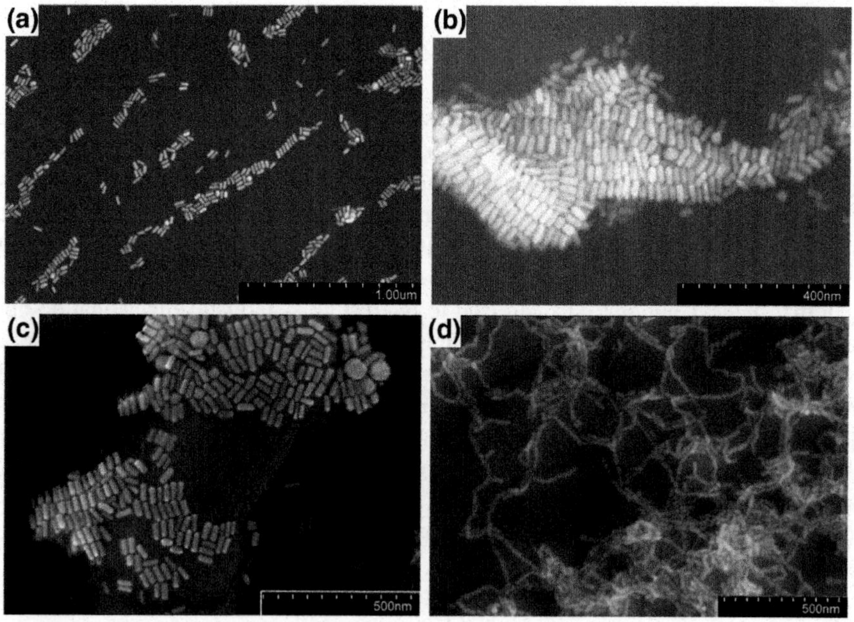

Fig. 3.16 SEM images of self-assembled Au nanorods via solvent evaporation

In solution, NPs can be spontaneously assembled into clusters, sheets or higher ordered assemblies through the minimization of attractive and repulsive forces [58–69]. The creation of superstructures, such as larger crystalline lattices, form as a result of long-range ordering [59–61, 75]. A highly specific route for self-assembly is through the use of oligonucleotide hybridization, which takes advantage of the Watson-Crick base pairing interactions to control assembly. Aptamers can also be used in these assemblies to incorporate sensing capabilities. In the work by Paul Rothemund, interesting nanoscale DNA structures were formed through DNA origami via self-assembly as shown in Fig. 3.17 [59]. This used a scaffolding approach; where one long DNA strand ($\sim$7000 bases) was assembled into various nanoscale structures by changing the smaller "staple" strands that link the scaffold together [59]. This research provided a route for the design of DNA origami into complex shapes with high yield [59].

The molecular recognition properties of nucleotides have been used to assemble NPs into various shapes, including three dimensional crystal lattice formations and clusters [60–66]. The strength and shape of the assembled clusters can be tuned through the size and shape of the NP building blocks, base-pair selection, the number of DNA strands on the NP surface and temperature, solvent and electrolyte concentration choices. The interparticle distances can be tuned through the length of the oligonucleotides and linker strands. For example, AuNP-DNA building blocks were assembled into different crystallographic patterns (face centered cubic vs body centered cubic) by changing the sequence of the linker strands, as shown in

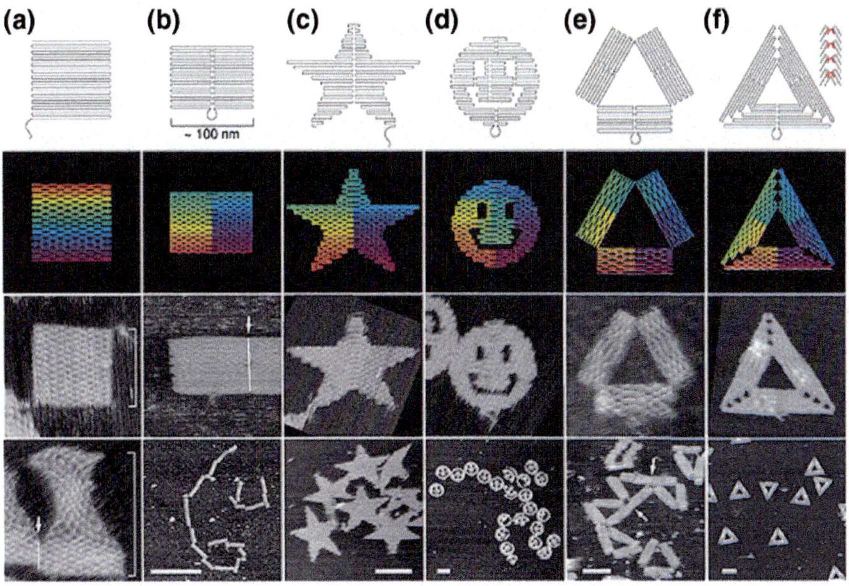

Fig. 3.17 Nanoscale DNA complexes created via self-assembly. Reprinted with permission from Ref. [59]

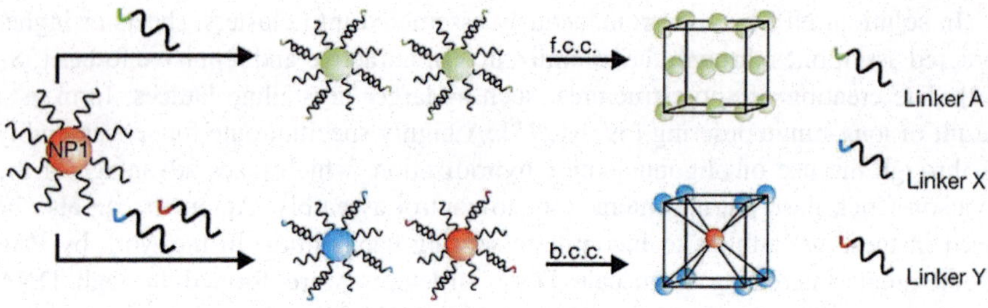

Fig. 3.18 Gold nanoparticle-DNA conjugates assembled into different crystallographic arrangements by changing the DNA linker sequence. Reproduced with permission from Ref. [60]

Fig. 3.18. Crystalline assemblies formed using DNA typically show well defined scattering rings in SAXS due to the formation of superlattices [60, 61].

Other small molecules have been used to create crystalline assemblies. For example, the *in situ* formation of benzoates during the "ligand-free" synthesis of ZrO_2 nanoparticles caused the creation of crystalline superlattices [75]. In this work, changing the reaction parameters led to the formation of different shapes [75]. Further heat treatment of these structures led to weight loss and the shrinking of the crystalline lattice with shape retention, indicating removal of the organic species [75]. This work indicates that, in some cases, the assembly directing ligands may be removed with preservation of the final structure. This is highly beneficial for applications that are affected by the presence of the ligand layer. The ability to create amorphous and crystalline assemblies allows for the ability to tune the synergistic properties for applications, including medical treatment and diagnostics, forensics testing, optical and electronic applications among others.

3.2 Top-Down Fabrication: The Engineering Approach

3.2.1 Overview

The first portion of this sub-chapter describes the many ways that chemical, kinetic, or thermodynamic forces within a solution can be harnessed to drive nanoparticle growth toward a desired shape or size. There is a wonderful elegance to coaxing such a wide array of beautiful structures out of spontaneous processes. In contrast, this next section describes a completely different approach to nanoparticle synthesis, one that attempts to circumvent natural ordering by using brute force to directly control the shape of matter at the nanoscale. The elegance of these so-called "top-down" techniques instead resides in the remarkable engineering that goes into the instruments and materials that enable these processes. By using top-down techniques, there are virtually no limits to the shapes and sizes of nanoparticles that can be produced through clever engineering. The iconic image of "IBM" written in xenon atoms is a classic representation of such ingenuity. In fact, the only real limits to the top down approach are the typical high cost and low throughput of producing nanoparticles in this manner. In spite of these drawbacks, top-down engineered nanoparticles are well represented in the research literature, and there is hope that the costs will be driven down in the future through innovation, allowing for the industrial production of matter that has been sculpted at the nanoscale.

3.2.2 Nano-Lithography

Many top-down techniques are grouped under the umbrella term of lithography. The name is borrowed from the traditional lithographic printing technique, where an image is drawn on a plate using a chemically resistant material. The plate is then immersed in a solution that etches away the portions of the plate not protected by the chemically resistant image. Initially, lithographic prints were created from wax images drawn on smooth limestone. Hence, the name "lithography," which comes from the ancient Greek words *lithos* and *graphein* meaning "stone" and "to write," respectively. Modern advanced lithographic nanofabrication techniques are conceptually very similar to the traditional method. However, now writing is done using beams of particles, and instead of writing into stone, images are written into modern polymer or dielectric materials called, "resists."

3.2.2.1 Photolithography

Photolithography is the most widely known and used top down micro-/nanofabrication technique, with a long history of use in the semiconductor industry for the production of integrated circuits [76–78]. The photolithographic process is very straight forward (Fig. 3.19): a light-sensitive polymer, called a photoresist, is coated onto a substrate. Next, an opaque mask with the desired pattern or image is placed on top of or in front of the photoresist; the photoresist is then exposed to UV light irradiation (typically, wavelengths: $\lambda = 193–436$ nm) that is collimated by the mask. Finally, the exposed film is placed in a developing solution, replicating the pattern from the mask in the photoresist. Whether or not the image is directly or inversely replicated in the photoresist depends on its chemical properties. If the photoresist material is *positive*, the UV illuminated portions of the photoresist will be removed by the developer—that is, light exposure increases the solubility of the photoresist. In comparison, *negative* photoresist materials become less soluble in the developing solution after UV light exposure, causing areas that were not exposed to light to be etched away during development and creating the inverse image of the photomask. In either case, the obtained patterned polymer layer can be used as is, or it can serve as a template for further processing, such as electroplating or thin film deposition.

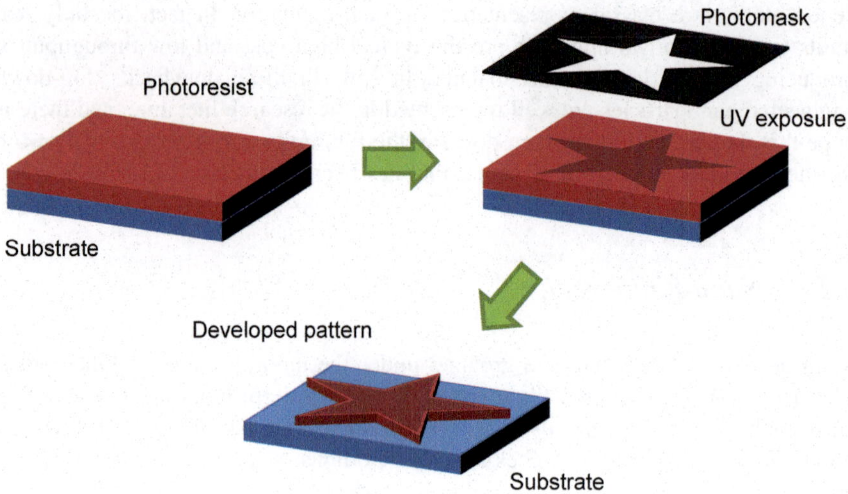

Fig. 3.19 Schematic of the photolithography process using negative tone photoresist

While photolithography is quite relevant for industrial micro-/nanofabrication, there are some challenges in the application of photolithography toward the development of anisotropic or shape-selective nanoparticles. The first challenge is that of resolution. As a general guideline, the minimum feature size that can be obtained within a plane using projection photolithography in ideal conditions is based on the Rayleigh criterion for imaging: $L_{min} = 0.61\ \lambda/NA$, where λ is the wavelength of light and NA is the numerical aperture of the lens imaging the mask onto the photoresist [79]. The numerical aperture of a lens is given by $NA = n \sin \theta$, where n is the refractive index in which the lens is working and θ is the half angle of the cone of light that enters the lens. This implies that the smallest feature size that can be obtained in air using $\lambda = 193$ nm is $L_{min} \approx 120$ nm, which is not quite in the true nanoscale regime (1–100 nm). Clever engineering has allowed researchers to push beyond this limit to yield films with <50 nm resolution [80–83]. However, further gains in resolution are non-trivial, requiring advanced photoresist materials [84], optics (e.g., hyperlenses) [85], light sources (e.g., extreme UV) [86], and patterning techniques (e.g., triple patterning) [87]. The second main challenge to the use of photolithography for fabrication of anisotropic nanoparticles is that photolithography is inherently a two-dimensional technique. Patterning is done in a single plane at a time. Of course, multiple planes can be built up to generate a three dimensional pattern, but given the resolution limits of the technique, patterning three-dimensional anisotropic nanoparticles by this method is impractical.

Photolithography does excel at chip patterning, as demonstrated by its dominance in the semiconductor industry. Thus, photolithography is an excellent method for patterning shape-selective and anisotropic nanoparticles that are synthesized using other methods. For example, thiol-stabilized gold nanoparticles are UV light-sensitive, and when exposed, the thiol molecules are oxidized, causing aggregation and densification of the nanoparticles. Thus, the photomasks can be employed to directly oxidize and aggregate these gold nanoparticles into patterns on different substrates [88]. Similarly, quantum dots can be fabricated into specific patterns through the incorporation of light-sensitive precursors into the photoresist [89]. Alternatively, photolithography fabricated patterns can serve as a foundation for further nanoparticle growth, which can create wafer scale patterns of anisotropic nanoparticles, such as nanowires grown by electrodeposition [90]. The ability to pattern anisotropic nanoparticles onto a wafer is especially beneficial for applications that interface with electronics or optics, such as biosensors or light-emitting diodes [91, 92].

3.2.2.2 Scanning Beam Lithography

Scanning beam techniques are similar to photolithography, but instead of using a photomask to shield portions of the resist, patterns are directly written into the resist using a beam of radiation that moves across the surface. The main advantages of these types of technique are enhanced resolution and greater design flexibility. These come at the expense of much lower throughput. However, these tools are extremely useful for research and development or other situations where long lead times are not as important as the ability to write arbitrary patterns with high resolution.

Light can be employed here as well, through direct laser writing. In this case, the resist is transparent to the laser's wavelength, but the non-linear process of multiphoton absorption becomes relevant at sufficiently high intensities, which changes the solubility of the resist in a confined region just around the focal point of the laser [93]. The focal point of the laser can trace out virtually any pattern, even in three dimensions. Therefore, direct laser writing can sculpt intricate shapes and patterns into a photoresist, creating micro-/nanostructures with any conceivable shape. Some of these remarkable structures are shown in Fig. 3.20. The resolution of direct laser writing can beat the diffraction limit, and resolutions of $\lambda/10$ and even $\lambda/50$ have been achieved [93]. However, the use of light as an energy source does limit the achievable resolution, but researchers can obtain even smaller structures by utilizing different beams of particles.

Electron beam lithography is the most widely recognized direct write lithography technique. The main advantage is the extremely small wavelength of electrons, $<0.4 \times 10^{-9}$ m for energies >100 keV, which greatly improves diffraction limits over optical methods. Indeed, resolutions <10 nm are consistently reported [94, 95]. Like direct laser writing above, electron beam lithography typically relies on a serial process, where a beam of electrons draws out a pre-determined pattern on a resist surface changing the solubility of the irradiated material, but higher energy electron beams can directly remove material from surface, as well [97]. Resists are organic or inorganic polymers, such as poly(methyl methacrylate) or hydrogen silsesquioxane, and may be negative or positive tone. The resolution of the electron beam lithography technique is controlled by the type of resist, environmental

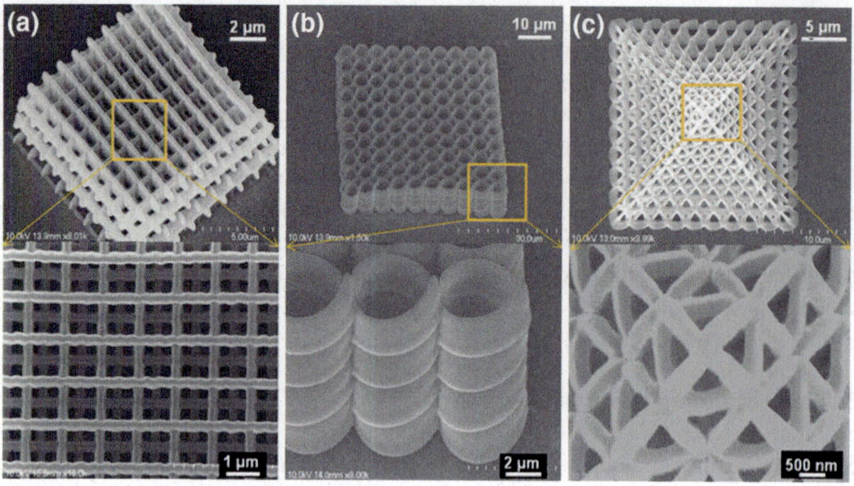

Fig. 3.20 SEM images of three-dimensional photonic crystals with different geometries, **a** woodpile, **b** spiral, and **c** pyramid, written by direct laser writing. Figure reproduced from Ref. [96] with permission from Springer

factors, resist thickness, and developing process [98]. In general, the minimum achievable feature sizes are not related to the diffraction limit, but to scattering and secondary electron generation within the resist layer.

A closely related technique is focused ion beam lithography, which relies on ion bombardment instead of an electron beam. In this case, ions may come from liquid metal ion sources (e.g., Ga, Au, Si, Ge) or from gas-field ionization sources (e.g., He, Ne, H, Ar). The focused ion beam interacts with materials in a number of different ways, all of which can be used to pattern micro-/nanostructures. In addition to changing the chemical properties of a resist through bond dissociation, amorphization, and/or intermixing, the beam of particles can mill or etch away portions of material, or it can deposit ions into a substrate and create patterns of alloyed or doped material [99]. It can also be used to build up materials from reactive precursors through a technique called focused ion beam deposition [100]. With gas field-ionization sources, the resolution can be better than <10 nm, which is comparable to electron beam lithography. However, due to the larger particle sizes of the beam, focused ion beam lithography processes require thinner resist layers than electron beams due to shallower penetration of the heavier ions. On the other hand, the direct milling and etching of material by ion beams reduces the number of steps and simplifies fabrication processing compared to electron beam lithography.

In general, scanning beam lithography techniques have the same benefits and challenges of photolithography techniques, only with greater resolution at the expense of slower throughput. Direct laser writing and focused ion beam milling and deposition have the added benefit of improved three-dimensional nanofabrication. Given their design flexibility, these fabrication techniques are well-suited to investigate how shape and size affect nanostructure properties. However, given their high cost and long fabrication times, this technique may not be readily available or relevant to many researchers. Additionally, the obtained sample sizes are typically relatively small and are generally confined to the surface of a substrate, which places limits on experimental characterization. Thus, scanning beam lithography might be an excellent choice for investigations into the optical and electrical properties of shape-designed nanoparticles, but the catalytic properties of such nanostructures will be more challenging to assess.

3.2.2.3 Scanning Probe Lithography

There is a direct relationship between different nanomaterial characterization techniques and different nanofabrication methods. In the sections above, we saw how light, electron beams, and X-rays all could be used to sculpt material at the nanoscale. These interactions also enable optical and electron microscopy and X-ray diffraction, among many other analytical techniques. Similarly, scanning probe microscopy techniques, such as scanning tunneling microscopy (STM), atomic force microscopy (AFM), and near-field scanning optical microscopy (NSOM), can also be used to pattern materials at the nanoscale and beyond. In fact, the well-known image of the IBM logo written in atoms was created using a STM in 1989. Researchers at IBM have recently taken this concept a step further and produced a stop-motion animated short film, *A Boy and His Atom*, which was created by moving carbon monoxide molecules around on a copper substrate using STM. Clearly, resolution is not an issue with such techniques.

A finely tipped scanning probe can interact with a surface through mechanical, thermal, electrical, and chemical processes, to enable many different patterning techniques [101]. Analogous to the lithography practices above, these probe-surface interactions can either physically remove or add material, induce chemical changes to the surfaces that are the end goal for an application, or they can be exploited for further processing in nanoscale pattern development (Fig. 3.21). An attractive feature of scanning probe lithography is the relatively simple technology, as the most common techniques are based on ubiquitous AFM instrumentation and can be conducted in open atmosphere on a variety of different materials [102]. Another benefit is that scanning probe techniques allow for simultaneous imaging and patterning, which greatly improves shape fidelity over different patterning areas. However, similar to the other direct write lithography methods, writing speeds and throughput are slow, especially for STM methods requiring ultra-high vacuum [97]. Additionally, the patterning by scanning probe techniques is confined to one surface plane at a time, making three-dimensional patterning cumbersome. These considerations indicate scanning probe techniques are not the ideal choice for the synthesis of anisotropic or shape-selective nanoparticles. On the other hand, given the relative simplicity and capabilities of AFM-based techniques, scanning probe techniques are attractive for patterning of nanoparticles into more complicated

Fig. 3.21 (*Left*) Original image of the Mona Lisa, scaled and pixelated for input into the model for thermochemical nanolithography experiment. (*Right*) Experimental rendition of the Mona Lisa with a total width of just ~32 μm produced by consecutive touchdowns of a scanning probe tip via the thermochemical nanolithography technique. *Scale bar* 10 μm. Reprinted with permission from Ref. [103]. Copyright 2013 American Chemical Society

architectures for electronics or other wafer-based applications, and they are also ideal for patterning on novel nanomaterials themselves, such as graphene and graphene oxide [104–106].

3.2.3 *Pattern Transfer and Templates*

The previous sections described techniques that allow researchers to sculpt and create new shapes and patterns to form anisotropic and shape selected nanoparticles from bulk materials. However, nanoscale patterns and particles can be found in natural systems or can be derived from spontaneous processes. Such patterns can serve as templates to engineer unique and complex nanostructures. Patterns can be generated on the original template, or alternatively, transferring the pattern to a different substrate can be done by a number of different ways, either by molding, embossing, imprinting, or contact printing [97].

3.2.3.1 Nanosphere Lithography

Perhaps the most pervasive and economical templating method is nanosphere lithography (NSL) [107]. NSL is a simple method that relies on colloidal self-assembly for producing arrays of nanoparticles on a substrate (Fig. 3.22). When at an interface between phases (e.g., air-water), colloidal spheres tend to aggregate into layered hexagonal-close-packed (*hcp*) crystalline structure, and by transferring this structure to a solid support by drying, dip-coating, etc., a *hcp* mask or template is formed on a substrate [108]. Colloidal spheres are generally composed of polystyrene or silica and can be found in a range of sizes from 20 nm to 50 μm plus, which allows nanoscale patterns to be easily reached. Typically, monolayers of spheres are used for patterning, while bilayers are less common [109]. Multi-layered structures are often backfilled for the production of inverse opal structures [110]. The most common technique of using the *hcp* mono-/bilayer templates is as sacrificial masks for material evaporation, where the gaps in between spheres creates patterns on the substrates below [108]. These patterns can be used directly, or they can be used as an additional mask for etching processes to produce ordered anisotropic Si nanopillar arrays [111]. Alternatively, self-shadowing by the *hcp* lattice itself can lead to the creation of anisotropic structures on the colloids themselves [112]. Such structures can be Janus structures or can even exhibit chirality [113–115]. NSL is a versatile and simple technique that allows for the production of a wide variety of nanoparticles with tunable shapes and sizes that can be used on the substrate or dispersed in solution. The shape of the template does place limits on the achievable structures, and there are some challenges to creating very large area single crystal monolayers. However, the technique continues to evolve and is very useful tool for researchers interested in anisotropic and shape-selected nanoparticles [116–118].

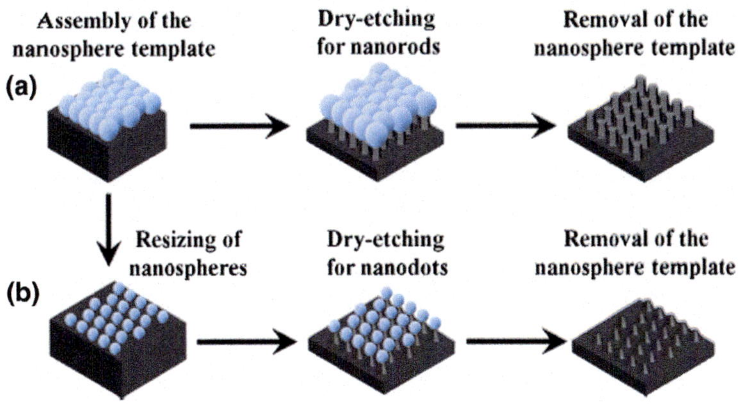

Fig. 3.22 Schematic of nanosphere lithography technique to produce nanorods from a Si-based substrate. Figure reproduced from Ref. [119] with permission

3.2.3.2 Spontaneously and Naturally Occurring Templates

Electrochemistry offers alternative routes for the creation of nanopatterned templates. A notable example is that of the spontaneous, self-ordering of porous oxide layers formed during anodization of metals [120]. Passing an anodic current through a metal in an electrolytic solution will cause an oxidation layer to grow and form on the surface, and by careful selection of the electrolyte and experimental parameters, the oxidation layer will form into a self-organized array of metal oxide nanopores or nanotubes. While this process can be done for many different metals, the process is most often seen in the literature for alumina on aluminum or titania on titanium [121]. Typically, titania nanotubes are used as active or photocatalytic materials [120], while alumina nanopores are often used as passive support materials or used as templates for nanopillar growth [122]. Pore diameters in anodic aluminum oxide can range from 10 to 400 nm, and pore lengths can be 10–150 μm [121]. Highly anisotropic nanoparticles can be formed readily by backfilling these pores with single or multiple materials and etching away the aluminum support. An interesting application of such structures is as catalytic bimetallic nanomotors that move autonomously in solution [123, 124].

Block copolymers can form spontaneous periodic patterns due to phase separation between the different chemically dissimilar components [125]. These patterns can consist of repeating domains of spheres, cylinders, gyroids, or lamellae, depending on synthetic conditions [126]. While the details of such self-assembled processes belong to the "bottom up" synthesis group, it is worthwhile to mention here, as very often these polymers are etched and backfilled with inorganic materials to engineer a new type of nanostructured film [127]. Due to their periodicity and nanoscale patterns, block copolymers are often used as templates for the formation of anisotropic and chiral optical components and photonic crystals [128–130].

There are no shortages of naturally occurring templates that can be used for patterning other materials. The natural ordering that occurs in living systems makes life possible and is a source for template-based nanofabrication of anisotropic nanoparticles. Most reports in the literature describe templates based on smaller life forms, such as bacteria and viruses [131], which is expected given the sizes of these microorganisms. Small and highly anisotropic nanoparticles have been derived using such techniques. For example, cobalt and nickel nanowires with a diameter of 3 nm and lengths up to 500 nm can be fabricated using the tobacco mosaic virus as a template [132]. Larger organisms can be useful, too. Plants and insects can have remarkable structures (e.g., hydrophobic lotus leaf surfaces, moth's eye, etc.) and are common sources of biotemplates [133–135]. Even the wings of cicadas have been used to pattern nanopillar arrays for anti-reflection coatings [136].

3.2.4 *Thin Film Growth*

Thin film growth is a bottom up nanofabrication method that has more of an engineering spirit than other bottom up techniques, and therefore, is included here. The engineering core of the technique is captured in the collection of heaters, vacuum pumps, lasers, electron and ion beams, microwave and plasma sources, and other instruments that can make up a thin film growth system [137]. The two major divisions of thin film growth are physical vapor deposition and chemical vapor deposition. These techniques were historically employed by semiconductor, optics, and other industries to create coatings of materials on a substrate. However, by judiciously changing the deposition parameters, a variety of different nanostructures can be grown on a surface.

3.2.4.1 Physical Vapor Deposition

Physical vapor deposition (PVD), as indicated by the name, uses a source of physical vapor to create thin films. The term "physical" here means that the constituents of the vapor are generally non-reactive, and merely physically deposit onto the substrate and accumulate into a larger structure of the same composition as the vapor. For a number of reasons, including vapor and gas purity, PVD is conducted in a vacuum chamber, usually under high vacuum. The vapor is generated from an initially solid source through a variety of methods, such as laser ablation, thermal evaporation, electron beam bombardment, or sputtering by plasma. Substrates are placed in the vacuum chamber such that the vapor plume is incident upon them, and since they are at a lower temperature than the vapor, the vaporized atoms condense on the substrate forming thin films. PVD is most commonly used in conjunction with some other lithographic technique, such as the ones described above, to generate micro-/nanopatterns on a surface by masking selected areas. However, PVD can also be used as a single technique to fabricate unique anisotropic nanostructures due to the self-shadowing effect (Fig. 3.23).

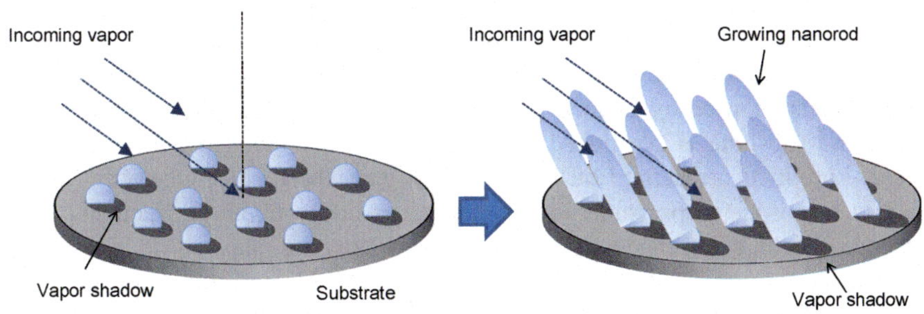

Fig. 3.23 Schematic illustrating the self-shadowing effect of physical vapor deposition at oblique vapor incidences

Self-shadowing arises from the use of high vacuum, where the vapor particles travel with mean free paths that are greater distances than the dimensions of the deposition chamber [138]. This ensures that vapor atoms travel in a ballistic, or line of sight, manner before impinging on the substrate. The first vapor atoms to arrive at the substrate surfaces condense and coalesce into nucleation centers. For a flat film that is placed perpendicular to the direction of vapor flux, these different nucleation "islands" grow in number and size until they coalesce into a continuous film. If the substrate is instead tilted at a large angle with respect to the vapor flux, the initial islands cast large vapor shadows that prevent the growth and development of additional islands, and instead, only the initial nucleation centers grow, creating an array of nanorods that are tilted in the direction of the vapor flux. If the tilted substrate is given a slow azimuthal rotation, the vapor shadow traces out a helical shape, creating micro-/nanohelices. If the substrate is instead rotated quickly, straight nanorods are formed. Discrete rotations can create additional shapes like chevrons and square helices. Thus, by controlling the substrate orientation (tilt and rotation) with respect to the vapor flux, a wide variety of different nanostructures can be sculptured from physical vapor (Fig. 3.24) [140]. This technique, known as glancing angle deposition (GLAD) or dynamic shadowing growth (DSG), applies to almost any material that can be evaporated, making it a powerful technique to investigate anisotropy and shape effects of different nanomaterials [114, 141, 142]. There are some material challenges in obtaining more complicated shapes beyond rods for nanostructure lengths less than 100 nm, especially for metals, but advanced techniques such as seeded growth and cryogenic substrate temperatures can circumvent these issues [143]. Additionally, the obtained nanostructures are typically amorphous or polycrystalline due to the non-equilibrium growth conditions, but again, advanced techniques have allowed greater control over the crystallinity of GLAD nanostructures, even permitting the growth of single crystals [144, 145].

3.2.4.2 Chemical Vapor Deposition

Chemical vapor deposition represents a collection of different methods, and like PVD, is primarily used for creating conformal thin film coatings [146]. In contrast to PVD, the CVD vapor phase precursors are volatile and are intended to react on the substrate surface to produce the desired coating. Typically, volatile vapor precursors are carried by an inert gas into a reaction chamber (e.g., tube furnace), where the substrate is positioned and coated, and any unreacted precursors and reactive products are carried off by the carrier gas. The reaction chamber may be held at atmospheric pressure, low pressure, or under ultrahigh vacuum. More advanced techniques can couple with microwaves and plasmas to assist with material deposition [147, 148]. Though, CVD may be performed as simply as evaporating silane monolayers onto Si wafers in a desiccator [149].

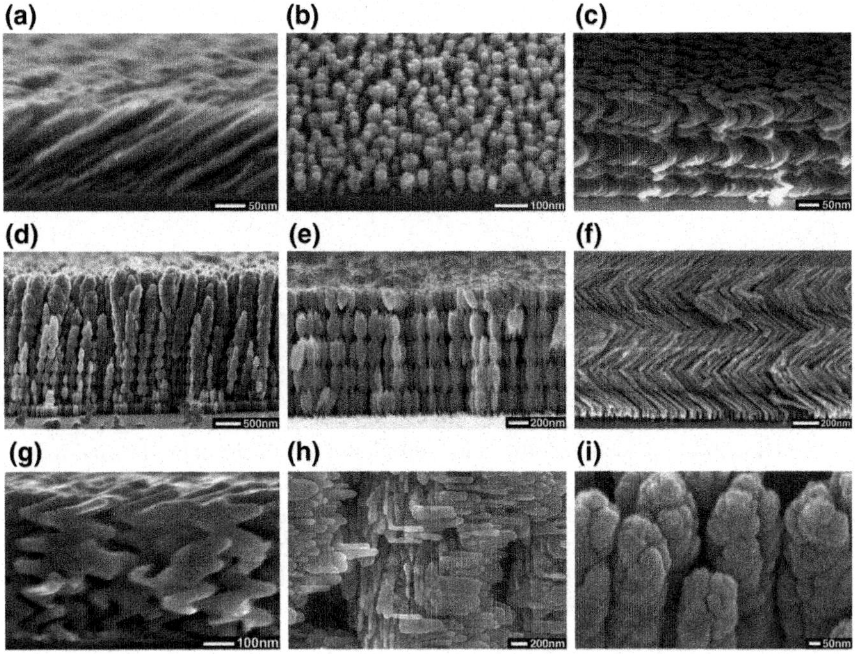

Fig. 3.24 SEM micrographs of thin films deposited with the UHV GLAD system **a** Si oblique columns, **b** Si pillars, **c** Si spirals, **d** Si rugate film on glass substrate, **e** Si rugate film with an antireflection layer on glass substrate, **f** Si square helix film, **g** Cu zig–zag film, **h** Si square spiral film, **i** Si envelope corrected rugate on glass substrate. Reprinted from Ref. [139], with the permission of AIP Publishing

Perhaps the most recognized use of CVD for anisotropic nanomaterials synthesis is the production of graphene and carbon nanotube, which is the preferred method of creating graphene films [150, 151]. Given the considerable interest in graphene, a great deal of information is available in the literature regarding this process and its variations [152, 153]. In addition to carbon-based structures, a host of different metal, oxide, and semiconductor nanowires have been grown using chemical vapor techniques [154–160]. Many, but not all, CVD nanowire growth processes require the use of a metal catalyst on the surface of the substrate in order to nucleate nanostructure growth. This method is known as the vapor-liquid-solid (VLS) method (Fig. 3.25) [161].

While the finer details of the VLS mechanism are still being investigated, it is generally understood that the liquid metal catalyst alloys with the incoming volatile vapor precursors (vapor-liquid interface), and as the alloy becomes supersaturated, material deposits at the liquid-solid interface creating a nanowire [162]. The balance

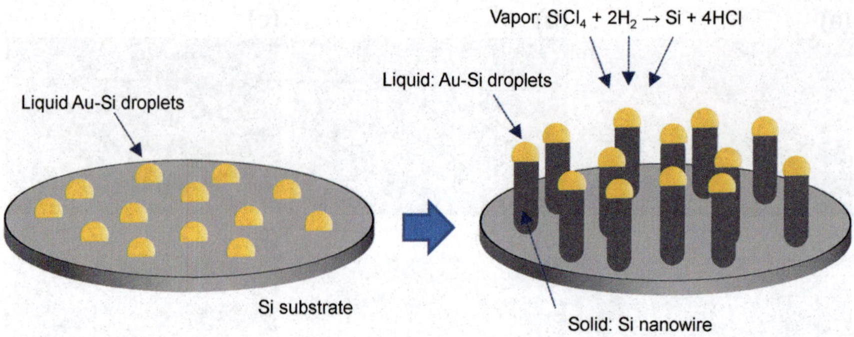

Fig. 3.25 Schematic illustrating Si nanowire growth via the vapor-liquid-solid method

between incoming vapor and the alloy droplet sizes determines the nanowire's morphological properties, and therefore by controlling these parameters, the structural parameters of the nanowire can be engineered along lengths less than 10 nm [163, 164]. Additionally, the directional growth of the wires can be modulated by changing the surface chemistry and introducing defects, which provides more control over the nanostructure [165–167]. However, in comparison to PVD, CVD provides much less control over the nanoparticle shape and is currently limited to the production of nanowires, nanotubes, or monolayer coatings. CVD processes do have a number of advantages for anisotropic nanomaterial synthesis: (i) the method is relatively simple and scalable; (ii) nanostructure growth occurs where the flowing vapor molecules react, allowing for conformal coating of non-planar substrates; and (iii) the obtained nanowires are generally single crystals.

3.3 Classification

3.3.1 Metal and Metal Oxides

Like bulk materials, nanoparticles can be classified by their material composition, and perhaps this the best way to classify nanoparticles, as physical properties of materials can change at the nanoscale. Metallic nanoparticles, especially noble metal nanoparticles, have received a large amount of attention, and have been found useful for a number of different applications in catalysis, sensors, electronics, and optics. Metallic nanoparticles are amenable to different synthetic methods. However, due to their greater reactivity and large surface areas, many metal nanoparticles are pyrophoric and require air-free techniques. Metal oxides represent

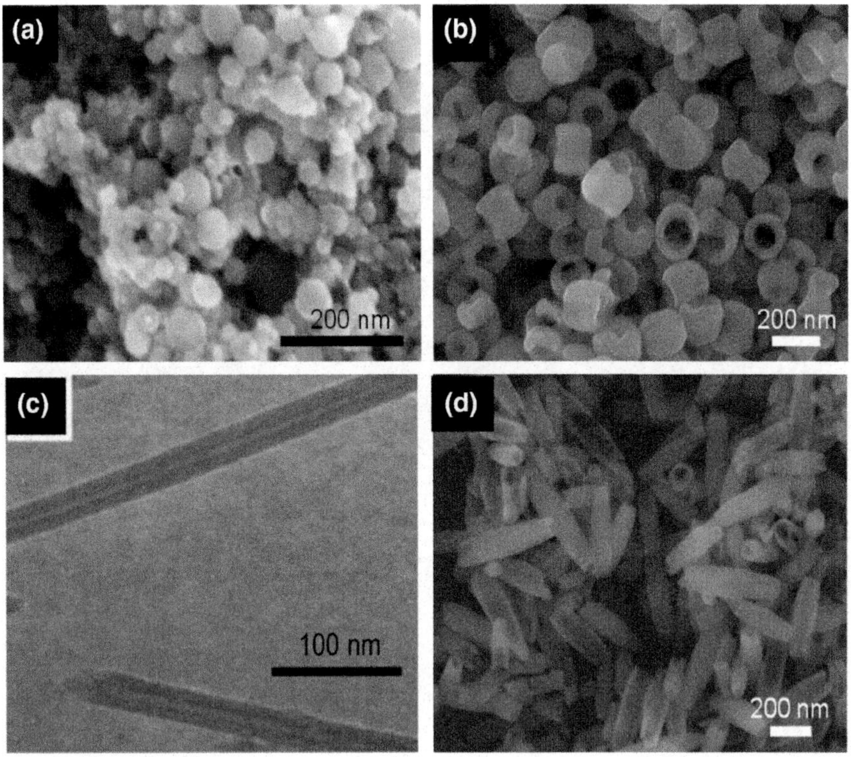

Fig. 3.26 SEM and TEM images of anisotropic metal oxides **a**, **b**, **d** Fe_2O_3 and **c** SiO_2. **a**, **b**, **d** Reproduced with permission from Ref. [170]. **c** Reprinted with permission from Ref. [23]. Copyright 2017 American Chemical Society

another large class of nanomaterials. Metal oxides (Fig. 3.26) are generally widely available, non-toxic, and stable, which make them excellent candidates for clean energy and biomedical applications, and an extensive amount of research has been devoted to such investigations. Since most metal oxides are well-behaved, a wide variety of different application methods have been used for their synthesis. Finally, it is important to note that the reduction/oxidation equilibria of metal oxides may change as a function of surface area, or nanoparticle size, which can alter the phase stability of the different oxides and make certain phases thermodynamically unstable at smaller sizes, hindering their synthesis [168, 169].

A variety of metal-metal oxide composite nanoparticles with a variety of structures have been created (Fig. 3.27). This allows one to couple the properties of

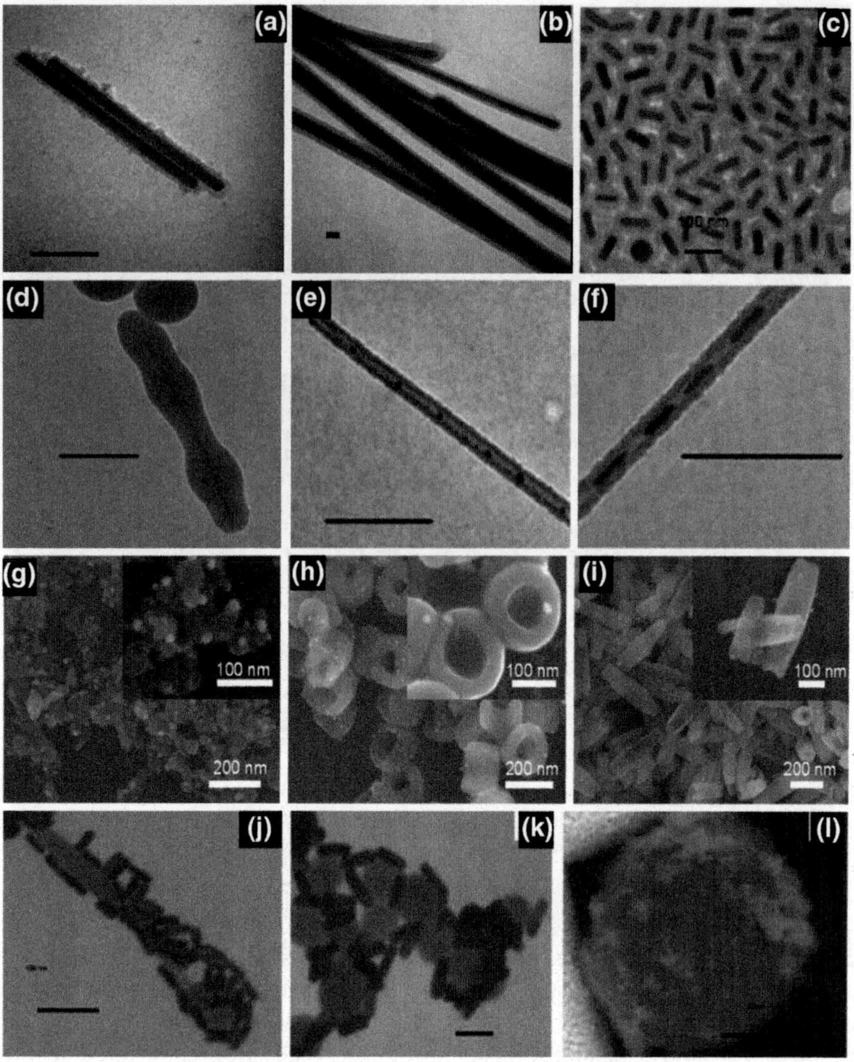
(a)
(b)
(c)
(d)
(e)
(f)
(g)
100 nm
200 nm
(h)
100 nm
200 nm
(i)
100 nm
200 nm
(j)
(k)
(l)

◂**Fig. 3.27** TEM and SEM images of **a** Au nanorods- coated with a layer of iron oxide. *Scale bar* 100 nm. **b** silica on silver nanowires. *Scale bar* 40 nm; **c** Au-titania rods. *Scale bar* 100 nm; **d** silver-silica dumbbell, *Scale bar* 100 nm; **e**, **f** Silver silica nano-peapods, *Scale bar* 100 nm; **g**, **h**, **i** Fe_2O_3 spheres, rings and tubes decorated with Au nanospheres. *Scale bar* 200 nm; **j** Silica nanowires decorated with Au nanorods. *Scale bar* 100 nm; **k** silica nanospheres decorated with Au nanorods. *Scale bar* 100 nm, **l** silica micro-sphere decorated with Au nanospheres. *Scale bar* 100 um. **a**, **b** Reproduced with permission from Ref. [23]. **c** Reproduced with permission from Ref. [33]. **d**, **e**, **f** Reproduced with permission from Ref. [171]. **g**, **h**, **i** Reproduced with permission from Ref. [170]. **j**, **k**, **l** Reproduced with permission from Ref. [33]. Copyright 2017 American Chemical Society

the metal and the metal-oxide for cost effectiveness, increased efficiency, interesting synergistic properties, among others [24, 170–172].

3.3.2 Semiconductor Nanostructures

Semiconductor nanomaterials, or quantum dots, are crystalline materials made up of inorganic compounds that have band gap energies (E_g) that are larger than a metal but smaller than an insulator. At room temperature, energy can be put into the system to excite the electrons across the bandgap, leading to the creation of an electron-hole pair. Once the excited electron and hole recombine, energy is released as photons (light) or phonons (heat) [173–175]. The energy of the excited electrons is dependent on E_g, which can be tuned by changing the size and material of the nanocrystal. As the size increases and more atoms are introduced, the density of orbitals increases and the size of the bandgap decreases. This leads to a red-shift in the photoluminescence with increasing size. Doping the semiconductor can also change the energies of the valence band and conduction band, changing the optical and conductive properties of the nanoparticles. Doping has also led to the creation of "hot electrons" which are electrons that have kinetic energy larger than E_g which can lead to improved photovoltaic and photocatalytic properties [176].

Quantum dots are favored over organic dyes due to their exceptional photostability against photobleaching, longer photoluminescent lifetimes and improved quantum yields. Unlike organic dyes, however, quantum emitting particles such as quantum dots, quantum wires and quantum rods display fluorescence intermittency called blinking [177, 178]. Although the physical origin of blinking is unknown, it has been suppressed by the addition of a thick layer of a higher E_g semiconductor around the quantum dot to prevent limitations in applications such as medical imaging and single particle tracking. Other applications take advantage of this fluorescence intermittency. For example, in super-resolution optical microscopy, the blinking is used to distinguish between individual quantum dot emitters to overcome the wavelength dependent limitations of traditional microscopy [177, 179]. Quantum dots are also employed in electronics or for solar energy conversion due to their photocatalytic properties [16, 180]. Quantum dots are typically made up of heavy metals including GaAs, InAs, PbS, PbSe, CdS, CdSe.

Quantum dots are similar to small proteins, and their photoluminescence is sensitive to the presence and nature of adsorbates. We have prepared CdS nanoparticles through an arrested precipitation method and deployed these nanomaterials as luminescent probes of DNA structure [181]. Sequence dependent conformational flexibility of DNA is of great interest due to its implications for drug–DNA and DNA–protein interactions. The counter-ion atmosphere surrounding DNA plays an important role in its structure, dynamics, and packaging. We investigated the effect that various monovalent and divalent cations have on the binding of 4.5 nm CdS quantum dots to oligonucleotides that have sequence-directed intrinsic structure. We have shown that the photoluminescence of 4.5 nm CdS nanoparticles surface-activated with Cd^{2+}, Mg^{2+}, and Zn^{2+} ions is quenched by "straight", "bent", and "kinked" oligonucleotides (Fig. 3.28). The relative binding constants of these DNAs are higher for the harder Mg^{2+} and Zn^{2+} ions compared to the softer Cd^{2+} ions, suggesting a higher affinity of these more biologically relevant ions for the DNA phosphate backbone.

3.3.3 Hybrid Nanostructures

Hybrid nanostructures are structures that couple the functionality of metal or semiconductor NPs with the unique capabilities of biological or organic molecules [2, 24, 172]. NPs can be coupled with biological molecules, such as DNA or liposomes, for diagnostic and therapeutic applications [182] or with polymers for drug delivery, sensing or electronic applications [183, 184]. NPs can either be passivated with the molecules or incorporated inside of porous structures [182–184]. By coupling antibodies or cell targeting agents to the surface of NPs, the NPs can gain the ability to penetrate cells or bind to toxins for disease treatment and immunology or they can accumulate in tumors for cellular imaging [143, 185–187]. In disease treatment, NPs are functionalized with small molecules that specifically target the therapeutic location to efficiently deliver a small amount of drug to increase treatment efficiency and decrease the negative impact drugs have on the patient [188, 189]. Hybrid NPs have also been reported to overcome multidrug resistance in cancer cells by encapsulating the drug into the NP [189, 190]. There are an abundance of examples of hybrid NPs in the literature with unique properties and applications due to the interesting coupled properties that arise.

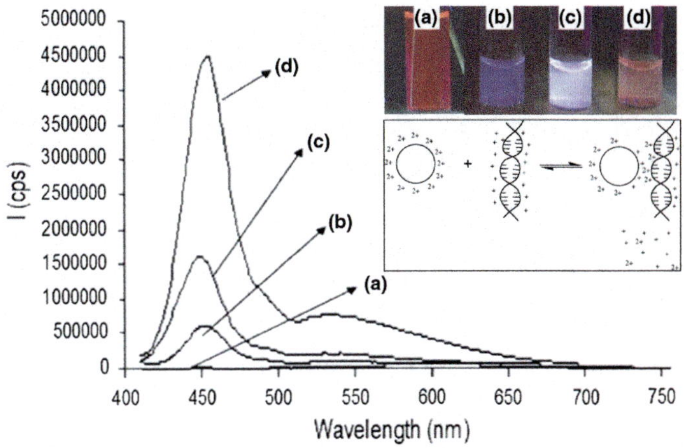

Fig. 3.28 Photoluminescence spectra of **a** "unactivated" CdS nanoparticles and **b–d** "activated" CdS nanoparticles. **b** CdS nanoparticles activated with Mg^{2+}; **c** CdS nanoparticles activated with Zn^{2+}; **d** CdS nanoparticles activated with Cd^{2+}. *Inset* Photograph of aqueous solutions of CdS nanoparticles upon irradiation with an ultraviolet lamp: **a** "unactivated" and **b–d** "activated" with **b** Zn^{2+}, **c** Cd^{2+}, and **d** Mg^{2+}. Proposed model of nanoparticle–oligonucleotide interactions mediated by ions. The nanoparticle (*circle*) initially has a loose web of divalent counterions surrounding it. The DNA (*helix*) has sodium counterions, initially, represented by the + symbols. The nanoparticle and DNA size are roughly to scale; only one DNA is shown for one nanoparticle for clarity. At equilibrium, after they bind, ions are expelled from the interface (shown as + and 2 + below the nanoparticle–DNA complex), and the DNA binds the divalent cations more tightly than the nanoparticle at the interface. The loss of divalent metal ions at the nanoparticle surface leads to a reverse of the activation process that resulted in more intense emission of the nanoparticles. Reproduced with permission from Ref. [181]. Copyright 2017 Elsevier

3.3.4 *Carbon Nanostructures*

Carbon nanostructures are nanosized allotropes of carbon atoms, including graphene, graphyne, buckyballs and carbon nanotubes. Carbon nanostructures can be made of the allotropes of graphite and diamond which are made up of sp^2 and sp^3 hybridized carbon, respectively [191]. Some of these carbon nanostructures, including carbon quantum dots and nanodiamonds, are photoluminescent. Compared to semiconductor quantum dots, carbon quantum dots are made up of more inert materials and do not display blinking behaviors [192]. Their photoluminescent lifetimes range in the picosecond to nanosecond range and quantum yield values of up to 80% have been reported [191]. The photoluminescence of carbon nanomaterials arises from several mechanisms impurity-vacancy centers, π-electron transitions, thermally activated fluorescence and surface defects.

Buckyballs are spherical fullerenes that are in the shape of a soccer ball and have the formula C_{60} (buckminsterfullerene) or C_{70}. Since these structures are in a

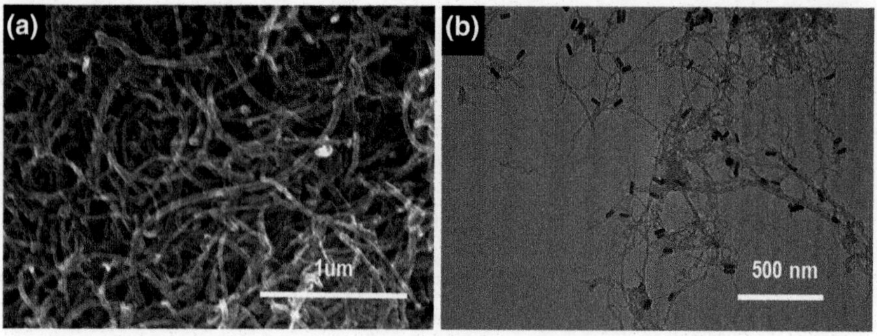

Fig. 3.29 SEM images of **a** CNTs and **b** Au nanorods decorated CNTs

spherical shape, there aren't any edges with dangling bonds or unpaired electrons and they're chargeless. Graphene is a 2 dimensional planar sheet of sp^2 hybridized carbon that forms a honeycomb structure. Many different derivatives of graphene have been synthesized including multi-layer graphenes, nanoribbons and nanoflakes as well as chemically modified graphene such as graphene oxide and hydrated graphene [191].

Carbon nanotubes are the more popularly used carbon structures due to their high mechanical stability and high electrical conductivity (Fig. 3.29) [24]. Single walled carbon nanotubes (SWCNTs) are made up of graphene like structures in the form of a tube. MWCNTs are made up on one or more concentric SWCNTs which leads to more mechanically stable nanotubes. These materials can act as a metal or a semiconductor, depending on the chirality, or the orientation of the carbon's crystal lattice compared to the tube's central axis. The metal SWCNTs do not display fluorescence and they quench the fluorescence from neighboring semiconductor SWCNTs [191].

3.4 Conclusions

A variety of NP shapes, structures and compositions have been synthesized via bottom-up and top-down techniques. High synthetic control, reproducibility and amenability to scale up are actively being pursued to be able to incorporate NPs into a myriad of applications. Each technique offers its own set of advantages and disadvantages and the synthetic route should be chosen according to application requirements. As nanomaterial synthesis progresses, more unique morphologies and compositions continue to arise with more interesting properties.

References

1. Murphy, C.J., et al. 2005. Anisotropic Metal Nanoparticles: Synthesis, Assembly, and Optical Applications (Feature Article; a Top Five ACS Article by Citations, National Chemistry Week, 2007). *The Journal of Physical Chemistry B* 109: 13857–13870.
2. Hunyadi, S.E. 2007. *Nanoengineered Materials: Synthesis, Design, Functionalization and Chemical Sensing Applications*. Columbia: University of South Carolina.
3. Liao, H., et al. 2014. Facet Development During Platinum Nanocube Growth. *Science* 345 (6199): 916–919.
4. Zhang, Q., L. Han, H. Jing, D.A. Blom, Y. Lin, H.L. Zin, and H. Wang. 2016. Facet Control of Gold Nanorods. *ACS Nano* 10 (2): 2960–2974.
5. Scarabelli, L., et al. 2015. A "Tips and Tricks" Practical Guide to the Synthesis of Gold Nanorods. *The journal of physical chemistry letters* 6: 4270–4279.
6. Carbone, L., and P.D. Cozzoli. 2010. Colloidal Heterostructured Nanocrystals: Synthesis and Growth Mechanisms. *Nano Today* 5: 449–493.
7. Yang, Y., W. Wang, X. Li, W. Chen, N. Fan, C. Zou, X. Chen, X. Xu, L. Zhang, and S. Huang. 2013. Controlled Growth of Ag/Au Bimetallic Nanorods through Kinetic Control. *Chemistry of Materials* 25: 34–41.
8. Meng, M., et al. 2016. Integration of Kinetic Control and Lattice Mismatch to Synthesize Pd@AuCu Core-shell Planar Tetrapods with Size-Dependent Optical Properties. *Nano Letters* 16: 3036–3041.
9. Biacchi, A.J., and R.E. Schaak. 2011. The Solvent Matters: Kinetic versus Thermodynamic Shape Control in the Polyol Synthesis of Rhodium Nanoparticles. *ACS Nano* 5: 8089–8099.
10. Wang, Y., et al. 2013. Shape-Controlled Synthesis of Palladium Nanocrystals: A Mechanistic Understanding of the Evolution from Octahedrons to Tetrahedrons. *Nano Letters* 13: 2276–2281.
11. Johnson, C.J., et al. 2002. Growth and Form of Gold Nanorods Prepared by Seed-Mediated Surfactant-Directed Synthesis. *Journal of Materials Chemistry* 12: 1765–1770.
12. Wang, Y., et al. 2015. Thermodynamics versus Kinetics in Nanosynthesis. *Angewandte Chemie International Edition* 54: 2022–2051.
13. Jang, K., H.J. Kim, and S.U. Son. 2010. Low-Temperature Synthesis of Ultrathin Rhodium Nanoplates via Molecular Orbital Symmetry Interactions Between Rhodium Precursors. *Chemistry of Materials* 22: 1273–1275.
14. Manna, L., et al. 2003. Controlled Growth of Tetrapod-Branced Inorganic Nanocrystals. *Nature Materials* 2: 382–385.
15. Wiley, B., S.H. Im, Z. Li, J. McLellan, and A. Siekkinen. 2006. XIa, Y, Maneuvering the Surface Plasmon Resonance of Silver Nanostructures through Shape-Controlled Synthesis. *The Journal of Physical Chemistry B* 110: 15666–15675.
16. Khon, E., K. Lambright, R. Khnayzer, P. Moroz, D. Perera, E. Butaeva, S. Lambright, F. Castellano, and M. Zamkov. 2013. Improving the Catalytic Activity of Semiconductor Nanocrystals through Selective Domain Etching. *Nano Letters* 13: 2016–2023.
17. Straney, P., C. Andolina, and J. Millstone. 2013. Seedless Initiation as an Efficient, Sustainable Route to Anisotropic Gold Nanoparticles. *Langmuir* 29: 4396–4403.
18. Thanh, N.T.K., N. Maclean, and S. Mahiddine. 2014. Mechanisms of Nucleation and Growth of Nanoparticles in Solution. *Chemical Reviews* 114: 7610–7630.
19. Wiley, B., et al. 2005. Shape-Controlled Synthesis of Metal Nanostructures: The Case of Silver. *Chemistry-A European Journal* 11: 454–463.
20. Clavijo-Chaparro, S.L., et al. 2016. Water Splitting Behavior of Copper-Cerium Oxide Nanorods and Nanocubes Using Hydrazine as a Scavenging Agent. *Journal of Molecular Catalysis A: Chemical* 423: 143–150.
21. Talapin, D.V., et al. 2007. Seeded Growth of Highly luminscent CdSe/CdS Nanoheterostructures with Rod and Tetrapod Morphologies. *Nano Letters* 7: 2951–2959.

22. Liu, X., et al. 2008. Capping Modes in PVP-Directed Silver Nanocrystal Growth: Multi-twinned Nanorods vs Single-Crystal Nano-Hexapods. *Crystal Growth & Design* 8: 1916–1923.
23. Murphy, C.J., et al. 2006. One-Dimensional Colloidal Gold and Silver Nanostructures. *Inorganic Chemistry* 45 (19): 7544–7554.
24. Hunyadi Murph, S.E., et al. 2012. Metallic and Hybrid Nanostructures: Fundamentals and Applications. In *Applications of Nanomaterials*, ed. J.N. Govil. Studium Press LLC: USA.
25. Near, R.D., S.C. Hayden, R.E. Hunter, D. Thackston, and M. El-Sayed. 2013. Rapid and Efficient Prediction of Optical Extinction Coefficients for Gold Nanospheres and Gold Nanorods. *Journal of Physical Chemistry C* 117: 23950–23955.
26. Ming, T., et al. 2009. Strong Polarization Dependence of Plasmon-Enhanced Fluorescence on Single Gold Nanorods. *Nano Letters* 9: 3896–3903.
27. Jackson, S.R., J.R. McBride, S.J. Rosenthal, and D.W. Wright. 2014. Where's the Silver? Imaging Trace Silver Coverage on the Surface of Gold Nanorods. *Journal of the American Chemical Society* 136 (14): 5261–5263.
28. DuChene, J.S., et al. 2013. Halide Anions as Shape-Directing Agents for Obtaining High-Quality Anisotropic Gold Nanostructures. *Chemistry of Materials* 25: 1392–1399.
29. Michel, J.A., W.H. Morris III., and C.M. Lukehart. 2015. Synthesis of Shaped Pt Nanoparticles Using Common Anions or Small Molecules as Shape-Directing Agents: Observation of a Strong Halide or Pseudo-Halide Effect. *Journal of Materials Chemistry A* 3: 2012–2018.
30. Lu, X., T.T. Tran, and W. Zhang. 2013. Shape-Selective Effect of Foreign Metal Ions on Growth of Noble Metal Nanocrystals with High-Index Facets. *Journal of Chemical Engineering and Process Technology* 1: 1009–1016.
31. Au, L., X. Lu, and Y. Xia. 2008. A Comparative Study of Galvanic Replacement Reactions Involving Ag Nanocubes and $AuCl_2^-$ or $AuCl_4^-$. *Advanced Materials* 20: 2517–2522.
32. Hunyadi, S.E., and C.J. Murphy. 2006. Bimetallic Silver-Gold Nanowires: Fabrication and Use in Surface- Enhanced Raman Scattering. *Journal of Materials Chemistry* 16 (Special Issue: Anisotropic Nanoparticles): 3929–3935.
33. Hunyadi Murph, S.E. et al. 2011. Synthesis, Functionalization, Characterization and Application of Controlled Shape Nanoparticles in Energy Production. In *Fluorine-Related Nanoscience with Energy Applications*, ed. D.J. Nelson and C.N. Brammer.
34. Hunyadi, S.E., and C.J. Murphy. 2009. Synthesis and Characterization of Silver-Platinum Bimetallic Nanowires and Platinum Nanotubes. *Journal of Cluster Science* 20: 319–330.
35. Wiley, B., Y. Sun, and Y. Xia. 2007. Synthesis of Silver Nanostructures with Controlled Shapes and Properties. *Accounts of Chemical Research* 40 (10): 1067–1076.
36. Menagen, G., et al. 2008. Selective Gold Growth on CdSe Seeded CdS Nanorods. *Chemistry of Materials* 20: 6900–6902.
37. Zheng, Y., et al. 2007. Ag/ZnO Heterostructured Nanocrystals: Synthesis, Characterization and Photocatalysis. *Inorganic Chemistry* 46: 6980–6986.
38. Mokari, T., et al. 2004. Selective Growth of Metal Tips onto Semiconductor Quantum Rods and Tetrapods. *Science* 304: 1787–1790.
39. Mokari, T., et al. 2005. Formation of Asymmetric One-Sided Metal-Tipped Semiconductor Nanocrystal Dots and Rods. *Nature Materials* 4: 855–863.
40. Hunyadi Murph, S.E. et al. 2011. Tuning of Size and Shape of Au-Pt Nanocatalyst for Direct Methanol Fuel Cells. *Journal of Nanoparticle Research* 13 (6347–6364): 6347.
41. Xia, Y., et al. 2009. Shape-Controlled Synthesis of Metal Nanocrystals: Simple Chemistry Meets Complex Physics? *Angewandte Chemie International Edition* 48: 60–103.
42. Wang, Z.L., T.S. Ahmed, and M.A. El-Sayed. 1997. Steps, Ledges, and Kinks on the Surfaces of Platinum Nanoparticles of Different Shapes. *Surface Science* 380: 302–310.
43. Hunyadi Murph, S.E., et al. 2015. A Possible Oriented Attachment Growth Mechanism for Silver Nanowire Formation. *Crystal Growth & Design* 15: 1968–1974.
44. Caswell, K.K., C.M. Bender, and C.J. Murphy. 2003. Seedless, Surfactantless Wet Chemical Synthesis of Silver Nanowires. *Nano Letters* 3 (5): 667–669.

45. Rajamathi, M., and R. Seshadri. 2002. Oxide and Chalcogenide Nanoparticles from Hydrothermal/Solvothermal Reactions. *Current Opinion in Solid State and Materials Science* 6: 337–345.
46. Han, H., G. Di Francesco, and M.M. Maye. 2010. Size Control and Photophysical Properties of Quantum Dots Prepared Via a Novel Hydrothermal Route. *Journal of Physical Chemistry C* 114: 19270–19277.
47. Yang, Y., et al. 2007. Solvothermal Synthesis of Multiple Shapes of Silver Nanoparticles and Their SERS Properties. *Journal of Physical Chemistry C* 111: 9095–9104.
48. Sasaki, T., et al. 2010. One-Step Solvothermal Synthesis of Cubic-Shaped ITO Nanoparticles Precisely Controlled in Size and Shape and their Electrical Resistivity. *Journal of Materials Chemistry* 20: 8153–8157.
49. Sochalski-Kolbus, L.M., et al. 2015. Solvothermal Synthesis and Surface Chemistry to Control the Size and Morphology of Nanoquartz. *Crystal Growth & Design* 15: 5327–5331.
50. Lidström, P., et al. 2001. Review of Microwave Assisted Organic Synthesis. *Tetrahedron* 57: 9225–9283.
51. Ribeiro, D.S.M. et al. 2017. Synthesis of distinctly thiol-capped CdTe quantum dots under microwave heating: multivariate optimization and characterization. *Journal of Materials Science* 52: 3208–3224.
52. Bilecka, I., and M. Niederberger. 2010. Microwave Chemistry for Inorganic Nanomaterials Synthesis. *Nanoscale* 2: 1358–1374.
53. Gerbec, J.A., et al. 2005. Microwave-Enhanced Reaction Rates for Nanoparticle Synthesis. *Journal of the American Chemical Society* 127: 15791–15800.
54. García, S., et al. 2014. Microwave Synthesis of Classically Immiscible Rhodium-Silver and Rhodium-Gold Alloy Nanoparticles: Highly Active Hydrogenation Catalysts. *ACS Nano* 8: 11512–11521.
55. Tsuji, M., et al. 2004. Microwave-Assisted Synthesis of Metallic Nanostructures in Solution. *Chemistry-A European Journal* 11: 440–452.
56. Mohamed, M.B., et al. 2010. Growth Mechanisms of Anisotropic Gold Nanocrystals Via Microwave Synthesis: Formation of Dioleamide by Gold Nanocrystals. *ACS Nano* 4: 2766–2772.
57. Zhu, H., et al. 2009. Microwave Synthesis of Fluorescent Carbon Nanoparticles with Electrochemiluminescent Properties. *Chemical Communications* 34: 5118–5120.
58. Nie, Z., A. Petukhova, and E. Kumacheva. 2010. Properties and Emerging Applications of Self-Assembled Structures Made from Inorganic Nanoparticles. *Nature Nanotechnology* 5: 15–25.
59. Rothemund, P.W.K. 2006. Folding DNA to Create Nanoscale Shapes and Patterns. *Nature* 440: 297–302.
60. Park, S.Y., et al. 2008. DNA-Programmable Nanoparticle Crystallization. *Nature* 451: 553–556.
61. Nykypanchuk, D., et al. 2008. DNA-Guided Crystallization of Colloidal Nanoparticles. *Nature* 451: 549–552.
62. Zhang, C., et al. 2013. A General Approach to DNA-Programmable Atom Equivalents. *Nature Materials* 12: 741–746.
63. Zhang, Y., et al. 2013. A General Strategy for the DNA-Mediated Self-Assembly of Functional nanoparticles into Heterogeneous Systems. *Nature Nanotechnology* 8: 865–872.
64. Auyeung, E., et al. 2012. Synthetically Programmable Nanoparticle Superlattices Using a Hollow Three-Dimensional Spacer Approach. *Nature Nanotechnology* 7: 24–28.
65. Mirkin, C.A., et al. 1996. A DNA-Based Method for Rationally Assembling Nanoparticles into Macroscopic Materials. *Nature* 382: 607–609.
66. Coopersmith, K., H. Han, and M.M. Maye. 2015. Stepwise Assembly and Characterization of DNA Linked Two-Color Quantum Dot Clusters. *Langmuir* 31: 7463–7471.
67. Chhabra, R., et al. 2010. DNA Self-Assembly for Nanomedicine. *Advanced Drug Delivery Reviews* 62: 617–625.

68. Abbas, A., et al. 2013. Molecular Linker-Mediated Self-Assembly of Gold Nanoparticles: Understanding and Controlling the Dynamics. *Langmuir* 29: 56–64.
69. Taylor, R.W., et al. 2011. Precise Subnanometer Plasmonic Junctions for SERS Within Gold Nanoparticle Assemblies Using CURCUBIT[n]uril "Glue". *ACS Nano* 5: 3878–3887.
70. Stobiecka, M., K. Coopersmith, and M. Hepel. 2010. Resonance Elastic Ligher Scattering (RELS) Spectroscopy of Fast Non-Langmuirian Ligand-Exchange in Glutathione-Induced Gold Nanoparticle Assembly. *Journal of Colloid and Interface Science* 350: 168–177.
71. Puig, H.d., et al. 2011. Quantifying the nanomachinery of the nanoparticle-biomolecule interface. *Small* **7**: 2477–2484.
72. Zhang, P., et al. 2012. Click-Functionalized Compact Quantum Dots Protected by Multidentate-Imidazole Ligands: Conjugation-Ready Nanotags for Living-Virus Labeling and Imaging. *Journal of the American Chemical Society* 134: 8388–8399.
73. Greybush, N.J., et al. 2014. Plasmon-Enhanced Upconversion Luminescence in Single Nanophosphor-Nanorod Heterodimers Formed Through Template-Assisted Self-Assembly. *ACS Nano* 8: 9482–9491.
74. Mallavajula, R., and L. Archer. 2011. Nanocrystal Self-Assembly Assisted by Oriented Attachment. *Angewandte Chemie International Edition* 50: 578–580.
75. Pucci, A., et al. 2013. One-Step Synthesis and Self-Assembly of Metal Oxide Nanoparticles into 3D Superlattices. *ACS Nano* 6: 4382–4391.
76. Broers, A.N. 1978. High Resolution Lithography Systems: A Review of the Current Status. *Physikalische Blätter* 34 (12): 704–712.
77. Smith, H.I. 1986. A Review of Submicron Lithography. *Superlattices and Microstructures* 2 (2): 129–142.
78. Wu, B., and A. Kumar. 2014. Extreme Ultraviolet Lithography and Three Dimensional Integrated Circuit—A Review. *Applied Physics Reviews* 1 (1): 011104.
79. Biswas, A., et al. 2012. Advances in Top–Down and Bottom–Up Surface Nanofabrication: Techniques, Applications & Future Prospects. *Advances in Colloid and Interface Science* 170 (1): 2–27.
80. Sanders, D.P. 2010. Advances in Patterning Materials for 193 nm Immersion Lithography. *Chemical Reviews* 110 (1): 321–360.
81. Wagner, C., and N. Harned. 2010. EUV Lithography: Lithography Gets Extreme. *Nature Photonics* 4 (1): 24–26.
82. Pease, R.F., and S.Y. Chou. 2008. Lithography and Other Patterning Techniques for Future Electronics. *Proceedings of the IEEE* 96 (2): 248–270.
83. Bailey, G.E., et al. 2007. Double Pattern EDA Solutions for 32 nm HP and Beyond. In *Advanced Lithography*. Bellingham: International Society for Optics and Photonics.
84. Naulleau, P.P., et al. 2011. Critical Challenges for EUV Resist Materials. In *SPIE Advanced Lithography*. Bellingham: International Society for Optics and Photonics.
85. Sun, J., T. Xu, and N.M. Litchinitser. 2016. Experimental Demonstration of Demagnifying Hyperlens. *Nano Letters* 16 (12): 7905–7909.
86. Bakshi, V. 2009. *EUV Lithography* (Vol. 178). Bellingham: Spie Press Bellingham.
87. Yu, B., et al. 2015. Layout Decomposition for Triple Patterning Lithography. *IEEE Transactions on Computer-Aided Design of Integrated Circuits and Systems* 34 (3): 433–446.
88. Sun, S., et al. 2006. Fabrication of Gold Micro- and Nanostructures by Photolithographic Exposure of Thiol-Stabilized Gold Nanoparticles. *Nano Letters* 6 (3): 345–350.
89. Bertino, M.F., et al. 2007. Quantum Dots by Ultraviolet and X-ray Lithography. *Nanotechnology* 18 (31): 315603.
90. Xiang, C., Y. Yang, and R.M. Penner. 2009. Cheating the Diffraction Limit: Electrodeposited Nanowires Patterned by Photolithography. *Chemical Communications* 8: 859–873.
91. Doan, N.M., et al. 2015. Low-Cost Photolithographic Fabrication of Nanowires and Microfilters for Advanced Bioassay Devices. *Sensors* 15 (3): 6091–6104.

92. Ayenew, G.T., et al. 2014. Self-Organized Nanoparticle Photolithography for Two-Dimensional Patterning of Organic Light Emitting Diodes. *Optics Express* 22 (S6): A1619–A1633.
93. Zhang, Y.-L., et al. 2010. Designable 3D Nanofabrication by Femtosecond Laser Direct Writing. *Nano Today* 5 (5): 435–448.
94. Hu, W., et al. 2004. Sub-10 nm Electron Beam Lithography Using Cold Development of Poly(methylmethacrylate). *Journal of Vacuum Science and Technology B* 22 (4): 1711–1716.
95. Lo, C.J., T. Aref, and A. Bezryadin. 2006. Fabrication of Symmetric sub-5 nm Nanopores Using Focused Ion and Electron Beams. *Nanotechnology* 17 (13): 3264.
96. Baldacchini, T., et al. 2016. Polymer-Based 3D Micro-/Nanofabrication by Laser Direct Writing. In *Encyclopedia of Nanotechnology*, ed. B. Bhushan, pp. 3349–3361. Dordrecht, Netherlands: Springer.
97. Gates, B.D., et al. 2005. New Approaches to Nanofabrication: Molding, Printing, and Other Techniques. *Chemical Reviews* 105 (4): 1171–1196.
98. Chen, Y. 2015. Nanofabrication by Electron Beam Lithography and its Applications: A Review. *Microelectronic Engineering* 135: 57–72.
99. Joshi-Imre, A., and S. Bauerdick. 2014. Direct-write ion beam lithography. *Journal of Nanotechnology* 2014.
100. Esposito, M., et al. 2014. Three Dimensional Chiral Metamaterial Nanospirals in Athe Visible Range by Vertically Compensated Focused Ion Beam Induced-Deposition. *Advanced Optical Materials* 2 (2): 154–161.
101. Garcia, R., A.W. Knoll, and E. Riedo. 2014. Advanced Scanning Probe Lithography. *Nature Nanotechnology* 9 (8): 577–587.
102. Martínez, R.V., et al. 2007. Patterning Polymeric Structures with 2 nm Resolution at 3 nm Half Pitch in Ambient Conditions. *Nano Letters* 7 (7): 1846–1850.
103. Carroll, K.M., et al. 2013. Fabricating Nanoscale Chemical Gradients with ThermoChemical NanoLithography. *Langmuir* 29 (27): 8675–8682.
104. Byun, I.-S., et al. 2011. Nanoscale Lithography on Monolayer Graphene Using Hydrogenation and Oxidation. *ACS Nano* 5 (8): 6417–6424.
105. Zhang, K., et al. 2012. Direct Writing of Electronic Devices on Graphene Oxide by Catalytic Scanning Probe Lithography. *Nature Communications* 3: 1194.
106. Lu, G., et al. 2010. Nanolithography of Single-Layer Graphene Oxide Films by Atomic Force Microscopy. *Langmuir* 26 (9): 6164–6166.
107. Hulteen, J.C., and R.P. Van Duyne. 1995. Nanosphere Lithography: A Materials General Fabrication Process for Periodic Particle Array Surfaces. *Journal of Vacuum Science and Technology A* 13 (3): 1553–1558.
108. Haynes, C.L., and R.P. Van Duyne. 2001. Nanosphere Lithography: A Versatile Nanofabrication Tool for Studies of Size-Dependent Nanoparticle Optics. *The Journal of Physical Chemistry B* 105 (24): 5599–5611.
109. Colson, P., C. Henrist, and R. Cloots. 2013. Nanosphere Lithography: A Powerful Method for the Controlled Manufacturing of Nanomaterials. *Journal of Nanomaterials* 2013: 21.
110. Wang, D., and F. Caruso. 2001. Fabrication of Polyaniline Inverse Opals Via Templating Ordered Colloidal Assemblies. *Advanced Materials* 13 (5): 350–354.
111. Peng, K., et al. 2007. Ordered Silicon Nanowire Arrays Via Nanosphere Lithography and Metal-Induced Etching. *Applied Physics Letters* 90 (16): 163123.
112. Larsen, G.K., et al. 2013. Hidden Chirality in Superficially Racemic Patchy Silver Films. *Nano Letters* 13 (12): 6228–6232.
113. Gibbs, J., and Y. Zhao. 2011. Catalytic Nanomotors: Fabrication, Mechanism, and Applications. *Frontiers of Materials Science* 5 (1): 25–39.
114. Larsen, G.K., et al. 2014. The Fabrication of Three-Dimensional Plasmonic Chiral Structures by Dynamic Shadowing Growth. *Nanoscale* 6 (16): 9467–9476.

115. Larsen, G.K., N. Stom, and Y. Zhao. 2015. Continuously Tuning the Spectral Response of Chiral Plasmonic Patchy Particles Through GALVANIC Replacement Reaction. *RSC Advances* 5 (123): 101257–101261.
116. Wu, Y., et al. 2013. Fabrication of Wafer-Size Monolayer Close-Packed Colloidal Crystals via Slope Self-Assembly and Thermal Treatment. *Langmuir* 29 (46): 14017–14023.
117. Zhang, J., et al. 2010. Colloidal Self-Assembly Meets Nanofabrication: From Two-Dimensional Colloidal Crystals to Nanostructure Arrays. *Advanced Materials* 22 (38): 4249–4269.
118. Park, C., et al. 2014. Quick, Large-Area Assembly of a Single-Crystal Monolayer of Spherical Particles by Unidirectional Rubbing. *Advanced Materials* 26 (27): 4633–4638.
119. Chang, H.-T., et al. 2013. Uniform SiGe/Si Quantum Well Nanorod and Nanodot Arrays Fabricated Using Nanosphere Lithography. *Nanoscale Research Letters* 8 (1): 349.
120. Ghicov, A., and P. Schmuki. 2009. Self-Ordering Electrochemistry: A Review on Growth and Functionality of TiO_2 Nanotubes and Other Self-Aligned MOx Structures. *Chemical Communications* 20: 2791–2808.
121. Jani, Md, A.M.D. Losic, and N.H. Voelcker. 2013. Nanoporous Anodic Aluminium Oxide: Advances in Surface Engineering and Emerging Applications. *Progress in Materials Science* 58 (5): 636–704.
122. Shingubara, S. 2003. Fabrication of Nanomaterials Using Porous Alumina Templates. *Journal of Nanoparticle Research* 5 (1): 17–30.
123. Wang, Y., et al. 2006. Bipolar Electrochemical Mechanism for the Propulsion of Catalytic Nanomotors in Hydrogen Peroxide Solutions. *Langmuir* 22 (25): 10451–10456.
124. Liu, R., and A. Sen. 2011. Autonomous Nanomotor Based on Copper-Platinum Segmented Nanobattery. *Journal of the American Chemical Society* 133 (50): 20064–20067.
125. Epps III, T.H., and R.K. O'Reilly. 2016. Block Copolymers: Controlling Nanostructure to Generate Functional Materials–Synthesis, Characterization, and Engineering. *Chemical Science* 7 (3): 1674–1689.
126. Mai, Y., and A. Eisenberg. 2012. Self-Assembly of Block Copolymers. *Chemical Society Reviews* 41 (18): 5969–5985.
127. Bigall, N.C., et al. 2015. High-Resolution Metal Nanopatterning by Means of Switchable Block Copolymer Templates. *ACS Applied Materials & Interfaces* 7 (23): 12559–12569.
128. Papalia, J.M., et al. 2010. Silicon Nanowire Polarizers for far Ultraviolet (sub-200 nm) Applications: Modeling and Fabrication. *Journal of Applied Physics* 107 (8): 084305.
129. Urbas, A.M., et al. 2002. Bicontinuous Cubic Block Copolymer Photonic Crystals. *Advanced Materials* 14 (24): 1850–1853.
130. Hur, K., et al. 2011. Three-Dimensionally Isotropic Negative Refractive Index Materials from Block Copolymer Self-Assembled Chiral Gyroid Networks. *Angewandte Chemie International Edition* 50 (50): 11985–11989.
131. Selvakumar, R., et al. 2014. Recent Advances in the Synthesis of Inorganic Nano/Microstructures Using Microbial Biotemplates and their Applications. *RSC Advances* 4 (94): 52156–52169.
132. Knez, M., et al. 2003. Biotemplate Synthesis of 3-nm Nickel and Cobalt Nanowires. *Nano Letters* 3 (8): 1079–1082.
133. Nguyen, S.H., et al. 2014. Natural Insect and Plant Micro-/Nanostructsured Surfaces: An Excellent Selection of Valuable Templates with Superhydrophobic and Self-Cleaning Properties. *Molecules* 19 (9): 13614–13630.
134. Mainwaring, D.E., et al. 2016. The Nature of Inherent Bactericidal Activity: Insights from the Nanotopology of Three Species of Dragonfly. *Nanoscale* 8 (12): 6527–6534.
135. Sun, M., et al. 2011. A Study of the Anti-Reflection Efficiency of Natural Nano-Arrays of Varying Sizes. *Bioinspiration & Biomimetics* 6 (2): 026003.
136. Xie, G., et al. 2008. The Fabrication of Subwavelength Anti-Reflective Nanostructures Using a Bio-Template. *Nanotechnology* 19 (9): 095605.
137. Ohring, M. 2001. *Materials science of thin films*. USA: Academic press.

138. Robbie, K., and M. Brett. 1997. Sculptured Thin Films and Glancing Angle Deposition: Growth Mechanics and Applications. *Journal of Vacuum Science and Technology A* 15 (3): 1460–1465.
139. Robbie, K., et al. 2004. Ultrahigh Vacuum Glancing Angle Deposition System for Thin Films with Controlled Three-Dimensional Nanoscale Structure. *Review of Scientific Instruments* 75 (4): 1089–1097.
140. Robbie, K., J. Sit, and M. Brett. 1998. Advanced Techniques for Glancing Angle Deposition. *Journal of Vacuum Science and Technology B* 16 (3): 1115–1122.
141. Barranco, A., et al. 2016. Perspectives on Oblique Angle Deposition of Thin Films: From Fundamentals to Devices. *Progress in Materials Science* 76: 59–153.
142. He, Y., and Y. Zhao. 2011. Advanced Multi-Component Nanostructures Designed by Dynamic Shadowing Growth. *Nanoscale* 3 (6): 2361–2375.
143. Mark, A.G., et al. 2013. Hybrid Nanocolloids with Programmed Three-Dimensional Shape and Material Composition. *Nature Materials* 12 (9): 802–807.
144. Chen, L., et al. 2013. Engineering Epitaxial-Nanospiral Metal Films Using Dynamic Oblique Angle Deposition. *Crystal Growth & Design* 13 (5): 2075–2080.
145. LaForge, J.M., et al. 2012. Flux Engineering To Control In-Plane Crystal and Morphological Orientation. *Crystal Growth & Design* 12 (7): 3661–3667.
146. Pierson, H.O. 1999. *Handbook of Chemical Vapor Deposition: Principles, Technology and Applications*. New York: William Andrew.
147. Yugo, S., et al. 1991. Generation of Diamond Nuclei by Electric Field in Plasma Chemical Vapor Deposition. *Applied Physics Letters* 58 (10): 1036–1038.
148. Bower, C., et al. 2000. Nucleation and Growth of Carbon Nanotubes by Microwave Plasma Chemical Vapor Deposition. *Applied Physics Letters* 77 (17): 2767–2769.
149. Dong, J., et al. 2006. Self-Assembly of Octadecyltrichlorosilane Monolayers on Silicon-Based Substrates by Chemical Vapor Deposition. *Thin Solid Films* 515 (4): 2116–2122.
150. Zhang, Y., L. Zhang, and C. Zhou. 2013. Review of Chemical Vapor Deposition of Graphene and Related Applications. *Accounts of Chemical Research* 46 (10): 2329–2339.
151. Kumar, M., and Y. Ando. 2010. Chemical Vapor Deposition of Carbon Nanotubes: A Review on Growth Mechanism and Mass Production. *Journal of Nanoscience and Nanotechnology* 10 (6): 3739–3758.
152. Li, X., L. Colombo., and R.S. Ruoff. 2016. Synthesis of Graphene Films on Copper Foils by Chemical Vapor Deposition. *Advanced Materials* 28 (29): 6247–6252.
153. Wang, H., and G. Yu. 2016. Direct CVD Graphene Growth on Semiconductors and Dielectrics for Transfer-Free Device Fabrication. *Advanced Materials* 28 (29): 4956–4975.
154. Xiang, B., et al. 2007. Rational Synthesis of p-type Zinc Oxide Nanowire Arrays Using Simple Chemical Vapor Deposition. *Nano Letters* 7 (2): 323–328.
155. Benson, J., et al. 2012. Chemical Vapor Deposition of Aluminum Nanowires on Metal Substrates for Electrical Energy Storage Applications. *ACS Nano* 6 (1): 118–125.
156. Kuykendall, T., et al. 2003. Metalorganic Chemical Vapor Deposition Route to GaN Nanowires with Triangular Cross Sections. *Nano Letters* 3 (8): 1063–1066.
157. Zhou, J., et al. 2006. Growth of Large-Area Aligned Molybdenum Nanowires by High Temperature Chemical Vapor Deposition: Synthesis, Growth Mechanism, and Device Application. *The Journal of Physical Chemistry B* 110 (21): 10296–10302.
158. Lo Nigro, R., et al. 2007. Template-Free and Seedless Growth of Pt Nanocolumns: Imaging and Probing Their Nanoelectrical Properties. *ACS Nano* 1 (3): 183–190.
159. Hwang, N.M., et al. 2000. Growth of Silicon Nanowires by Chemical Vapor Deposition: Approach by Charged Cluster Model. *Journal of Crystal Growth* 218 (1): 33–39.
160. Ermez, S., et al. 2015. Self-Seeded Growth of GaAs Nanowires by Metal-Organic Chemical Vapor Deposition. *Crystal Growth & Design* 15 (6): 2768–2774.
161. Wu, Y., and P. Yang. 2001. Direct Observation of Vapor-Liquid-Solid Nanowire Growth. *Journal of the American Chemical Society* 123 (13): 3165–3166.

162. Klamchuen, A., et al. 2015. Rational Concept for Designing Vapor–Liquid–Solid Growth of Single Crystalline Metal Oxide Nanowires. *Nano Letters* 15 (10): 6406–6412.
163. Pinion, C.W., J.D. Christesen, and J.F. Cahoon. 2016. Understanding the Vapor-Liquid-Solid Mechanism of Si Nanowire Growth and Doping to Synthetically Encode Precise Nanoscale Morphology. *Journal of Materials Chemistry C* 4 (18): 3890–3897.
164. Ergen, O., et al. 2010. Shape-Controlled Synthesis of Single-Crystalline Nanopillar Arrays by Template-Assisted Vapor-Liquid-Solid Process. *Journal of the American Chemical Society* 132 (40): 13972–13974.
165. Shin, N., et al. 2013. Rational Defect Introduction in Silicon Nanowires. *Nano Letters* 13 (5): 1928–1933.
166. Shin, N., and M.A. Filler. 2012. Controlling Silicon Nanowire Growth Direction via Surface Chemistry. *Nano Letters* 12 (6): 2865–2870.
167. Shin, N., M. Chi, and M.A. Filler. 2014. Interplay between Defect Propagation and Surface Hydrogen in Silicon Nanowire Kinking Superstructures. *ACS Nano* 8 (4): 3829–3835.
168. Navrotsky, A., L. Mazeina, and J. Majzlan. 2008. Size-Driven Structural and Thermodynamic Complexity in Iron Oxides. *Science* 319 (5870): 1635–1638.
169. Navrotsky, A., et al. 2010. Nanophase Transition Metal Oxides Show Large Thermodynamically Driven Shifts in Oxidation-Reduction Equilibria. *Science* 330 (6001): 199–201.
170. Larson, G.K., W. Farr, and S.E. Hunyadi Murph. 2016. Multifunctional Fe_2O_3-Au Nanoparticles with Different Shapes: Enhanced Catalysis, Photothermal Effects and Magnetic Recycling. *Journal of Physical Chemistry C* 120: 15162–15172.
171. Hunyadi Murph, S.E., and C.J. Murphy. 2006. Tunable One-Dimensional Silver-Silica Nanopeapod Architectures. *The Journal of Physical Chemistry B* 110: 7226–7231.
172. Hunyadi Murph, S.E., and C.J. Murphy. 2013. Patchy Silica-Coated Silver Nanowires as SERS Substrates. *Journal of Nanoparticle Research* 15 (6): 1607.
173. Giblin, J., and M. Kuno. 2010. Nanostructure Absorption: A Comparative Study of Nanowire and Colloidal Quantum Dot Absorption Cross Sections. *The Journal of Physical Chemistry Letters* 1: 3340–3348.
174. Pons, T., I.L. Medintz, K.E. Sapsford, S. Higashiya, A.F. Grimes, D.S. English, and H. Mattoussi. 2007. On the Quenching of Semiconductor Quantum Dot Photoluminescence by Proximal Gold Nanoparticles. *Nano Letters* 7: 3157–3164.
175. Krishnan, R., M. Hahn, Z. Yu, J. Silcox, P. Fauchet, and T. Krauss. 2004. Polarization Surface-Charge Density of Single Semiconductor Quantum Rods. *Physical Review Letters* 92 (21): 216803.
176. Dong, Y., D. Parobek., D. Rossi., and D.H. Son. 2016. Photoemission of Energetic Hot Electrons Produced via Up-Conversion in Doped Quantum Dots. *Nano Letters*, 16 (11): 7270–7275.
177. Wang, Y., G. Fruhwirth, E. Cai, T. Ng, and P.R. Selvin. 2013. 3D Super-Resolution Imaging with Blinking Quantum Dots. *Nano Letters* 12 (11): 5233–5241.
178. Galland, C., et al. 2012. Lifetime Blinking in Nonblinking Nanocrystal Quantum Dots. *Nature Communications* 3: 908.
179. Xu, J., K.F. Tehrani, and P. Kner. 2015. Multicolor 3D Super-resolution Imaging by Quantum Dot Stochastic Optical Reconstruction Microscopy. *ACS Nano* 9 (3): 2917–2925.
180. Liqiang, J., Q. Yichun, W. Baiqi, L. Shudan, J. Baojiang, Y. Libin, F. Wei, F. Honggang, and S. Jiazhong. 2006. Review of Photoluminscence Perfomance of Nano-sized Semiconductor Materials and its Relationship with Photocatalytic Activity. *Solar Energy Materials and Solar Cells* 90 (12): 1773–1787.
181. Mahtab, R., et al. 2007. Influence of the Nature of Quantum Dot Surface Cations on Interactions with DNA. *Journal of Inorganic Biochemistry* 10: 559–564.
182. Sailor, M.J., and J. Park. 2013. Hybrid Nanoparticles for Detection and Treatment of Cancer. *Advanced Materials* 24: 3779–3802.

183. Zheng, Y., et al. 2015. Reversible Gating of Smart Plasmonic Molecular Traps Using Thermoresponsive Polymers for Single-Molecule Detection. *Nature Communications* 6: 8797.
184. Wu, S., et al. 2006. Electrodeposition of Silver-DNA Hybrid Nanoparticles for Electrochemical Sensing of Hydrogen Peroxide and Glucose. *Electrochemistry Communications* 8: 1197–1203.
185. Cavadas, M.A.S., et al. 2016. Unravelling Malaria Antigen Binding to Antibody-Gold Nanoparticle Conjugates. *Particle & Particle Systems Characterization* 33: 906–915.
186. Huang, X., et al. 2010. The Effect of the Shape of Mesoporous Silica Nanoparticles on Cellular Uptate and Cell Function. *Biomaterials* 31 (3): 438–448.
187. Elzoghby, A.O., A.L. Hemasa, and M.S. Freag. 2016. Hybrid Protein-Inorganic Nanoparticles: From Tumor-Targeted Drug Delivery to Cancer Imaging. *Journal of Controlled Release* 243: 303–322.
188. Asadishad, B., M. Vossoughi, and I. Alemzadeh. 2010. Folate-Receptor-Targeted Delivery of Doxorubicin Using Polyethylene Glycol-Functionalized Gold Nanoparticles. *Industrial and Engineering Chemistry Research* 49: 1958–1963.
189. Wang, F., et al. 2011. Doxorubicin-Tethered Responsive Gold Nanoparticles Facilitate Intracellular Drug Delivery for Overcoming Multidrug Resistance in Cancer Cells. *ACS Nano* 5: 3679–3692.
190. Soma, C.E., et al. 2000. Reversion of Multidrug Resistance by Co-Encapsulation of Doxorubicin and Cyclosporin A in Polyalkylcyanoacrylate Nanoparticles. *Biomaterials* 21: 1–7.
191. Kozák, O., M. Sudolská, G. Pramanik, P. Cígler, M. Otyepka, and R. Zbořil. 2016. Photoluminescent Carbon Nanostructures. *Chemistry of Materials* 28: 4085–4128.
192. Lim, S.Y., W. Shen, and Z. Gao. 2015. Carbon Quantum Dots and their Applications. *Chemical Society Reviews* 44: 362–381.

Chapter 4
Characterization of Anisotropic and Shape-Selective Nanomaterials: Methods and Challenges

George K. Larsen

Abstract Research into shape-selective and anisotropic nanoparticles is generally motivated by the desire to create better materials for a specific application, and therefore, it is critical to understand how and why shape affects nanoscale properties. Such information can be revealed through analytical experimentation, and this chapter describes characterization methods and challenges associated with analyzing anisotropic and shape-selective nanoparticles. Researchers can typically employ commonly available techniques used in materials characterization. However, in the case of anisotropic and shape-selective nanoparticles, greater concern for orientational and/or shape effects and artifacts should be shown during analyses.

Keywords Anisotropic metallic nanostructures · Anisotropic metallic oxide nanostructures · Multifunctional nanostructures · Gold nanoparticles · Silver nanoparticles · Silica · Iron oxide · Titania · Sensing · Imaging · Photodegradation · Environmental applications · Catalysis · Electrocatalysis · Solar energy to fuel conversion · Photothermal · Self-assemblies

4.1 Overview

Similar to their fabrication, the characterization of anisotropic and shape-selective nanomaterials can present unique challenges. These challenges are mostly due to implicit assumptions during experimental measurements that an instrument is probing a region and orientation of sample that is representative of the whole (*i.e.*, the sample is homogeneous and isotropic). The rationality behind these assumptions is more readily apparent for local, precise measurements, but anisotropic artifacts can arise during bulk measurement too, due to alignment and preferential

G.K. Larsen (✉)
National Security Directorate, Savannah River National Laboratory, Aiken, SC, USA
e-mail: George.Larsen@srnl.doe.gov

S.E. Hunyadi Murph et al. (eds.), *Anisotropic and Shape-Selective Nanomaterials*, Nanostructure Science and Technology, DOI 10.1007/978-3-319-59662-4_4

orientation of aggregates. In spite of these considerations, the characterization tools of anisotropic and shape-selective nanomaterials are generally no different than typical, isotropic nanomaterials. The only difference is that a researcher should show greater concern for orientational and/or shape effects and artifacts during measurements and analyses. After all, it is the influence of shape and orientation on material properties that generally motivates research into anisotropic and shape-selective nanomaterials in the first place.

4.2 Structural and Chemical Characterization

Generally, morphological and compositional analyses are the first and most important characterizations conducted on nanomaterials. The small sizes of nanoparticles prevent direct observation, and therefore, researchers must employ analytical instruments that irradiate, shoot particle beams, or scratch and bump across surfaces in order to interrogate the nanoparticles. By analyzing the subtleties of these interactions, precise nanoscale information regarding shape and chemistry can be obtained.

4.2.1 Microscopy

Microscopy is very often the first technique that is conducted after synthesis in order to visually confirm the production of the desired nanoparticles. Nanoparticles are too small to be observed by optical microscopy, and therefore, electron and scanning probe microscopy are generally employed. Although, other types of microscopes, such as X-ray and scanning helium ion microscopes, are occasionally used as well. These techniques as applied to nanomaterial characterization are well covered in the literature and in textbooks [1–3], and there are no dramatic differences in extending application to anisotropic or shape-selective nanoparticles. However in practice, care should be taken to ensure that the anisotropic nature of the nanoparticles do not obscure measurements. For example, nanoparticles may aggregate in preferential orientations when drop-cast onto transmission electron microscopy grids (Fig. 4.1). Being aware of such shape-dependent effects will prevent sampling errors in statistical morphological analyses. Careful selection of dilute samples may also eliminate these issues. Electron microscopy is a powerful technique that allows visual inspection of growth, dissolution, and other crystallographic properties of anisotropic nanomaterials [4–6]. However, the ability of scanning probe methods to contact and interact with the surface of nanoparticles gives even greater functionality to these techniques. For example, the measurement of the friction between an atomic force microscopy (AFM) probe tip and a surface has enabled much progress in the field of nanotribology [7, 8]. Analyses of such measurements can provide additional insights beyond just frictional properties and have been used to explain the differences between the theoretical and actual behavior of ideal and exfoliated graphene [9]. Because the probe tip localizes measurements to a sub-nanometer location on the surface, AFM can be coupled with other analytical measurements to create precise chemical, thermal, and electrical property maps of nanoparticles and nanostructured surfaces, including anisotropic ones [11–15]. A good example of is the combination of AFM with Raman spectroscopy, through a technique called tip-enhanced Raman spectroscopy (TERS), which uses the AFM tip as resonance location for Raman enhancement to provide local chemical information (Fig. 4.2) [16, 17]. The combination of simultaneous imaging with property measurement improves the overall accuracy of such measurements, and the ability of AFM to manipulate a nanoparticle's position and orientation gives additional control during characterization.

Clearly, scanning probe microscopy techniques offer several benefits for characterization of anisotropic and shape-selective nanoparticles, especially AFM, which is widely available. That being said, there are some issues that need to be considered during scanning probe characterization of anisotropic or shape-selective nanomaterials. Since the probe must be rastered across the surface plane, directionality is induced into the measurement. Thus, any shape anisotropy in the probe tip, or any hysteresis in the probe motion, will induce directional artifacts in the obtained image (Fig. 4.3). Creating images by scanning in several different directions is one way to account for such artifacts. Long duration scans over the surface may also be subject to thermal drift inducing distortions into the image, though there are some techniques to account for such effects [18]. Therefore, when using scanning probe microscopy, it is important to take steps to ensure that the observed shapes and patterns are accurate and not distorted by anisotropy or artifacts from the measurements.

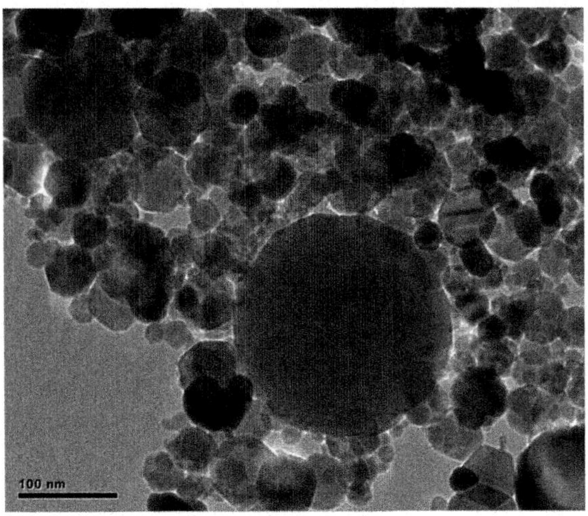

Fig. 4.1 Transmission electron micrograph of a drop-cast aggregation of nanoparticles having different *sizes* and *shapes*

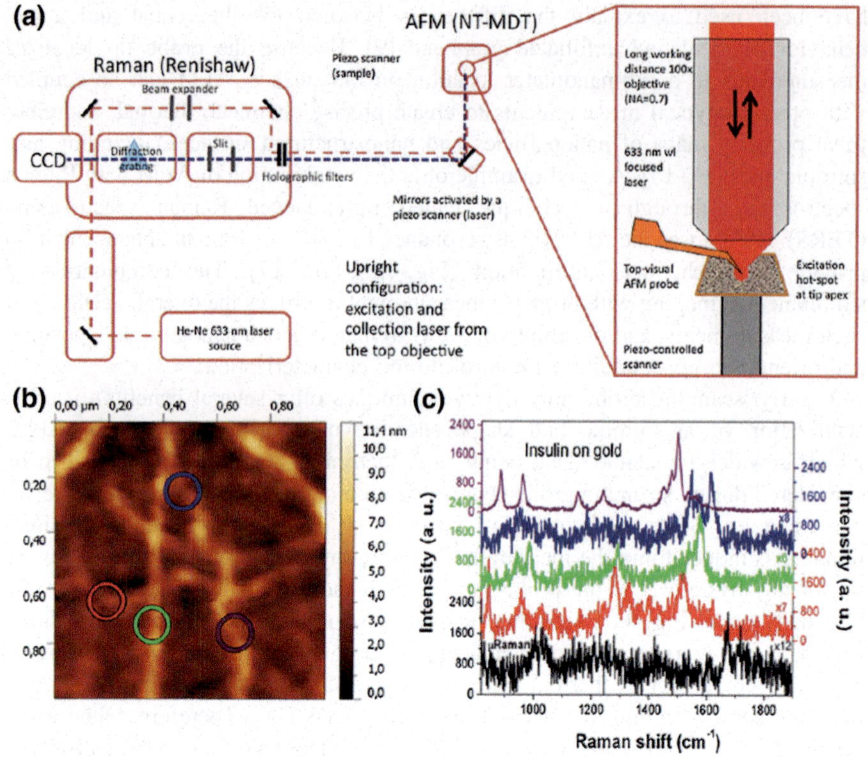

Fig. 4.2 **a** Representative experimental setup for a TERS instrument. **b** AFM map and **c** corresponding TERS spectra of insulin fibrils; the *color* of the spectra correspond with the measurement location designated by the *colored circles*. Figure reproduced from Ref. [10] with permission

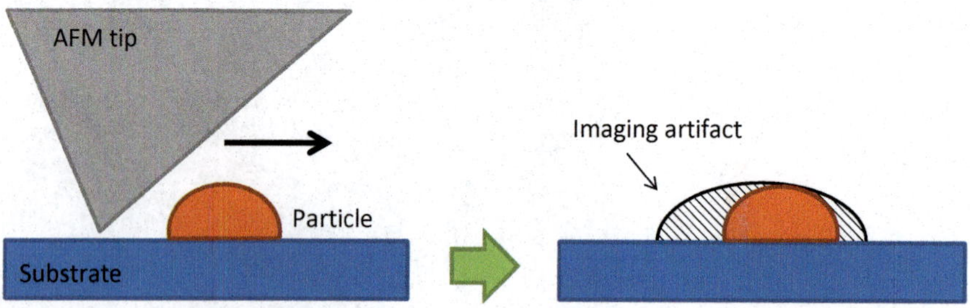

Fig. 4.3 Schematic illustrating possible imaging artifacts associated with a non-symmetric AFM tip

4.2.2 *Diffraction and Scattering Techniques*

Microscopy often uses transmitted or reflected light or particles to recreate planar images based on intensity patterns. However, other information can be obtained by measuring light or particles that have been diffracted or scattered by nanomaterials. Most common diffraction and scattering techniques use light, X-rays, and electrons as probes, but neutron and proton scattering methods also exist.

4.2.2.1 Dynamic Light Scattering

Dynamic light scattering (DLS) is a common technique used to characterize the size distribution of nanoparticles dispersed in a solution. The most common form of this technique is called photon correlation spectroscopy (PCS), and it is relatively straightforward. Polarized laser light is scattered by a dilute solution of small particles (<250 nm), creating a "speckle pattern" of scattered light. The speckle pattern is the result of the constructive and destructive interference of light scattered by the particles in the solution, and the particular pattern at any given time is a function of the location of the individual scatterers. Thus, the intensity of scattered light within the speckle pattern fluctuates over time as the individual small particles undergo Brownian motion and move about. The timescale of the pattern change, as quantified by the autocorrelation function, can be related to the sizes of the particles (Fig. 4.4). More specifically, the translational diffusion coefficient can be obtained, and this can be related to the hydrodynamic radius of a sphere through the Stokes-Einstein equation. An interested reader can pursue the finer details of PCS and the Stokes-Einstein equation from various sources in the literature [19].

For the purposes of this chapter, it is important to realize that the underlying assumptions of PCS, the most common form of DLS, which requires a monodisperse collection of optically isotropic spherical particles in order to obtain an accurate size distribution. Polydispersity can be accounted for in the determination of the translational diffusion coefficient by choosing an appropriate form factor [20]. However, these factors still depend on the particular shape of the particles being

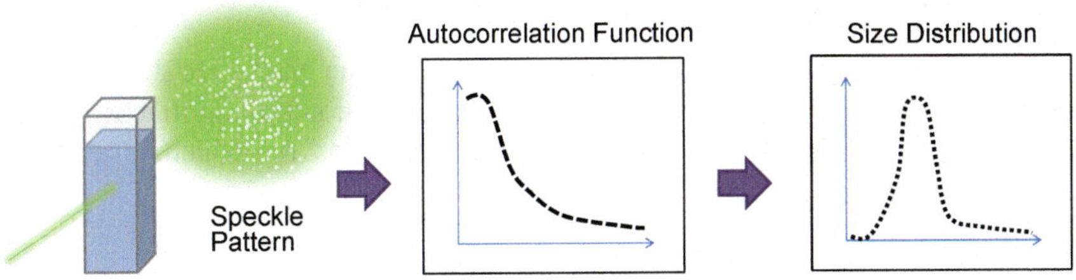

Fig. 4.4 Schematic illustrating the basic concept of dynamic light scattering: a speckle pattern is generated from a solution, which changes in time as quantified by the autocorrelation function, and the rate of change is related to size distribution of particles in the solution

investigated. DLS can still be used to characterize anisotropic nanoparticles, and the correct diffusion coefficient is generally measured, possibly with a higher polydispersity. Additionally, careful data analysis can allow the inference of geometric parameters for anisotropic nanoparticles from PCS data. For example, an investigation of Au nanorods using a standard DLS instrument yielded both the translational and rotational diffusion coefficients, which could then be related to the aspect ratio of the rods through theoretical calculations [21]. A DLS technique that uses depolarized light, called depolarized Fabry-Perot interferometry (FPI) can be directly applied to optically anisotropic nanoparticles [22], but it is far less common. Finally, it is very common for DLS instruments to also characterize electrophoretic light scattering in order to obtain a particle's Zeta potential. Since the measurement of the electrophoretic mobility is not generally affected by particle shape, Zeta potential measurements of anisotropic nanoparticles are accurate, though measurements might exhibit greater polydispersity.

4.2.2.2 X-ray Scattering and Diffraction

As with imaging, moving away from visible light to higher energy electromagnetic fields, such as X-rays, allows researchers to probe smaller regions of space. X-rays commonly used in the laboratory have wavelengths around 0.1–0.2 nm, and therefore, allow access to sub-nanometer information. A host of different scattering techniques have evolved from measuring the scattering of X-rays from materials, including X-ray reflectivity (XRR), X-ray diffraction (XRD), and inelastic scattering techniques. XRR measures the specular reflection of X-rays from surfaces and is sensitive to the electron density from the surface down to several tens of nanometers in depth [23]. The dependence on electron density allows XRR to provide information on porosity, roughness, and other structural parameters of surface layers [23, 24], and XRR has been useful in studying the size distributions of anisotropic nanoparticles, such as nanowires and nanotubes [25]. While the inference of size distributions from X-ray scattering data is more abstract than imaging, XRR measures a larger sample population, which can improve statistical accuracy. In general, data obtained from XRR must be fit to a model using regression techniques in order to obtain structural information, and therefore, it may be necessary to perform complementary analyses in order to ensure the relevancy of the model used for extrapolation.

XRD is the most common X-ray scattering technique encountered in the laboratory and relies on diffraction instead of specular reflection [26]. Diffraction results from constructive interference of waves, the conditions for which is given by Bragg's law:

$$m\lambda = 2d\sin\theta, \tag{4.1}$$

where m is the integer diffraction order, λ is the incident wavelength, d is the distance between diffraction planes, and θ is the angle of the diffracted wave.

The angstrom size of X-ray wavelengths, gives access to diffraction by crystalline planes. A fixed X-ray wavelength (*e.g.*, commonly used Cu $K\alpha$ radiation, $\lambda = 1.54184$ Å) incident on a crystal will produce an angular spread of intensity peaks, and this angular pattern is defined by Bragg's law and the distances between the lattice planes that are unique to each crystal. Therefore, the angular diffraction patterns of materials can serve as a fingerprint to identify the presence of a specific crystal phase. For nanoparticles, single crystal diffraction measurements are not currently practical, and instead, XRD analyses of nanomaterials rely on powder diffraction techniques. As the name implies, powder diffraction measures the X-ray diffraction pattern from a solid sample as a function of detector angle, which when scanned results in a series of intensity peaks at different angles, corresponding to different lattice planes (Fig. 4.5). By comparing the obtained pattern with known patterns of materials in a reference database, the phase or phases present in a sample can be readily identified. Reference patterns have been obtained from experimental measurements or have been calculated based on first principles techniques.

Ideally, the obtained diffraction peaks will be extremely sharp, and the peak angular locations and relative intensities will perfectly match those of the corresponding reference pattern. For anisotropic nanoparticles, this is almost never the case. This is because most reference patterns, especially those based on first principles calculations, assume large, randomly ordered, isotropic, perfect crystallites. For most materials, peak locations may shift due to instrumental alignment errors, but shifts in peak locations also result from faults, vacancies, and intentional or unintentional impurities, which all can cause changes in the lattice spacing. Peak broadening is caused by a number of factors including instrumental effects, crystalline imperfections, inhomogeneous strain, and small crystallite sizes. This latter effect is often employed to convert diffraction peak width into nanocrystal size via the Scherrer equation [27, 28]. It is important to note that, unless the other effects

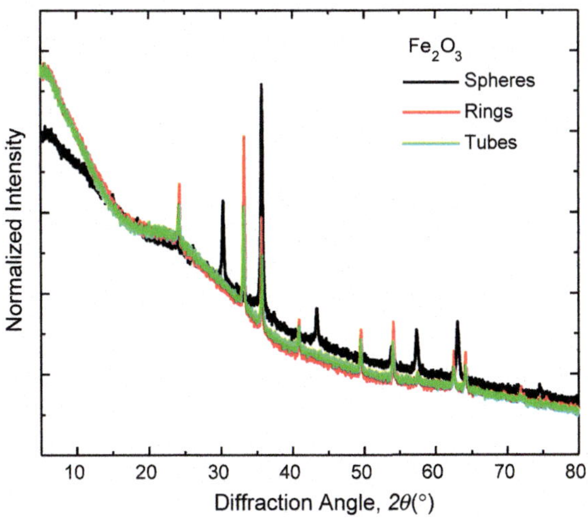

Fig. 4.5 Representative X-ray diffraction data, showing the diffraction patterns obtained from iron oxide (Fe_2O_3) powders of nanoparticle with different shapes, *spheres*, *rings*, and *tubes*

can be negated or taken into account, Scherrer calculations only provide a lower bound on size. Nevertheless, by analyzing the shape of the diffraction pattern and peaks, XRD data can provide structural information in addition to phase information. The relative intensities of the individual diffraction peaks of an anisotropic nanoparticle may not agree with the reference diffraction pattern of the same material. Again, this is because reference patterns are generally based on perfectly random, isotropically organized crystallites, and anisotropic nanoparticle typically aggregate and align into preferred orientations. This is especially common if the nanoparticles are directly grown on a substrate through physical or chemical vapor depositions, but is also often the case for anisotropic or shape-controlled nanoparticles that are drop-cast or spin-coated onto a substrate [29]. A possible way to minimize this alignment is to mix the nanoparticles in an amorphous matrix material prior to measurement. However, analyzing the orientation-dependent diffraction pattern can provide insights into the growth mechanisms of the nanoparticles [29], and therefore, XRD texture measurements, which measure a diffraction peak intensity versus polar and azimuthal rotations [30], are a useful tool for characterizing anisotropic and shape-selective nanoparticles [31–33].

In summary, X-ray scattering and diffraction techniques are useful tools for studying anisotropic and shape-selective nanoparticles. In particular, X-ray methods are common, well-established, and can provide chemical, crystalline, and structural information of a statistically large ensemble of nanoparticles. Some limitations are that X-ray methods often require relatively large amounts (gram quantities are often consider large for nanomaterials) of solid material for characterization, and that data analysis and refinement can be complicated for non-ideal materials.

4.2.2.3 Electron Diffraction

Similar to X-rays, electrons are diffracted by the atomic lattices of materials and can be used to characterize the crystalline properties of materials. The main differences between the two techniques are the smaller wavelengths of electrons compared to X-rays and the negative charge carried by electrons versus neutral X-rays. These factors change the nature of the diffraction. For example, X-rays are typically influenced by the valence electrons in a lattice, while electrons interact with both the positively-charged nucleus and the negatively charged electrons in the lattice. There are several different electron scattering and diffraction techniques. Some surface sensitive electron techniques, such as low-energy electron diffraction (LEED) or reflective high-energy diffraction (RHEED), have been used to characterize anisotropic or shape-selective nanomaterials [34–36]. However, electron diffraction imaging using a TEM is by far the most common method of using electron diffraction to characterize nanoparticles (Fig. 4.6) [37]. Typically, this is done using selected area electron diffraction (SAED), which images the diffraction of a parallel beam of electrons from a thin (<100 nm) sample. The SAED image is a series or pattern of spots that correspond with a collection of electrons that have satisfied the diffraction criterion upon transmission through the sample. Electron

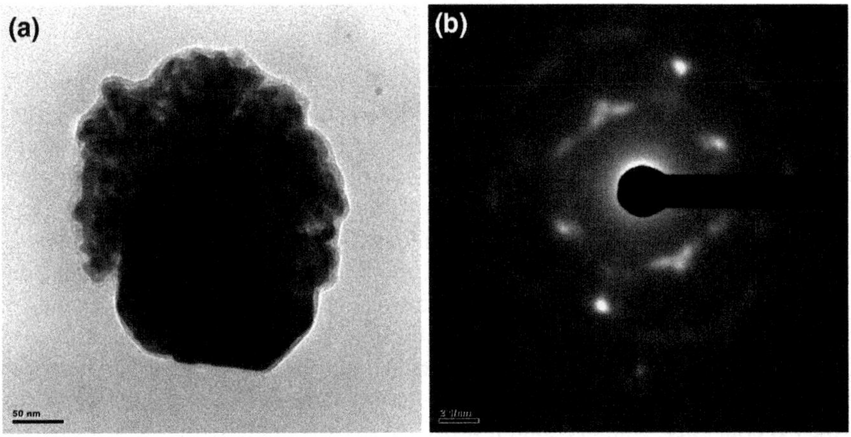

Fig. 4.6 Representative **a** TEM image and corresponding **b** SAED pattern from a gold- zinc alloy nanoparticle cluster

diffraction combined with TEM has been widely used to characterize and study anisotropic and shape-selective nanoparticle growth and properties, and numerous references are available in the literature [38–41]. This popularity is motivated by the combination of imaging with crystal structure interrogation, which is a great advantage and allows for real time imaging of very targeted and precise crystalline information. The limitations of electron diffraction characterization are similar to those of TEM - data is typically collected from a statistically small sample number, and the use of high energy electrons and vacuum environments limits sample types and observation conditions. Though, environmental TEM techniques are becoming more common and allow access to a wider variety of experimental conditions [42–45]. In spite of the limitations, electron diffraction is a useful and important characterization technique for understanding the structural properties and growth mechanisms of anisotropic and shape-selective nanoparticles.

4.2.3 Spectroscopic Techniques

Spectroscopy represents a broad category of techniques used to study nanomaterials. While historically the term emerged in reference to the study of light-matter interactions as a function of incident wavelength or frequency (energy), spectroscopy now encompasses a wide range of techniques that study the effects of radiative energy on matter versus wavelength or frequency. Similarly, a wide range of different information can be obtained from spectroscopic techniques, including composition, electronic structure, molecular bonding, and structure. Given the vastness of this topic, the discussion below will address a few of the more common techniques with applications toward anisotropic and shape-selective nanoparticles.

4.2.3.1 Optical Spectroscopy

Absorption spectroscopy is the oldest and most widely known optical spectroscopic technique. There are many different configurations, but the most basic conceptualization of optical absorption spectroscopy involves passing light of a specific wavelength range through a sample and measuring what wavelengths are absorbed. The wavelength region used corresponds with the information desired. For example, UV-visible spectroscopy provides information on the electronic structure of materials, infrared spectroscopy reveals molecular vibrational information, and moving out to the far-infrared and microwave regions provides information on molecular rotations. Other optical spectroscopies, such as photoluminescence or Raman spectroscopy, measure how incident light energy is changed after interacting with a sample. Photoluminescence measures the spectrum of light re-emitted from material after light absorption, which is typically in the visible-NIR region and corresponds with electronic configurations [46]. Raman spectroscopy measures small changes in the wavelengths of light induced after scattering off of a sample, and these smaller energy changes correspond with vibrational modes of molecules and lattices [47]. In general, different optical spectroscopic techniques complement each other, and by using several different analyses, a coherent picture of a nanomaterial's structure and composition can be obtained.

For anisotropic and shape-selective nanoparticles, it is important to consider how shape and anisotropy can influence spectroscopic measurements. In most cases, this is determined by how shape and anisotropy affect electronic configurations and vibrational or rotational movement. A notable example is that of quantum confinement. The quantum confinement effect has been well-established and describes how nanoscale dimensions change the energies of electron levels within semiconductors [48]. In particular, the energy gap between the valence and the conduction bands increases with decreasing dimensions, similar to the "particle in a box" concept (Fig. 4.7). For some anisotropic nanoparticles such as nanowires and nanorods, quantum confinement effects are one dimensional [49–51], and for others such as nanoplatelets and nanodisks, two-dimensional confinement is observed [52]. In comparison, quantum dots exhibit three-dimensional confinement. The electronic density of states is highly dependent on the shape and size of semiconducting nanoparticles in the confinement regime, and controlling for volume, nanospheres exhibit higher exciton binding energies than nanowires [54], which leads to different absorption bands, and corresponding emission bands, in the UV-visible spectrum. Similar confinement effects are seen in lattice vibrations (phonons) of nanoparticles [55–59], which affects infrared absorption and Raman scattering.

Localized surface plasmon resonances (LSPRs) are another well-known example of how shape and size of nanoparticles can alter optical properties and measurements [60]. As discussed in previous chapters, LSPRs are a collective oscillation of free electrons in nanoparticles that can be excited by coupling with incoming light. The energetics of this coupling is highly dependent on nanoparticle shape, size, composition, and ambient conditions, and are, therefore, tunable. Some generalities

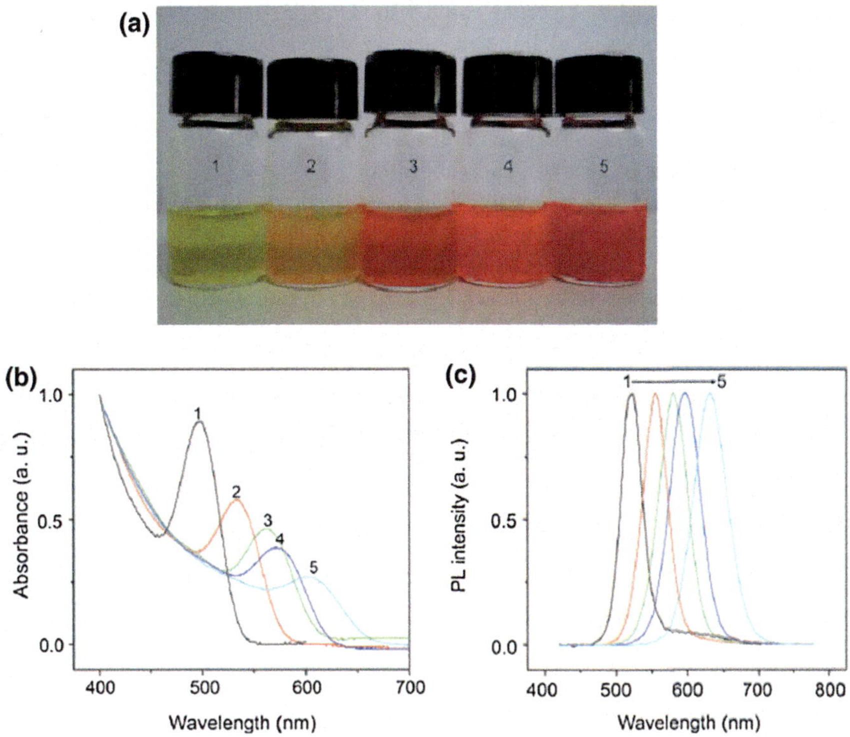

Fig. 4.7 **a** Photograph of CdTe quantum dots of different sizes (increasing diameters from 1 to 5), and corresponding **b** absorbance and **c** photoluminescence spectra. Figure reproduced from Ref. [53] with permission

exist. For example, LSPRs in noble metal nanoparticles tend to couple with visible frequencies of light [61], while LSPRs in doped semiconducting nanoparticles are usually excited in the infrared region [62].

The illustrative phenomena described above would not necessarily change how a spectroscopic experiment is performed, but it could affect how the measurements are analyzed. The variation of properties with shape and size means that reference spectra for bulk materials may no longer be applicable to nanomaterials. Furthermore, reference spectra from isotropic nanomaterials may no longer be similar to what is obtained from measurements of anisotropic materials. Knowledge of such changes is important when trying to use reference spectra for phase identification. For example, the changes in peak profile and position with particle size and shape caused by quantum confinement of lattice phonons will make phase identification using Raman spectroscopy more challenging. Anisotropy will cause additional notable effects in Raman spectra, such as the appearance of new peaks or the increase or decrease in relative intensities of other peaks. General trends are difficult to establish, but being aware of the underlying physical principles or concepts (*e.g.*, particle in a box) can help guide analyses. More rigorous comparisons can include first principles calculations [58]. On the other hand, if the phase of the material is already known, changes in Raman spectral features can be used to identify structural features of isotropic and anisotropic nanostructures [63–67]. For example, Raman analyses of peak shape and position has become an indispensable method for characterizing the quality and structure of graphene and carbon nanotubes [68–70].

In the same way, an identifying characteristic of LSPR in noble metal nanoparticles is the appearance of new peaks in the UV-visible spectrum, and the number and nature of these peaks can provide structural information regarding the nanoparticles that produced them [71, 72]. For example, gold nanorods typically exhibit two distinct LSPR bands in UV-visible-NIR spectra. One of these peaks is due to the oscillation of free electrons along the transverse axis of the rod, and the other corresponds with the longitudinal oscillation of electrons. Shifts in the energy of the longitudinal LSPR have been used to quantify the aspect ratios of Au nanorods in solution [73, 74]. Likewise, the aggregation state of monodisperse Au nanorods in solution causes shifts and changes in the profile and position of the transverse and longitudinal LSPR bands, and monitoring these shifts allows for remote monitoring of the ordering of nanorods in solution [75]. Such results are not only interesting from a materials characterization point of view, but can also be used as a basis for chemical and biological sensing [76].

4.2.3.2 Polarization-Dependent Measurements

There is an implicit assumption that the optical spectroscopy techniques described above rely on the use and measurement of unpolarized light, and this is often the case. However, many optical spectroscopies can be extended to polarization-dependent measurements, and additional challenges and advantages arise in the characterization of anisotropic and shape-selective nanoparticles when polarization effects are considered. In general, it is important to be aware of polarization effects for any optical spectroscopy because such effects can still arise in unpolarized measurements due to light being partially polarized by a highly anisotropic sample or the instrument being used out of tolerance.

As is well known, light is described by a transverse wave, where the direction of oscillation is orthogonal to the direction of energy transfer. That is, if an electromagnetic wave is propagating in the z-direction, the electric component of the field will be oscillating in the x-y plane (Fig. 4.8). The magnetic component of the field will also be oscillating in the x-y plane, only it will also be orthogonal to the electric component [77]. Linear polarized light occurs when light has been generated or filtered such that the components only oscillate in one direction (*e.g.*, along the x-axis for the electric and y-axis for the magnetic). In general, any fully polarized electromagnetic wave can be described by a linear combination of orthogonally oriented polarization states [78]. While it is often more convenient to think in linear terms (*e.g.*, light polarized in vertical and horizontal directions with respect to an incident plane), fully polarized light can also be expressed as linear combinations of circular

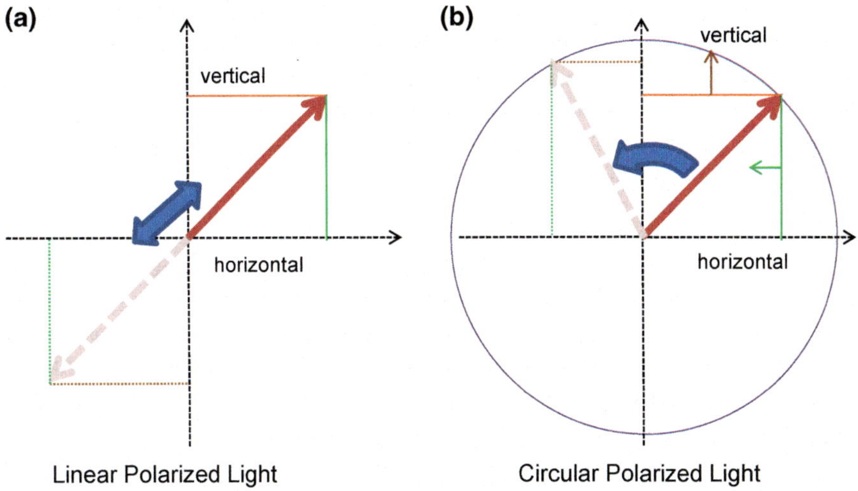

Fig. 4.8 Diagram illustrating the **a** oscillation and **b** rotation of the electric field amplitude of *linear* and *circular polarized light*, respectively

polarization states, where instead of the electric field amplitude oscillating in one direction, the electric field amplitude remains constant and rotates either clockwise or counter-clockwise in the x–y plane for a wave propagating in the z-direction. For circular polarization states, the x- and y-polarizations have a quarter wave phase shift between them. Different phase shifts will cause elliptical polarization states.

Materials, nano or otherwise, can exhibit anisotropy with respect to these two different types of polarizations, linear and circular. Linear and circular *birefringence* measurements characterize the difference in the refractive index between the different polarization directions (*i.e.*, vertical versus horizontal or right circular versus left circular), and linear and circular *dichroism* measurements characterize the difference in absorption between the different polarization directions. Bulk crystalline materials are known to exhibit birefringence and dichroism due to their crystal structure because light may propagate more freely when polarized along a particular crystal axis. Nanomaterials, on the other hand, may exhibit birefringence and dichroism due to their crystal structure and/or their particular shape. An easy way to visualize this phenomenon is to consider a plasmonic (gold or silver) nanorod. If light is linearly polarized along the major axis of the nanorod, it will more efficiently couple with the longitudinal LSPR mode of the nanorod and couple less efficiently with the transverse mode, leading to increased and decreased absorbance, respectively, or linear dichroism [79]. In general, it is expected that light will show absorption maxima/minima when polarized along the symmetry axes of plasmonic nanoparticles of different shapes [80]. Now imagine a plasmonic helix instead of a plasmonic rod [81–83]. If light is circularly polarized such that rotation of the electric field matches the winding of the helix, it will couple efficiently with the LSPR modes of the plasmonic helix [84–86]. On the other hand, the opposite rotation will couple less efficiently. This leads to circular dichroism and birefringence. While the discussion has focused on absorption and transmission effects (dichroism and birefringence), similar effects are observed in the emission and scattering of light. Quantum nanorods can emit highly polarized light during photoluminescence [87–89], and helical and other chiral nanostructures have been observed to emit circularly polarized light [90–93]. The effect of polarization of incident light on the polarization and intensity of Raman scattered light is also nanoparticle shape dependent [94–97].

Measurements from polarization-based spectroscopy can be directly related to the structure of the material under investigation and are also often done as a means to quantify the anisotropy of a sample. However, the symmetry of the measurement will affect whether or not polarization effects will be observed in anisotropic nanoparticles. Consider the case of a sample of nanorods dispersed in solution. If the dispersion is stable and the nanorods are randomly distributed and oriented throughout the solution, the sample will not display any linear dichroism or birefringence due to the symmetry of the perfectly random distribution. On the other hand, if the nanorods have been grown on a substrate through vapor deposition techniques or manipulated in such a way that they preferentially align, then linear polarization effects will occur. In comparison, nano-helices will exhibit circular dichroism and birefringence when aligned on a substrate and when randomly distributed in a solution. This is because helices have a particular type of anisotropy known as *chirality*. The defining characteristic of a chiral object is that it lacks reflective symmetry (Fig. 4.9). The most common examples are right and left hands. Just like right/left hands and right/left circular polarized light, a true chiral object can be oriented in any way without changing its "*handedness*." Chirality is a

Fig. 4.9 Diagram comparing the reflective symmetries and asymmetries of *achiral* and *chiral* objects, respectively

pervasive feature of living organisms, and circular dichroism spectroscopy has become an invaluable tool for the characterization of the secondary structure of proteins and other macromolecular structures, including ones that incorporate nanoparticles [98–101]. It is important to note that circular birefringence/dichroism measurements are very often susceptible to linear birefringence/dichroism artifacts, and vice versa [102]. Therefore, additional techniques are needed to separate circular and linear effects if the nanostructured sample exhibits alignment and both linear and chiral asymmetries [103–106].

4.2.3.3 Other Spectroscopies

The list of other spectroscopic techniques not described above is extensive, and a single chapter, or even a book in some cases, could be devoted to each technique and its application toward anisotropic nanoparticles. An experimental technique's absence from this chapter is not a judgment on its utility for the characterization of anisotropic nanoparticles. Each spectroscopic technique has its own advantages and drawbacks, and its use is determined by a specific sample and the information desired. For example, nuclear magnetic resonance (NMR) spectroscopy is a well-known technique for the characterization of organic molecules. However, its sensitivity to carbon and hydrogen bonding make NMR very useful in the characterization of anisotropic carbon-based nanomaterials, such as carbon nanotubes, graphene, and graphene oxide [107–110]. Similarly, Mössbauer spectroscopy is sensitive to magnetic fields and the presence of iron, and therefore, it has proven to be an advantageous technique for the investigation of magnetic anisotropy in iron nanowires [111]. Features such as quantum confinement and polarization sensitivity are ubiquitous to anisotropic nanoparticles, and their effects will arise in any measurement that intentionally or unintentionally probes these conditions. For example, band gap variation with size will be seen in UV-visible absorbance spectroscopy, as well as in photoacoustic spectroscopy [112]. Similarly, polarization and quantum size effects will be seen in ultrafast spectroscopic characterization of anisotropic nanoparticles [113–116]. However, new phenomena may also emerge in different experimental regimes. For example, time-resolved fluorescence spectroscopy measurements of the ultrafast emission of gold nanorods and nanospheres found that the intensities of fluorescence of nanorods are enhanced with respect to nanospheres due to an enhanced local field [117, 118]. Such effects can be attributed to the reduction of dephasing rate with volume in nanorods [119].

4.3 "Bulk" Property Characterization

Generally, materials are developed with a specific application in mind, and after structural and chemical characterization, the next step is to determine the bulk properties of the materials to see how well they will perform at certain tasks.

For example, in the development of a new catalyst, materials might be characterized by thermal gravimetric analysis for thermal stability and by chemical adsorption analysis to determine active surface area prior to actually testing the materials for the catalytic process. Similarly, the bulk optical and electrical properties of a material should be characterized before testing its use in a photodiode, phototransistor, or other opto-electronic device. For bulk materials composed of anisotropic nanostructures, experimental considerations for bulk analyses are similar to those above; the symmetries of both the materials and the measurement system need to be taken into account during inspection, and care must be taken to avoid negating any built-in assumptions of the technique.

The designation of "bulk properties" is a bit of a misnomer, as generally the hope is that unique nanoscale phenomenon will emerge when bulk materials are nanostructured or are composed of nanoparticle constituents. Therefore, there has been a great deal interest in characterizing the "bulk properties" (*e.g.*, electrical, optical, thermal, mechanical, magnetic) properties of individual nanoparticles in order to assess how they would perform in miniaturized applications, as well as to compare with bulk aggregates and verify theory [120–125]. The "bulk" property characterization of single nanoparticles is typically quite challenging and requires sophisticated equipment and clever micro-/nanoengineering and/or experimental design. Shape effects and anisotropy could further complicate such measurements on a case-by-case basis. Nevertheless, such single nanoparticle characterizations have become numerous in the literature, with single nanowire conductivity measurements being a common example (Fig. 4.10) [127–133]. An alternative to single nanoparticle experiments is to infer nanoscale properties from macroscale measurements. This can be done using effective medium theories [134], for example. However, most theoretical solutions are usually only exactly solvable for simple systems, which excludes most anisotropic and complex-shaped nanoparticles.

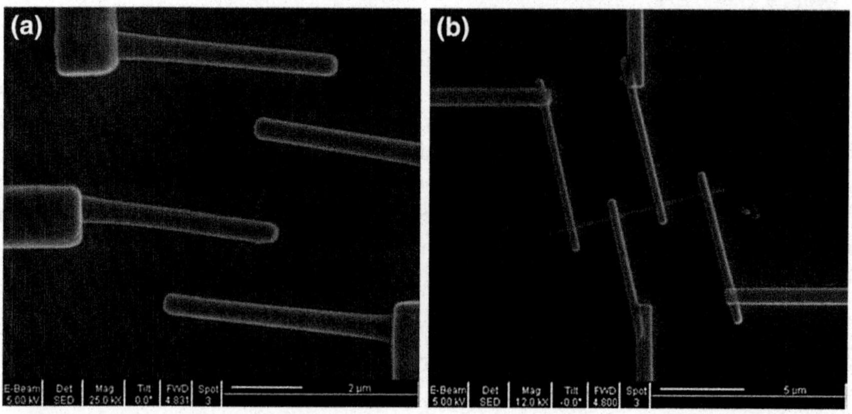

Fig. 4.10 SEM images of template-synthesized PEDOT nanowires and attached four Pt microleads for two-point and four-point probe analyses of single nanowires. Figure reproduced from Ref. [126] with permission

Though, anisotropic effective medium theories do exist and can be applied to the thermal, electrical, elastic, and optical properties [135–140]. Numerical solutions to theoretical equations are also possible, but it is not always clear how and if these calculations can be homogenized into bulk properties [141]. For such intractable problems, computational models, such as those based on molecular dynamics, are becoming more important, especially as computer technology improves [142].

4.4 Conclusion

Research into shape-selective and anisotropic nanoparticles is generally motivated by the desire to create better materials for a specific application, and therefore, it is critical to understand how and why shape affects nanoscale properties. As discussed above, in order to gain this understanding through experimentation, researchers can employ commonly available techniques used in materials characterization. However, in the case of anisotropic and shape-selective nanoparticles, greater concern for orientational and/or shape effects and artifacts should be shown during analyses.

References

1. Zhou, W., and Z.L. Wang. 2007. *Scanning Microscopy for Nanotechnology: Techniques and Applications*. Berlin: Springer science & business media.
2. Pennycook, S.J., and P.D. Nellist. 2011. *Scanning Transmission Electron Microscopy: Imaging and Analysis*. Berlin: Springer Science & Business Media.
3. Bhushan, B., and O. Marti. 2010. Scanning Probe Microscopy–Principle of Operation, Instrumentation, and Probes. In *Springer Handbook of Nanotechnology*, pp. 573–617. Berlin: Springer.
4. Cordeiro, M.A.L., P.A. Crozier, and E.R. Leite. 2012. Anisotropic Nanocrystal Dissolution Observation by in Situ Transmission Electron Microscopy. *Nano Letters* 12 (11): 5708–5713.
5. Angshuman, P., et al. 2016. Observation of the Formation of Anisotropic Silver Microstructures by Evanescent Wave and Electron Microscopy. *Nanotechnology* 27 (7): 075708.
6. Lu, W., et al. 2015. Probing the Anisotropic Behaviors of Black Phosphorus by Transmission Electron Microscopy, Angular-Dependent Raman Spectra, and Electronic Transport Measurements. *Applied Physics Letters* 107 (2): 021906.
7. Carpick, R.W., and M. Salmeron. 1997. Scratching the Surface: Fundamental Investigations of Tribology with Atomic Force Microscopy. *Chemical Reviews* 97 (4): 1163–1194.
8. Rota, A., et al. 2016. AFM-Based Tribological Study of Nanopatterned Surfaces: The Influence of Contact Area Instabilities. *Journal of Physics: Condensed Matter* 28 (13): 134008.
9. Choi, J.S., et al. 2011. Friction Anisotropy-Driven Domain Imaging on Exfoliated Monolayer Graphene. *Science* 333 (6042): 607.
10. Moretti, M., et al. 2013. Reflection-Mode TERS on Insulin Amyloid Fibrils with Top-Visual AFM Probes. *Plasmonics* 8 (1): 25–33.

11. Lintao, C., T. Hitoshi, and K. Tomoji. 2001. Probing Electrical Properties of Oriented DNA by Conducting Atomic Force Microscopy. *Nanotechnology* 12 (3): 211.
12. Nardes, A.M., et al. 2007. Microscopic Understanding of the Anisotropic Conductivity of PEDOT: PSS Thin Films. *Advanced Materials* 19 (9): 1196–1200.
13. Stöckle, R.M., et al. 2000. Nanoscale Chemical Analysis by Tip-Enhanced Raman Spectroscopy. *Chemical Physics Letters* 318 (1): 131–136.
14. Schönherr, H., Z. Hruska, and G.J. Vancso. 2000. Toward High Resolution Mapping of Functional Group Distributions at Surface-Treated Polymers by AFM Using Modified Tips. *Macromolecules* 33 (12): 4532–4537.
15. Majumdar, A., J. Carrejo, and J. Lai. 1993. Thermal Imaging Using the Atomic Force Microscope. *Applied Physics Letters* 62 (20): 2501–2503.
16. Zhang, Z., et al. 2016. Tip-Enhanced Raman Spectroscopy. *Analytical Chemistry* 88 (19): 9328–9346.
17. Sonntag, M.D., et al. 2014. Recent Advances in Tip-Enhanced Raman Spectroscopy. *The Journal of Physical Chemistry Letters* 5 (18): 3125–3130.
18. Rostislav, V.L. 2004. Feature-Oriented Scanning Methodology for Probe Microscopy and Nanotechnology. *Nanotechnology* 15 (9): 1135.
19. Pecora, R. 2000. Dynamic Light Scattering Measurement of Nanometer Particles in Liquids. *Journal of Nanoparticle Research* 2 (2): 123–131.
20. Aragon, S., and R. Pecora. 1976. Theory of Dynamic Light Scattering from Polydisperse Systems. *The Journal of Chemical Physics* 64 (6): 2395–2404.
21. Rodríguez-Fernández, J., et al. 2007. Dynamic Light Scattering of Short Au Rods with Low Aspect Ratios. *The Journal of Physical Chemistry C* 111 (13): 5020–5025.
22. Lehner, D., H. Lindner, and O. Glatter. 2000. Determination of the Translational and Rotational Diffusion Coefficients of Rodlike Particles Using Depolarized Dynamic Light Scattering. *Langmuir* 16 (4): 1689–1695.
23. Parratt, L.G. 1954. Surface Studies of Solids by Total Reflection of X-rays. *Physical Review* 95 (2): 359–369.
24. Holy, V., et al. 1993. X-ray Reflection From Rough Layered Systems. *Physical Review B* 47 (23): 15896–903.
25. Ito, Y. 2009. Grazing-Incidence Small-Angle X-ray Scattering Technique for Nanostructure Determination of Surfaces and Interfaces of Thin Films. *Rigaku Journal* 25 (1): 1–6.
26. Warren, B.E. 1969. *X-ray Diffraction*. USA: Courier Corporation.
27. Scherrer, P. 1918. Bestimmung der Größe und der inneren Struktur von Kolloidteilchen mittels Röntgenstrahlen. *Nachrichten von der Gesellschaft der Wissenschaften zu Göttingen, Mathematisch-Physikalische Klasse* 1918: 98–100.
28. Holzwarth, U., and N. Gibson. 2011. The Scherrer Equation Versus the 'Debye-Scherrer Equation'. *Nature Nanotechnology* 6 (9): 534.
29. Gordon, T.R., et al. 2015. Characterization of Shape and Monodispersity of Anisotropic Nanocrystals through Atomistic X-ray Scattering Simulation. *Chemistry of Materials* 27 (7): 2502–2506.
30. Wagner, F. 1999. Texture Determination by Using X ray Diffraction. In *Characterization Techniques of Glasses and Ceramics*, ed, J.M. Rincon and M. Romero, pp. 169–186. Berlin: Springer.
31. Melanie, K., et al. 2007. Solid Au Nanoparticles as a Catalyst for Growing Aligned ZnO Nanowires: A New Understanding of the Vapour–Liquid–Solid Process. *Nanotechnology* 18 (36): 365304.
32. Baker, J.L., et al. 2010. Quantification of Thin Film Crystallographic Orientation Using X-ray Diffraction with an Area Detector. *Langmuir* 26 (11): 9146–9151.
33. Kruszynska, M., et al. 2010. Synthesis and Shape Control of CuInS2 Nanoparticles. *Journal of the American Chemical Society* 132 (45): 15976–15986.
34. Gu, E., et al. 1999. Two-Dimensional Paramagnetic-Ferromagnetic Phase Transition and Magnetic Anisotropy in Co(110) Epitaxial Nanoparticle Arrays. *Physical Review B* 60 (6): 4092–4095.

35. Shima, T., et al. 2006. Formation of Octahedral FePt Nanoparticles by Alternate Deposition of FePt and MgO. *Applied Physics Letters* 88 (6): 063117.
36. Hanada, T., et al. 2001. Anisotropic Shape of Self-Assembled in As Quantum Dots: Refraction Effect on Spot Shape of Reflection High-Energy Electron Diffraction. *Physical Review B* 64 (16): 165307.
37. Williams, D.B., and C.B. Carter. 1996. The Transmission Electron Microscope. In *Transmission Electron Microscopy*, pp. 3–17. Berlin: Springer.
38. Kumar, C.S. 2013. *Transmission Electron Microscopy Characterization of Nanomaterials*. Berlin: Springer Science & Business Media.
39. Lee, S.-M., et al. 2002. Single-Crystalline Star-Shaped Nanocrystals and Their Evolution: Programming the Geometry of Nano-Building Blocks. *Journal of the American Chemical Society* 124 (38): 11244–11245.
40. Guiton, B.S., et al. 2005. Single-Crystalline Vanadium Dioxide Nanowires with Rectangular Cross Sections. *Journal of the American Chemical Society* 127 (2): 498–499.
41. Wang, Z.-l., and C. Hui. 2013. *Electron Microscopy of Nanotubes*. Berlin: Springer Science & Business Media.
42. Bogner, A., et al. 2005. Wet STEM: A New Development in Environmental SEM for Imaging Nano-Objects Included in a Liquid Phase. *Ultramicroscopy* 104 (3–4): 290–301.
43. de Jonge, N., and F.M. Ross. 2011. Electron Microscopy of Specimens in Liquid. *Nature Nanotechnology* 6 (11): 695–704.
44. Sutter, E., et al. 2014. In Situ Liquid-Cell Electron Microscopy of Silver–Palladium Galvanic Replacement Reactions on Silver Nanoparticles. *Nature communications* 5.
45. De Yoreo, J.J., and N. Sommerdijk. 2016. Investigating Materials Formation with Liquid-Phase and Cryogenic TEM. *Nature Reviews Materials* 1: 16035.
46. Kumar, C.S. 2013. *UV-VIS and Photoluminescence Spectroscopy for Nanomaterials Characterization*. Berlin: Springer.
47. Kumar, C.S. 2012. *Raman Spectroscopy for Nanomaterials Characterization*. Berlin: Springer Science & Business Media.
48. Haug, H., and S.W. Koch. 2004. *Quantum Theory of the Optical and Electronic Properties of Semiconductors* (Vol. 5). Singapore: World Scientific.
49. Krahne, R., et al. 2013. Optical Properties of Semiconductor Nanorods. In *Physical Properties of Nanorods*, pp. 7–55. Berlin: Springer.
50. Li, L.-S., et al. 2001. Band Gap Variation of Size- and Shape-Controlled Colloidal CdSe Quantum Rods. *Nano Letters* 1 (7): 349–351.
51. Yu, H., et al. 2003. Two- Versus Three-Dimensional Quantum Confinement in Indium Phosphide Wires and Dots. *Nature Materials* 2 (8): 517–520.
52. Bouet, C., et al. 2013. Flat Colloidal Semiconductor Nanoplatelets. *Chemistry of Materials* 25 (8): 1262–1271.
53. Duan, J., L. Song, and J. Zhan. 2009. One-Pot Synthesis of Highly Luminescent CdTe Quantum Dots by Microwave Irradiation Reduction and Their Hg^{2+} -sensitive properties. *Nano Research* 2 (1): 61–68.
54. Scher, J.A., J.M. Elward, and A. Chakraborty. 2016. Shape Matters: Effect of 1D, 2D, and 3D Isovolumetric Quantum Confinement in Semiconductor Nanoparticles. *The Journal of Physical Chemistry C* 120 (43): 24999–25009.
55. Krahne, R., et al. 2006. Confinement Effects on Optical Phonons in Polar Tetrapod Nanocrystals Detected by Resonant Inelastic Light Scattering. *Nano Letters* 6 (3): 478–482.
56. Nobile, C., et al. 2007. Confinement Effects on Optical Phonons in Spherical, Rod-, and Tetrapod-Shaped Nanocrystals Detected by Raman Spectroscopy. *physica status solidi (a)* 204 (2): 483–6.
57. Holger, L., et al. 2009. Geometry Dependence of the Phonon Modes in CdSe Nanorods. *Nanotechnology* 20 (4): 045705.
58. Saviot, L., and D.B. Murray. 2009. Acoustic Vibrations of Anisotropic Nanoparticles. *Physical Review B* 79 (21): 214101.

59. Wheaton, S., R.M. Gelfand, and R. Gordon. 2015. Probing the Raman-Active Acoustic Vibrations of Nanoparticles with Extraordinary Spectral Resolution. *Nature Photonics* 9 (1): 68–72.
60. Maier, S.A. 2007. *Plasmonics: Fundamentals and Applications.* Berlin: Springer Science & Business Media.
61. Link, S., and M.A. El-Sayed. 1999. Spectral Properties and Relaxation Dynamics of Surface Plasmon Electronic Oscillations in Gold and Silver Nanodots and Nanorods. *The Journal of Physical Chemistry B* 103 (40): 8410–8426.
62. Hsu, S.-W., K. On, and A.R. Tao. 2011. Localized Surface Plasmon Resonances of Anisotropic Semiconductor Nanocrystals. *Journal of the American Chemical Society* 133 (47): 19072–19075.
63. Zuo, J., et al. 1994. Study of the Raman spectrum of nanometer SnO_2. *Journal of Applied Physics* 75 (3): 1835–1836.
64. Spanier, J.E., et al. 2001. Size-Dependent Properties of ${\mathrm{CeO}}_{2-y}$ Nanoparticles as Studied by Raman Scattering. *Physical Review B* 64 (24): 245407.
65. Hearne, G.R., et al. 2004. Effect of Grain Size on Structural Transitions in Anatase TiO_2: A Raman Spectroscopy Study at High Pressure. *Physical Review B* 70 (13): 134102.
66. Alim, K.A., V.A. Fonoberov, and A.A. Balandin. 2005. Origin of the Optical Phonon Frequency Shifts in ZnO Quantum Dots. *Applied Physics Letters* 86 (5): 53103.
67. Kim, Y.K., and H.M. Jang. 2003. Raman Line-Shape Analysis of Nano-Structural Evolution in Cation-Ordered $ZrTiO_4$-Based Dielectrics. *Solid State Communications* 127 (6): 433–437.
68. Ferrari, A.C., et al. 2006. Raman Spectrum of Graphene and Graphene Layers. *Physical Review Letters* 97 (18): 187401.
69. Ferrari, A.C. 2007. Raman Spectroscopy of Graphene and Graphite: Disorder, Electron–Phonon Coupling, Doping and Nonadiabatic Effects. *Solid State Communications* 143 (1–2): 47–57.
70. Elias, D.C., et al. 2009. Control of Graphene's Properties by Reversible Hydrogenation: Evidence for Graphane. *Science* 323 (5914): 610–613.
71. Lee, K.-S., and M.A. El-Sayed. 2005. Dependence of the Enhanced Optical Scattering Efficiency Relative to That of Absorption for Gold Metal Nanorods on Aspect Ratio, Size, End-Cap Shape, and Medium Refractive Index. *The Journal of Physical Chemistry B* 109 (43): 20331–20338.
72. Jain, P.K., S. Eustis, and M.A. El-Sayed. 2006. Plasmon Coupling in Nanorod Assemblies: Optical Absorption, Discrete Dipole Approximation Simulation, and Exciton-Coupling Model. *The Journal of Physical Chemistry B* 110 (37): 18243–18253.
73. Schmucker, A.L., et al. 2010. Correlating Nanorod Structure with Experimentally Measured and Theoretically Predicted Surface Plasmon Resonance. *ACS Nano* 4 (9): 5453–5463.
74. Near, R.D., S.C. Hayden, and M.A. El-Sayed. 2013. Thin to Thick, Short to Long: Spectral Properties of Gold Nanorods by Theoretical Modeling. *The Journal of Physical Chemistry C* 117 (36): 18653–18656.
75. Umar, A., and S.-M. Choi. 2013. Aggregation Behavior of Oppositely Charged Gold Nanorods in Aqueous Solution. *The Journal of Physical Chemistry C* 117 (22): 11738–11743.
76. Wang, C., et al. 2007. Biorecognition-Driven Self-Assembly of Gold Nanorods: A Rapid and Sensitive Approach Toward Antibody Sensing. *Chemistry of Materials* 19 (24): 5809–5811.
77. Jackson, J.D. 1999. *Classical Electrodynamics.* New Jersey: Wiley.
78. Jones, R.C. 1941. *A New Calculus for the Treatment of Optical SystemsI. Description and Discussion of the Calculus.* JOSA 31 (7): 488–93.
79. Sönnichsen, C., and A.P. Alivisatos. 2005. Gold Nanorods as Novel Nonbleaching Plasmon-Based Orientation Sensors for Polarized Single-Particle Microscopy. *Nano Letters* 5 (2): 301–304.
80. Nehl, C.L., H. Liao, and J.H. Hafner. 2006. Optical Properties of Star-Shaped Gold Nanoparticles. *Nano Letters* 6 (4): 683–688.

81. Gansel, J.K., et al. 2009. Gold Helix Photonic Metamaterial as Broadband Circular Polarizer. *Science* 325 (5947): 1513–1515.
82. Larsen, G.K., et al. 2014. Scalable Fabrication of Composite Ti/Ag Plasmonic Helices: Controlling Morphology and Optical Activity by Tailoring Material Properties. *Advanced Optical Materials* 2 (3): 245–249.
83. Titus, J., et al. 2016. Large Circular Dichroism and Optical Rotation in Titanium Doped Chiral Silver Nanorods. *Annalen der Physik* 528 (9–10): 677–683.
84. Zhang, Z.-Y., and Y.-P. Zhao. 2007. Optical Properties of Helical Ag Nanostructures Calculated by Discrete Dipole Approximation Method. *Applied Physics Letters* 90 (22): 221501.
85. Schäferling, M., et al. 2014. Helical Plasmonic Nanostructures as Prototypical Chiral Near-Field Sources. *ACS Photonics* 1 (6): 530–537.
86. Gansel, J.K., et al. 2010. Gold Helix Photonic Metamaterials: A Numerical Parameter Study. *Optics Express* 18 (2): 1059–1069.
87. Hu, J., et al. 2001. Linearly Polarized Emission From Colloidal Semiconductor Quantum Rods. *Science* 292 (5524): 2060–2063.
88. Hikmet, R.A., et al. 2005. Polarized-Light-Emitting Quantum-Rod Diodes. *Advanced Materials* 17 (11): 1436–1439.
89. Bruhn, B., J. Valenta, and J. Linnros. 2009. Controlled Fabrication of Individual Silicon Quantum Rods Yielding High Intensity, Polarized Light Emission. *Nanotechnology* 20 (50): 505301.
90. Hrudey, P.C.P., K.L. Westra, and M.J. Brett. 2006. Highly Ordered Organic Alq3 Chiral Luminescent Thin Films Fabricated by Glancing-Angle Deposition. *Advanced Materials* 18 (2): 224–228.
91. Naito, M., et al. 2010. Circularly Polarized Luminescent CdS Quantum Dots Prepared in a Protein Nanocage. *Angewandte Chemie International Edition* 49 (39): 7006–7009.
92. Singh, H.J., et al. 2016. Circular Differential Two-Photon Luminescence from Helically Arranged Plasmonic Nanoparticles. *ACS Photonics* 3 (5): 863–868.
93. Kumar, J., T. Kawai, and T. Nakashima. 2017. Circularly Polarized Luminescence in Chiral Silver Nanoclusters. *Chemical Communications*.
94. Rao, A., et al. 2000. Polarized Raman Study of Aligned Multiwalled Carbon Nanotubes. *Physical Review Letters* 84 (8): 1820.
95. Cançado, L., et al. 2004. Anisotropy of the Raman Spectra of Nanographite Ribbons. *Physical Review Letters* 93 (4): 047403.
96. Schäfer-Nolte, E., et al. 2010. Highly Polarized Raman Scattering Anisotropy in Single GaN Nanowires. *Applied Physics Letters* 96 (9): 091907.
97. Acevedo, R., et al. 2009. Plasmonic Enhancement of Raman Optical Activity in Molecules Near Metal Nanoshells. *The Journal of Physical Chemistry A* 113 (47): 13173–13183.
98. Fasman, G.D. 2013. *Circular Dichroism and the Conformational Analysis of Biomolecules.* Berlin: Springer Science & Business Media.
99. Fan, Z., and A.O. Govorov. 2010. Plasmonic Circular Dichroism of Chiral Metal Nanoparticle Assemblies. *Nano Letters* 10 (7): 2580–2587.
100. Slocik, J.M., A.O. Govorov, and R.R. Naik. 2011. Plasmonic Circular Dichroism of Peptide-Functionalized Gold Nanoparticles. *Nano Letters* 11 (2): 701–705.
101. Kuzyk, A., et al. 2012. DNA-Based Self-Assembly of Chiral Plasmonic Nanostructures with Tailored Optical Response. *Nature* 483 (7389): 311–314.
102. Schellman, J., and H.P. Jensen. 1987. Optical Spectroscopy of Oriented Molecules. *Chemical Reviews* 87 (6): 1359–1399.
103. Lu, S.-Y., and R.A. Chipman. 1994. Homogeneous and Inhomogeneous Jones Matrices. *Journal of the Optical Society of America A* 11 (2): 766–773.
104. Arteaga, O., and A. Canillas. 2009. Pseudopolar Decomposition of the Jones and Mueller-Jones Exponential Polarization Matrices. *Journal of the Optical Society of America A* 26 (4): 783–793.

105. Arteaga, O., and A. Canillas. 2010. Analytic Inversion of the Mueller-Jones Polarization Matrices for Homogeneous Media. *Optics Letters* 35 (4): 559–561.
106. Larsen, G.K., and Y. Zhao. 2014. Extracting the Anisotropic Optical Parameters of Chiral Plasmonic Nanostructured Thin Films Using Generalized Ellipsometry. *Applied Physics Letters* 105 (7): 071109.
107. Frota, H.O., and A. Ghosh. 2012. NMR of Localized Magnetic States in Graphene. *Physica B: Condensed Matter* 407 (7): 1170–1174.
108. Nelson, D.J., and R. Kumar. 2013. Characterizing Covalently Sidewall-Functionalized Single-Walled Carbon Nanotubes by Using 1H NMR Spectroscopy. *The Journal of Physical Chemistry C* 117 (28): 14812–14823.
109. Freitas, J.C., et al. 2015. Determination of the Hyperfine Magnetic Field in Magnetic Carbon-Based Materials: DFT Calculations and NMR Experiments. *Scientific reports*, 2015. **5**.
110. MacIntosh, A.R., K.J. Harris, and G.R. Goward. 2016. Structure and Dynamics in Functionalized Graphene Oxides through Solid-State NMR. *Chemistry of Materials* 28 (1): 360–367.
111. Zhan, Q.-F., et al. 2004. Applied Field Mössbauer Study of Shape Anisotropy in Fe Nanowire Arrays. *Applied Physics Letters* 85 (20): 4690–4692.
112. Liu, F.X., et al. 1999. Quantum Size and composition Effects in Photoacoustic Spectra of FexZr1 − xO_2 clusters. *Journal of Applied Physics* 85 (2): 734–738.
113. Mohamed, M.B., C. Burda, and M.A. El-Sayed. 2001. Shape Dependent Ultrafast Relaxation Dynamics of CdSe Nanocrystals: Nanorods vs Nanodots. *Nano Letters* 1 (11): 589–593.
114. Seo, M.A., et al. 2011. Polarization Anisotropy of Transient Carrier Dynamics in Single Si Nanowires. In *CLEO: 2011—Laser Science to Photonic Applications*.
115. Wu, K., H. Zhu, and T. Lian. 2015. Ultrafast Exciton Dynamics and Light-Driven H2 Evolution in Colloidal Semiconductor Nanorods and Pt-Tipped Nanorods. *Accounts of Chemical Research* 48 (3): 851–859.
116. Ge, S., et al. 2015. Dynamical Evolution of Anisotropic Response in Black Phosphorus under Ultrafast Photoexcitation. *Nano Letters* 15 (7): 4650–4656.
117. Varnavski, O.P., et al. 2003. Relative Enhancement of Ultrafast Emission in Gold Nanorods. *The Journal of Physical Chemistry B* 107 (14): 3101–3104.
118. Varnavski, O.P., et al. 2005. Femtosecond Excitation Dynamics in Gold Nanospheres and Nanorods. *Physical Review B* 72 (23): 235405.
119. Sönnichsen, C., et al. 2002. Drastic Reduction of Plasmon Damping in Gold Nanorods. *Physical Review Letters* 88 (7): 077402.
120. Klein, D.L., et al. 1996. An Approach to Electrical Studies of Single Nanocrystals. *Applied Physics Letters* 68 (18): 2574–2576.
121. Wernsdorfer, W., D. Mailly, and A. Benoit. 2000. Single Nanoparticle Measurement Techniques. *Journal of Applied Physics* 87 (9): 5094–5096.
122. Johnson, J.C., et al. 2001. Single Nanowire Lasers. *The Journal of Physical Chemistry B* 105 (46): 11387–11390.
123. Gambardella, P., et al. 2003. Giant Magnetic Anisotropy of Single Cobalt Atoms and Nanoparticles. *Science* 300 (5622): 1130–1133.
124. Zijlstra, P., et al. 2008. Acoustic Oscillations and Elastic Moduli of Single Gold Nanorods. *Nano Letters* 8 (10): 3493–3497.
125. Doerk, G.S., C. Carraro, and R. Maboudian. 2010. Single Nanowire Thermal Conductivity Measurements by Raman Thermography. *ACS Nano* 4 (8): 4908–4914.
126. Long, Y., et al. 2009. Electrical Conductivity Studies on Individual Conjugated Polymer Nanowires: Two-Probe and Four-Probe Results. *Nanoscale Research Letters* 5 (1): 237.
127. Li, D., et al. 2003. Thermal Conductivity of Individual Silicon Nanowires. *Applied Physics Letters* 83 (14): 2934–2936.
128. Zhang, D., et al. 2004. Detection of NO_2 Down to ppb Levels Using Individual and Multiple $In2O_3$ Nanowire Devices. *Nano Letters* 4 (10): 1919–1924.

129. Li, Q.H., et al. 2004. Electronic Transport Through Individual ZnO Nanowires. *Applied Physics Letters* 84 (22): 4556–4558.
130. Samitsu, S., et al. 2005. Conductivity Measurements of Individual poly (3,4-ethylenedioxythiophene)/poly(styrenesulfonate) Nanowires on Nanoelectrodes Using Manipulation with an Atomic Force Microscope. *Applied Physics Letters* 86 (23): 233103.
131. Fujii, M., et al. 2005. Measuring the Thermal Conductivity of a Single Carbon Nanotube. *Physical Review Letters* 95 (6): 065502.
132. Boukai, A., K. Xu, and J.R. Heath. 2006. Size-Dependent Transport and Thermoelectric Properties of Individual Polycrystalline Bismuth Nanowires. *Advanced Materials* 18 (7): 864–869.
133. Kuang, Q., et al. 2007. High-Sensitivity Humidity Sensor Based on a Single SnO_2 Nanowire. *Journal of the American Chemical Society* 129 (19): 6070–6071.
134. Choy, T.C. 2015. *Effective Medium Theory: Principles and Applications* (Vol. 165). Oxford: Oxford University Press.
135. Berthier, S. 1994. Anisotropic Effective Medium Theories. *Journal de Physique I* 4 (2): 303–318.
136. Zhou, X.F., and L. Gao. 2006. Effective Thermal Conductivity in Nanofluids of Nonspherical Particles with Interfacial Thermal Resistance: Differential Effective Medium Theory. *Journal of Applied Physics* 100 (2): 024913.
137. Garcia-Vidal, F., J. Pitarke, and J. Pendry. 1997. Effective Medium Theory of the Optical Properties of Aligned Carbon Nanotubes. *Physical Review Letters* 78 (22): 4289.
138. Hornby, B.E., L.M. Schwartz, and J.A. Hudson. 1994. Anisotropic Effective-Medium Modeling of the Elastic Properties of Shales. *Geophysics* 59 (10): 1570–1583.
139. Song, C., G.K. Larsen, and Y. Zhao. 2013. Anisotropic Resistivity of Tilted Silver Nanorod Arrays: Experiments and Modeling. *Applied Physics Letters* 102 (23): 233101.
140. Bradley, L., G. Larsen, and Y. Zhao. 2016. Designed to Fail: Flexible, Anisotropic Silver Nanorod Sheets for Low-Cost Wireless Activity Monitoring. *The Journal of Physical Chemistry C* 120 (27): 14969–14976.
141. Hornung, U. 2012. *Homogenization and Porous Media* (Vol. 6). Berlin: Springer Science & Business Media.
142. Che, J., T. Cagin, and W.A. Goddard III. 2000. Thermal Conductivity of Carbon Nanotubes. *Nanotechnology* 11 (2): 65.

Part II
Effect of the Morphology and the Nanometric Dimension of Materials on Their Physico-Chemical Properties

Chapter 5
Anisotropic Metallic and Metallic Oxide Nanostructures-Correlation Between Their Shape and Properties

Simona E. Hunyadi Murph

Abstract In this chapter, we highlight recent innovations from our laboratory by featuring uniquely shaped nanostructures and how their morphology and dimension affect their physico-chemical properties and subsequently their applications. We aim to cover a wide range of applications including optical and plasmonic applications, sensing and imaging, catalytic and photocatalytic applications, bio-medical and environmental implications as well as energy related applications.

Keywords Anisotropic metallic nanostructures · Anisotropic metallic oxide nanostructures · Multifunctional nanostructures · Gold nanoparticles · Silver nanoparticles · Silica · Iron oxide · Titania · Sensing · Imaging · Photodegradation · Environmental applications · Catalysis · Electrocatalysis · Solar energy to fuel conversion · Photothermal · Self-assemblies

5.1 Sensing and Optical Imaging

In the last several decades, a great deal of interest has been directed toward the accurate detection of ultra-low levels of analytes. Trace detection of analytes and single molecule detection [1] in real time are essential in virtually every scientific discipline as sensors find applications in many industries; among them transportation, communications, medicine, safety, and national security, including both homeland defense and military operations. There is a constant need to design sensors with substantially smaller size, lower weight, faster detection, greater sensitivity, better specificity, and modest power requirements that can be deployed in the field. These tasks are technically complex and require innovative and advanced technologies to achieve success. Recent advances in microscopy and

S.E. Hunyadi Murph (✉)
National Security Directorate, Savannah River National Laboratory, Aiken, SC, USA
e-mail: Simona.Murph@srnl.doe.gov

S.E. Hunyadi Murph
Department of Physics and Astronomy, University of Georgia, Athens, GA, USA

S.E. Hunyadi Murph et al. (eds.), *Anisotropic and Shape-Selective Nanomaterials*, Nanostructure Science and Technology, DOI 10.1007/978-3-319-59662-4_5

analytical techniques coupled with researchers' ability to produce nanoscale materials with unique geometries promise to solve a variety of analytical, biomedical, environmental and global problems.

Anisotropic gold and silver nanoparticles (rods, wires, stars, cubes, triangles, etc.) are uniquely suited for sensing applications as they address many of these tasks. Metal nanoparticles are particularly attractive materials because they can be easily synthesized and chemically transformed [2]. They are significantly different from the same materials in the bulk because as their size decreases, they exhibit quantum size effects [2]. Metallic nanoparticles interact strongly with light waves, even though the wavelength of the light may be much larger than the particle. In metal nanoparticles, "plasma oscillations" driven by external electromagnetic fields are localized and lead to strong resonances at specific wavelengths that are dependent on the particle size, shape and the local dielectric environment (Fig. 5.1) [3]. Plasmonic properties of metallic nanoparticles find applications in plasmon-enhanced spectroscopy, near-field imaging, sensing, and nanophotonic devices.

The plasmon band of gold and silver nanoparticles is tunable throughout the visible and near-infrared region of the spectrum as a function of particle shape, size and the local refractive index of the medium (Fig. 5.1). Isotropic gold nanoparticles display only one plasmon band around 520 nm. Anisotropic metal nanoparticles can absorb and scatter light along multiple axes; therefore, metal nanorods and nanowires display both longitudinal and transverse plasmon bands (Fig. 5.1). In the case of nanorods, two plasmon bands emerge, corresponding to light being absorbed (and scattered) along the short axis (the transverse plasmon band) and the long axis (the longitudinal plasmon band). In general, the longer the rod, the more red-shifted the longitudinal band will be (Fig. 5.1). Thus, metal nanoparticle shape

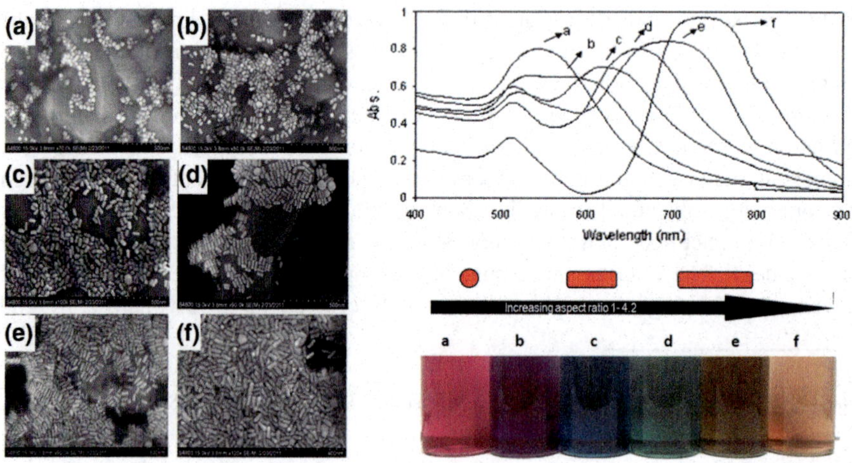

Fig. 5.1 Electron micrographs images (**a–f**), photographs, and UV-Vis spectra of Au nanoparticles (**a**) spheres, (**b-f**) different aspect ratio rods

can be tuned to match the frequency of common lasers (e.g., 633, 785, or 1064 nm) critical for detection of trace quantities of analytes. Their tunable optical properties, including light absorption, elastic light scattering, and inelastic light scattering (e.g., surface-enhanced Raman scattering) [2–5], makes them extremely valuable as components for incorporation in macro-devices.

5.1.1 Sensing via Inelastic Light Scattering-Surface-Enhanced Raman Scattering

The optical properties of the anisotropic and isotropic nanomaterials become very important for examining surface-enhanced Raman scattering (SERS) effects [6]. SERS can be used for detection of trace amount of substances for a large variety of molecules of environmental [7], biomedical [8], and pharmaceutical interest [9]. Single molecule detection of analytes has been reported with silver and gold colloids [1, 10].

SERS is a vibrational technique in which the Raman signatures of molecules that are near metal nanoscale surfaces are greatly enhanced in intensity compared to the molecules alone (Fig. 5.2) [4, 11]. While the Raman vibrations of molecules are in general very weak, in the presence of metals (copper, silver, gold) with nanoscale roughness, the molecular Raman vibrations excited by visible light are enhanced by orders of magnitude. In the classical measurement, roughly 10^{-6} of the incident electromagnetic (EM) field (laser) is radiated at a shifted frequency and detected as Raman scattering [4–6, 11, 12]. There are two mechanisms to describe the overall SERS enhancement: the electromagnetic (EM) and chemical (CHEM) enhancement mechanisms. EM enhancement is due to the increased local electric field incident on an adsorbed molecule at a metallic surface, due to visible light absorption by the metal. CHEM enhancement results from electronic resonance/charge transfer

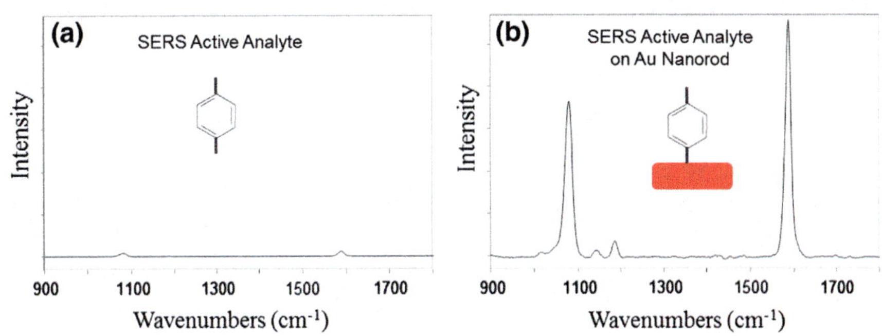

Fig. 5.2 **a** Raman signature of molecule alone and **b** Raman signatures of molecules that are near metal nanoscale surfaces are greatly enhanced in intensity compared to the molecules alone

between a molecule and a metal surface, which leads to an increase in the polarizability of the molecule [4–6, 11–13].

Anisotropic gold and/or silver nanoparticles with dimensions less than 100 nm that have highly curved, sharp surface features (i.e., rods, wires, stars, triangles, etc.) are widely used for the detection of gas and liquid analytes. They have enhanced sensing capabilities compared to isotropic nanospheres [1] due to an increased electric field, or "hot spots" at the tips of the particles [4–6, 10–13]. This is of particular interest if the molecules to be detected are located at the tips of the sharp features, rather than the sides [3, 4, 6, 12, 15]. An enhanced electric field at the tips compared to spheres generates an increased SERS response that is due to the localized plasmon band of the nanoparticles. Up to 100-fold increase of the local electric field at the nanoparticle's tip has been reported [3–14].

Surface enhanced Raman scattering (SERS) spectra of 4-mercaptobenzoic acid (4-MBA) in the presence of various shapes of gold nanoparticles was investigated by Orendorff et al. [16]. Cubes, blocks, and dogbones exhibited the largest surface enhancement factors on the order of 10^9, while spheres showed the smallest enhancement with enhancement factors around 10^7. The dependence of enhancement factor on nanoparticle shape was attributed to the chemical contributions to SERS, specifically, the surface structure and sharpness of structural features of the gold nanocrystals or SERS "hot spot" (Fig. 5.3) [3, 6, 12].

There are many factors that affect the SERS response including the nanomaterial's plasmonic response, Raman cross-section, affinity of the analyte to the nanomaterial, analyte surface oxidation, hot-spot density, as well as nanomaterial geometry and morphology. Careful selections of the aforementioned parameters are critical for achieving best SERS activities. Additional improvements can be generated by using laser frequencies that are resonant with the electronic and vibrational transitions of the analyte, thereby increasing the efficiency of the Raman process [3, 4]. When the incident radiation is tuned to overlap with the nanoparticle's absorption maxima, even greater SERS enhancements can be observed [5, 6]. This creates locally intense electric fields due to the excitement of plasmons. Nanoparticle surfaces can be tuned for the excitation and Raman scattering frequencies. For example, nanorods with longitudinal plasmon bands in resonance with the excitation source exhibited a 10^2 increase in EM enhancement for both

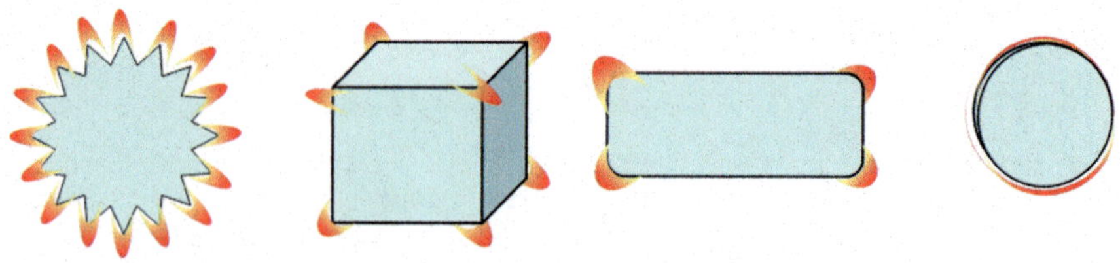

Fig. 5.3 Nanoparticles with sharp features, "hot spots", concentrate the local electromagnetic field into a small volume and are considered superior to spheres

gold and silver nanorods compared to nanorods with plasmon bands in the off-resonance condition [3, 17, 18].

Surface-enhanced Raman scattering studies were reported by us for 4-MBA attached to Ag nanowires, Ag–Au bimetallic nanowires and Au nanotubes in colloidal solutions [19]. Silver nanowires that were synthesized using an aqueous chemical reduction process, were reacted with gold salt to create tunable Ag–Au bimetallic nanowires (Fig. 5.4a, b). A continuous red-shift of the transverse plasmon band of the Ag nanowires, from ~ 400 to ~ 650 nm was observed during the production of the Ag-Au bimetallic nanowires. Upon incubation of the bimetallic products in aqueous ammonia, gold nanotubes were produced. By using bimetallic nanowires instead of Au nanotubes or even Ag nanowires in solution, enhancement factors of ~ 10^7 were observed compared to the analyte alone, which, once again, are contingent on the composition of the substrate (Fig. 5.4c) [4].

In our studies [19], the enhancements calculated for bimetallic nanowires resulted in a factor of 40, at best, compared to silver nanowires alone (Fig. 5.4c).

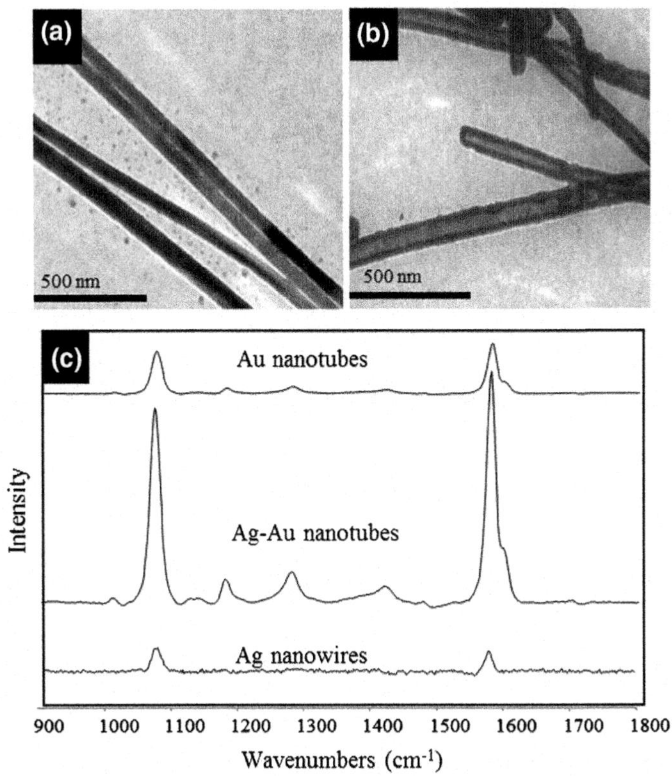

Fig. 5.4 TEM images of **a** Ag nanowires and **b** Ag-Au bimetallic nanotubes; **c** Surface enhanced Raman spectra of 4-MBA using silver nanowires, Ag-Au nanotubes and Au nanotubes. Adapted with permission from Ref. [19]. Copyright 2017 American Chemical Society

These results suggest that matching the laser wavelength with the material composition and plasmon band resonance is beneficial for chemical detection. Au nanotubes, while not providing enhancement factors as significant as those with the bimetallic Ag-Au nanotwires, are surprisingly better than solid long gold nanorods in similar colloidal SERS experiments. Analogous experiments with nanospheres showed SERS enhancement on the order of 10 times larger for Ag-Au bimetallic compared to Ag alone [19]. This shows that nanowires provide an additional advantage compared to spheres, as their plasmon band is tunable in the Vis- near IR region of the spectrum.

The SERS response is highly dependent on the nature of analyte investigated, the substrate's composition and experimental conditions, chemical or physical affinity for the substrate, phase of analyte (liquid or gas phase), etc. [4, 6, 10, 12]. Larger SERS effects have been reported for analytes that are electrostatically attracted to nanoparticle substrates than those samples without electrostatic interactions [20].

Ordered arrays of silver nanorods prepared by physical vapor deposition were explored by us as "lab-on-a-chip" type SERS substrates (Fig. 5.5). Liquid and gas phase analytes were investigated. SERS enhancement factors greater than 10^8 were recorded for liquid analytes compared to five to ten order of magnitude enhancement for gas analytes [11]. Since silver nanostructures are more prone to oxidation,

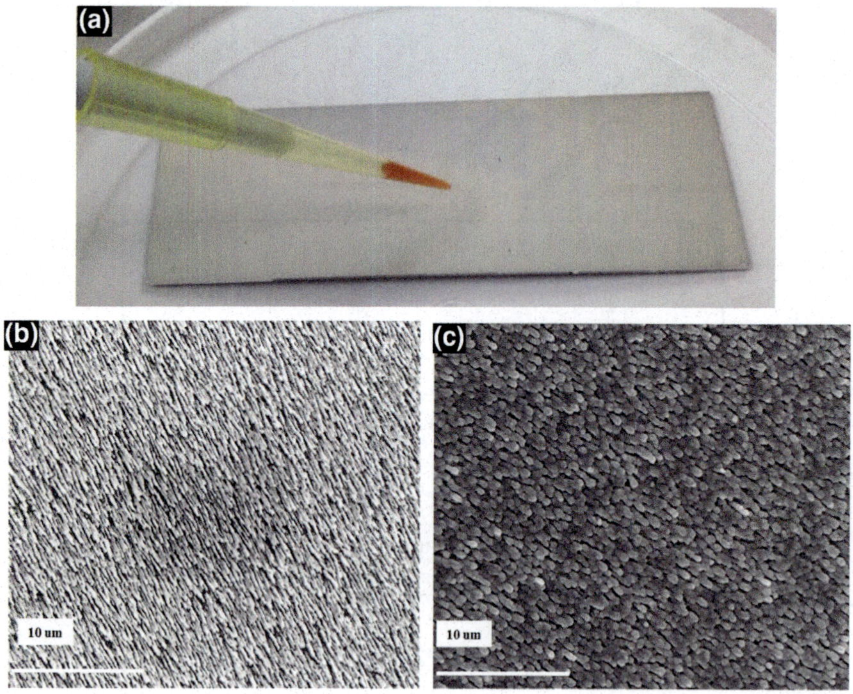

Fig. 5.5 **a** Photograph of ordered arrays of silver nanorods—lab-on-a-chip system, and SEM images of **b** Ag nanorods and **c** Ag-Au nanorods

the silver nanorods were incubated in gold salt to create ordered arrays of bimetallic Ag–Au and Au nanostructures to increase the material's lifetime (Fig. 5.5). The SERS response tested using 4-mercaptophenol decreased exponentially with reaction time due to the compositional evolution of the nanostructure from pure Ag to Au–Ag alloy type with increasing Au content. However, the stability of the SERS substrate increases significantly which makes them valuable for use in harsh environments. Additionally, the Ag-Au nanorods are more efficient than the Ag nanorods for detection of gas adsorbates in the ambient air than their counterparts making them suitable for gas detection applications [11].

Large enhancement factors and even single-molecule SERS are expected for molecules at junctions between aggregated nanoparticles [21]. This is a result of localized surface plasmon (LSP) coupling between nanoparticles and enhanced electromagnetic field intensity localized at nanoparticle junctions activated by absorption of visible light [1–23].

An optical cell that is suitable for the use of surface-enhanced Raman spectroscopy (SERS) for gaseous samples was developed by us (Fig. 5.6) [22, 23]. The technology is based on a specifically modified glass capillary and holder that simultaneous allows: (a) gas flow through the capillary, (b) exposure of the capillary interior to laser light, and (c) collection of light scattering arising from the SERS effect. The capillary has an optically thick base layer of metal to support waveguide-like propagation of light down the long axis of the tube. The layer is topped with a self-assembled (ordered) array of nanoparticles which support plasmonic resonances required for the SERS effect (Fig. 5.6b). The combination of the two components allows the SERS effect to be measured over a much wider surface area [22, 23]. The improved Raman method described here uses molecular fingerprints to identify the presence of analytes for fluidic analysis or gas analysis in humid environments. The improved detector sensitivity and reduced sample analysis make this an excellent option for medical field applications, analysis of gases (e.g. flue/stack monitoring), monitors for controlled environments such as glove

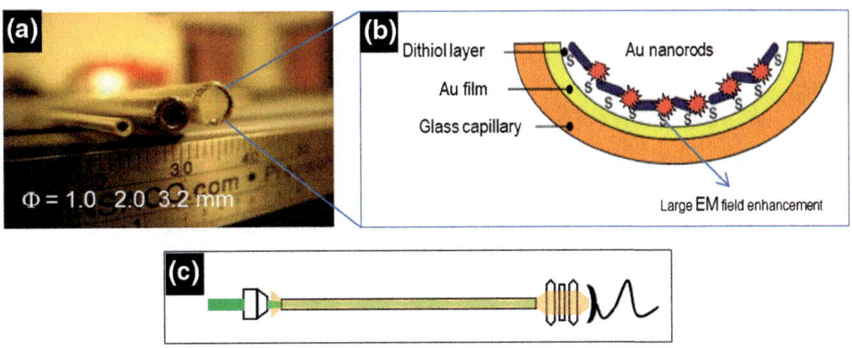

Fig. 5.6 **a** Photograph of optical cell of different diameter, **b** glass capillary modified with Au nanorods, **c** schematic of the optical cell [22]

Fig. 5.7 Schematic showing different anisotropic nanostructures used for sensing applications

boxes or confined spaces, and as a detector for certain types of analytical instrumentation (e.g. identification of effluents from a gas chromatography column).

Nanoparticle's surface tailoring can be used as way to explore chemical specificity for sensing, labeling and environmental applications [14]. Through a convenient and straightforward wet-chemical route, we produced "nanopeapod" architectures in which silver "peas" are embedded in silica "pod" [14]. The dimensions of the peas, and their spacing within the pod, are controllable on the nanoscale. The nanopeapods were investigated as surface-enhanced Raman scattering of a model analyte promoter, as the tips of metallic nanoparticle can lead to significant enhancements [4–8, 11, 12, 14]. Our results show enhancement factors of $\sim 10^7$ were calculated compared to the analyte alone, which are determined by the nanopeas' dimensions and spacing [12, 14].

Patchy silica islands were produced by us on silver nanowire substrates and used for nanomolar concentration detection of 4-MBA as an analyte of interest (Fig. 5.7) [17].

This study showed that the patchy silica coated silver nanowires are better SERS substrate than silver nanowires since nanomolar concentration of 4-MBA can be detected by employing these architectures [17]. In addition, "nano-matryoshka" configurations were used to quantify the effect of the electromagnetic field at the tips of the nanowire ("hot spots") in the Raman scattering experiment. Measurements performed with 4-MBA show a linear calibration detection from 260 to 0.08 μM 4-MBA (Fig. 5.8). Silver nanowire "hot spots" encapsulated in full silica shell showed a 65% increase in Raman intensity [17].

5.1.2 Sensing Based on Surface-Enhanced Fluorescence (SEF)

Surface-enhanced fluorescence (SEF) bears some analogies with SERS and relies on the photoexcitation of the metal nanoparticle to induce an altered response in the presence of the analyte molecule [3–5]. While for SERS applications, analytes are placed in the close proximity of metal nanoparticle, in the case of SEF, the optimum fluorescence signal occurs when the molecule is $\sim$ 10–100 nm away from the metal surface [24–26]. Such structures are able to produce desirable effects such as increased fluorescence quantum yield, decreased lifetime, increased fluorophore

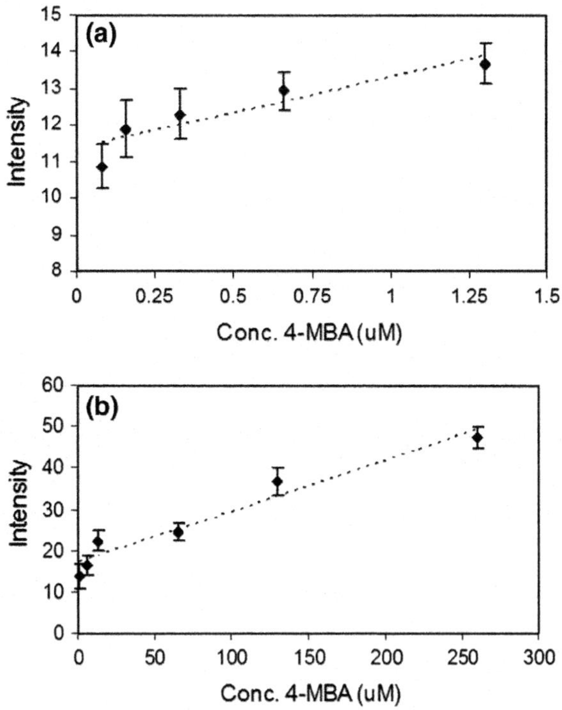

Fig. 5.8 Calibration curve for surface enhanced Raman spectra of 4-MBA using silver nanowires engineered with patchy silica coats by capturing various concentrations of 4-MBA **a** from 260 to 1.3 μM 4-MBA and **b** from 1.3 to 0.08 μM 4-MBA. Reproduced from Ref. [17] with permission from Springer

photo-stability, and improved energy transfer [27]. However, when fluorophores are very close to the metal surface, the fluorescence quenching effect competes with these favorable effects and it dominates within 5 nm from the surface of metallic particles. At larger distances, the enhancement starts to override the quenching and reaches its maximum at about 100 nm from the metal surface [28]. Theoretically, such surface-enhanced effects are larger for nanorods compared to nanospheres [25, 26]. This process offers promise for a range of applications, including LEDs, sensor technologies, drug delivery and imaging, separation and/or transport of molecules (nanofluidics) and single-molecule studies [6, 24, 27, 28].

Metal induced fluorescence enhancement has been investigated in a variety of multi-compositional nanoparticles. In one example, high aspect ratio silica nanotubes and silver-silica nanoparticles doped with Eu(III) were prepared by a facile and flexible two step sol-gel approach with different silica shell thickness [6, 24]. The optical properties of europium doped silica nanocomposite and europium doped silver-silica nanocomposite materials were interrogated for metal induced fluorescence enhancement. Photoluminescence spectra of Eu(III)-doped silica nanotubes, and Ag-silica-Eu nanocomposite materials showed stronger emission peaks in the presence of silver and increased with silica shell thickness from 10 to 100 nm for identical Eu(III) doping (Fig. 5.9). The enhanced emission was attributed to the

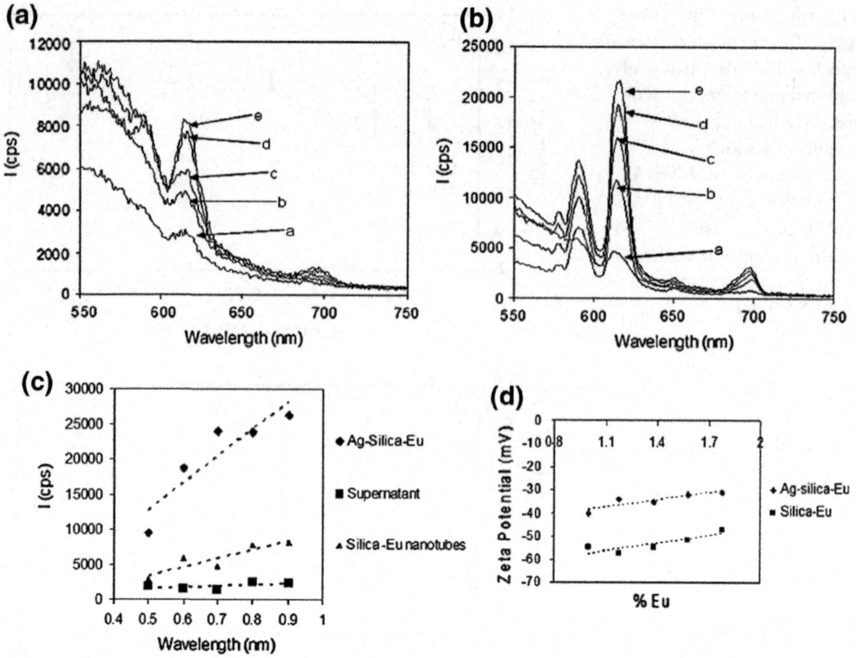

Fig. 5.9 Emission spectra of **a** Eu doped-silica nanotubes and **b** Eu doped-Ag-silica nanowires for different Eu loadings: a-1.0%, b-1.18%, c-1.38%, d-1.58%, e-1.78%; f-Eu(III) in solution; **c** calibration curve at 615 nm for Eu doped-silica nanotubes and Eu doped-Ag-silica nanowires; **d** Zeta-Pals measurement for (*square*) Eu doped-silica nanotubes and (*circle*) Eu doped-Ag-silica nanowires for different Eu loadings: a-1.0%, b-1.18%, c-1.38%, d-1.58%, e-1.78%. Reproduced from Ref. [24]with permission from Cambridge University Press

surface enhanced fluorescence displayed by the nanocomposite materials in the vicinity of metallic surfaces in agreement with theoretical and experimental reports [24–28].

5.1.3 *Sensing Based on Nanoparticle's Aggregation-Colorimetric Sensors*

The optical properties of gold and silver nanoparticles are also useful in aggregation based colorimetric sensing as a strategy for molecular detection. We have exploited the negative surface charge characteristics of the citrate modified Au nanospheres for the detection of metal ions on the surface of the nanoparticles. Nanoparticle aggregation resulted in visible color changes from ruby red (well separated nanoparticles) to blue-gray color (Fig. 5.10). These colorimetric changes arise from a combination of absorption and scattering of light by the nanoparticle solutions. Through a combination of techniques, one could monitor the "loading" of analytes

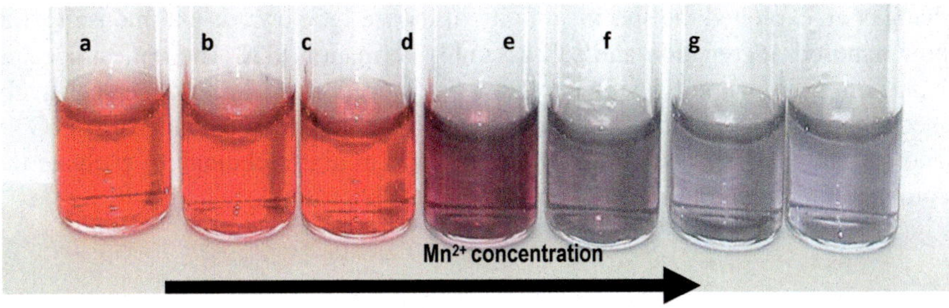

Fig. 5.10 Photograph of Mn–Au nanospheres. Mn^{2+} concentration varies from (from *left*) **a** 0 mM, **b** 0.1 mM, **c** 0.2 mM, **d** 0.4 mM, **e** 0.6 mM, **f** 0.8 mM and **g** 1 mM

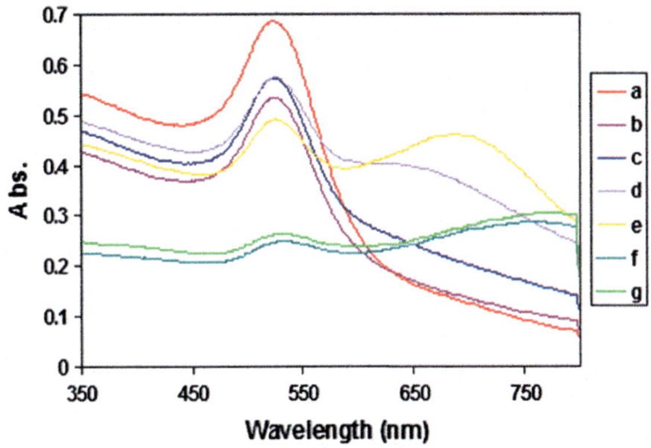

Fig. 5.11 UV-Vis spectra of Au–Mn nanomaterials used as MRI contrast agents. Mn^{2+} ions concentration varies from **a** 0 mM, **b** 0.1 mM, **c** 0.2 mM, **d** 0.4 mM, **e** 0.6 mM, **f** 0.8 mM and **g** 1 mM. As the Mn ions concentration increases a red shift of the peaks is observed –from 520 to ~ 770 nm

on the nanoparticle's surface [29]. Changes in the plasmon band peak maxima of gold nanoparticles are indicative of aggregation and can be correlated with aggregate size through light scattering measurements (Fig. 5.11). It is believed that the metal cations are bound to the nanoparticle through some combination of nonspecific electrostatic interactions and complexation to the negatively charged carboxyl groups on the particle surface. A lithium-ion sensor based on gold nanoparticle aggregation that occurs specifically in the presence of lithium ions was developed by Obare et al. [30]. Mirkin's group also explored the detection of DNA with gold nanoparticles. In their study, nanoparticle aggregation resulted upon the addition of target oligonucleotides that were complementary to the ssDNA attached to the gold nanosphere surfaces [31].

While the assembly of spherical metal particles is rather well studied, the aggregation-type sensing system of anisotropic particles is comparatively poorly investigated. The transverse and longitudinal plasmon bands of the nanorods are excellent probes for the detection of chemicals, as they are responsive to aggregation by surface plasmon band broadening and red-shifting in two separate wavelength regimes [5]. This makes them superior candidates to spheres, as one could tailor the nanoparticle shape and create chemical specificity for signatures of interest. This is critical for the creation of successful sensing and imaging platforms.

Chang et al. exploited the specific affinity of mouse IgG towards anti-mouse IgG to show nanorod aggregation and self-assembly formation [32]. The optical spectral changes in gold nanorods were exploited by Sudeep et al. through the selectively detection of micro molar concentrations of cysteine and glutathione [33]. Gold nanorod aggregation was used by Murphy's group as a technique to evaluate the protonation and deprotonation of adipic acid [34].

5.1.4 Sensing Based on Plasmon Shifts with Local Refractive Index

For molecular specificity, one could transform the nanomaterial's surface chemically. One of the advantages of using gold and silver nanoparticles is their ability to be surface-engineered with a large variety of functional groups. Particularly, their high affinity to thiols and amine allows one to change their charge and functionality while increasing their chemical stability and dispersability [4, 5]. Surface engineering of metallic nanoparticles is a promising route for the creation of complex architectures with multimodal functionalities and anisotropies. Advanced anisotropic structures with precisely defined properties and functionalities can be generated, including monometallic/bimetallic, metal/metal oxide nanoparticles with core-shell geometries, nano-peapods, decorated structures, and solid and hollow architectures. Surface modification of inorganic nanoparticles can also improve the interactions between nanoparticles and biomolecules or can help to target specific cells and even the nucleus [35]. Rational assemblies of individual nanostructures in multidimensional configurations have become increasingly attractive because they demonstrate improved physical and chemical properties over their single-component counterparts, so they are potentially useful in a broader range of applications [4, 6, 12, 23].

Surface engineering of noble metal nanoparticles with materials that have very different dielectric constants can lead to dramatic plasmon band shifts. This process can be easily spectroscopically monitored and exploited for analyte detection. For example, van Duyne's group demonstrated femtomolar detection of proteins, and zeptomolar detection of thiols based on the plasmon band resonance shifts of individual silver nanoscale triangles in a dark-field optical microscope [36]. In our group, we modified anisotropic nanostructures, particularly one dimensional silver and gold nanostructures, with a variety of inorganic and organic moieties, namely silica, iron oxide, titania, Pt, Pd, Au, Mn, polymers, and Raman reporters, and investigated their use for sensing, imaging, catalytic, environmental and energy related applications.

Silica and silica-titania shells were grown onto gold nanorods and silver nanowires by an indirect sol-gel approach (Fig. 5.12) [3–6, 12, 14, 17]. The presence of a silica shell around Au nanorods and silver nanowires was monitored spectroscopically. A red shift of the transverse and longitudinal plasmon bands was

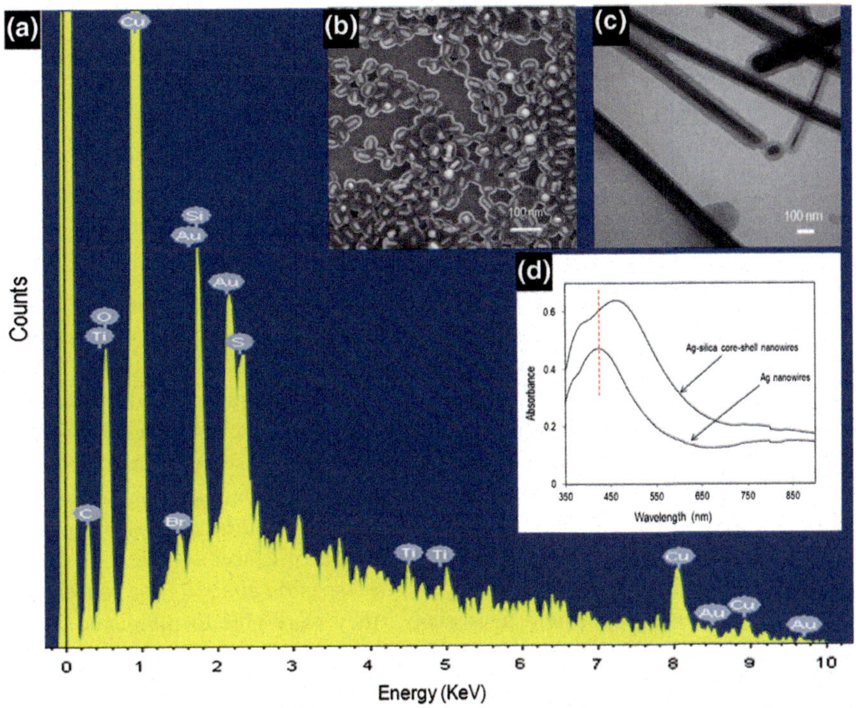

Fig. 5.12 **a** EDS data on Au-silica-titania nanorods; **b** SEM data on Au-silica-titania nanorod. Adapted with permission from Ref. [67]. Copyright 2017 American Chemical Society; **(c)** TEM image of Ag-silica nanowires and **d** UV-Vis spectra of Ag nanowires before and after coating with a silica shell. Adapted with permission from Ref. [14]. Copyright 2017 American Chemical Society

observed due to an increase in the dielectric constant surrounding the metallic nanoparticles [3, 4, 6, 12]. This is consistent with the theoretical predictions in the literature. Silica is an excellent material for various applications because it is highly porous, easy to functionalize, non-toxic, biocompatible, and can be further exploited for sensing, imaging and biological applications [12].

5.2 Medical and Biological Applications

The increasing availability of nanomaterials with tunable physico-chemical and surface properties has created an interest in the use of anisotropic nanoparticles in biological systems [29]. A staggering number of nanostructures have been investigated, namely metals, oxide, quantum dots, carbon based nanostructures, polymers, or hybrid nanostructures. From drug delivery systems, to theranostics, bioimaging, tissue repair, tracking and manipulation, bioengineering, cancer therapy, etc., the field is rich in opportunities [37].

5.2.1 *Metallic Nanostructures*

One advantage of using anisotropic gold and silver nanoparticles for biological and medical applications is their elastic light scattering (Rayleigh scattering) [5, 6, 12]. When anisotropic gold or silver nanoparticles are imaged with a dark-field microscope (white light illuminate the sample), they scatter light elastically. Their elastic light scattering (Rayleigh scattering) properties are intense, and can be used to visualize nanoparticle location, even to the single nanoparticle level in cells [5]. The color of the scattered light is highly dependent on nanoparticle size, shape, composition, orientation, and local environment. For example, gold nanocubes "glow" orange, while gold hexagons "glow" yellow-green (Fig. 5.13).

Mostafa and El-Sayed showed that the detection of cancer cells could become much easier if gold nanoparticles are bound to a specific antibody for cancer cells [6, 38, 39]. By using dark-field microscopy, they were able to illuminate cancer

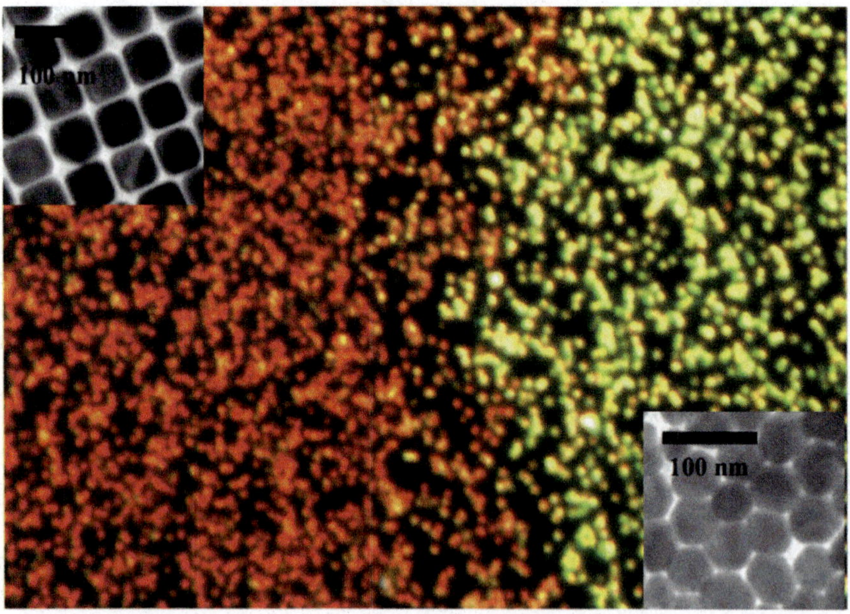

Fig. 5.13 Dark field microscopy image of gold nanocubes (*left*) and gold "nanohexagons" (*right*), taken on solutions dried down on a microscope slide. The field of view is ~1 mm. *Insets* transmission electron micrographs of the nanoparticles, *scale bars* 100 nm. Reprinted with permission from Ref. [3]. Copyright 2017 American Chemical Society

cells when bound to nanoparticles. On the other side, healthy cells were not illuminated because they did not bind to gold nanoparticles. This technique is promising because it can provide an instantaneous response, is simple and inexpensive, non-toxic to human cells and can detect single particles [3, 6, 38, 39].

Compared to their spherical counterparts, anisotropic particles are more attractive, as their dimensionality brings another level of complexity and benefits to their inherent nanoscale properties. A unique feature of the anisotropic nanostructures, including gold and silver nanorods, is that they display multiple plasmon bands and light absorption in the near-infrared (NIR) regions, which open up a variety of applications. The use of optically tunable gold nanoparticles in conjunction with a near infrared (NIR) laser is an attractive option for biomedical applications, as it capitalizes on the plasmonic heating of gold nanoparticles tuned to absorb light strongly in the NIR region. Basically, because the gold strongly absorbs light in the near infrared, these particles can be remotely heated via near-infrared light, thereby releasing the "payloads" remotely. Placing these nanomaterials in the close proximity of bacteria or cancer cells would destroy these cells [40–42]. There are numerous excellent papers that use the photothermal heating process in cancer therapy to damage cancerous cells and tissues [42–44]. One of the benefits of the photothermal therapies compared with the traditional cancer therapies is that the heating is targeted and only occurs in the area directly around the gold nanoparticles. Temperatures can rise to tens or hundreds of degrees above physiological temperature within seconds and minutes [42–44].

Gold nanorods (AuNR) and nanospheres (AuNS) were incorporated into a polyurethane-based shape-memory polymer (SMP) EG-72D matrix, SMP nanocomposite films capable of being remotely triggered by low-power laser. Comparative studies on the laser irradiation response of EG-72D/AuNS and EG-72D/AuNR nanocomposite films suggest that Au nanorods have significantly higher photothermal conversion efficiency than Au nanospheres and on-resonance laser irradiation, where the incident laser wavelength matches with the longitudinal plasmon resonance of Au nanorods, is necessary to induce the fast response of gold nanoparticle enabled SMP nanocomposites [45].

Nanoparticle size and shape are important parameters that dictate the uptake and bioavailability of nanoparticles in cells, plants and animals. Numerous reports have investigated the role of shape and size on cellular internalization with metallic nanoparticles, Au and Ag, due to their ease of preparation, surface modification, and stability [4, 6]. Recently, it was reported that different gold shaped nanoparticles entered the tumors in mice at different rates and are also distributed differently according to their shape. Nanorods and hollow gold nanocubes were also found to enter more readily to the core of the tumors than discs and sphere of similar sizes [46].

We also investigated the bioavailability of different size Au and Cu nanoparticles, albeit spherical nanoparticles only, in soil and demonstrated their use as bio-tracers to evaluate environmental behavior of nanoparticles [47–49]. In our studies, the size range of 20–55 nm Au and Cu nanoparticle did not show any clear

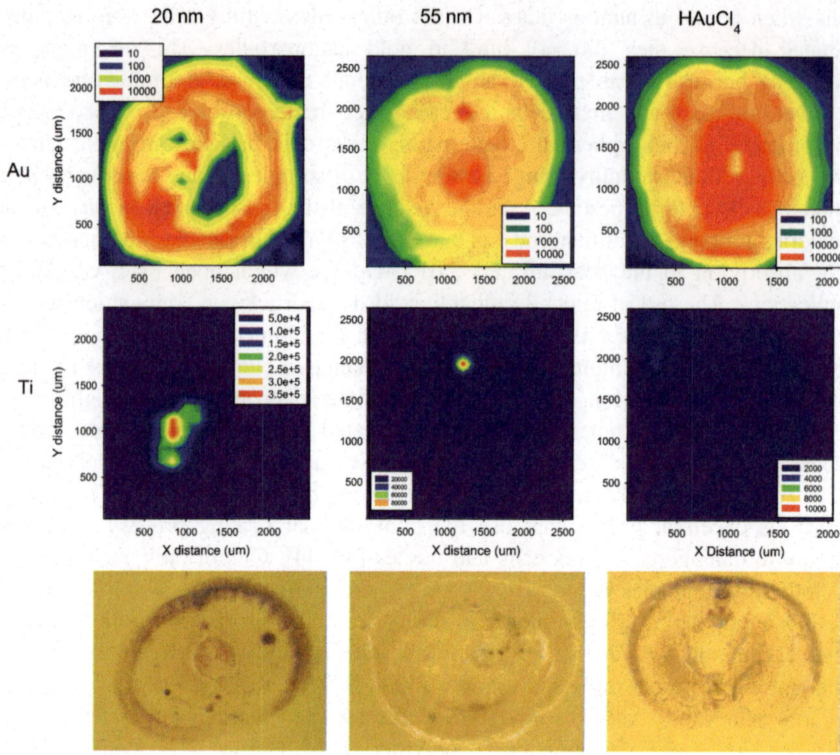

Fig. 5.14 Example contour plots showing the spatial localization of Au (*top row*) in cross sections of earthworms exposed to either $HAuCl_4$, 20 nm Au NPs or 55 nm Au NPs. Plots of Ti intensity are also shown (*middle row*) to indicate the location of soil particles remaining within the gut lumen. *Color bars* indicate intensity of target analyte in counts per second. Reprinted with permission from Ref. [48]. Copyright 2017 American Chemical Society

particle size dependence for uptake and accumulation in earthworms (Fig. 5.14); however, the uptake was significantly higher when *Eisenia fetida* earthworms or even mice were exposed to 4 nm Au nanoparticles [49, 50]. These results suggest different mechanisms of gastrointestinal absorption based on the size and environment.

Silver nanostructures exhibit antimicrobial activities and skin penetration capabilities that are dependent on the nanoparticle's shape and their crystallographic facets. Recent proof-of-concept studies revealed that silver nanorods have the highest penetration and accumulation in the dermal layer compared with triangles or spheres [51]. Cheng's group [52] also reported the preparation of anisotropic nanostructures with multimodal theranostic capabilities, chemotherapy, photothermal therapy, magnetic resonance imaging (MRI), and photoacoustic imaging (PAI) for image-guided cancer therapy.

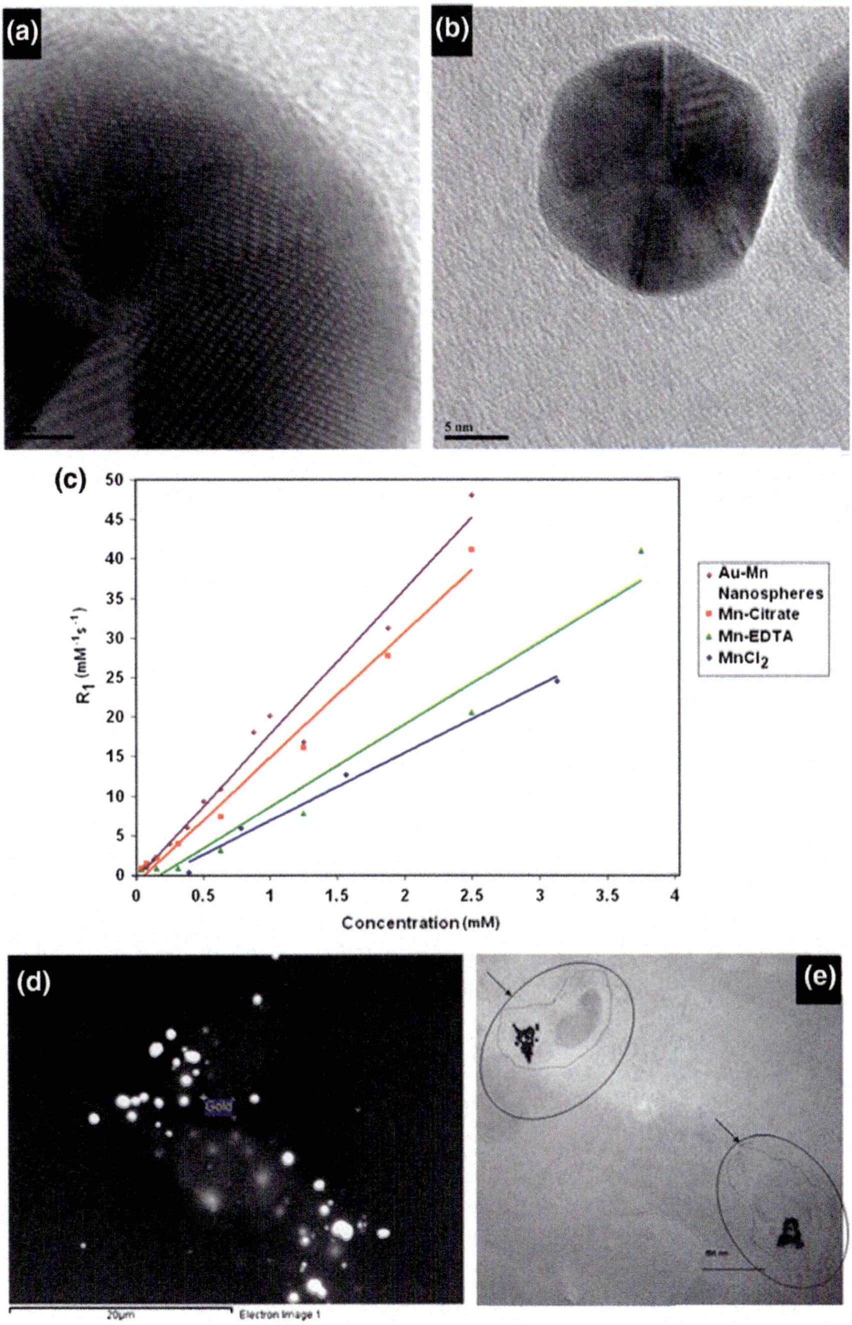
(a)
(b)
5 nm
(c)
R1 (mM-1s-1)
Concentration (mM)
0
0.5
1
1.5
2
2.5
3
3.5
4
0
5
10
15
20
25
30
35
40
45
50
Au-Mn Nanospheres
Mn-Citrate
Mn-EDTA
MnCl2
(d)
Gold
20µm
Electron Image 1
(e)

◂**Fig. 5.15** **a** HRTEM of Au nanosphere seed (scale bar 2 nm); **b** HRTEM of Mn–Au nanoparticle (scale bar 5 nm) SEM images of live MSCs **a** without and **b** with Mn–Au nanoparticles (*backscatter image*); **c** Relaxation rate versus Mn concentration for agar phantoms; Au–Mn Nanoparticles, Mn-Citrate, Mn–EDTA, Mn–Cl_2; **d** SEM images of live MSCs incubated with nanoparticles; **e** TEM images of Mn–Au NPs incubated in MSCs. Mn–Au are located in intracellular compartments of the stem cells. *Scale* 500 nm. Adapted from Ref. [29] with permission from Springer

Heterogeneous spatial arrangements of two or more metallic/non-metallic nano-entities of various compositions and modalities are a great way to create hybrid nanostructures with fine-tuned functional and directional anisotropy that are of great relevance for medical applications. Of particular interest for biological applications is the use of metallic nanoparticles as contrast agents in microscopy, as they are electron dense and can be easily detected [3–6]. We have taken advantage of the unique chemical properties of Mn in nanoscale environments to develop a "tunable" positive contrast agent for magnetic resonance imaging applications (Fig. 5.15) [29, 53, 54]. For example, Mn contrast agents have become useful to study myocardial imaging processes due to their T1 shortening effect and relatively long half-life in cells. Because Mn^{2+} can enter excitable cells through the voltage-gated L-type calcium channels, manganese-enhanced MRI can be used to determine the viability and the inotropic state of the heart [55].

We have developed several nanomaterial syntheses routes that provide a means to control the metal speciation and particle size and, thus, the MRI relaxation rates and biological activity (e.g., metabolic uptake and toxicity) [29]. Gold-manganese multifunctional nanostructures showed excellent plasmonic and paramagnetic properties. HRTEM reveals that the internal lattice of the nanoparticles is twinned crystals (Fig. 5.15a, b). The core Au nanoparticles exhibited structures with a lattice constant of 2.24 Å which is suggestive of (111) facets. Addition of Mn (II) ions resulted in an in-plane expansion on the gold (111) surface of approximately 2.40 Å, indicating that Mn ions are substituting into the gold lattice at the nanoparticle surface. T1 data, a measure of the effectiveness of the contrast agent, for several Mn–Au containing nanoparticles resulted in measured relaxation rates greater (four folds higher) than the ones used in imaging agents currently used for clinical application (Fig. 5.15c) [29, 53, 54]. For example, magnetic resonance imaging (MRI) relaxation studies performed using Mn–Au NPs exhibited a high longitudinal relaxivity ($18.26 \pm 1.04\ s^{-1}\ mmol^{-1}$) and low stem cell mortality, both of which are beneficial properties for use as Mn-based contrast agents [29, 53, 54]. This relaxivity value is higher than the Mn–DPDP currently applied in clinical settings with a reported value of $2.8\ s^{-1}\ mmol^{-1}$ [55]. Controlled incubation of these nanoparticles with mesenchymal stem cells (MSCs) was used to study cellular uptake of these particles for cell tracking (Fig. 5.15d, e). Additionally, we demonstrated that these nanoparticles can be taken up by stem cells, thus allowing for targeted delivery of these contrast agents in biological systems.

5.2.2 Non-metallic Nanostructures

The overall bioavailability, internalization and transport of NPs by cells and membranes are also determined by the nanomaterials' composition and surface properties. Anisotropic polymers, hydrogels, oxides, carbon nanotubes, and quantum dots, have been theoretically and experimentally investigated. One study probed the rate of macrophage uptake in rodents of nano-polymer micelle assemblies with different shapes. According to the published results, nanoparticles with very high aspect ratios (~ 20) that align with their long axis parallel to the cell membrane internalize more slowly than particles that align with their short axis parallel to the cell membrane [56]. Additionally, long rod nanoparticles infiltrate into cells at four times the rate of short nanorods with similar volumes [57]. A team of researchers found that hydrogel rod-shaped nanoparticles are much more likely to penetrate cells than spheres. The research, published by DeSimone [57], indicates that internalization of rod-shaped particles of high aspect ratio, 150 nm in diameter and 450 nm in length, in human cells is about four times faster and traveled farther into the cells than particles with symmetric dimensions (200 nm in diameter spheres), even with the same volume. The shape of nanoparticles dictates the interaction of nanoparticles with cell membranes. Recent findings suggest that non-spherical nanoparticles (e.g. rods, discs, hemispheres, ellipsoids) may target tumors more effectively than the spherical counterparts [58].

Uptake studies performed in vitro in different cancerous cell lines show that, along with particle shape and surface functionalization, cellular origin and features may also influence the uptake of particles in cells. One study shows that both rod-like and spherical porous silica nanoparticles are readily internalized by HeLa cells with slight shape and charge-induced differences [59]. In the case of Caco-2 cells, rod shaped particles are internalized more efficiently than the spheres.

Ferrari's group investigated the biodistribution of different shape silicon-based particles, quasi-hemispherical, cylindrical and discoidal shapes, with diameters from 700 nm to 3 μm injected into tumor bearing mice [60]. The binding dynamics of nanoparticles from transport, margination, to adhesion are also dependent on the nanoparticle's shape. Computational studies revealed that rod-shaped NPs have higher binding rates compared with their spherical counterparts due to the tumbling motion of NPs and larger adhesion area once contact [61]. The binding probability of a nanorod under a shear rate of 8 s^{-1} is found to be three times higher than that of a nanosphere with the same volume.

The effect of particle geometry (i.e., the interplay between nanoscale shape and size) is also important for the nanoparticle uptake by mammalian cells. Recently, Roy's group [62] conducted a comprehensive in vitro study showing the complex interplay between shape and size of anionic nanohydrogels on their uptake in epithelial, endothelial, and immune cells. Their studies reveal that the nanoparticle's uptake and mechanisms of uptake were shape- and cell type-specific. Specifically, disc-shaped hydrophilic nanoparticles were internalized more efficiently than nanorods, while larger nanodiscs and rods had higher uptake compared with the smallest particles tested.

Nanoparticle geometry plays a major role in extravasation in tumors. High-aspect ratio single-walled carbon nanotubes (SWNTs) display extravasational behavior different from spherical nanoparticles with similar surface coatings, area, and charge. The study shows that quantum dots extravasate more than SWNTs in colon adenocarcinoma, while SWNTs extravasate much more than quantum dots in human glioblastoma tumors [63].

A nanoparticle's shape dramatically affects how effectively it delivers gene therapy to the cells. DNA-containing micellar nanoparticles with distinctly different and highly uniform morphologies are paving the way for targeted gene delivery [64]. Nanoparticle uptake by cells is dependent on particle's shape, size, and surface functionalization, as well as on the particle orientation and membrane properties [65]. Nanostructures can be tailored with multifunctional capabilities for multi-modal selective theranostics [66].

5.3 Catalysis and Electrocatalysis

The larger surface-to-volume ratios of nanoscale catalysts compared to their bulk counterparts makes them attractive candidates for many chemical reactions. Since over 80% of all chemical reactions involve the use of catalysts, state-of-the-art advances in this field is highly desired [67, 68]. Nanoscale materials are more chemically/electrochemically active than their bulk counterpart catalysts [69–73]. Much work has focused on investigating the catalytic behavior of nanoscale materials, as catalytic reactions are surface phenomena with significant contributions at the microscopic and molecular levels.

Palladium and platinum are some of the most widely used materials for industrial applications because they have extraordinary physical and chemical properties and withstand harsh environments [69, 74]. For example, they find applications as catalysts for electrochemical processes for proton exchange membrane fuel cells [67, 68, 75], organic reactions (Heck and Suzuki reactions) [76], synthesis of nitric acid, and reduction of pollutant gases emitted from automobiles, among many others [77, 78]. Staggering amounts of theoretical and experimental work have been conducted in this field [6, 67, 68, 70, 79].

The pioneering work conducted by El-Sayed in this field opened new doors for exploring different shape nanostructures for catalytic applications [80, 81]. For the first time, in 1996, they produced tetrahedral, cubic and truncated octahedral platinum nanostructures. El-Sayed [80] and his team investigated the activity of cubic, tetrahedral, and nearly spherical Pt nanoparticles in catalyzing the electron-transfer reaction between hexacyanoferrate (III) ions and thiosulfate ions and found that tetrahedral particles had almost half the activation energy as cubic particles of similar size. They proposed that such high activities are due to the presence of valence-unsatisfied surface atoms and correlated to the large number of atoms present at the corners and edges of the nanoparticles. These studies generated a lot of interested in the scientific community and soon it was postulated that metallic nanoparticles with sharp edges, sharp corners, or rough surfaces are more active than the spherical counterparts [6, 67, 68, 70, 81]. Over the years, it has been reported that catalytic/electrocatalytic activity and selectivity of nanomaterials depends highly on their morphology (size, shape), spatial distribution, surface or support composition and electronic structure among others [6, 67–70, 72, 73, 76, 79, 82, 83].

It is generally accepted today that, depending on the interfacial atomic arrangement of the crystal and which crystallographic facets are exposed to the environment, different catalytic activities may be obtainable [84, 85]. Subsequently, different crystals facets have different surface energies that are directly related to the materials catalytic activities and stabilities [67]. Numerous theoretical and experimental studies show that higher index planes with a high density of atomic steps, ledges, and kinks serves as active sites for breaking chemical bonds; generally, they exhibit much higher catalytic activity than the low index planes [67, 81, 85]. A higher percentage of atoms on the corners and edges make them more catalytically active. Moreover, different shape nanocatalysts could lead to different products. We recommend several excellent review papers published in the literature [81, 86, 87] that discuss the catalytic dependence of different shape Pt and Pd nanostructures in detail.

Electrocatalytic reactions are structure sensitive and their rates vary dramatically with shape or crystallographic facets of the catalyst. Pt-based catalysts are the most rigorously studied electrocatalysts for fuel cells applications, as they are extremely efficient in these processes [67, 68, 70]. Fuel cells (FCs) are electrochemical devices that convert a fuel, such as hydrogen, methanol or natural gas, directly into electricity through an electro-catalytic process. Direct methanol fuel cells are considered ideal fuel cell systems for field operations and military applications since they provide high energy power densities and instantaneous refueling time with the simple exchange of the fuel cartridge [67].

We developed a new class of shape-selective platinum-gold nanomaterials (Fig. 5.16) with high electrocatalytic activity and stability for direct-metanol fuel cell applications. We investigated the Au–Pt nanocatalysts electrochemical performance in the oxygen reduction reaction (ORR) and the methanol oxidation reaction of direct methanol fuel cells (DMFC) and demonstrated its sensitivity to the nanoscale shape of a catalyst particle [88]. The presence of Au in bimetallic electrocatalysts has shown to alter the Pt electronic band structure leading to a synergistic catalytic effect, which modifies the strength for oxygen surface adsorption in the ORR (Fig. 5.17) [89].

Carbon monoxide often poisons platinum nanocatalysts, significantly reducing DMFC efficiency by slowing down the highly irreversible methanol oxidation reaction and the oxygen reduction reaction at the cathode. We determined that the gold-platinum nanocatalysts that we developed can effectively suppress carbon monoxide adsorption onto the active platinum surface. Depending on the geometry and the Pt/Au ratio, Pt-based catalysts can be prepared that reduce and, in some cases, eliminate the detrimental effects of methanol crossing over to the cathode. The Au–Pt dogbones are partially, and in some cases completely, unaffected by methanol poisoning during the evaluation of the ORR. The ORR performance of

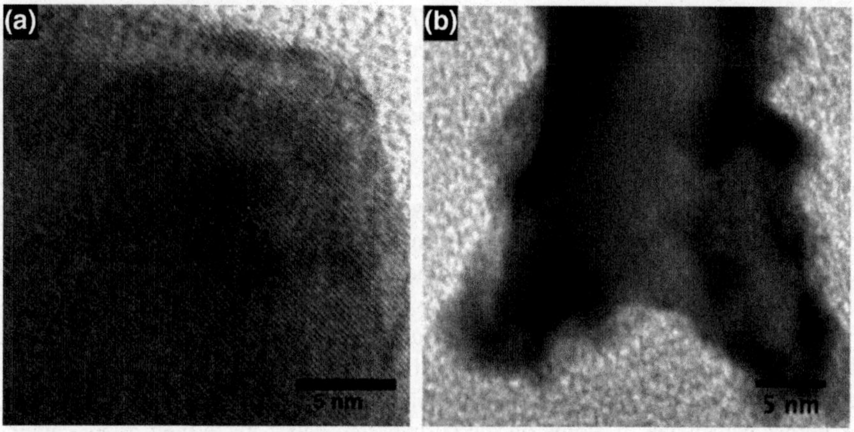

Fig. 5.16 HRTEM images of Au–Pt nanocatalysts: **a** octahedron and **b** “dogbone” shaped. Reprinted from Ref. [88] with permission from Springer

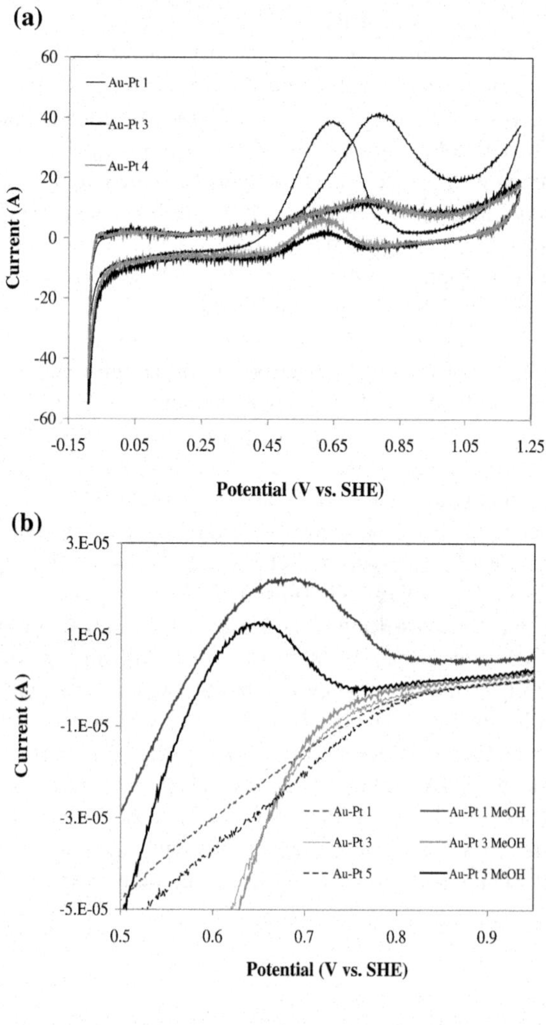

Fig. 5.17 **a** Representative cyclic voltammetric profiles (100 mV s−1) between −100 and 1200 mV in Ar-saturated 0.5 M H_2SO_4 and 0.1 M MeOH; **b** Linear sweep voltammetry on the different Au–Pt nanocatalysts in O_2 saturated 0.5 M H_2SO_4 with and without 0.1 M MeOH. Current density evaluated using Pt electrochemical active surface area from the CVs. Sweep rate 10 mV s-1 and a rotation rate 800 rpm. Reprinted from Ref. [88] with permission from Springer

the octahedron particles in the absence of MeOH is superior to that of the Au–Pt dogbones and Pt-black; however, its performance is affected by the presence of MeOH (Fig. 5.17) [88].

A large number of commercial catalytic reactions require regeneration of catalysts at high temperatures. One way to circumvent issues regarding the stability, sintering and durability of the catalysts is by placing nanoparticles on "inactive" supports. The synergistic effects between catalysts and supports have been demonstrated by many research groups [12].

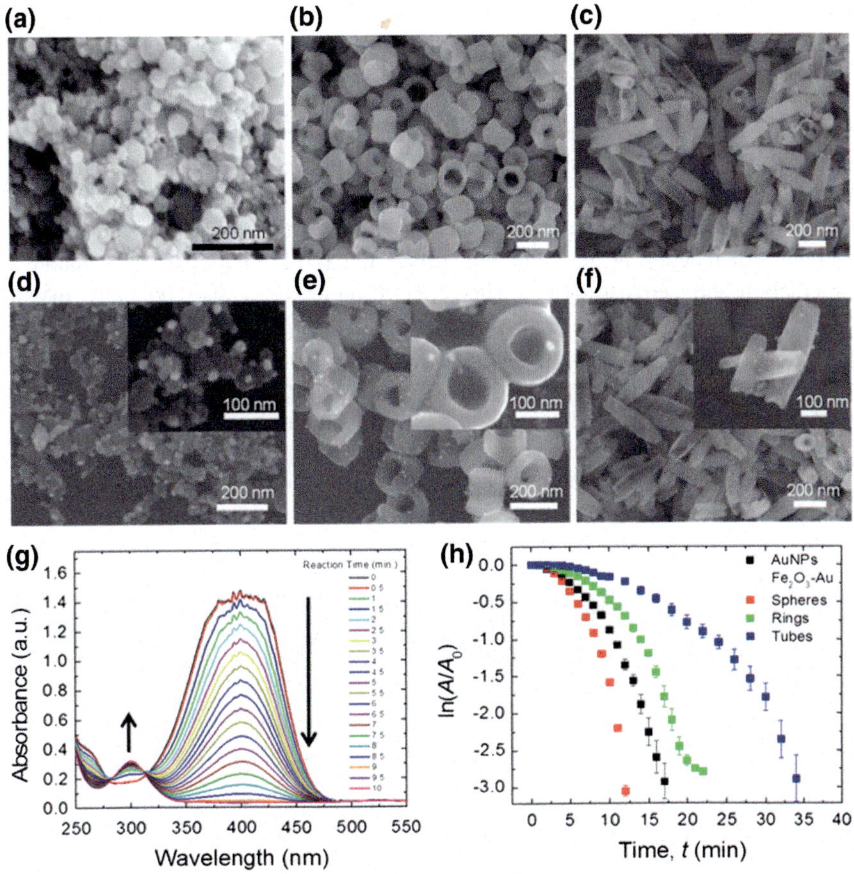

Fig. 5.18 Plot of the change in the 4-nitrophenol peak intensity versus time during the catalytic reduction experiments using the AuNPs and the Fe_2O_3−Au spheres, rings, and tubes. SEM micrographs of the **a** Fe_2O_3 spheres, **b** Fe_2O_3 rings, **c** Fe_2O_3 tubes, **d** Fe_2O_3-Au spheres, **e** Fe_2O_3−Au rings, and **f** Fe_2O_3−Au tubes. The insets in panels (**d**)–(**f**) show zoomed in images at different magnifications of the same particle types. Note that the size of the scale bar decreases from (**d**) to (**f**). **g** Representative 4-nitrophenol reduction spectra obtained during a Fe_2O_3-Au spheres catalytic decay experiment. **h** Plot of the change in the 4-nitrophenol peak intensity versus time during the catalytic reduction experiments using the Au NPs and the Fe_2O_3-Au spheres, rings, and tubes. Adapted with permission from Ref. [90]. Copyright 2017 American Chemical Society

We evaluated different shape Fe_2O_3−Au nanoparticles and AuNPs for their ability to catalytically reduce 4-nitrophenol, a pollutant found in wastewater, as a model catalytic reaction. It was found that Fe_2O_3−Au nanoparticles are more efficient catalysts than AuNPs because they can achieve the same, or better, catalytic reaction rates using significantly smaller quantities of the catalytically active and significantly more expensive Au. Taking into account the Au-loadings, the Fe_2O_3 rings and tubes are superior to the Fe_2O_3 spheres as catalytic supports due to their γ-Fe_2O_3 crystal phase (Fig. 5.18) [90, 91].

5.4 Environmental Applications

Nanomaterials have received widespread attention for environmental stewardship and remediation applications. At the nanometer scale, these materials have very high surface energy, making them extremely reactive and susceptible to the environment. The high surface area to volume ratio leads to a larger fraction of atoms available for reaction at the particle's surface making the nanoparticle valuable as remediation and sequestration agents [92]. Nanoparticles also typically display faster reaction kinetics, higher remediation capacities, and higher adsorption activity compared to conventional macroscale/bulk materials [93–95]. The scattering component of the extinction spectra of noble metal nanoparticles is often used for rapid environmental and chemical screening (Fig. 5.19) [3–6, 96]. Detection of analytes in real time is essential for optimal performance of treatment systems at contaminated sites.

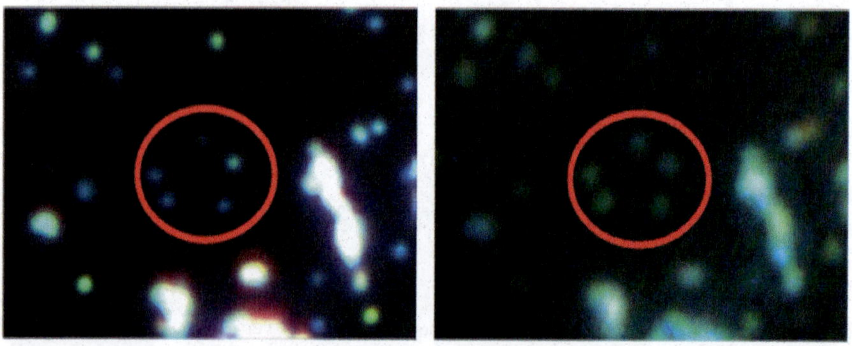

Fig. 5.19 Darkfield optical micrograph of light elastically scattered from silver nanoparticles, ~50 nm diameter. The field of view is approximately 1 mm. *Left panel* Micrograph taken in air (refractive index 1). *Right panel* Micrograph, same region, taken in mineral oil (refractive index ~1.5). The *red-circled* region highlights five spots that shift in scattered color from blues to greens. Reproduced from Ref. [5] with permission from The Royal Society of Chemistry

5.4.1 Detection and Sequestration of Environmental Contaminants

Hybrid nanoparticles take advantage of the physico-chemical properties of two or more materials and are often preferred to their single components. Hybrid nanostructures with multifunctional components were developed by our group and used for two parallel applications (a) as a sequestering agent and (b) reporter for uptake of radioactive contaminants [97]. The hybrid Raman sensor-ion exchanged materials promoted sequestration of radioactive contaminants existing in radioactive waste solutions and allowed the process efficiency to be evaluated via Raman spectroscopy. The sequestering inorganic ion-exchange agent, namely monosodium titanate (MST), was used for the uptake and removal of radioactive elements from waste solutions. Gold nanostars were attached on the sequestering agent and used as a Raman sensing/reporter component. The use of gold star nanostructures was instrumental, as the tips of the stars are considered "hot-spots" that enhance sensing capabilities. Matching the optical adsorption peak of the gold nanostars to the laser wavelength resulted in an additional enhancement effect as observed in the Raman response. A key advantage of this approach is that it minimizes radiation exposure of the workers, reduces the processing time, and limits the amount of radioactive waste generated.

We also developed anisotropic nanostructures for the sensing and capture of radioactive gas compounds. Gold nanomaterials were functionalized on stainless steel filters (SSF) for radioactive zinc vapor sequestration (Fig. 5.20) [92]. Without nanoparticle modification, stainless steel coupons do not react or alloy with Zn. Different shape nanostructures were investigated for zinc vapor sequestration, namely nano-urchins, cubes and spheres on bronze supports (Fig. 5.21). Among the three different shape nanostructures, the Au nano-urchins showed the highest Zn gettering capacity. Specifically, the loading capacity shows that the Au nanourchins

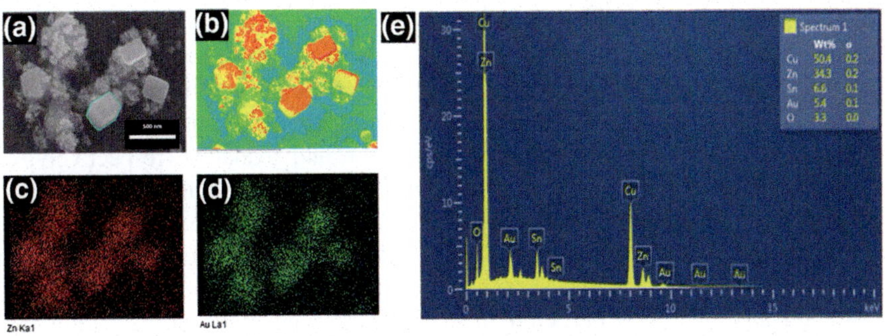

Fig. 5.20 Zn Deposits "captured" by Au nanoparticle-treated SSW: **a** SEM; **b** EDS mapping analysis; **c** Au NPs EDS mapping **d** Zn EDS mapping; **e** EDS analysis. Reprinted from Ref. [92] with permission from Springer

Fig. 5.21 SEM images of gold-based **a** nano-urchins and **b** nanocubes

retained twice the amount of radioactive Zn65 compared to cubes and spheres due to a larger surface area available for sequestration. This demonstrates once again the importance of the nanomaterial's shape and morphology.

5.4.2 Detection and Destruction of Environmental Contaminants

Titania has received increased attention in photo-decontamination applications due to its powerful oxidation capability, superior charge transport, and corrosion resistance [67]. Titania proved to be an efficient nanomaterial for the degradation of organic pollutants found in wastewater. For example, a large number of organic azo dyes used in textile, paper, leather, ceramic, cosmetics, ink, and food processing industries pose a hazard to human health and the environment. Unfortunately, 15% of the dyes produced are lost with wastewater during the synthesis and processing of these dyes worldwide [98]. TiO_2 is also inexpensive, has reasonable activity, and is abundant [12, 67, 84, 99, 100]. Despite these attributes, the efficiency of TiO_2 for photocatalytic applications is severely limited by its large band gap ($\sim$ 3.2 eV) and rapid charge carrier recombination dynamics, which means that anatase titania can use less than approximately 1% (only UV light) of the solar spectrum.

The absorption of photons from natural light in photocatalytic materials generates electrons and holes that enable the reduction and oxidation, respectively, of chemical species in the surrounding environment [100]. An efficient photocatalytic process of a semiconductor must also allow a large fraction of the photo-generated electrons and holes to separate in bulk and to transport to the redox reaction sites at surface before they recombine. Thus, a fair mobility for photoinduced electron–hole separation and their transportation in crystal lattice and the corresponding low probability of electron– hole recombination are demanded in order to achieve a high photocatalytic performance of a semiconductor [84, 88, 99, 100].

Photoinduced charge separation in a semiconductor nanoparticle can be greatly improved by coupling it with another semiconductor particle having favorable energetics. By selecting a short band gap semiconductor, one can also harvest photons in the visible wavelength region. One strategy employed by us and others was the combination of high surface area nanostructured semiconductor arrays, e.g. TiO_2 with visible light-absorbing properties of metallic nanoparticles or quantum dots [12, 67, 99–101]. The idea of the formation of heterojunction structure between two semiconductors provides an effective way to accelerate photo-stimulated electron–hole separation and consequently improve the photocatalytic activity [102].

The use of titania based photocatalysts for the selective reduction of nitrates to N_2 has been used in groundwater remediation efforts to clean up nitrates from high agricultural areas [103]. Noble metals, such as Pd metal, have been shown to have high N_2 selectivity (100%) and improve the efficiency of the process [103]. Photocatalytic destruction of nitrate ions in drinking water has also been demonstrated [104]. Recently, we demonstrated that nitrate ions from different sources, $NaNO_3$ and HNO_3, at extremely low pH (<1) can be effectively and completely removed via photocatalytic processes using titania nanospheres [105].

A major shortcoming of spherical nanostructures for photocatalytic applications is their lack of directionality for efficient electron transport subsequent to light harvesting. One dimensional nanostructures, namely nanorods, nanowires, nanotubes, are favorable for photocatalytic destruction of organic dyes because they provide a direct electrical pathway for photogenerated charges and an additional surface area for analyte absorption. This results in superior charge transport properties [106]. Light scattering and trapping is sensitive to particle shape. The distribution of an electromagnetic field around the particles is dependent upon the size and shape of the nanoparticles, the interparticle distance, and the dielectric function of the surrounding medium. The light-trapping effects are most pronounced at the peak of the plasmon resonance spectrum, and can be tuned by engineering the dielectric constant of the surrounding media [107]. For example, light scattering from a small metal nanoparticle embedded in a homogeneous medium is nearly symmetric in all directions. When the particle is placed close to the interface between two dielectrics, light will scatter preferentially into the dielectric with the larger permittivity. The scattered light will then acquire an angular spread in the dielectric that effectively increases the optical path length [108]. Ultimately, designing efficient photocatalysts requires careful selection of the nanoparticle size, geometry, composition, and local dielectric environments.

By using simple wet chemical synthesis approaches we designed and produced a series of core-shell nano-photocatalysts, namely Au rods-silica-titania or titania-Au rods (Fig. 5.12a, b) [12, 67, 84]. The photo-catalytic activity of these novel nanostructures was tested for the decomposition of an organic dye, methyl orange, and compared to bare titania nanoparticles. When such metal core-semiconductor shell composite particles were subjected to UV irradiation, the decomposition of methyl orange in the presence of Au-SiO_2-TiO_2 or TiO_2-Au particles was 20% and 32%, respectively [12, 67]. In another study, photocatalytic activity of Au-silica/titania nanomaterials under visible and UV illumination was measured via degradation of methyl orange under visible and UV illumination. The results indicate $\sim$3 fold improvement in the photocatalytic decomposition rate of methyl orange under visible illumination versus UV illumination [84]. This enhancement is attributed to the electric field enhancement near the Au nanorods and subsequent energy transfer to the semiconductor. By integrating strongly plasmonic Au nanoparticles with strongly catalytic TiO_2, Cronnin's group reports enhanced photocatalytic decomposition of methyl orange under visible illumination [109]. Irradiating Au nanoparticles at their plasmon resonance frequency creates intense electric fields that can be used to increase electron–hole pair generation rates in semiconductors.

Watanabe's group [110] reported a new type of plasmonic photocatalyst that employed the enhanced electric field amplitude on the surface of Ag nanoparticles in the spectral vicinity of their plasmon resonances. The measured photocatalytic activity under UV illumination of methyl orange onto these plasmonic photocatalysts, consisting of titania deposits onto silver core, was enhanced by a factor of 7 and was highly dependent on the silica shell thickness [110].

A unique oblique angle co-deposition technique was used to fabricate well aligned arrays of Ag nanoparticle-embedded TiO_2 composite nanorods (Fig. 5.22)

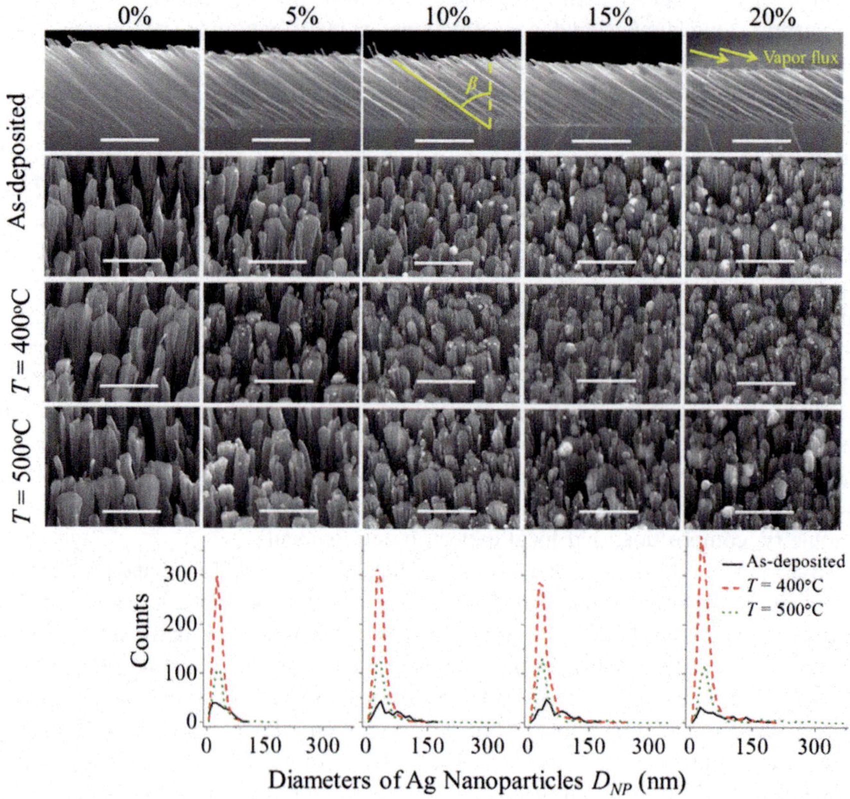

Fig. 5.22 The representative SEM images of TiO_2 and Ag NP embedded TiO_2 composite nanorod arrays and the size distributions of AgNPs. The *scale bar* in each SEM image represents 1 μm. Reprinted with permission from Ref. [100]. Copyright 2017 American Chemical Society

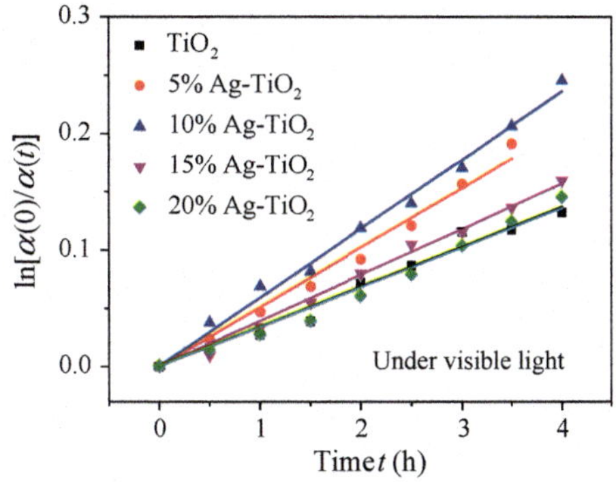

Fig. 5.23 Photocatalytic MB degradation kinetics under visible light illumination for the samples annealed at T = 500 °C. Reprinted with permission from Ref. [100]. Copyright 2017 American Chemical Society

and their photodegradation response for a dye contaminant was investigated [100]. The addition of Ag in TiO_2 results in enhanced photocatalytic activity compared to pure TiO_2. Ag NPs play different roles in enhancing photocatalytic activity under different light sources (Fig. 5.23) [100]. Under visible illumination, the Ag NPs are excited due to LSPR and the electron transfer from Ag to TiO_2 is the primary contribution to the enhancement in photocatalytic activity. Under UV illumination, TiO_2 is activated to generate electron-hole pairs and Ag NPs work as electron traps, which facilitates the charge separation, thus enhancing the photocatalytic activity.

5.5 Energy Related Applications

5.5.1 Conversion of Solar Energy to Fuel

The increasing economic and environmental dependence on oil and electricity has created a worldwide effort toward developing alternative fuels [111, 112]. Among renewable energy sources at our disposal (wind, geothermal, hydroelectric, etc.), solar energy is the most abundant of all carbon-neutral energy sources [112, 113]. Solar energy does not pose any dangers to the environment or climate, unlike harvesting and burning fossil fuels.

Conversion of solar energy and waste chemicals, such as CO_2, into stored chemical energy is one of the most promising and environmentally friendly technologies for future energy generation. Nature already provides us with an outstanding solar energy conversion system that exquisitely manages photon capture and conversion processes to drive water-splitting and carbon fixation. However, limited availability of biomaterial and cultivatable land assets restricts the use of bioenergy conversion as a fuel source on a very large scale. Therefore, in similar fashion, bio-inspired artificial photosynthetic strategies are attractive approaches for the production of molecular fuels, such as hydrogen from the splitting of water and/or high energy organic molecules by the reduction of CO_2.

In spite of its enormous potential, the efficient harnessing, conversion, and storage of energy from sunlight has thus far proven to be difficult and costly tasks. This is due to high stability of CO_2. For example, CO_2 requires at least 8 photons to photo-catalytically convert CO_2 and H_2O to CH_4 [114, 115]. To be advantageous and efficient, the artificial photosynthetic devices must (a) directly capture CO_2 and convert it into liquid solar fuels that can be easily stored, (b) have a higher solar energy-to-chemical fuel conversion efficiency, and (c) be robust, cheap and amenable to scaling up. On the quest for both highly efficient and cost effective solar conversion devices, researchers have been exploring new material designs for decades, including novel hetero-structures, morphologies and crystallinities, while improving their photon-free electron generation rate, or shifting the absorption band of the photocatalysts to lower energy.

The first successful photo-electrochemical reduction of CO_2 to hydrocarbons was reported by Halmann in 1978 [116]. Several other studies followed, but the majority of them used electrochemical cells, which clearly limits any energy advantage of converting CO_2 to fuels [111]. In contrast, the photocatalytic conversion of CO_2 to hydrocarbon fuels using sunlight only received traction in recent decades as better solution to this problem. A material that can efficiently catalyze the photo reduction of atmospheric CO_2 and H_2O in sunlight would essentially make our diminishing supply of naturally occurring fossil fuels recyclable while, at the same time, reducing CO_2 emissions. The most important solar energy conversion technologies, whether direct photovoltaic devices or solar driven catalytic reactions (CO_2 to fuels and water-splitting), depend on the photoelectronic properties of nanoscale semiconductors (TiO_2 nano-particles/rods/tubes, for example, or semiconductor quantum dots.

The absorption of photons from natural light in photocatalytic materials generates electrons and holes which could drive energy conversion processes in the solar-driven photoreduction of CO_2 and/or water-splitting for hydrogen fuels [101, 102]. Despite these attributes, the efficiency of TiO_2 for photovoltaic and photocatalytic applications is severely limited by its large band gap ($\sim$ 3.2 eV) and rapid charge carrier recombination dynamics, which means that titania can use less than approximately 1% of the solar spectrum [67, 84, 101, 102]. Grimes' group used nitrogen-doped titania nanotube array loaded with both Cu and Pt nanoparticles as the catalysts to convert CO_2, and reported a hydrocarbon production rate of 160 μL/(hr) [117]. Other novel spherical catalysts, such as $ZnAl_2O_4$-modified mesoporous ZnGaNO with Pt co-catalyst or SiC nanoparticles modified by Cu_2O, have also been reported for CO_2 conversion with similar efficiencies [114, 115]. With billions spent in the R&D effort, the conventional semiconductor technologies have achieved a lot. So far, the reported highest efficiency by TiO_2-based catalysts is around 0.1% [114, 115].

One-dimensional nanostructures have proven to be the most promising technique for combatting the charge transfer and recombination issues prevalent in traditional nanocomposite films. Such structures, including nanowires, nanotubes, nanorods, nanobelts, and nanofibers, provide a direct pathway for electron transport between the photon-excited surface and the surrounding media. One dimensional hollow nanostructures are favored to other shapes as they allow for the reactant gases to occupy both the inner and outer surfaces of the TiO_2 nanotubes, maximizing the surface area for photocatalytic reactions [99–102, 117]. Additionally, if the nanotube wall thickness is about half of the hole diffusion length in TiO_2, the surface species have access to charge carriers generated near the surface and also deep within the walls via diffusion. This is advantageous because regardless of the nanotube length, charge carriers are never generated far from the semiconductor-reactant surface. Therefore, the nanotubes can be lengthened to increase reactive surface area and facilitate light absorption without compromising the photocatalytic properties processes taking place in the array [114].

Semiconductor nanoparticles, metallic nanoparticles, and quantum dots possess unique electrical and optical properties that can significantly enhance solar energy conversion efforts. When used in conjunction with photoactive semiconductor materials, nanoparticles can increase the surface-to-volume ratio and the spectral window of such devices. The superior light-absorbing and charge-separating capabilities of quantum dots and plasmonic nanoparticles in conjunction with the electron-transport properties of nanostructured semiconductor materials are a promising avenue for new solar conversion devices. The development of such hybrid multifunctional structures has also led to major advances in solar-to-fuel conversion applications as discussed in the next section.

We demonstrated solar conversion of carbon dioxide and water vapor to carbon monoxide, hydrogen and hydrocarbons by exposure to titania-based nanophotocatalysts under ambient light only. TiO_2 and WO_3-core TiO_2-shell nanorod arrays fabricated by glancing angle deposition were functionalized with quantum dots (CdSe) prepared by solution chemistry (Fig. 5.24) [118]. A gas chromatographic analysis shows that the primary products are carbon monoxide, hydrogen and hydrocarbons (methanol and methane) after several hours of exposure to sunlight [99, 118]. The overall CO_2 conversion efficiency of such core-shell and quantum dot-titania nanostructures is significantly higher than that of pure TiO_2 nanorod array photocatalysts. The improved conversion efficiency may result from the additional absorbance of visible light by WO_3 and CdSe quantum dots and the charge separation at the WO_3-TiO_2 and CdSe-TiO_2 interfaces. The conversion efficiency of such an artificial photosynthesis process remains to be improved for practical applications.

Solar conversion of carbon dioxide and water vapor in the presence of Fe_2O_3-TiO_2 core-shell nanorod arrays was also investigated. Carbon monoxide, hydrogen, methane, and methanol along with other hydrocarbons were produced after only several hours' exposure under ambient illumination (Figs. 5.25 and 5.26) [99]. Core-shell Fe_2O_3-TiO_2 nanostructures (Fig. 5.25a) were chosen because they have high surface areas for reactions. The titania-iron composite also maximizes the interfacial area between the two different materials, which should improve charge separation and transport and, therefore, increase the overall efficiency of the material. Coupling TiO_2 with a lower band gap material (Fe_2O_3) created a composite structure that is active over a larger region of the spectrum, allowing one to take advantage of the excellent photocatalytic properties of TiO_2 under visible light

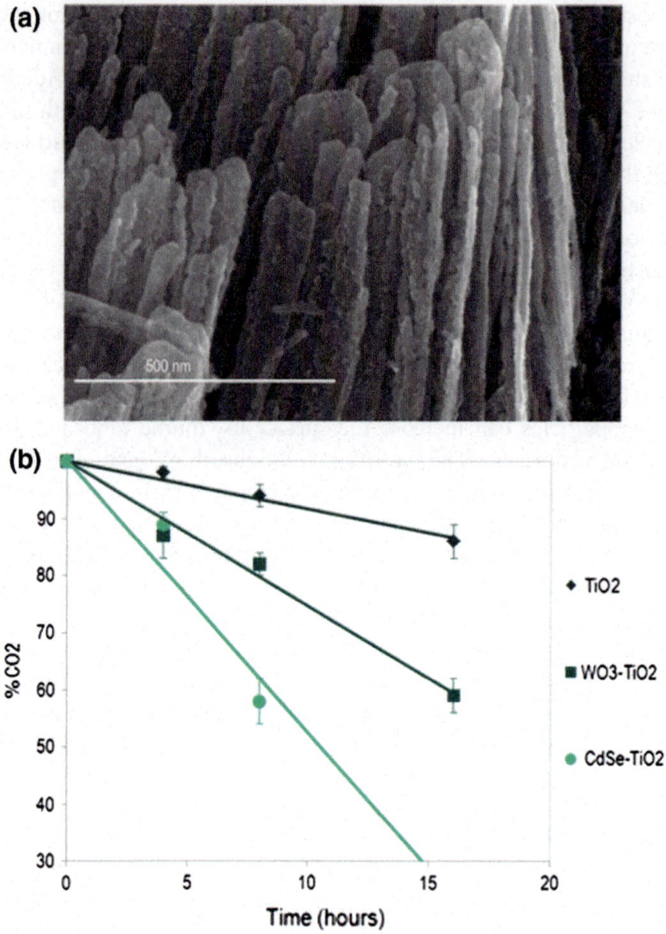

Fig. 5.24 **a** Scanning electron micrograph of representative CdSe-WO_3-TiO_2 nanoparticles; **b** CO_2 reduction process with TiO_2, WO_3-TiO_2 or CdSe-TiO_2 nanophotocatalysts after prolonged exposure in sun light

absorbance. Fe_2O_3 is an excellent choice as a coupling material for TiO_2 because it is stable, non-toxic, widely available, and catalytically active. Furthermore, it has a band gap of 2.2 eV so it can absorb visible light.

In order to utilize solar radiation as the sole energy input for CO_2 reduction, sunlight sensitive materials need to be further developed. The combination of high surface area nanostructured semiconductor arrays with the visible light-absorbing properties of metallic nanoparticles and quantum dots have led to significant advances in CO_2 photoconversion in recent years [114, 115, 117].

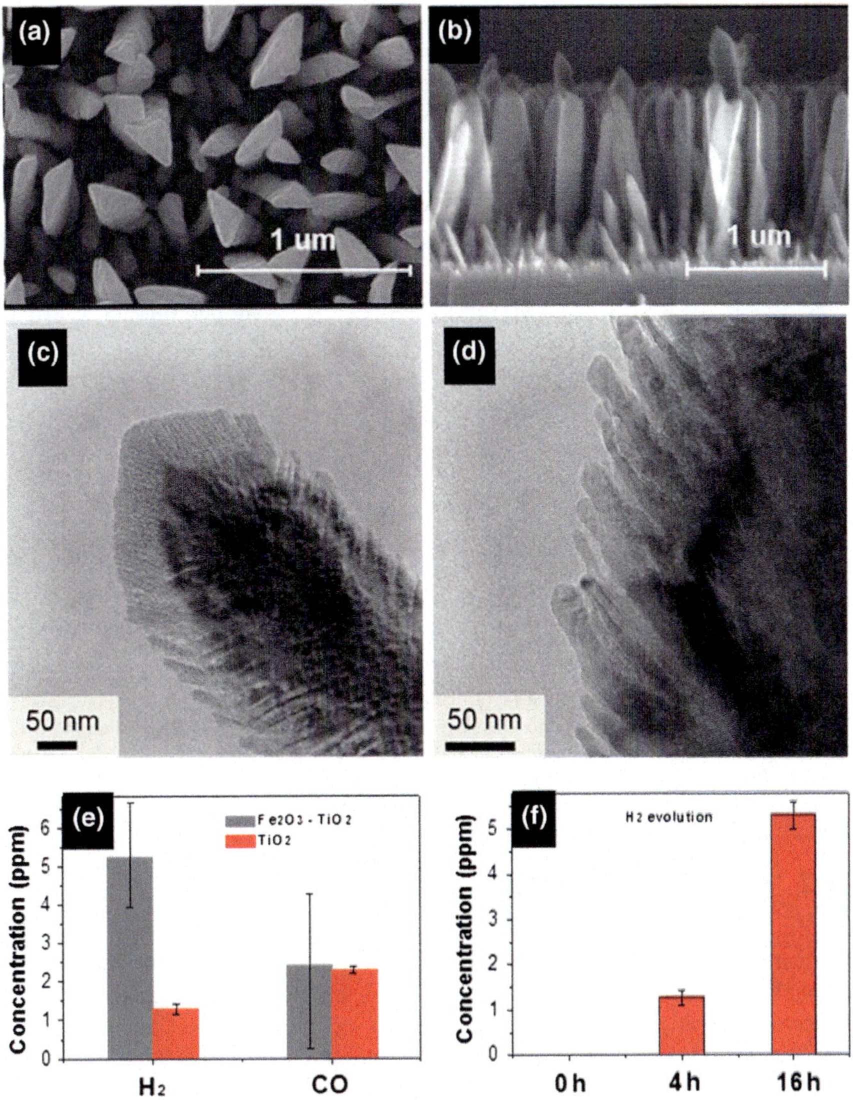

Fig. 5.25 **a**, **b** SEM images of the Fe_2O_3–TiO_2 core-shell nanorod array from top view and cross section view. **c**, **d** HRTEM image of the as-deposited Fe_2O_3–TiO_2 core-shell nanorods. **b**–**d** HRTEM images of the Fe_2O_3–TiO_2 core-shell nanorods annealed at 500 °C. **e** H_2 and CO production in the presence of Fe_2O_3–TiO_2 core-shell or TiO_2 nanorod array photocatalysts after 4 h exposure. **f** H_2 evolution after 16 h exposure in the presence of TiO_2. Adapted from Ref. [99] with permissions from Elsevier

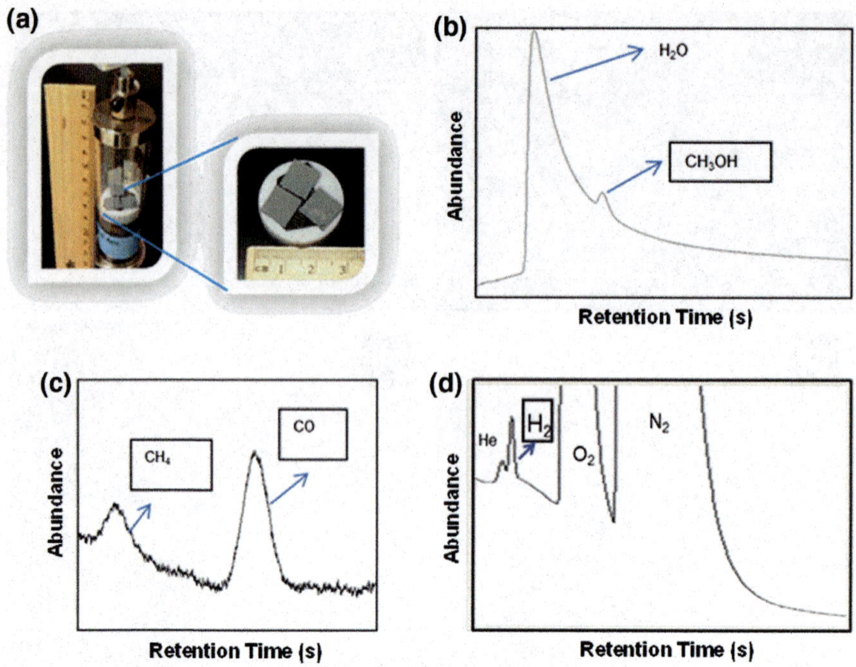

Fig. 5.26 **a** Pictures of the syringe-type chamber with the catalysts. **b–d** Representative chromatographs collected on a micro-GC showing CO, CH_3OH, CH_4 and H_2 evolution, in the presence of the core-shell nanorod array photocatalysts exposed under ambient light for 4 hours. Reprinted from Ref. [99] with permissions from Elsevier

5.5.2 Energy Storage Materials

Nanoparticle mediated energy storage is also highly dependent on nanoparticle morphology. Researchers at Stanford showed that Pd nanoparticles of various shapes, specifically cubes and pyramids, may be more efficient energy storage materials than icosahedra shaped Pd. Dionne's group examined how different shapes and crystalline structures of palladium nanoparticles affected their ability to absorb and release hydrogen atoms [119]. Multiply twinned icosahedral particles show hydrogen intercalation behavior markedly different from that of their single-crystalline counterparts. Notably, icosahedra absorb hydrogen gradually over several pressure steps and apparently exhibit reduced hydrogen storage capacity compared to single-crystalline cubes and prisms [119]. Generally, nanoparticles with highly compressively strained, defective central regions have a reduced energy storage capacity compared to their single-crystalline counterparts.

5.6 Photothermal Applications

As described in previous chapters, gold and silver nanoparticles exhibit localized surface plasmon resonance (LSPR) that occur when the incident electromagnetic radiation resonantly drives free electrons into collective oscillations in metallic materials leading to highly intense electromagnetic fields [90, 91]. The plasmonic properties of gold and silver nanostructures are tunable throughout the visible and near-infrared region of the spectrum as a function of nanoparticle size, shape, aggregation state, and local environment, allowing for increased tunability [2–6]. This phenomenon has been exploited for photothermal applications as the large amount of near-IR light absorbed by gold nanoparticles of appropriate size and shape can be released into the surroundings as heat [38, 42, 120, 121]. The temperature profiles generated by shining light of appropriate wavelength can be optimized by simply matching the laser light wavelength to the nanoparticle's plasmon band. For example, in the presence of near infrared (NIR) laser excitation, the plasmonic induced heating of 100 nm gold nanorods with longitudinal plasmon band at around 800 nm is superior to spheres with plasmon band at 525 nm.

While these properties have been largely exploited for cancer studies [40–44], there are numerous other applications where these phenomena are easily exploited for drug delivery, uptake and release of analytes of interest, catalysis, among others. Coupling of optically tunable gold nanoparticles with other nanoparticles or media, specifically oxides, polymers, metals, bacteria, cells, etc. [49] are attractive options for photothermal applications [90, 91].

Plasmonic heating has been employed to study temperature gradients in catalytic reactions [90, 91]. The photothermal properties of shape selective multifunctional Fe_2O_3-Au nanoparticles, namely, Fe_2O_3 rings, spheres and tubes (Fig. 5.18a–c), were harnessed to increase temperatures using light and tested as catalysts for 4-nitrophenol reduction reactions. Compared to pure AuNPs, Fe_2O_3-Au nanoparticles were superior catalysts, not only by having greater efficiencies but also by using significantly much smaller quantities of gold (Fig. 5.27). The Fe_2O_3-Au spheres were the most catalytically active structure compared with the Au-Fe_2O_3 rings and tubes. However, the Au loadings on the Fe_2O_3-Au rings and tubes were significantly smaller than Au loading on the Fe_2O_3 sphere. Specifically, Fe_2O_3-Au rings and tubes were more than twice as efficient as the spheres when normalized against the amount of Au loaded onto the Fe_2O_3 support. Fe_2O_3-Au nanoparticles can also be recovered and reused via magnetic collection. The Fe_2O_3-Au nanoparticles and AuNPs are found to efficiently transduce heat from light through plasmonic absorbance phenomenon that was demonstrated using the photothermal catalytic reduction of 4-nitrophenol. The reaction rates increase with increasing laser intensities/solution temperatures for both, as expected. Linear fits of the Arrhenius data result in measured activation energies, $E_A = 106 \pm 7$ kJ/mol and $E_A = 180 \pm 10$ kJ/mol for the AuNPs and the Fe_2O_3-Au spheres, respectively (Fig. 5.27h–i) [90].

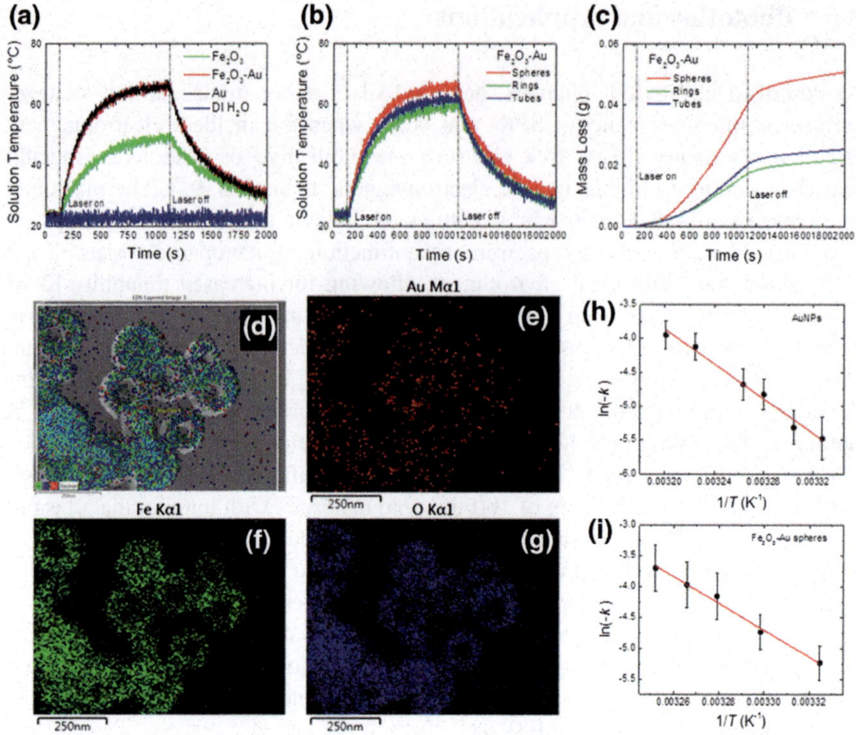

Fig. 5.27 **a** Temperature traces of the solutions containing the AuNPs, Fe_2O_3–Au and Fe_2O_3 spheres, and deionized water under laser irradiation. **b** Temperature traces and **c** solution mass loss comparing the photothermal heating effects of the Fe_2O_3–Au spheres, rings, and tube. **d–h** EDX mappings respectively showing the distribution of **e–g** Au, Fe, and O in the Fe_2O_3–Au rings. **h**, **i** Arrhenius style plots of the photothermally induced temperature-rise versus the catalytic reaction constant for the **h** AuNPs and the **i** Fe_2O_3–Au spheres. Adapted with permission from Ref. [90]. Copyright 2017 American Chemical Society

Nanoparticle size and shape are demonstrated to affect the specific heat, melting entropy and enthalpy of nanomaterials. The specific heat increases with a decrease in particle size, whereas the melting entropy and enthalpy decrease as the particle size decreases for Ag, Cu, In, Se nanospheres, nanwires and nanofilms. The amount of reduction in the melting entropy is increasing in the following order: nanofilm, nanowire, and nanosphere. The melting entropy difference for nanofilm to nanowire to nanosphere increases for the same particle size [122].

5.7 Self-assembled Nanostructures

The organization and manipulation of one dimensional nanostructures (nanorods and nanowires) have attracted a lot of attention. Further scientific and technological advances in the application of 1-D nanostructures in functional devices depend strongly on the ability to assemble them into ordered and complex architectures. Self-assembly of nanostructures into optimized geometries can lead to interesting synergistic effects and are of particular interest for molecular-based sensors, components for optical and electronic devices, photovoltaics, among others. One dimensional nanostructures are superior to nanospheres for self-assembly applications due to the unique spatial properties. For example, gold nanorods showed unidirectional oriented self-assembly at high concentration in nematic liquid crystals while no alignment was observed for shape-isotropic nanoparticles [123].

Whether through spontaneous or directed self-assembly, nanoparticles form secondary architectures with properties dictated not only by the individual building blocks but also by their spatial arrangement. Spatial assembly of nanoparticles in 1-D nanoarchitectures can be achieved by surface functionalization/biorecognition, template directed self-assembly, external electric or magnetic fields, Langmuir-Blodgett technique, or solvent directed assemblies, among others [124–127]. It has been reported that the collective behavior and interparticle coupling in well-organized 1-D nanostructures results in new functions or significant improvement of their single optoelectronic properties [123–127]. The long-range unidirectional alignment of gold nanorods gives rise to strong polarization-dependent absorption and scattering, abnormal dispersions of refractive index, and enhanced optical birefringence with sign reversal in the spectral vicinity of the longitudinal SPR peak [123]. Optimal dimension ratios for anisotropic nanostructures, rods, platelets, bowls and dumbbells, have been reported for particle alignment using an external electric field [127].

Theoretical and experimental work suggests that one-dimensional metal nanoparticle arrays can be utilized to transport electromagnetic energy ("plasmonics") [128, 129]. Maier et al. demonstrated that periodic arrays of metallic structures (specifically, lithographically prepared silver nanorods) embedded in dielectric media are able to guide and modulate light transmission (even through corners and tee structures) in a regime dominated by near-field coupling [130].

We developed a convenient and straightforward wet-chemical route to produce "nanopeapod" architectures in which silver "peas" are embedded in silica "pod" in an ordered fashion (Fig. 5.28). Since a wet-chemical approach was employed for their fabrication, the synthesis is amenable to scaling up [14].

Optical manipulation of nanoparticles is of particular interest in the plasmonics field. The shape of an optically trapped gold NP strongly determines the stable position and orientation in the optical tweezers for a particular trapping wavelength. Theoretical analysis focused on the three dimensional optical trapping of non-spherical gold nanoparticles using a tightly focused laser beam (i.e. optical

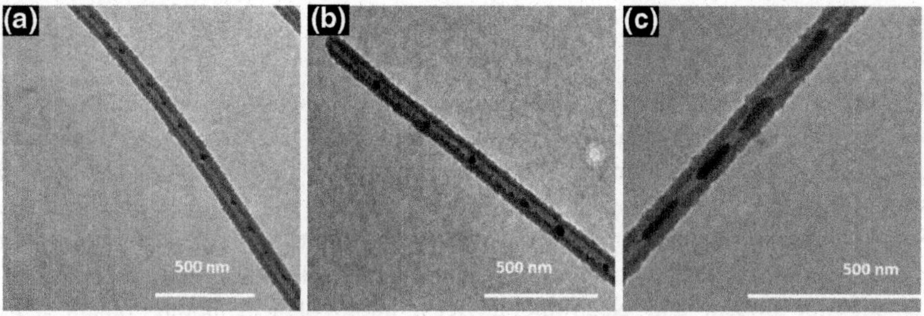

Fig. 5.28 Transmission electron micrographsof representative "peapod" architectures obtained under conditions of incomplete dissolution of the silver with ammonia, showing silver "peas" in silica "pods". *Scale bar* 500 nm. Adapted with permission from Ref. [14]. Copyright 2017 American Chemical Society

tweezers) shows that thin triangular gold prisms can be trapped in 3D even if spheres of the same volume cannot be confined [131].

Anisotropic nanoparticles, unlike spherical particles, can align and aggregate in different configurations. Solvent-mediated interactions between gold nanorods with uniform shape and size often leads to end-to-end, parallel or 'nondirectional' orientations. Figure 5.29 *s*hows an arrangement of gold nanorods with localized arrangements based on solvent-mediated interactions. The solvent mediated aggregation is used for identifying a change in local environments. It is important to note that control of the self-assembled structures is not possible when the self-assembly is induced by surfactant-mediated interactions and drying alone. Assemblies of nanorods are known to produce a shift of the localized plasmon bands in the UV-vis spectra (Table 5.1) [132]. In our experiment, a red shift in the localized plasmon band was recorded and, in some cases, appearance of additional peaks due to aggregate formation.

The self-assembly of gold nanorods of different aspect ratios (AuNRs) on carbon nanotubes (CNTs) has been studied by us (Fig. 5.30). The gold nanoparticles have a positive surface charge due to the CTAB surfactant bilayer whereas the CNTs have an intrinsic negative surface charge. Simply mixing solutions of these two species together in the appropriate solvents can lead to electrostatic assembly of the Au nanoparticles on the CNT surface. Bifunctional linkers can also be used to covalently attach Au nanoparticles to CNTs by exploiting the strong affinity of Au to thiol or amino groups [12].

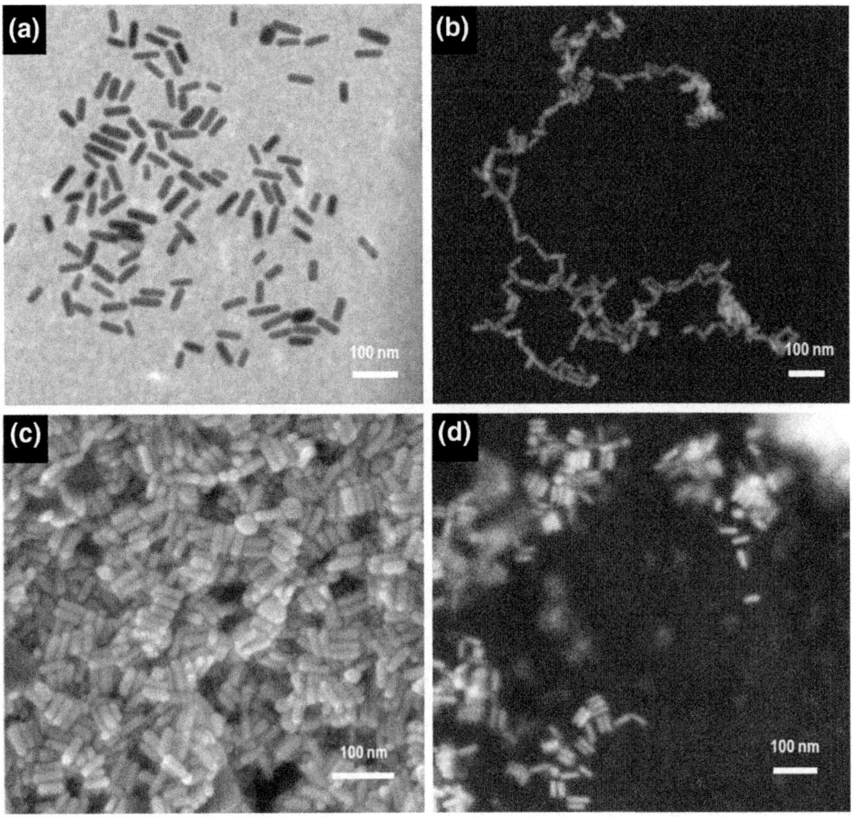

Fig. 5.29 Electron micrographs of Au nanorod aggregates in different solvent: **a** water, **b** methanol, **c** acetone, **d** pyridine

Table 5.1 Au nanorods' plasmon band shifts with the environment

Solvent name	Longitudinal plasmon band(s) (nm)	Aggregated assemblies
Water	711	No aggregation; individual nanoparticles
Methanol	914 1046	End-to-end aggregates; ~100 nanorods
Acetone	909 1025	Large aggregates; ~Hundreds of nanorods with 3–7 side-to-side aggegates
Pyridine	737 870	Small nanorod aggregates; side-to-side aggregates; ~3–5 nanorods

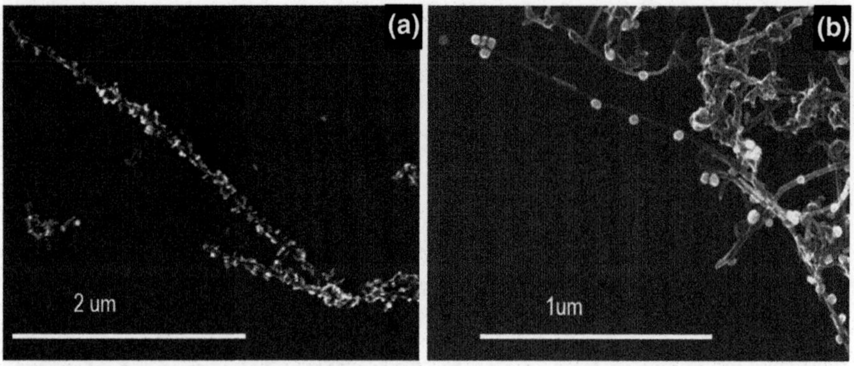

Fig. 5.30 SEM images of self-assembled gold nanorods (**a**) or gold nanosphere (**b**) onto carbon nanotubes

5.8 Conclusions

The promise of viable future applications for anisotropic nanomaterials lies in the scientists' ability to understand, generate, and control materials at the nanoscale. Since nanomaterial's properties can be precisely tuned by manipulating atomic growth, a high degree of control is vital. In this chapter, the reader was immersed in the latest state-of–the-art developments in nanotechnology with a focus on the broad applications of these technologies at the laboratory and/or industrial scale. Specifically, the reader was introduced to a plethora of current and emerging applications of anisotropic nanomaterials, including sensing and imaging, catalysis, biomedical imaging and drug delivery, environmental remediation, plasmonics, and energy- related missions. The effect of morphology and nanometric dimensions of metallic, metallic oxide and nonmetallic and materials on their physico-chemical properties have been described. It is our hope that the reader recognizes the inimitability of anisotropic nanostructures and acquired a greater appreciation for the vast degree of excitement in this area.

References

1. Doering, W.E., and S.M. Nie. 2002. Single-Molecule and Single-Nanoparticle SERS: Examining the Roles of Surface Active Sites and Chemical Enhancement. *The Journal of Physical Chemistry B* 106: 311–317.
2. Fedlheim, D.L., and C.A. Foss. 2001. *Metal Nanoparticles: Synthesis, Characterization, and Applications*. New York: Marcel Dekker Inc.
3. Murphy, C.J., T.K. Sau, A.M. Gole, C.J. Orendorff, J. Gao, L. Gou, S.E. Hunyadi, and T. Li. 2005. Anisotropic Metal Nanoparticles: Synthesis, Assembly, and Optical Applications (Feature Article; a Top Five ACS article by citations, National Chemistry Week, 2007). *The Journal of Physical Chemistry B* 109: 13857–13870.

4. Murphy, C.J., A.M. Gole, S. Hunyadi Murph, and C.J. Orendorff. 2006. One-Dimensional Colloidal Gold and Silver Nanostructures. *Inorganic Chemistry* 45 (19): 7544–7554.
5. Murphy, C.J., A.M. Gole, S.E. Hunyadi, J.W. Stone, P.N. Sisco, A. Alkilany, B.E. Kinard, and P. Hankins. 2008. Chemical Sensing and Imaging with Metallic Nanorods. *Chemical Communications* 4 (554–557): 554.
6. Hunyadi, S.E. 2007. *Nanoengineered Materials: Synthesis, Design Functionalization and Chemical Sensing Applications*. Columbia: University of South Carolina.
7. Ruan, C.M., W. Wang, and A.H. Gu. 2006. Surface-Enhanced Raman Scattering for Perchlorate Detection Using Cystamine-Modified Gold Nanoparticles. *Analytica Chimica Acta* 567: 114–120.
8. Vo-Dinh, T., F. Yan., and M.B Wabuyele. 2006. Surface-Enhanced Raman Scattering for Biomedical Diagnostics and Molecular Imaging. In *Surface-Enhanced Raman Scattering: Physics and Applications*, ed, K. Kneipp, M. Moskovits and H. Kneipp, pp. 409–26. Berlin: Springer.
9. Cîntă Pînzaru, S., I. Pavel, N. Leopold, and W. Kiefer. 2004. Identification and Characterization of Pharmaceuticals Using Raman and Surface-Enhanced Raman Scattering. *Journal of Raman Spectroscopy* 35 (5): 338–346.
10. Kneipp, K., Y. Wang, H. Kneipp, L.T. Perelman, I. Itzkan, R.R. Dasari, and M.S. Feld. 1997. Single Molecule Detection Using Surface-Enhanced Raman Scattering (SERS). *Physical Review Letters* 78 (9): 1667–1670.
11. Song, C., J. Abbell, Y. He, S.E. Hunyadi Murph, Y. Cui, and Y. Zhao. 2012. Gold-Modified Silver Nanorod Arrays: Growth Dynamics and Improved SERS Properties. *Journal of Materials Chemistry* 22: 1150–1159.
12. Hunyadi Murph, S.E., K. Heroux., C. Turick, and D. Thomas. 2012. Metallic and Hybrid Nanostructures: Fundamentals and Applications. In *Applications of Nanomaterials* (Vol. 4), ed, J.N. Govil. USA: Studium Press LLC.
13. Valden, M., X. Lai, and D.W. Goodman. 1998. Onset of Catalytic Activity of Gold Clusters on Titania with the Appearance of Nonmetallic Properties. *Science* 281 (5383): 1647–1650.
14. Hunyadi Murph, S.E., and C.J. Murphy. 2006. Tunable One-Dimensional Silver-Silica Nanopeapod Architectures. *The Journal of Physical Chemistry B* 110: 7226–7231.
15. Zou, S., and G.C. Schatz. 2005. Silver Nanoparticle Array Structures that Produce Giant Enhancements in Electromagnetic Fields. *Chemical Physics Letters* 403 (1–3): 62–67.
16. Orendorff, C.J., A. Gole, T.K. Sau, and C.J. Murphy. 2005. Surface-Enhanced Raman Spectroscopy of Self-Assembled Monolayers: Sandwich Architecture and Nanoparticle Shape Dependence. *Analytical Chemistry* 77 (10): 3261–3266.
17. Hunyadi Murph, S.E., and C.J. Murphy. 2013. Patchy Silica-Coated Silver Nanowires as SERS Substrates. *Journal of Nanoparticle Research* 15 (6): 1607.
18. Orendorff, C.J., L. Gearheart, N.R. Jana, and C.J. Murphy. 2006. Aspect Ratio Dependence on Surface Enhanced Raman Scattering Using Silver and Gold Nanorod Substrates. *Physical Chemistry Chemical Physics* 8 (1): 165–170.
19. Hunyadi, S.E., and C.J. Murphy. 2006. Bimetallic Silver-Gold Nanowires: Fabrication and Use in Surface- Enhanced Raman Scattering. *Journal of Materials Chemistry.* 16 (Special Issue: Anisotropic Nanoparticles): 3929–35.
20. Alvarez-Puebla, R.A., E. Arceo, P.J.G. Goulet, J.J. Garrido, and R.F. Aroca. 2005. Role of Nanoparticle Surface Charge in Surface-Enhanced Raman Scattering. *The Journal of Physical Chemistry B* 109 (9): 3787–3792.
21. Nie, S., and S.R. Emory. 1997. Probing Single Molecules and Single Nanoparticles by Surface-Enhanced Raman Scattering. *Science* 275 (5303): 1102–1106.
22. Lascola, R., S. McWhorter, and S.H. Murph. 2010. Advanced Gas Sensors Using SERS-Activated Waveguides. *AIP Conference Proceedings* 1267 (1): 1095–1096.
23. Lascola, R.J., C.S. McWhorter, and S.H. Murph. 2015. Surface Enhanced Raman Scattering Spectroscopic Waveguide.U.S. Patent 4/15/2015, serial # 501665408S.
24. Murphy, C.J., and S.E. Hunyadi. 2006. Europium-Doped Silica Nanotubes: Synthesis and Optical Properties. *Materials Research Society Symposium Proceedings* 922E: U01–U03.

25. Malicka, J., I. Gryczynski, and J.R. Lakowicz. 2003. Enhanced Emission of Highly Labeled DNA Oligomers near Silver Metallic Surfaces. *Analytical Chemistry* 75 (17): 4408–4414.
26. Sokolov, K., G. Chumanov, and T.M. Cotton. 1998. Enhancement of Molecular Fluorescence near the Surface of Colloidal Metal Films. *Analytical Chemistry* 70 (18): 3898–3905.
27. Aslan, K., M. Wu, J.R. Lakowicz, and C.D. Geddes. 2007. Metal Enhanced Fluorescence Solution-based Sensing Platform 2: Fluorescent Core-Shell Ag@SiO_2 Nanoballs. *Journal of Fluorescence* 17 (2): 127–131.
28. Aslan, K., M. Wu, J.R. Lakowicz, and C.D. Geddes. 2007. Fluorescent Core—Shell Ag@SiO_2 Nanocomposites for Metal-Enhanced Fluorescence and Single Nanoparticle Sensing Platforms. *Journal of the American Chemical Society* 129 (6): 1524–1525.
29. Hunyadi Murph, S.E., S. Jacobs., M. Siegfried., T. Hu., S. Serkiz., and J. Hudson. 2012. Manganese- Doped Gold Nanoparticles as Positive Contrast Agents for Magnetic Resonance Imaging (MRI). *Journal of Nanoparticle Research* 14: 658–659.
30. Obare, S.O., R.E. Hollowell, and C.J. Murphy. 2002. Sensing Strategy for Lithium Ion Based on Gold Nanoparticles. *Langmuir* 18 (26): 10407–10410.
31. Mirkin, C.A., R.L. Letsinger, R.C. Mucic, and J.J. Storhoff. 1996. A DNA-Based Method for Rationally Assembling Nanoparticles into Macroscopic Materials. *Nature* 382: 607–609.
32. Chang, J.-Y., H. Wu, H. Chen, Y.-C. Ling, and W. Tan. 2005. Oriented Assembly of Au Nanorods Using Biorecognition System. *Chemical Communications* 8: 1092–1094.
33. Sudeep, P.K., S.T.S. Joseph, and K.G. Thomas. 2005. Selective Detection of Cysteine and Glutathione Using Gold Nanorods. *Journal of the American Chemical Society* 127 (18): 6516–6517.
34. Orendorff, C.J., P.L. Hankins, and C.J. Murphy. 2005. pH-Triggered Assembly of Gold Nanorods. *Langmuir* 21 (5): 2022–2026.
35. Bhadra, D., S. Bhadra, S. Jain, and N.K. Jain. 2003. A PEGylated Dendritic Nanoparticulate Carrier of Fluorouracil. *International Journal of Pharmaceutics* 257 (1–2): 111–124.
36. Haes, A.J., and R.P. Van Duyne. 2002. A Nanoscale Optical Biosensor: Sensitivity and Selectivity of an Approach Based on the Localized Surface Plasmon Resonance Spectroscopy of Triangular Silver Nanoparticles. *Journal of the American Chemical Society* 124 (35): 10596–10604.
37. McFarland, A.D., and R.P. Van Duyne. 2003. Single Silver Nanoparticles as Real-Time Optical Sensors with Zeptomole Sensitivity. *Nano Letters* 3 (8): 1057–1062.
38. Huang, X., I.H. El-Sayed, W. Qian, and M.A. El-Sayed. 2006. Cancer Cell Imaging and Photothermal Therapy in the Near-Infrared Region by Using Gold Nanorods. *Journal of the American Chemical Society* 128 (6): 2115–2120.
39. El-Sayed, I.H., X. Huang, and M.A. El-Sayed. 2005. Surface Plasmon Resonance Scattering and Absorption of anti-EGFR Antibody Conjugated Gold Nanoparticles in Cancer Diagnostics: Applications in Oral Cancer. *Nano Letters* 5 (5): 829–834.
40. El-Sayed, I.H., X. Huang, and M.A. El-Sayed. 2006. Selective Laser Photo-Thermal Therapy of Epithelial Carcinoma Using Anti-EGFR Antibody Conjugated Gold Nanoparticles. *Cancer Letters* 239 (1): 129–135.
41. Pissuwan, D., S.M. Valenzuela, and M.B. Cortie. 2006. Therapeutic Possibilities of Plasmonically Heated Gold Nanoparticles. *Trends in Biotechnology* 24 (2): 62–67.
42. Hirsch, L.R., R.J. Stafford, J.A. Bankson, S.R. Sershen, B. Rivera, R.E. Price, J.D. Hazle, N. J. Halas, and J.L. West. 2003. Nanoshell-Mediated Near-Infrared Thermal Therapy of Tumors Under Magnetic Resonance Guidance. *Proceedings of the National Academy of Sciences of the United States of America* 100: 13549–13554.
43. Abadeer, N.S., and C.J. Murphy. 2016. Recent Progress in Cancer Thermal Therapy Using Gold Nanoparticles. *The Journal of Physical Chemistry C* 120 (9): 4691–4716.
44. Day, E.S., J.G. Morton., and J.L. West. 2009. Nanoparticles for Thermal Cancer Therapy. *Journal of Biomechanical Engineering* 131 (7): 074001-074001-5.

45. Xiao, Z., Q. Wu, S. Luo, C. Zhang, J. Baur, R. Justice, and T. Liu. 2013. Shape Matters: A Gold Nanoparticle Enabled Shape Memory Polymer Triggered by Laser Irradiation. *Particle & Particle Systems Characterization* 30 (4): 338–345.
46. Black, K.C.L., Y. Wang, H.P. Luehmann, X. Cai, W. Xing, B. Pang, Y. Zhao, C.S. Cutler, L.V. Wang, Y. Liu, and Y. Xia. 2014. Radioactive 198Au-Doped Nanostructures with Different Shapes for In Vivo Analyses of Their Biodistribution, Tumor Uptake, and Intratumoral Distribution. *ACS Nano* 8 (5): 4385–4394.
47. Unrine, J.M., O.V. Tsyusko, S.E. Hunyadi, J. Judy, and P.M. Bertsch. 2010. Effects of Particle Size on Chemical Speciation and Bioavalability of Copper to Earthworms. *Journal of Environmental Quality* 39: 1942–1953.
48. Unrine, J.M., S.E. Hunyadi, O.V. Tsyusko, W. Rao, W.A. Shoults-Wilson, and P.M. Bertsch. 2010. Evidence for Bioavailability of Au Nanoparticles from Soil and Biodistribution within Earthworms (Eisenia fetida). *Environmental Science and Technology* 44 (21): 8308–8313.
49. Unrine, J., P. Bertsch., and S. Hunyadi. 2008. Bioavailability, Trophic Transfer, and Toxicity of Manufactured Metal and Metal Oxide Nanoparticles in Terrestrial Environments. In *Nanoscience and Nanotechnology* (pp. 345–66). New Jersey: Wiley.
50. Hillyer, J.F., and R.M. Albrecht. 2001. Gastrointestinal Persorption and Tissue Distribution of Differently Sized Colloidal Gold Nanoparticles. *Journal of Pharmaceutical Sciences* 90 (12): 1927–1936.
51. Tak, Y.K., S. Pal, P.K. Naoghare, S. Rangasamy, and J.M. Song. 2015. Shape-Dependent Skin Penetration of Silver Nanoparticles: Does It Really Matter? *Scientific Reports* 5: 16908.
52. Zhang, R., K. Cheng, A.L. Antaris, X. Ma, M. Yang, S. Ramakrishnan, G. Liu, A. Lu, H. Dai, M. Tian, and Z. Cheng. 2016. Hybrid Anisotropic Nanostructures for Dual-Modal Cancer Imaging and Image-Guided Chemo-Thermo Therapies. *Biomaterials* 103: 265–277.
53. Jacobs, S., S. Murph, M. Siegfried, S. Serkiz, and T.C.C. Hu. 2009. Manganese-Gold Nanospheres as Positive Contrast Agents for Magnetic Resonance Imaging (MRI). *Proceedings of the International Society for Magnetic Resonance in Medicine* 17: 1275.
54. Hunyadi Murph, S.E., M. Siegfried., S. Jacobs., T. Hu., and S. Serkiz. 2009. Manganese-Doped Gold Nanoparticle as Positive Contrast Agents for MRI. *Proceedings of Front Characterization and Control of Magnetic Carriers*.
55. Nagel, E., N. Al-Saadi, and E. Fleck. 2000. Cardiovascular Magnetic Resonance: Myocardial Perfusion. *Herz* 25 (4): 409–416.
56. Geng, Y., P. Dalhaimer, S. Cai, R. Tsai, M. Tewari, T. Minko, and D.E. Discher. 2007. Shape Effects of Filaments Versus Spherical Particles in Flow and Drug Delivery. *Nature Nanotechnology* 2 (4): 249–255.
57. Gratton, S.E.A., P.A. Ropp, P.D. Pohlhaus, J.C. Luft, V.J. Madden, M.E. Napier, and J.M. DeSimone. 2008. The Effect of Particle Design on Cellular Internalization Pathways. *Proceedings of the National Academy of Sciences* 105 (33): 11613–11618.
58. Toy, R., P.M. Peiris, K.B. Ghaghada, and E. Karathanasis. 2013. Shaping Cancer Nanomedicine: The Effect of Particle Shape on the In Vivo Journey of Nanoparticles. *Nanomedicine* 9 (1): 121–134.
59. Karaman, D.S., D. Desai, R. Senthilkumar, E.M. Johansson, N. Råtts, M. Odén, J.E. Eriksson, C. Sahlgren, D.M. Toivola, and J.M. Rosenholm. 2012. Shape Engineering vs Organic Modification of Inorganic Nanoparticles as a Tool for Enhancing Cellular Internalization. *Nanoscale Research Letters* 7 (1): 358.
60. Decuzzi, P., B. Godin, T. Tanaka, S.Y. Lee, C. Chiappini, X. Liu, and M. Ferrari. 2010. Size and Shape Effects in the Biodistribution of Intravascularly Injected Particles. *Journal of Controlled Release* 141 (3): 320–327.
61. Shah, S., Y. Liu, W. Hu, and J. Gao. 2011. Modeling Particle Shape-Dependent Dynamics in Nanomedicine. *Journal of Nanoscience and Nanotechnology* 11: 919–928.
62. Agarwal, R., V. Singh, P. Jurney, L. Shi, S.V. Sreenivasan, and K. Roy. 2013. Mammalian Cells Preferentially Internalize Hydrogel Nanodiscs Over Nanorods and Use Shape-Specific

Uptake Mechanisms. *Proceedings of the National Academy of Sciences* 110 (43): 17247–17252.
63. Smith, B.R., P. Kempen, D. Bouley, A. Xu, Z. Liu, N. Melosh, H. Dai, R. Sinclair, and S.S. Gambhir. 2012. Shape Matters: Intravital Microscopy Reveals Surprising Geometrical Dependence for Nanoparticles in Tumor Models of Extravasation. *Nano Letters* 12 (7): 3369–3377.
64. Jiang, X., W. Qu., D. Pan., Y. Ren., J.-M. Williford., H. Cui., E. Luijten., and H.-Q. Mao. 2013. Plasmid-Templated Shape Control of Condensed DNA–Block Copolymer Nanoparticles. *Advanced Materials* 25 (2): 227–32.
65. Dasgupta, S., T. Auth, and G. Gompper. 2014. Shape and Orientation Matter for the Cellular Uptake of Nonspherical Particles. *Nano Letters* 14 (2): 687–693.
66. Liu, Y., J. Tan, A. Thomas, D. Ou-Yang, and V.R. Muzykantov. 2012. The Shape of Things to Come: Importance of Design in Nanotechnology for Drug Delivery. *Therapeutic Delivery* 3 (2): 181–194.
67. Hunyadi Murph, S.E., S. Serkiz., E. Fox., H. Colon-Mercado., L. Sexton., and M. Siegfried. 2011. Synthesis, Functionalization, Characterization and Application of Controlled Shape Nanoparticles in Energy Production. In *Fluorine-Related Nanoscience with Energy Applications*, ed, D.J. Nelson, and C.N. Brammer (Vol. 1064).
68. Murph, S. 2012. Nanotechnology and Fuel Cells. *Innovation: America's Journal of Technology Commercialization*.
69. Narayanan, R., and M.A. El-Sayed. 2003. Effect of Catalytic Activity on the Metallic Nanoparticle Size Distribution: Electron-Transfer Reaction between Fe(CN)6 and Thiosulfate Ions Catalyzed by PVP—Platinum Nanoparticles. *The Journal of Physical Chemistry B* 107 (45): 12416–12424.
70. Hunyadi, S.E., and C.J. Murphy. 2009. Synthesis and Characterization of Silver-Platinum Bimetallic Nanowires and Platinum Nanotubes. *Journal of Cluster Science* 20: 319–330.
71. Baker, L.A., P. Jin, and C.R. Martin. 2005. Biomaterials and Biotechnologies Based on Nanotube Membranes. *Critical Reviews in Solid State and Materials Sciences* 30 (4): 183–205.
72. Liu, Z., B. Zhao, C. Guo, Y. Sun, F. Xu, H. Yang, and Z. Li. 2009. Novel Hybrid Electrocatalyst with Enhanced Performance in Alkaline Media: Hollow Au/Pd Core/Shell Nanostructures with a Raspberry Surface. *The Journal of Physical Chemistry C* 113 (38): 16766–16771.
73. Crooks, R.M., M. Zhao, L. Sun, V. Chechik, and L.K. Yeung. 2001. Dendrimer-Encapsulated Metal Nanoparticles: Synthesis, Characterization, and Applications to Catalysis. *Accounts of Chemical Research* 34 (3): 181–190.
74. El-Sayed, M.A. 2001. Some Interesting Properties of Metals Confined in Time and Nanometer Space of Different Shapes. *Accounts of Chemical Research* 34 (4): 257–264.
75. Reddington, E., A. Sapienza, B. Gurau, R. Viswanathan, S. Sarangapani, E.S. Smotkin, and T.E. Mallouk. 1998. Combinatorial Electrochemistry: A Highly Parallel, Optical Screening Method for Discovery of Better Electrocatalysts. *Science* 280 (5370): 1735–1737.
76. Roucoux, A., J. Schulz, and H. Patin. 2002. Reduced Transition Metal Colloids: A Novel Family of Reusable Catalysts? *Chemical Reviews* 102 (10): 3757–3778.
77. Steele, B.C.H., and A. Heinzel. 2001. Materials for Fuel-Cell Technologies. *Nature* 414 (6861): 345–352.
78. Liang, H.P., H.M. Zhang, J.S. Hu, Y.G. Guo, L.J. Wan, and C.L. Bai. 2004. Pt Hollow Nanospheres: Facile Synthesis and Enhanced Electrocatalysts. *Angewandte Chemie International Edition* 43 (12): 1540–1543.
79. Bell, A.T. 2003. The Impact of Nanoscience on Heterogeneous Catalysis. *Science* 299 (5613): 1688–1691.
80. Burda, C., X. Chen, R. Narayanan, and M.A. El-Sayed. 2005. Chemistry and Properties of Nanocrystals of Different Shapes. *Chemical Reviews* 105 (4): 1025–1102.

81. Mahmoud, M.A., R. Narayanan, and M.A. El-Sayed. 2013. Enhancing Colloidal Metallic Nanocatalysis: Sharp Edges and Corners for Solid Nanoparticles and Cage Effect for Hollow Ones. *Accounts of Chemical Research* 46 (8): 1795–1805.
82. Williams, K.R., and G.T. Burstein. 1997. Low Temperature Fuel Cells: Interactions Between Catalysts and Engineering Design. *Catalysis Today* 38 (4): 401–410.
83. Murph, S. *Gold-Manganese Nanoparticles for Targeted Diagnostic and Imaging*. http://www.nanowerk.com/spotlight/spotid=41828.php.
84. Hunyadi Murph, S.E. 2011. One-Dimensional Plasmonic Nano-Photocatalysts: Synthesis, Characterization and Photocatalytic Activity. In *Proceedings of SPIE*, ed, Y. Tachibana, pp 1–11.
85. Narayanan, R., and M.A. El-Sayed. 2005. Catalysis with Transition Metal Nanoparticles in Colloidal Solution: Nanoparticle Shape Dependence and Stability. *The Journal of Physical Chemistry B* 109 (26): 12663–12676.
86. Wu, B., and N. Zheng. 2013. Surface and Interface Control of Noble Metal Nanocrystals for Catalytic and Electrocatalytic Applications. *Nano Today* 8 (2): 168–197.
87. Semagina, N., and L. Kiwi-Minsker. 2009. Recent Advances in the Liquid-Phase Synthesis of Metal Nanostructures with Controlled Shape and Size for Catalysis. *Catalysis Reviews* 51 (2): 147–217.
88. Hunyadi Murph, S.E., C.J. Murphy., H. Colon-Mercado., R. Torres., K. Heroux., E. Fox., L. Thompson., and R. Haasch. 2011. Tuning of Size and Shape of Au-Pt Nanocatalyst for Direct Methanol Fuel Cells. *Journal of Nanoparticle Research* 13: 6347–6364.
89. Chuan-Jian, Z., L. Jin, F. Bin, N.W. Bridgid, N.N. Peter, L. Rameshwori, and Y. Jun. 2010. Nanostructured Catalysts in Fuel Cells. *Nanotechnology* 21 (6): 062001.
90. Larsen, G.K., W. Farr, and S.E.H. Murph. 2016. Multifunctional Fe_2O_3-Au Nanoparticles with Different Shapes: Enhanced Catalysis, Photothermal Effects, and Magnetic Recyclability. *Journal of Physical Chemistry C* 120 (28): 15162–15172.
91. Murph, S.E.H., G.K. Larsen., and R.J. Lascola. 2016. Multifunctional Hybrid Fe_2O_3-Au Nanoparticles for Efficient Plasmonic Heating. *Journal of Visualized Experiments* (108): e53598.
92. Murph, S.E.H., G.K. Larsen, P. Korinko, K.J. Coopersmith, A.J. Summer, and R. Lewis. 2017. Nanoparticle Treated Stainless Steel Filters for Metal Vapor Sequestration. *JOM Journal of the Minerals Metals and Materials Society* 69 (2): 162–172.
93. Kar, M., M. Pauline, K. Sharma, G. Kumaraswamy, and S. Sen Gupta. 2011. Synthesis of Poly-l-glutamic Acid Grafted Silica Nanoparticles and Their Assembly into Macroporous Structures. *Langmuir* 27 (19): 12124–12133.
94. Li, X.q., and W.x. Zhang. 2006. Iron Nanoparticles: The Core—Shell Structure and Unique Properties for Ni(II) Sequestration. *Langmuir* 22 (10): 4638–42.
95. Wang, H., Y.F. Yu, Q.W. Chen, and K. Cheng. 2011. Carboxyl-Functionalized Nanoparticles with Magnetic Core and Mesopore Carbon Shell as Adsorbents for the Removal of Heavy Metal Ions From Aqueous Solution. *Dalton Transactions* 40 (3): 559–563.
96. Raschke, G., S. Kowarik, T. Franzl, C. Sönnichsen, T.A. Klar, J. Feldmann, A. Nichtl, and K. Kürzinger. 2003. Biomolecular Recognition Based on Single Gold Nanoparticle Light Scattering. *Nano Letters* 3 (7): 935–938.
97. Fondeur, F.F., S.E. Murph., K. Taylor-Pashow., and D.T. Hobbs. 2014. *Hybrid Ion Exchange-SERS Sensors for Determining Distribution Coefficients*. USA: Savannah River Site.
98. Badr, Y., M.G. Abd El-Wahed, and M.A. Mahmoud. 2008. Photocatalytic Degradation of Methyl Red Dye by Silica Nanoparticles. *Journal of Hazardous Materials* 154 (1–3): 245–253.
99. Yao, K., P. Basnet, H. Sessions, G.K. Larsen, S.E.H. Murph, and Y. Zhao. 2016. Fe_2O_3–TiO_2 core–shell nanorod Arrays for Visible Light Photocatalytic Applications. *Catalysis Today* 270: 51–58.

100. He, Y., P. Basnet, S.E.H. Murph, and Y. Zhao. 2013. Ag Nanoparticle Embedded TiO_2 Composite Nanorod Arrays Fabricated by Oblique Angle Deposition: Toward Plasmonic Photocatalysis. *ACS Applied Materials & Interfaces* 5 (22): 11818–11827.
101. Smith, W., and Y. Zhao. 2008. Enhanced Photocatalytic Activity by Aligned WO_3/TiO_2 Two-Layer Nanorod Arrays. *The Journal of Physical Chemistry C* 112 (49): 19635–19641.
102. Smith, W., and Y.P. Zhao. 2009. Superior Photocatalytic Performance by Vertically Aligned Core–Shell TiO_2/WO_3 Nanorod Arrays. *Catalysis Communications* 10 (7): 1117–1121.
103. Shand, M., and J.A. Anderson. 2013. Aqueous Phase Photocatalytic Nitrate Destruction Using titania Based Materials: Routes to Enhanced Performance and Prospects for Visible Light Activation. *Catalysis Science & Technology* 3 (4): 879–899.
104. Doudrick, K., T. Yang, K. Hristovski, and P. Westerhoff. 2013. Photocatalytic Nitrate Reduction in Water: Managing the Hole Scavenger and Reaction By-Product Selectivity. *Applied Catalysis, B: Environmental* 136–137: 40–47.
105. S.E. Hunyadi Murph., and G.K. Larsen. 2017. Titania Nanoparticles for Environmental Applications. *Journals of Materials Chemistry.* (*Manuscript in preparation*).
106. Su, J., X. Feng, J.D. Sloppy, L. Guo, and C.A. Grimes. 2011. Vertically Aligned WO_3 Nanowire Arrays Grown Directly on Transparent Conducting Oxide Coated Glass: Synthesis and Photoelectrochemical Properties. *Nano Letters* 11 (1): 203–208.
107. Dupont, P.-H., C. Couteau, D.J. Rogers, F.H. Téhérani, and G. Lérondel. 2010. Waveguiding-assisted random lasing in epitaxial ZnO thin film. *Applied Physics Letters* 97 (26): 261109.
108. Kreibig, U., and M. Vollmer. 1995. *Optical Properties of Metal Clusters* (Vol. 25). Berlin: Springer.
109. Hou, W., Z. Liu, P. Pavaskar, W.H. Hung, and S.B. Cronin. 2011. Plasmonic Enhancement of Photocatalytic Decomposition of Methyl Orange Under Visible Light. *Journal of Catalysis* 277 (2): 149–153.
110. Awazu, K., M. Fujimaki, C. Rockstuhl, J. Tominaga, H. Murakami, Y. Ohki, N. Yoshida, and T. Watanabe. 2008. A Plasmonic Photocatalyst Consisting of Silver Nanoparticles Embedded in Titanium Dioxide. *Journal of the American Chemical Society* 130 (5): 1676–1680.
111. Roy, S.C., O.K. Varghese, M. Paulose, and C.A. Grimes. 2010. Toward Solar Fuels: Photocatalytic Conversion of Carbon Dioxide to Hydrocarbons. *ACS Nano* 4 (3): 1259–1278.
112. Lewis, N.S., G. Crabtree., A.J. Nozik., M.R. Wasielewski., P. Alivisatos., H. Kung., J. Tsao, E. Chandler., W. Walukiewicz., M. Spitler., R. Ellingson., R. Overend., J. Mazer., M. Gress., J. Horwitz., C. Ashton., B. Herndon., L. Shapard., and R.M. Nault. 2005. *Basic Research Needs for Solar Energy Utilization*; DOESC (USDOE Office of Science (SC)): Report of the Basic Energy Sciences Workshop on Solar Energy Utilization, p Medium: ED.
113. Kamat, P.V. 2007. Meeting the Clean Energy Demand: Nanostructure Architectures for Solar Energy Conversion. *The Journal of Physical Chemistry C* 111 (7): 2834–2860.
114. Yan, S., H. Yu, N. Wang, Z. Li, and Z. Zou. 2012. Efficient conversion of CO_2 and H_2O into hydrocarbon fuel over $ZnAl_2O_4$-modified mesoporous ZnGaNO under visible light irradiation. *Chemical Communications* 48 (7): 1048–1050.
115. Li, H., Y. Lei, Y. Huang, Y. Fang, Y. Xu, L. Zhu, and X. Li. 2011. Photocatalytic Reduction of Carbon Dioxide to Methanol by Cu2O/SiC Nanocrystallite Under Visible Light Irradiation. *Journal of Natural Gas Chemistry* 20 (2): 145–150.
116. Halmann, M. 1978. Photoelectrochemical Reduction of Aqueous Carbon Dioxide on p-type Gallium Phosphide in Liquid Junction Solar Cells. *Nature* 275 (5676): 115–116.
117. Varghese, O.K., M. Paulose, T.J. LaTempa, and C.A. Grimes. 2009. High-Rate Solar Photocatalytic Conversion of CO_2 and Water Vapor to Hydrocarbon Fuels. *Nano Letters* 9 (2): 731–737.
118. Hunyadi Murph, S.E., H. Sessions., Y. Kun., and Y.Zhao. 2014. Nanocomposite Photocatalysts, Conversion of CO_2 to Fuel. In *American Chemical Society National Meeting* (Vol. C1 Chemistry), Denver.

119. Narayan, T.C., A. Baldi., A.L. Koh., R. Sinclair., and J.A. Dionne. 2016. Reconstructing Solute-Induced Phase Transformations Within Individual Nanocrystals. *Nature Materials* 15: 768–774.
120. Huang, X., P.K. Jain, I.H. El-Sayed, and M.A. El-Sayed. 2006. Determination of the Minimum Temperature Required for Selective Photothermal Destruction of Cancer Cells with the Use of Immunotargeted Gold Nanoparticles. *Photochemistry and Photobiology* 82 (2): 412–417.
121. Zharov, V.P., K.E. Mercer, E.N. Galitovskaya, and M.S. Smeltzer. 2006. Photothermal Nanotherapeutics and Nanodiagnostics for Selective Killing of Bacteria Targeted with Gold Nanoparticles. *Biophysical Journal* 90 (2): 619–627.
122. Singh, M., S.o. Lara., and S. Tlali. 2016. Effects of Size and Shape on the Specific Heat, Melting Entropy and Enthalpy of Nanomaterials. *Journal of Taibah University for Science.*
123. Liu, Q., J. Tang, Y. Zhang, A. Martinez, S. Wang, S. He, T.J. White, and I.I. Smalyukh. 2014. Shape-Dependent Dispersion and Alignment of Nonaggregating Plasmonic Gold Nanoparticles in Lyotropic and Thermotropic Liquid Crystals. *Physical Review E* 89 (5): 052505.
124. van der Beek, D., A.V. Petukhov, P. Davidson, J. Ferré, J.P. Jamet, H.H. Wensink, G. J. Vroege, W. Bras, and H.N.W. Lekkerkerker. 2006. Magnetic-Field-Induced Orientational Order in the Isotropic Phase of Hard Colloidal Platelets. *Physical Review E* 73 (4): 041402.
125. Ryan, K.M., A. Mastroianni, K.A. Stancil, H. Liu, and A.P. Alivisatos. 2006. Electric-Field-Assisted Assembly of Perpendicularly Oriented Nanorod Superlattices. *Nano Letters* 6 (7): 1479–1482.
126. Caswell, K.K., J.N. Wilson, U.H.F. Bunz, and C.J. Murphy. 2003. Preferential End-to-End Assembly of Gold Nanorods by Biotin—Streptavidin Connectors. *Journal of the American Chemical Society* 125 (46): 13914–13915.
127. Kwaadgras, B.W., M. Dijkstra., and R.v. Roij. 2012. Communication: Bulkiness Versus Anisotropy: The Optimal Shape of Polarizable Brownian Nanoparticles for Alignment in Electric Fields. *The Journal of Chemical Physics* 136 (13): 131102.
128. Maier, S.A., M.L. Brongersma, P.G. Kik, S. Meltzer, A.A.G. Requicha, and H.A. Atwater. 2001. Plasmonics—A Route to Nanoscale Optical Devices. *Advanced Materials* 13 (19): 1501–1505.
129. Maier, S.A., P.G. Kik, H.A. Atwater, S. Meltzer, E. Harel, B.E. Koel, and A.A.G. Requicha. 2003. Local Detection of Electromagnetic Energy Transport Below the Diffraction Limit in Metal Nanoparticle Plasmon Waveguides. *Nature Materials* 2 (4): 229–232.
130. Atwater, H.A., S. Maier, A. Polman, J.A. Dionne, and L. Sweatlock. 2011. The New "p–n Junction": Plasmonics Enables Photonic Access to the Nanoworld. *MRS Bulletin* 30 (5): 385–389.
131. Brzobohatý, O., M. Šiler, J. Trojek, L. Chvátal, V. Karásek, and P. Zemánek. 2015. Non-Spherical Gold Nanoparticles Trapped in Optical Tweezers: Shape Matters. *Optics Express* 23 (7): 8179–8189.
132. Jain, P.K., S. Eustis, and M.A. El-Sayed. 2006. Plasmon Coupling in Nanorod Assemblies: Optical Absorption, Discrete Dipole Approximation Simulation, and Exciton-Coupling Model. *The Journal of Physical Chemistry B* 110 (37): 18243–18253.

Chapter 6
Putting Nanoparticles to Work: Self-propelled Inorganic Micro- and Nanomotors

Kaitlin J. Coopersmith

Abstract The development of nanomotors (nano- and micron sized particles that convert energy into mechanical movement) is an exciting endeavor. Nanomotors have been crafted in an extensive variety of sizes, morphologies and compositions for applications such as drug delivery, cargo transport, sensing, and lithography. Inspired by nature's elegant use of chemical gradients and cellular tracks for independently driven molecular processes, a variety of machines have been created. With the recent bestowment of the Nobel Prize for molecular machines, this concept is being actively pursued to create inorganic nano- and microparticles that independently move for a gamut of applications.

Keywords Anisotropic nanoparticle · Bimetallic nanoparticle · Nanomotor · Micromotor · Propulsion · Autonomous movement · Sensing · Acoustic · Optical · Magnetic · Electrophoresis · Biofuel

6.1 Introduction

As technological creations become smaller, machines that perform work expand into the micro- and nano-sized regimes. With the advancement of nanotechnology, the creation of nanoscopic machinery is becoming a popular endeavor. Molecular machines have been crafted using various molecules that interact with light, sound, and biological entities to create miniscule moving parts that are highly controlled [1–3]. More recently, nanomotors have been fabricated from micro- and nanoparticles rather than molecules [4–10]. This exploits the unique optical and physical properties of micro and nanoparticles that are highly dependent on size, composition, and morphology. Typically, nanomotors do not have moving parts, although, carbon nanotube based machines with moving parts have been created [9].

K.J. Coopersmith (✉)
National Security Directorate, Savannah River National Laboratory, Aiken, SC, USA
e-mail: kaitlin.lawrence@srnl.doe.gov

S.E. Hunyadi Murph et al. (eds.), *Anisotropic and Shape-Selective Nanomaterials*, Nanostructure Science and Technology, DOI 10.1007/978-3-319-59662-4_6

Nanomotors harvest energy from different forms of chemical or physical stimuli, including acoustic, optical, magnetic, and electrical energies and convert it into mechanical work [4–9]. Self-propelled nanoparticles have an array of applications related to medical treatment and diagnostics [4], biological and chemical sensing [7], energy storage [11], lithography [12], mass transportation [13, 14], and chemical degradation [15]. Self-powered nanomachines can cause controlled agitation without external energy, which can be useful in applications such as chemical degradation in ecosystems, where external agitation is not feasible but mixing is highly desirable. Compared to the diffusion controlled counterparts, nanomotors have higher reactivity due to the increased fluid transport [15]. The successful integration of nanomotors requires highly efficient propulsion using fuel at concentrations relevant to the applications. A variety of motors exists but with low efficiencies and high fuel concentration requirements that are not suitable for most biological applications. Propulsion speed is typically expressed in body lengths per second (bl/s), with speeds up to 176 bl/s having been reported for biological motors [16], 75 bl/s for inorganic nanomotors [8], and 375 bl/s for microtube nanomotors [17]. For comparison, a cheetah can only move with a speed of about 32 bl/s [18] and the record for the world's fastest land animal currently belongs to the *Paratarsotomus macropalpis*, a mite found in California that moves with a speed of 192 bl/s [19].

The sophistication found in biological entities have inspired the creation of nanomotors [20, 21]. Cells have an internal structure that is anisotropic and non-homogeneous and their complex makeup leads to directional flow and transport of materials [20]. Various molecular motors inside of the cells carry out the directional processes that overcome Brownian motion using stored chemical energy [20]. Flagella, which are tubular protein appendages, are present on most prokaryotic and some eukaryotic cells and move in response to ion gradients [22–24]. In prokaryotic cells, flagella move in sinusoidal wave-like motions and in eukaryotic cells, they move in whipping motions. Bacterial flagella are driven by a protein rotary engine that is powered by a proton gradient that comes about during the cell's metabolism [20]. Fig. 6.1i shows video micrographs of flagellar movement that resulted from the oscillatory sliding motion in bull sperm flagella in the presence of ATP and Ca^{2+} ions [24]. Transverse movement was attributed to the two sliding mechanisms that occur between adjacent double microtubules [24].

Cellular motor proteins can also move along a track. For example, dynein and kinesin convert energy from adenosine triphosphate (ATP) to movement along cytoskeletal tracks to generate forces and transport cargos (Fig. 6.1ii) [21]. Kinesin moves along a predefined path composed of microtubule (MT) filaments toward the positive end to transport cellular cargo as well as help the cell carry out mitosis and meiosis. Dyneins move along the MT's negative end to transport cellular cargo, position organelles in the cell, and play a role in other important cellular functions [21]. A more sophisticated biological motor is ATP synthase, the enzyme that produces ATP. It is made up of two rotary molecular motors that move in opposite directions depending on cell conditions [20].

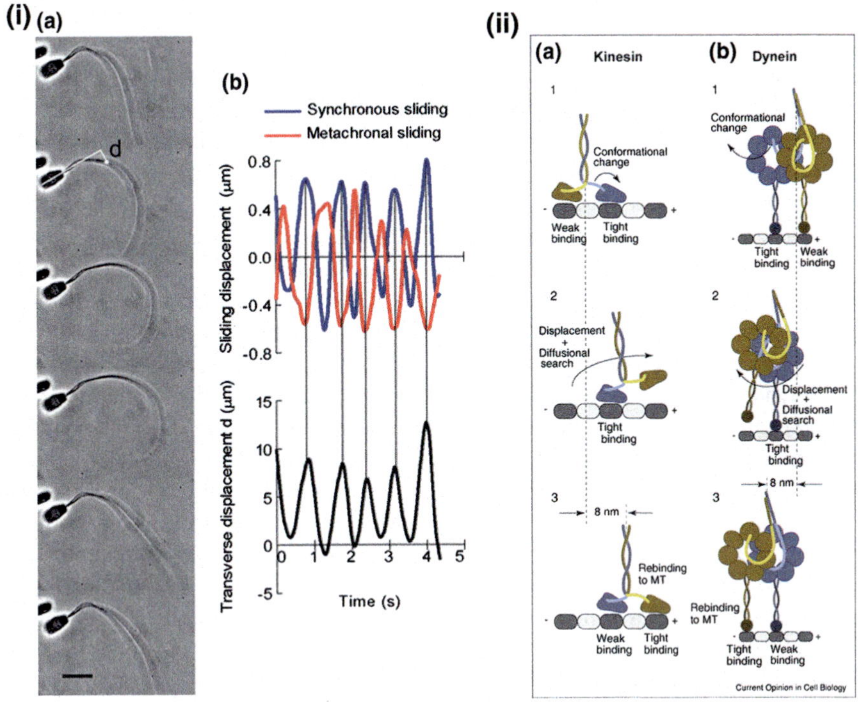

Fig. 6.1 Movement of biological motors: **i** (*a*) Phase contrast micrograph of the effect of sliding displacement on transverse displacement in bull sperm models (time interval between successive images t = 0.15 s; beat frequency of flagellar bending was f = 1.2 Hz); (*b*) profiles of microtubule sliding displacement and of the transverse displacement of the flagellum shown in (*a*). Scale bar = 10 μm. Open access from Ref. [24]. **ii** Models for (*a*) kinesin and (*b*) dynein stepping. Reprinted with permission from Ref. [23]

Nanomotors have been created from organic compounds, biological molecules such as DNA and enzymes, carbon nanoparticles, metals, metal oxides, and semi-conducting materials, both with and without moving parts [4–10]. The altered movement of nanomotors in response to changes in the local environment can be exploited for chemical sensing, environmental remediation, electronic self-repair, among others [4–7]. In medical applications, nanomotors can be used to pick up and deliver chemical payloads or as intracellular gene silencers in genetic therapy [25]. Most of these applications focus on the creation of non-toxic and environmentally friendly nano- and micromotors. The speed, direction of movement, and trajectory is directly related to the concentration of fuel, which can be acoustic, optical, magnetic, electrophoretic/electrocatalytic, chemical, and enzymatic/catalytic. Movement can be varied with the addition of fuel, namely ions or secondary moieties/molecules/components, to the nanomotors. The speed can be dramatically increased or decreased when a combination of fuels are used [7]. For environmental or biological applications, nanomotors are employed in complex sample matrices (e.g. high ionic strength), so high selectivity is imperative.

In liquid suspensions, the solvent's viscosity and the particle's drag force in the solvent have profound effects on motor movement. Locomotion in fluids is typically described by Reynolds number, which is the ratio of inertial forces to viscous forces. Due to their small sizes, the movement of nano- and micromotors takes place at low Reynolds numbers where viscous forces dominate [26]. Since inertial forces do not play a role, motors in this regime must continuously apply force to move. Particle diffusion can be understood using Stokes' Law, which calculates the drag force experienced by a spherical particle with a certain radius in low Reynolds' number regimes. Diffusion coefficients are also size dependent, so it is harder for smaller motors to move in response to a field. Nanomotors have to overcome hydrodynamic forces (e.g. diffusion and viscous drag), Brownian motion and rotation, and unfavorable interactions with the energy source (e.g. magnetic or plasmonic heating) or physical environment. Non-equilibrium forces are required to create directional movement. Thus, careful design of the nanomotor and power source must be carried out to ensure the movement is related to the power source rather than hydrodynamic forces and Brownian motion. The interactions between the motors, such as repulsive or attractive forces or long-range ion gradients, can lead to assemblies and collective interactions that will also affect the movement of the motors [27]. Depending on the morphology, composition, and fuel used, the nano- and micromotors can have oscillatory or linear motion.

This chapter focuses on metallic inorganic nanomotors. These nanomotors are not typically made up of moving parts but rather are created in a way to facilitate movement through their anisotropic structure [7]. The biomimetic capabilities of nanomotors have also been applied to electronic applications, where nanomotors were used to autonomously seek and repair nano-and micro-sized cracks [6]. The ability to repair damage based on changes in the chemical or physical environment can be extended into the medical field, where nanomotors can be applied to help "heal" damage to the body. The ability of nanomotors to change speed and trajectory in the presence of certain chemicals is beneficial in a wide array of applications including chemical sensing and transport.

6.2 Synthetic Nanomotor Design

6.2.1 Synthesis and Characterization

A variety of approaches have been used to fabricate a wide array of nanomotors, including Janus particles, anisotropically shaped nanoparticles, and protein and polymer based particles. The movement of nano and micromotors in solution is typically tracked through optical spectroscopy. Videos and time lapse images are obtained and single particle tracking statistics are used to determine the speed and trajectory of the motors [5, 7, 11]. Temporal particle tracks are used to graphically demonstrate the motor's trajectory [11]. Diffusion coefficients, which are typically on the order of $\sim 1\ \mu m^2/s$ for nanomotors, can be calculated from the slope from mean squared displacement plots or extrapolated from dynamic light scattering (DLS) measurements [28, 29]. Diffusion coefficients are used to quantify the mobility of the motors and are dependent on viscosity. Optical tweezers have also been used to measure the force required for propulsion [29].

The presence of ligands on nanomotor's surfaces will affect the movement and efficiency of the nanomotor, either hindering or enhancing the movement [30]. When ligands hinder the movement, anisotropic rods and wires can be synthesized without surface capping molecules via physical routes such as sequential electrodeposition onto templates [7, 10, 11, 30] or thermal evaporation [5]. In electrodeposition routes, the rod or wire length can be tuned via electrodeposition time where a longer deposition time leads to longer rods or wires. The fast movements of propelled nanomotors may also lead to a decrease in physically adsorbed species that can interfere with the movement or reactivity.

Anisotropy can be appended to particles that may not otherwise be anisotropic, such as spherical particles. For example, Janus nanoparticles have two or more domains with distinct physical or chemical properties. The different domains are typically fabricated by creating a separate domain on a spherical nanoparticle at various types of interfaces, such as air-liquid, liquid-liquid or gas-liquid interfaces. In the most common approach, nanoparticles are deposited onto substrates and the exposed side is coated with a different composition, either through wet chemistry [31], electron beam deposition [29] or altered sputtering processes [15]. In the thermal evaporation technique, single composition nanoparticles are attached to a substrate and metal vapor condenses onto the substrate in a vacuum, leading to the creation of a sphere with two different metallic domains [5]. Other synthetic methods, including the Kirkendall effect, may also be utilized to create anisotropic nanoparticles. For the Kirkendall effect, particles are synthesized from two incompatible metals that phase separate into individual domains. The ideal technique for nanomotor fabrication depends on the nanomotor's requirements.

6.2.2 Efficiency

Nanomotor efficiency is based on the conversion of fuel into mechanical motion. A variety of mechanisms contribute to efficiency loss, including consumption of the motor, depletion of the fuel, and side reactions, such as the decomposition of the fuel via paths that do not lead to movement [10, 11, 30]. Nanomotors should be able to be employed in complex sample matrices, where contaminants may be present. Various routes have been studied to increase the efficiency, including the photochemical decomposition to regenerate the nanomotor and fuel [30]. The efficiency is typically calculated using Eq. 6.1:

$$\eta = \frac{power\ output}{total\ power\ output}, \tag{6.1}$$

where η is the efficiency of the nanomotor as a function of power output (speed and trajectory) and total power input (chemical concentrations, etc.) [30]. The efficiency of nanomotors have been reported on the order of 10^{-9}–10^{-5}, and external field driven motors typically have higher efficiency than chemically driven motors [30].

6.3 Propulsion Routes

Synthetic nanomotors can be powered non-autonomously by external fields, such as magnetic [12, 13, 32–35], acoustic [25, 36–40], and optical [41–44] energies, or autonomously through self-generated chemical gradients [7, 11, 28, 45] or bubble propulsion [4, 15]. For motors powered via external fields, the speed and directionality may be easier to control. In contrast, chemically powered nanomotors do not require direct manipulation but they typically require high concentrations of chemicals, since the instantaneous efficiency depends on the amount of fuel remaining in solution. Chemically powered motors also have fine-tuned sensing capabilities, where movement as a response to changes in the environment can translate into lower detection limits and energy conservation for industrial applications such as water detoxification. The propulsion route and nanomotor design can be tailored depending on the desired application.

6.3.1 External Propulsion

External energy can be transduced into movement via magnetic, acoustic or optical energy, with speed and efficiency related to the strength and power of the energy input. These energy routes are non-contact and highly targeted methods that do not require additional chemical changes to the environment. External manipulation and control also gives one the ability to tune the route and location of the nanomotors with longer motor lifetimes.

6.3.1.1 Acoustic

Acoustic energy has been developed for medical treatment and diagnostics, such as ultrasonography. Due to the minimal adverse side effects on biological systems, the use of acoustic energy to power nanomotors for biologically relevant applications is desired. Nanorods, nanowires, and microrods have been accelerated in megahertz (MHz) acoustic fields without the addition of chemical fuel [25, 37–40]. In these studies, standing waves are used to acoustically confine particles to a levitation plane, where the acoustic pressure is at a minimum (nodal plane). Experimental geometry can also be tailored to create arrays of particles. In the presence of acoustic radiation, spherical particles create linearly and spherically assembled shapes that are difficult to control [38]. Though anisotropic metal rods also displayed linear and spherical assembly, they displayed directional motion, in-plane rotation and pattern formations, as shown in Fig. 6.2 [38]. The directional motion was dependent on their location in the levitation plane and the frequency and the speed of the rods were tuned by changing the amplitude of the acoustic fields [38].

The mechanism that is believed to be responsible for movement in acoustic fields is the acoustic streaming mechanism, where particles oscillate in the plane of the acoustic field and experience a stress that causes them to move [37]. Particle asymmetry causes streaming effects that create a net force along the axis of the rods,

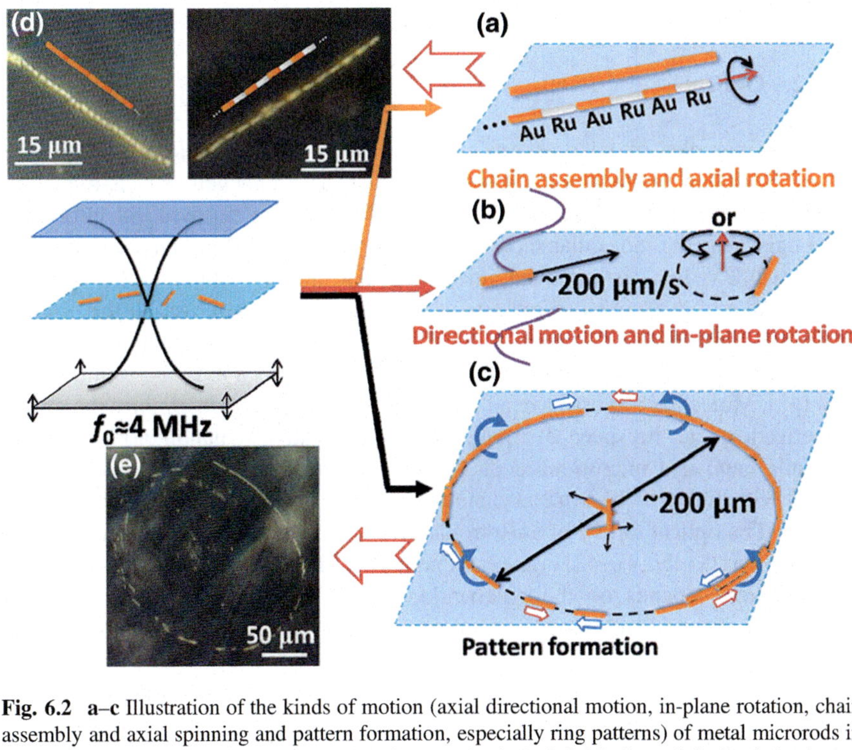

Fig. 6.2 **a–c** Illustration of the kinds of motion (axial directional motion, in-plane rotation, chain assembly and axial spinning and pattern formation, especially ring patterns) of metal microrods in a 3.7 MHz acoustic field. AuRu rods (gold-silvery color in dark field) showed similar behaviour to the Au rods, except that they moved with their Ru ends (the silvery end in the image) forward and aligned head-to-tail into chains. **d** and **e** Dark field images of typical chain structures and ring patterns formed by Au and AuRu rods. Note that the cartoons superimposed on **d** are intended to show the alignment of the rods and are not to scale or in proportion to the aspect ratio of the Au or AuRu rods. Reproduced with permission from Ref. [38] Copyright © 2012 American Chemical Society

leading to directional movement [37, 38]. For bimetallic nanorods, lighter density material led the movement of the nanomotor but when the materials had similar densities, the movement occurred in the direction of the concave end of the rod as opposed to the convex end due to shape asymmetry rather than material asymmetry [37].

Both cells and nanomotors can be manipulated using acoustic fields. When exposed to ultrasonic radiation, cells aggregate at pressure nodes that are perpendicular to the direction of the acoustic waves without damage to the cells or drug loaded nanomotors [46]. Acoustic fields can be used to increase the contact between cells and drug loaded nanomotors that can subsequently deliver their drugs [25, 46]. Enhanced propulsion under the influence of the acoustic field allows nanomotors to pierce through the cell membrane [25]. Drug therapy has been found to be more efficient in the presence of acoustic radiation [25]. More efficient drug therapy leads to lower drug concentration requirements, which can decrease the prevalence of unwanted side effects.

6.3.1.2 Optical

The first examples of nanomachines were created from organic molecules that change based on photochemical processes, such as photochemical cis-trans isomerization and thermal isomerization [47, 48]. When exposed to ultraviolet (UV) light, organic molecular motors rotate in response to the change in chirality. When the UV light is removed, the molecular motors rotate in the opposite direction due to thermal isomerization as the molecule relaxes back to the original structure. Organic molecular motors have been created that can rotate objects that are on the order of 10,000 times larger than their size when exposed to ultraviolet light [48]. Molecular motors have a large electronic effect that determines the rate of isomerization and the speed of rotation [47]. Alternatively, nanomotors based on inorganic nano and micron sized particles have been created from materials that interact with light, such as metal nanoparticles that have a surface plasmon band [41–44]. The optical torque of anisotropic gold nanostructures using light has been studied theoretically [41, 42] and experimentally [44]. Generally, the optical torque and direction depends on the polarization of the light, location of the surface plasmon resonance band, the efficiency of light absorption, solution viscosity, and magnitude of the light torque [41].

Figure 6.3 shows the interactions between AuNRs and circularly polarized light as well as the effect of torque on wavelength. At the surface plasmon resonance band, rotational torque is at a maximum; however, no rotation was observed with linearly polarized light [41]. Nanomotors that rotate in response to light can be used as nano stir bars or as probes measure biological processes or determine the viscosity of nano environments [44].

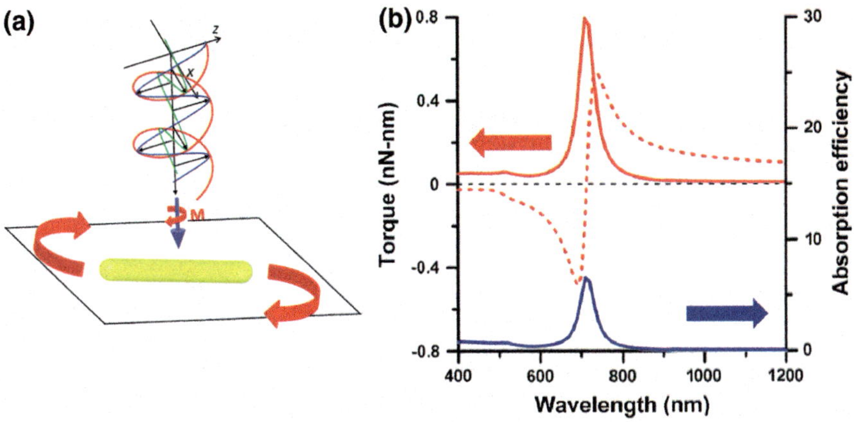

Fig. 6.3 **a** AuNR rotation as a response to circularly polarized light; **b** effect of absorption efficiency on torque for a rotating AuNR. Open access from Ref. [41]

6.3.1.3 Magnetic

Another external actuation mechanism involves the use of magnetic fields [12, 13, 32–35]. Propulsion has been achieved using homogenous rotating magnetic fields and magnetic field gradients, though as the size of the nanomotor decreases, actuation by the former becomes more reliable [12, 33]. Magnetically propelled machines typically contain helical or chiral components that allow them to be steered through a matrix. Independent control over propulsion in a mixture of different nano and micromotors has been achieved in a variety of different ways [12, 33]. In one route, two different nanomotors can be combined that respond to different types of magnetic fields. Alternatively, various nanomotors can be fabricated with different magnetic moments or speed-frequency relationships [12, 33]. For the latter, the ratios between speeds of the different propellers are related to the ratios between the critical frequencies, allowing for predictable motor control [12]. Nanomotors propelled via magnetic fields have been used for assembly, material delivery, and nanomotor lithography. Magnetic manipulation is useful for biological applications because most biological systems can tolerate magnetic field energy. As a result of this benefit, magnetic nanomotors have been used for applications such as the magnetic delivery of drugs using a predetermined route [13], biosensing [49], among others.

6.3.2 *Chemical Propulsion*

Chemically powered motors use stored chemical energy as the power source, leading to a fully autonomous movement without the need for external energy.

In these systems, there is an imbalance in the chemical environment that drives the motion. Chemical mechanisms that have been used to describe nanomotor acceleration in the presence of fuel or ions include underpotential deposition (UPD), where ions are absorbed and reduced on the surface [7], self-generated ion gradients, catalyzed reactions, and gas propulsion as a result of gas release from chemical reactions. The speed and directionality of chemically powered nanomotors depends on the nano-engineered motor as well as the amount of the fuel.

6.3.2.1 Diffusiophoresis

Movement can result from a gradient of uncharged solutes (electrophoretic movement) or a gradient of charged solutes (chemophoretic movement) on motor's surfaces [7, 11, 28, 45]. These mechanisms direct movement in particles by creating a gradient of particles that generates a change in osmotic pressure, leading to fluid flow in the direction of higher concentrations of ions or chemicals [28]. In self-electrophoresis, or the movement in the presence of a self-generated ion gradient, the change in ion concentrations results in movement via electro-osmotic flow through the change in the electrical double layer. Since the movement is dependent on ion concentrations, speed typically decreases with solution conductivity [7]. Nanoparticles that are propelled by self-generated chemical concentration gradients are propelled by self-diffusiophoresis and are effective even in high ionic conditions; however, these nanomotors tend to have slower speeds compared to electrophoretic conditions [27]. The fastest diffusiophoretic mechanisms occur in systems that release protons since the diffusion of protons is an order of magnitude faster than the diffusion of anions [27].

One of the most well studied self-electrophoretic fuel mechanisms is the decomposition of H_2O_2 in the presence of bimetallic rods and wires [7]. In one example, Au–Pt nanowires were used, where the Pt–Ag system acted as an anode–cathode for the decomposition of hydrogen peroxide. This system had an unusual increase in speed with an increase in solution conductivity when the added ion was Ag^+. In the presence of Ag^+ ions, there was a dramatic increase in speed, from 10 $\mu m\ s^{-1}$ to 52 $\mu m\ s^{-1}$, whereas the addition of other metal ions led to a decrease in speed, ranging from 0.3 to 7.1 $\mu m\ s^{-1}$ [7]. This increase in speed was attributed to underpotential deposition (UPD), where the Ag was reduced at the Pt surface of the nanowires, leading to Au–Pt–Ag compositions. Control experiments indicated that the acceleration was due to the addition of the Ag domain. This added domain possibly led to an increased difference in chemical potentials between the anode and the cathode, an increase in the catalytic activity, and an increase in concentration gradient of reaction products around the rod (self-diffusiophoresis).

Although hydrogen peroxide is one of the more well studied fuels, other materials such as halogen liquids have been investigated. Figure 6.4 shows an example of a bimetallic nanowire that moved in response to the addition of Br_2 and

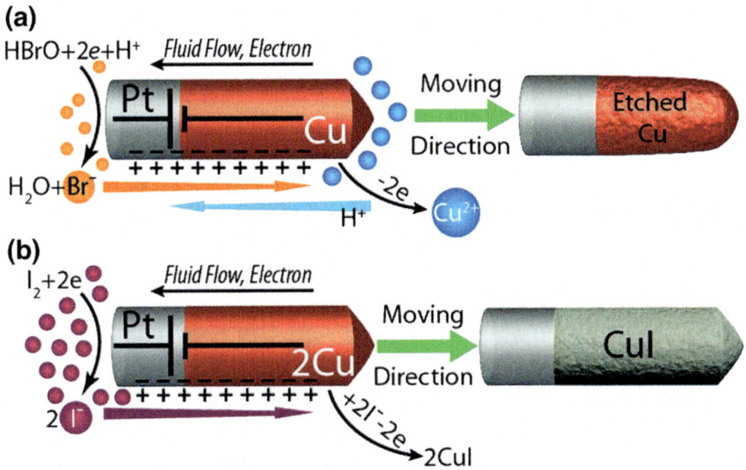

Fig. 6.4 Effect of aqueous solutions of Br_2 (a) and I_2 (b) on Cu morphology in Pt–Cu bimetallic nanowires. Reproduced with permission from Ref. [11]. Copyright © 2012 American Chemical Society

I_2 solutions due to the etching of the Cu domain in Br_2 and the conversion of Cu to CuI in I_2 solutions [11]. The movement of the nanowire is in the opposite direction of the fluid flow. The shorter the Cu segment, the faster the rods moved but the shorter the lifetime due to the consumption of the Cu metal [11].

In this study, asymmetric Cu nanorods were also synthesized into a ratchet shape and introduced into I_2 to create a rotor. The different redox reaction rates at the anisotropic end of the Cu nanorod led to the torque for the rotational movement [11].

Enzyme catalytic reactions are specific and efficient reactions that can occur even in the highly molecularly crowded locations inside of cells [50]. There is a large library of known enzymatic reactions, with over 2000 known enzymes. Since enzymes are found in living cells, they have a high biocompatibility that is beneficial for in vivo cargo delivery, biological sensing and other biomedical applications [29, 50]. In one example, Janus nanoparticles were created and functionalized with three different enzymes as shown in Fig. 6.5 [29]. The enzyme functionalized nanoparticles had directional movement as a result of the chemical gradient from the enzyme catalyzed reactions [29]. An increase in diffusion coefficient was measured with increasing fuel concentrations up to a saturation point (Fig. 6.5c), which was attributed to the Michaelis-Menten enzyme kinetic values and the viscosity of the solution due to increased glucose fuel concentration. Unlike the H_2O_2 fueled motors, these motors used glucose and urea, which are biologically benign [29].

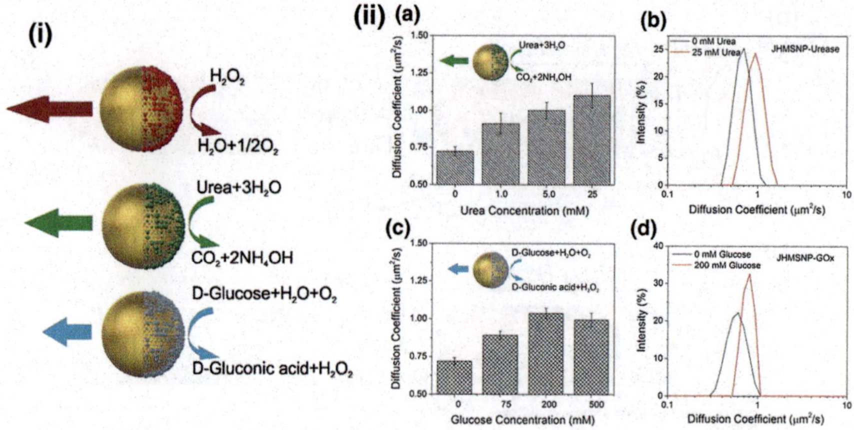

Fig. 6.5 **i** Overview of the different enzymes used to create self-propelled nanomotors. **ii** (*a*) Diffusion coefficient as a function of urea concentration for urease functionalized nanoparticles; (*b*) effect of urea on DLS measured diffusion coefficient; (*c*) diffusion coefficient as a function of glucose concentration for glucose oxidase functionalized nanoparticles. Open access from Ref. [29]

6.3.2.2 Bubble Propulsion

Micro- and nano sized jets have been created that are propelled via bubble recoil (Fig. 6.6). In these systems, bubbles are created due to a chemical reaction on one side of the nanomotor. For example, in the creation of H_2 gas from a metallic reaction with water or acid [15, 51], the hydrogen bubble thrust generated as a result of the reaction led to particle propulsion in the opposite direction. Controlled and gradual dissolution leads to an increase in nanomotor lifetime. This has been accomplished through various routes, including in core/shell architectures, where the presence of a small "hole" in the shell allows gradual exposure of the reactive species [15]. Bubble formation has also been used to create nanomachines that move in response to the ejection of single layer graphene sheets as a result of H_2 release from the sodium-water reaction [51].

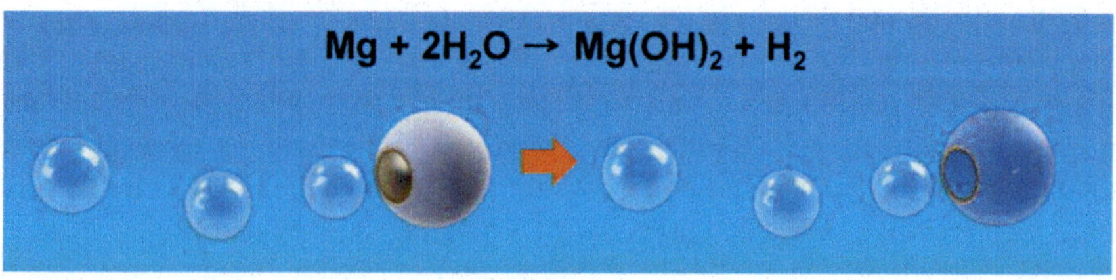

Fig. 6.6 Nanomotors powered via bubble propulsion as a result of the manganese reaction with water. Reproduced with permission from Ref. [15]. Copyright © 2014 American Chemical Society

6.3.3 Multiple Energy Sources

Dual-fuel and hybrid motors have also been realized for increased flexibility, control, and lifetime [5]. In one example, Ag/Mg Janus microparticles could be propelled in two directions based on the fuel; silver was propelled through electrocatalytic reactions with hydrogen peroxide and the magnesium component was propelled through the water–magnesium reaction with sodium bicarbonate as fuel. These nanomotors moved with an average velocity of 90 $\mu m\ s^{-1}$ in sodium bicarbonate (~3.6 bl/s) and 67 $\mu m\ s^{-1}$ in hydrogen peroxide (~2.7 bl/s) and traveled for 15 and 30 min in sodium bicarbonate and hydrogen peroxide, respectively, before consuming all of the fuel [5]. When exposed to *E. coli* with and without fuel, these motors killed 90% of the bacteria when the fuel was added, compared to 10% for the static motors. The more efficient antibacterial properties of the moving nanomotors was a result of the increased movement of the Janus particles in solution, allowing more bacteria to be exposed to the antimicrobial silver portion. These findings can lead to more efficient antimicrobial delivery and a decrease in antibiotic resistance.

Nanomotors that are powered by two different energy routes, also known as hybrid nanomotors, have been created [34–36]. These can be used to overcome the limitations of certain fuel sources or to reverse motion [36]. In one example, nanowires were created with Ag, Ni and Pt segments for catalytic and magnetic powered motors [34]. Individually, the motors experienced similar speeds in the presence of either catalytic fuel or magnetic energy; however, in the presence of both, the speeds were slower due to the increased fluid drag resistance from opposing propulsion forces [34].

Figure 6.7 shows the design of a hybrid nanomotor that can be powered using either acoustic or magnetic energy [35]. In these nanomotors, the concave Au nanorod segment responded to acoustic fields and the Ni coated Pd segment responded to a magnetic field. The magnetic propulsion was achieved using a rotating magnetic field with speeds of 12.2 μm/s and the acoustic propulsion, under an ultrasound field, led to speeds of 16.8 μm/s [35]. Creating nanomotors that can be powered using two different routes offers a new degree of flexibility for their use in dynamic environments [35].

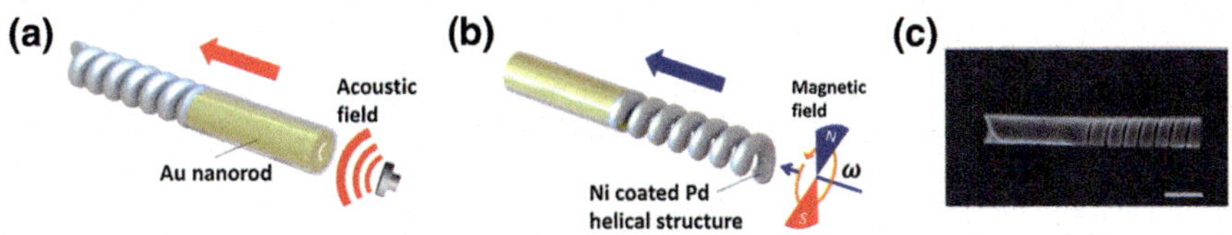

Fig. 6.7 Hybrid nanomotor that can be powered by **a** acoustic energy, which is absorbed by the Au nanorod segment, or **b** magnetic energy, which is absorbed by the Ni coated Pd segment; **c** SEM image of the magneto-acoustic nanomotor (scale bar = 500 nm). Reproduced with permission from Ref. [35]. Copyright © 2015 American Chemical Society

6.3.4 Conclusions and Future Outlook

The employment of micro- and nanomotors for medical treatment and diagnostics, chemotaxis, electronic repair, and environmental remediation has been realized. A wide array of machines has been created and their interactions with various chemical and external fields have been studied. While each propulsion system has their benefits and detriments, careful pairing of particle size, morphology and composition with the input energy route can lead to machines that are highly efficient for the job they are employed to do. It is hopeful that in the future, nanomotors are created with an enhanced ability to work in complex sample matrices and ability to move in biologically and environmentally relevant fuel concentrations.

References

1. Badjić, J.D., et al. 2004. A molecular elevator. *Science* 303: 1845–1849.
2. Bruns, C.J., and J.F. Stoddart. 2014. Rotaxane-based molecular muscles. *Accounts of Chemical Research* 47: 2186–2199.
3. Balzani, V., et al. 2006. Autonomous artificial nanomotor powered by sunlight. *Proceedings of the National Academy of Sciences* 103 (5): 1178–1183.
4. Duan, W., et al. 2015. Synthetic nano- and micromachines in analytical chemistry: Sensing, migration, capture, delivery and separation. *Annual Review of Analytical Chemistry* 8: 311–333.
5. Ge, Y., et al. 2016. Dual-fuel-driven bactericidal micromotor. *Nano-Micro Letters* 8 (2): 157–164.
6. Li, J., et al. 2015. Self-propelled nanomotors autonomously seek and repair cracks. *Nano Letters* 15: 7077–7085.
7. Kagan, D., P. Calvo-Marzal, S. Balasubramanian, S. Sattayasamitsathit, K.M. Manesh, G. Flechsig, and J. Wang. 2009. Chemical sensing based on catalytic nanomotors: Motion-based detection of trace silver. *Journal of the American Chemical Society* 131: 12082–12083.
8. Demirok, U.K., et al. 2008. Ultrafast catalytic alloy nanomotors. *Angewandte Chemie International Edition* 47 (48): 9349–9351.
9. Cai, K., et al. 2016. A method for measuring rotation of a thermal carbon nanomotor using centrifugal effect. *Scientific Reports* 6: 27338.
10. Wang, W., et al. 2013. Understanding the efficiency of autonomous nano- and microscale motors. *Journal of the American Chemical Society* 135: 10557–10565.
11. Liu, R., and A. Sen. 2011. Autonomous nanomotor based on copper-platinum segmented nanobattery. *Journal of the American Chemical Society* 133: 20064–20067.
12. Vach, P.J., S. Klumpp, and D. Faivre. 2016. Steering magnetic microproplers along independent trajectories. *Journal of Physics D: Applied Physics* 49: 065003.
13. Gao, W., et al. 2012. Cargo-towing fuel-free magnetic nanowsimmers for targeted drug delivery. *Small* 8 (3): 460–467.
14. Chen, J., et al. 2015. Impeded mass transportation due to defects in thermally driven nanotube nanmotor. *Journal of Physical Chemistry C* 119: 17362–17368.
15. Li, J., et al. 2014. Water-driven micromotors for rapid photocatalytic degradation of biological and chemical warfare agents. *ACS Nano* 8: 11118–11125.

16. Abdelmohsen, L.K.E.A., M. Nijemeisland, G.M. Pawar, G.A. Janssen, R.J.M. Nolte, J.C.M. van Hest, and D.A. Wilson. 2016. Dynamic loading and unloading of proteins in polymeric stomatocytes: Formation of an enzyme-loaded supramolecular nanomotor. *ACS Nano* 10: 2652–2660.
17. Gao, W., S. Sattayasamitsathit, and J. Wang. 2012. Catalytically propelled micro-/nanomotors: How fast can they move? *The Chemical Record* 12 (1): 224–231.
18. Young, L.E. 2009. Equine athletes, the equine athlete's heart and racing success. *Experimental Physiology* 88 (5): 659–663.
19. Rubin, S., M.H. Young, J.C. Wright, D.L. Whitaker, and A.N. Ahn. 2016. Exceptional running and turning performance in a mite. *Journal of Experimental Biology* 219: 676–685.
20. Bustamante, C., D. Keller, and G. Oster. 2001. The physics of molecular motors. *Accounts of Chemical Research* 34: 412–420.
21. Roberts, A.J., et al. 2013. Functions and mechanisms of dynein motor proteins. *Nature Reviews Molecular Cell Biology* 14: 713–726.
22. Bayly, P.V., and S.K. Dutcher. 2016. Steady dynein forces induce flutter instability and propagating waves in mathematical models of flagella. *Journal of the Royal Society, Interface* 13 (123): 20160523.
23. Gennerich, A., and R.D. Vale. 2009. Walking the walk: How kinesin and dynein coordinate their steps. *Current Opinion in Cell Biology* 21: 59–67.
24. Ishijima, S. 2016. Self-sustained oscillatory sliding movement of doublet microtubules and flagellar bend formation. *PLoS ONE* 11 (2): e0148880.
25. Esteban-Fernández de Ávila, B., et al. 2016. *Acoustically propelled nanomotors for intracellular siRNA delivery. ACS Nano*. **10**: p 4997–5005.
26. Purcell, E.M. 1977. Life at low reynolds number. *American Journal of Physics* 45: 3–11.
27. Wang, W., et al. 2015. From one to many: Dynamic assembly and collective behavior of self-propelled colloidal motors. *Accounts of Chemical Research* 48: 1938–1946.
28. Pavlick, R.A., et al. 2011. A polymerization-powered motor. *Angewandte Chemie International Edition* 50 (40): 9374–9377.
29. Ma, X., et al. 2015. Enzyme-powered hollow mesoporous janus nanomotors. *Nano Letters* 15: 7043–7050.
30. Wong, F., and A. Sen. 2016. Progress toward light-harvesting self-electrophoretic motors: Highly efficient bimetallic nanomotors and micropumps in halogen media. *ACS Nano* 10: 7172–7179.
31. Perro, A., et al. 2009. Production of large quantities of "Janus" nanoparticles using wax-in-water emulsions. *Colloids and Surfaces A* 332 (1): 57–62.
32. Walker, D., et al. 2015. Optimal length of low reynolds number nanopropellers. *Nano Letters* 15 (7): 4412–4416.
33. Mandal, P., V. Choptra, and A. Ghosh. 2015. Independent positioning of magnetic nanomotors. *ACS Nano* 9 (5): 4717–4725.
34. Gao, W., et al. 2011. Hybrid nanomotor: A catalytically/magnetically powered adaptive nanowire swimmer. *Small* 7 (14): 2047–2051.
35. Li, J., et al. 2015. Magneto-acoustic hybrid nanomotor. *Nano Letters* 15: 4814–4821.
36. Wang, W., et al. 2015. A tale of two forces: Simultaneous chemical and acoustic propulsion of bimetallic micromotors. *Chemical Communications* 51: 1020–1023.
37. Ahmed, S., et al. 2016. Density and shape effects in the acoustic propulsion of bimetallic nanorod motors. *ACS Nano* 10: 4763–4769.
38. Wang, W., et al. 2012. Autonomous motion of metallic microrods propelled by ultrasound. *ACS Nano* 6: 6122–6132.
39. Nadal, F., and E. Lauga. 2014. Asymmetric steady streaming as a mechanism for acoustic propulsion of rigid bodies. *Physics of Fluids* 26: 082001.
40. Rao, K.J., et al. 2015. A force to be reckoned with: A review of synthetic microswimmers powered by ultrasound. *Small* 11 (24): 2836–2846.
41. Liaw, J., Y. Chen, and M. Kuo. 2014. Rotating Au nanorod and nanowire driven by circularly polarized light. *Optics Express* 22 (21): 26005–26015.

42. Liaw, J., Y. Chen, and M. Kuo. 2016. Spinning gold nanoparticles driven by circularly polarized light. *Journal of Quantitative Spectroscopy and Radiative Transfer* 175: 46–53.
43. Guix, M., C.C. Mayorga-Martinez, and A. Merkoçi. 2014. Nano/micromotors in (bio) chemical science applications. *Chemical Reviews* 114: 6285–6322.
44. Bonin, K.D., B. Kourmanov, and T.G. Walker. 2002. Light torque nanocontrol, nanomotors and nanorockets. *Optics Express* 10 (19): 984–989.
45. Sundararajan, S., et al. 2008. Catalytic motors for transport of colloidal cargo. *Nano Letters* 8: 1271–1276.
46. Lee, Y., and Z. Wu. 2015. Enhancing macrophage drug delivery efficiency via co-localization of cells and drug-loaded microcarriers in a 3D resonant ultrasound field. *PLoS ONE* 10 (8): e0135321.
47. Pijper, D., et al. 2005. Acceleration of a nanomotor: Electronic control of the rotary speed of a light-driven molecular rotor. *Journal of the American Chemical Society* 127 (50): 17612–17613.
48. Eelkema, R., et al. 2006. Nanomotor rotates microscale objects. *Nature* 440: 163.
49. Chałupniak, A., E. Morales-Narváez, and A. Merkoçi. 2015. Micro and nanomotors in diagnostics. *Advanced Drug Delivery Reviews* 95: 104–116.
50. Küchler, A., et al. 2016. Enzymatic reactions in confined environments. *Nature Nanotechnology* 11: 409–420.
51. Akhavan, O., M. Saadati, and M. Jannesari. 2016. Graphene jet nanomotors in remote controllable self-propulsion swimmers in pure water. *Nano Letters* 15: 5619–5630.

Chapter 7
Prospects for Rational Control of Nanocrystal Shape Through Successive Ionic Layer Adsorption and Reaction (SILAR) and Related Approaches

Andrew B. Greytak, Rui Tan and Stephen K. Roberts

Abstract This chapter describes the use of colloidal successive ionic layer addition as an additive growth method to form inorganic colloidal nanocrystals with controlled shapes. Rational design of nanocrystal dimensions and layer thicknesses in nanocrystal heterostructures is important to many current and anticipated uses of nanocrystals in optoelectronics, energy conversion, and fluorescence imaging. One approach to shape control is the use of a series of self-limiting surface reactions to build up crystals one atomic layer at a time. This approach is especially applicable to nanocrystals made of binary ionic or polar-covalent crystalline compounds, as are found in colloidal quantum dots (QDs). The intrinsic symmetry present in colloidal nanocrystal nuclei can be suppressed to enforce conformal layer growth or harnessed to promote regioselective growth. We specifically discuss two families of methods that are colloidal analogues of vapor-source atomic layer deposition. In colloidal successive ionic layer adsorption and reaction (colloidal SILAR), reagents are introduced to a nanocrystal solution in metered doses corresponding to the total surface area. In colloidal atomic layer deposition (colloidal ALD), reagents are added in excess, and unreacted reagent is subsequently separated and removed. An extensive literature exists on the use of colloidal SILAR to form nominally isotropic core/shell quantum dots (QDs) with high photoluminescence quantum yield. Colloidal ALD has been introduced more recently. There is increasing interest in applying both methods to the formation of anisotropic nanocrystal heterostructures, both through deposition of conformal layers on anisotropic substrates and through controlled anisotropic growth. We review the historical development of these methods, common precursors, and recent developments in monitoring of reaction progress and mechanisms. We also present contemporary examples of isotropic and anisotropic growth, and prospects for future development in the context of several representative applications including cell membrane voltage measurements and fluorescence anisotropy.

A.B. Greytak (✉) · R. Tan · S.K. Roberts
Department of Chemistry and Biochemistry, University of South Carolina, 631 Sumter Street, 29208 Columbia, SC, USA
e-mail: greytak@mailbox.sc.edu

S.E. Hunyadi Murph et al. (eds.), *Anisotropic and Shape-Selective Nanomaterials*, Nanostructure Science and Technology, DOI 10.1007/978-3-319-59662-4_7

7.1 Overview

When the sizes of semiconductor and metal nanoparticles are reduced below characteristic length scales for physical processes in these materials, the properties begin to diverge from the bulk, resulting in a host of size-dependent phenomena that can be used to tune the properties to advantage in applications. However, the size-dependence means that a great deal of control must be exercised over size and shape in order to have narrowly distributed properties among particles in the ensemble. A particularly interesting challenge is the definition of anisotropic particles in template-free colloidal syntheses. Such anisotropic particles are potentially useful for measuring or imposing directional physical conditions such as static electric fields and absorption and emission of polarized light in order to determine orientations in biological samples.

Shape control in inorganic nanocrystals can be achieved via surface interactions and crystal symmetry in colloidal growth. These factors do not require a pre-existing template and are therefore highly scalable and can exert control at the nanometer and sub-nanometer level. We focus on the case of nanocrystals composed of binary semiconducting compounds that have shown considerable promise as colloidal quantum dots. Growth of these crystals can lead to anisotropic shapes based on interfacial energies and/or growth kinetics in epitaxial growth as well as due to relaxation of lattice strain in formation of heterostructures. Decades of research have identified solvent conditions, ligands, and temperatures under which colloidal nanocrystals of common compound semiconductors preferentially grow into quasi-spherical nanocrystals, nanorods, nanoplatelets, and a variety of other shapes. However, these conditions do not necessarily allow critical dimensions such as length and width to be defined independently or for conformal coatings to be deposited on anisotropic cores. Additionally, to achieve samples with well-defined properties, it is necessary to eliminate sources of structural inhomogeneity, including the nucleation of undesired particles due to cross-reaction of precursors in solution.

Colloidal successive ionic layer addition encompasses a family of processes that are designed to improve the control of surface growth in colloidal nanocrystals by using a series of self-limiting surface reactions that each supply one of the complementary ions. These processes are analogous to atomic layer deposition (ALD) [1–4] and successive ion layer adsorption and reaction (SILAR) [5–8] techniques for conformal thin film growth. To date, such processes have primarily been used for the formation of quasi-spherical core/shell quantum dots. Though successful in this objective, the potential value of colloidal successive ionic layer addition in the synthesis of anisotropic nanocrystal shapes with precisely programmed dimensions is just beginning to be recognized.

This chapter begins by describing the significance of shape control to spectroscopy and applications of semiconductor nanocrystals, and processes that lead to anisotropic growth in representative colloidal materials. Successive ionic layer addition approaches to nanocrystal growth are then introduced, including assessments of the effectiveness of such methods in enforcing conformal and isotropic growth of core/shell heterostructures. By combining successive ionic layer addition with kinetically- or thermodynamically-controlled continuous growth steps, it is possible to form selected classes of anisotropic nanocrystal heterostructures. We then discuss the prospects for adapting successive ionic layer addition methods to form anisotropic structures directly via sub-monolayer addition cycles, and some emerging application areas where such shape control is likely to be especially useful.

7.2 Influence of Shape on Electronic Properties of Colloidal Nanocrystals

Colloidal semiconductor nanocrystals, which can constitute one type of quantum dot [9, 10], have been widely explored over the past 30 years and a number of good reviews are available [11–13]. They have been studied as the basis for photocatalysts [14], components of solution-processable solar cells [12, 15] and photodetectors [16], as emitters for lighting and display applications [17–19], and as fluorescent probes and sensors for biomedical applications [20–28]. Many, but not all, of the foregoing applications make use of the unique electronic structure of semiconductor nanocrystals; namely, delocalized band-edge states whose energies are subject to quantum confinement imposed by the boundaries of the crystal. As a result of their electronic structure, quantum dots exhibit size-tunable absorption and emission spectra, large molar extinction coefficients [29–31] and 2-photon cross sections [21, 32], and high photostability compared to most molecular fluorophores [33].

In what follows, the term nanocrystal (NC) will be used to refer to nanocrystals in general and quantum dot (QD) to refer specifically to those whose relevant properties can be reasonably understood from spherical-well excited states, such as are typically employed as fluorophores; the terms are in some cases interchangeable. Nanocrystal samples are typically inhomogeneous ensembles, formed via kinetically-controlled, high-order reactions [34, 35]. In well-represented cases such as CdSe, advances in NC growth have permitted the formation of samples with good crystallinity and narrow size distributions, as well as core/shell heterostructures that achieve high quantum yields by enclosing the core within a semiconductor having a larger bandgap [36–41]. The ensemble linewidths of the lowest-energy electronic transition (LEET, or band-edge exciton) in QD samples can approach the single-particle limit at room temperature [42]. Advances in QD surface chemistry have enabled the formation of biocompatible nanoparticles with low non-specific binding, high brightness, and the ability to attach additional molecular components in a modular fashion to design interactions as sensors and labels [43–48].

Despite many successes in the use of QDs in fluorescence applications, it has been challenging to achieve homogenous properties among QDs in a sample and to reconcile single particle results with ensemble properties. For example, in samples displaying less than unity photoluminescence quantum yield (QY), it is important to understand the origin of non-radiative decay and whether it is a homogenous property of the sample or is associated with a subset of particles. Fluorescence intermittency, or blinking, is ubiquitous among QDs, but structural variations among particles also play a key role. It has long been known that one contribution to diminished QY is the homogenous nucleation of particles of the shell material during procedures intended to form core/shell heterostructures [49]. These shell particles will absorb incident light at energies above their effective bandgaps, but will not transfer the excitation efficiently to the intended product QDs. Lattice mismatch between core and shell can pose difficulties in achieving conformal, epitaxial shell growth. Indeed, recent experiments that permit registry of single-particle fluorescence microscopy images to high-resolution scanning transmission electron microscopy (scanning TEM, STEM) have shown that many QDs that show poor QY also display inhomogeneous or incomplete shell coverage [50].

Alternating layer approaches to core/shell QDs were introduced by Peng and co-workers with the goal of suppressing homogenous nucleation and enforcing conformal shell growth [51]. The techniques have since been widely adopted in many variations, and are now a standard preparative approach to core/shell QDs. More recent studies have investigated the intermediate stages of these reactions to determine the conditions under which the mechanistic goals of the approach can be met. Sections 7.4–7.7 of this chapter review alternating layer methods designed to form approximately spherical core/shell QDs, along with several references to shell growth procedures under simultaneous reagent addition for comparison.

In bioimaging, QDs have been employed as isotropic fluorophores in labeling, tracking [52], biodistribution [53], and energy transfer sensing applications [24]. Yet, there is a crowded field of small-molecule or genetically-encoded fluorophores and alternative luminescent nanoparticles [54] that are suitable for many applications. As such, there may be a limited set of cases in which the advantages of QDs outweigh drawbacks such as blinking and incorporation of toxic elements.

The success of colloidal QDs in fluorescence applications can be extended to achieve an even greater impact by research to alleviate these concerns, but also by harnessing the unique electronic properties afforded by NCs, including NCs with anisotropic shapes. NCs can be synthesized that display significant fluorescence anisotropy due to dielectric confinement and/or exciton fine structure [55–58], or that display significant responses to applied electric fields at room temperature [59]; such NCs offer a way to probe anisotropic biological processes and structures if the crystal orientation in the biological environment can be controlled. The application of semiconductor NCs as probes of membrane potential for neuroscience studies in particular has been considered for several years [60, 61] and recently discussed theoretically [62], but to realize this goal will require advances in the ability to form anisotropically-functionalized NCs.

7.3 Mechanisms of Anisotropic Growth and Erosion

Figure 7.1 illustrates several means by which continued growth of single-crystal colloidal nanoparticles may lead to anisotropic structures. Firstly, the surfaces of NCs necessarily include regions that do not share the same symmetry with respect to the crystal lattice and are therefore structurally distinct [11, 63–65]. Thermodynamic effects, such as a lower surface free energy, may favor the preponderance of low-index facets with greater coordination numbers, while these structurally distinct regions (facets, edges, corners) may naturally possess different reaction rates toward ligand exchange reactions or further crystal growth [66]. In this sense colloidal nanocrystals are no different than larger crystals, except that surface coating more strongly affects the relative stability of the system and the exchange kinetics at different sites. If growth is allowed to continue towards thermodynamic equilibrium with dissolved species, and/or if growth is conducted under conditions in which large differences in growth rates among crystal orientations exist, the resulting NC shapes will be anisotropic, and the anisotropies that are achieved will be derived from the symmetry of the crystal lattice.

Among the most common compound semiconductor NC lattice structures are cubic rocksalt (e.g. PbS, PbSe) and zincblende (e.g. InAs, InP, CdSe) . The wurtzite structure is also widely found (e.g. CdS, CdSe, ZnO) and is an example of a hexagonal lattice. A range of nanocrystal shapes can be derived from each of these structures without reducing their intrinsic crystal symmetries.

Zinc blende and rock salt can each be thought of as a pair of interpenetrating face-centered cubic lattices. In the zincblende structure, the atoms of the second lattice occupy a set of tetrahedral voids in the first; each atom has a tetrahedral bonding geometry, exactly as in diamond but with each element in the compound occupying half of the sites. In a zincblende crystal, six {100} facets may be terminated by cation or anion layers, but are all symmetrically equivalent. There are eight possible {111} facets, but {111} facets on opposite sides of the crystal are inequivalent. If the {111} facets are terminated so that each surfaceatom has only one dangling bond, then one of the opposing {111} surfaces will be cation-terminated (commonly defined as the A face, pointing in the [111] direction), while the other will be anion-terminated (commonly defined as the B face, pointing in the [1′1′1′] direction). This creates a dipole that, in a symmetric crystal, is canceled out by equivalent dipoles in the other three tetrahedrally equivalent directions, much as in a molecular compound such as CF_4. The {110} faces are charge-neutral, having equal numbers of cations and anions in each layer. As a result of these considerations, zincblende nanocrystals commonly appear as cubes, octahedrons, truncated octahedrons, or tetrahedral shapes.

In the rocksalt structure, the atoms of the second lattice occupy the octahedral voids in the first; each atom has an octahedral bonding geometry. In rocksalt crystals, the {100} and {110} faces are charge-neutral, while all eight {111} faces are symmetrically equivalent, with alternating layers of cations and anions.

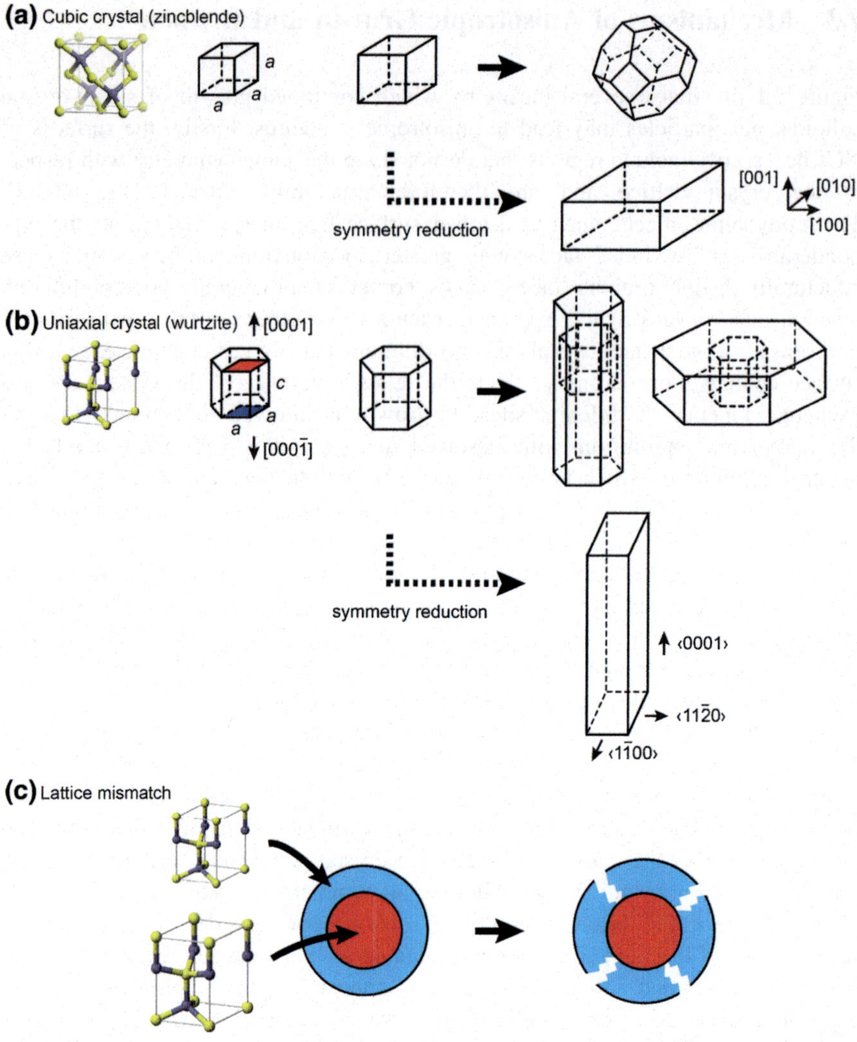

Fig. 7.1 Factors driving anisotropic growth in nanocrystals. **a** Growth of a *cubic structure*, such as zincblende. Preference for low-index facets may favor non-spherical shapes that share the symmetry of the lattice. Breaking the symmetry between crystallographically equivalent facets can lead to 2D structures with *square* or *rectangular shapes*. **b** The wurtzite crystal structure is hexagonal and lacks inversion symmetry. *Red* and *blue* patches represent dissimilar *c* faces of the unit cell; they are chemically distinct and exhibit different growth rates. Reduction in symmetry between equivalent sidewall facets can lead to 2D structures with *rectangular shapes* and distinct edge facets. **c** In core/shell heterostructures, lattice strain can be relieved by formation of defect structures leading to inhomogeneous or incomplete coverage of the core

As a result, cubes, octahedrons, and truncated octahedrons can be formed without loss of symmetry. Note that charge-neutral {111} faces can only be achieved with a partial monolayer at the surface.

The wurtzite structure is uniaxial. It is derived from a hexagonal close packed cation lattice, with anions occupying tetrahedral sites such that each atom has a tetrahedral bonding geometry, but with an eclipsed arrangement for successive pairs of atoms along the hexagonal axis, rather than the staggered arrangement found in zincblende and diamond. It does not possess inversion symmetry about the axis of hexagonal symmetry. As such, the two hexagonal facets ({0001}, normal to the *c* axis) are chemically distinct from each other and from the orthogonal surfaces. If the {0001} facets are terminated such that each atom has only one dangling bond, one face will be cation terminated (commonly used to define the [0001] direction, homologous to the ZB A face), while face in the opposing [0001′] direction will be anion-terminated (homologous to the ZB B face). Low-index sidewalls include the six {112′0} facets, and the six {11′00} facets rotated by 30° from these, all of which present charge-neutral monolayers. As a result of these considerations, when wurtzite NCs are extended through growth (or eroded through etching) the process can be characterized by at least three distinct growth rates (two axial, plus at least one orthogonal to the axis). Nanorods with tips of differing lengths, as well as hexagonal platelets, have been observed.

In each of these structures, reductions in symmetry have also been observed that are attributed to strong ligand interactions. Examples include the formation of quasi-2D nanoplatelets featuring extended (001) and (001′) facets in the case of zincblende, and nanoplatelets (or "nanoribbons") with {112′0} facets with rectangular symmetry in the case of wurtzite. The monolayer structure orthogonal to the <111> direction in zincblende is identical to that orthogonal to the <0001> direction in wurtzite; only the stacking sequence varies. As such, stacking faults and polytypism among wurtzite and zincblende are common and this has been harnessed to form complex structures such as tetrapods with zincblende cores and wurtzite arms extending in each of four equivalent <111> directions. These nanocrystal shapes strongly influence properties through quantum confinement and dielectric contrast.

Nanocrystal heterostructures can be formed by heterogeneous nucleation of a shell material on pre-existing cores (sometimes referred to as seeds). The growth of such shells may be regioselective based on the underlying anisotropy of the crystal structure as described above. For example, for wurtzite NC heterostructures, dot-in-rod and dot-in-plate morphologies are seen. Additionally, strain due to lattice mismatch can interfere with the ability to form epitaxial heterostructures. Above a critical thickness, the strain energy exceeds the cost of forming defects and/or increasing surface area, such that roughened and discontinuous layers are formed (Fig. 7.1c) [67]. Such islanding is well-known in thin-film growth, and tolerances are often more lenient in nanoscale junctions than in planar interfaces, but there can be a tradeoff between lattice mismatch and the desire to form heterojunctions with large band offsets to confine the band-edge wavefunctions. In some cases, strain can be used to repeatably obtain branched shapes such as tetrapods [68], but strain-driven anisotropy can also be accompanied by considerable variation in shapes among NCs in ensemble samples.

The above processes provide an avenue to template-free formation of anisotropic colloidal nanocrystals, which can sometimes be highly desirable. The formation of CdSe/CdS rod/tetrapod core/shell structures [69–74], $CdS_{1-x}Se_x$ nanorods with axial anisotropy [75–77], multi-component nanobarbells [78], and CdSe/CdS core/shell nanoplatelets [79] has led to NCs displaying highly polarized excited states [80–82], high sensitivity to non-isotropic external stresses [72], multiexcitonic dual emission [83, 84], and the capability to engineer charge and energy flows that are valuable in applications [70, 85, 86]. We emphasize that while nanocrystals with the wurtzite and cubic lattices have been widely studied, nanocrystal shape control via crystal symmetry is of course not limited to these materials: for example, 2D growth in transition metal sulfide colloidal nanocrystals has recently been reported [87].

In other cases, these tendencies toward anisotropic growth are detrimental. For example, a conformal and isotropic core/shell heterostructure can increase the photoluminescence (PL) performance of QDs by isolating the QD core from the surface, chemically and electronically. Quasi-spherical CdSe/CdS or CdSe/ZnS core/shell QDs [37, 41, 88–91] display much higher PL QY, [40, 92–94] excitation rate (absorption cross-section) [95], and photo- and chemical stability than QDs of homogeneous composition ("core-only" QDs) [33, 96]. The alternating layer addition processes that are the focus of this chapter have largely been applied in an effort to enforce isotropic growth and suppress homogenous nucleation in forming core/shell heterostructures, by suppressing the pathways to anisotropic growth outlined in Fig. 7.1.

In many cases the selection between isotropic and anisotropic core/shell growth is made by adjusting the solvent/surfactant and temperature conditions to achieve a desirable ratio between the growth rates along the different crystal axes [97]. Some examples of specific conditions for anisotropic growth of chalcogenide nanocrystals are mentioned in Sect. 7.10. This leads to only a limited degree of control as the length and diameter of resulting particles cannot be independently selected. Additional control can be achieved through etching following growth [86]. Nonetheless, the development of a controllable method for growing shells with different morphology remains a critical goal. In addition, an understanding of the shell growth mechanisms at play in existing procedures should aid in the design of a new generation of NC heterostructures.

This chapter will describe nanocrystal surface chemistry pertinent to shape control and alternating layer growth methods; several good reviews are available for readers interested in a more general comprehensive overview [98], including measurements of ligand binding and the effect on photophysics [99], and structural aspects of ligand binding [100].

7.4 Enforcing Isotropic Growth with Alternating Layer Approaches

Successive Ionic Layer Adsorption and Reaction (SILAR) is a solution-phase technique analogous to atomic layer deposition that can be used to grow binary films from solution, in which addition of each ion type is nominally self-limiting (Fig. 7.2). The goals of such an approach are to: (1) saturate available surface binding sites in each half-cycle in order to enforce conformal growth; and (2) avoid the simultaneous presence of both precursors in the solution or vapor so as to prevent uncontrolled surface growth or homogeneous nucleation of film material. While SILAR was originally developed for use on planar substrates [5, 6], over the last 14 years alternating-layer methods have been extensively applied to growth of shells on colloidal NCs.

In discussing alternating layer addition approaches for colloidal particles, it is important to make a distinction between the process and the mechanism that is achieved by it. In typical thin film growth by SILAR (or ALD), an excess of reagent is used in each half-cycle step to drive the surface reaction to completion at all available sites by mass action. The excess is easily removed before the next step, for example by lifting the solid substrate out of the reagent solution and rinsing (Fig. 7.3a). As long as the adsorbed ions do not rearrange quickly on introduction of the film into the next reagent solution, conformal growth can be achieved. However, separation of colloidal particles from excess reagent in each half-cycle poses an extra challenge, and several methods have been adopted in the effort to

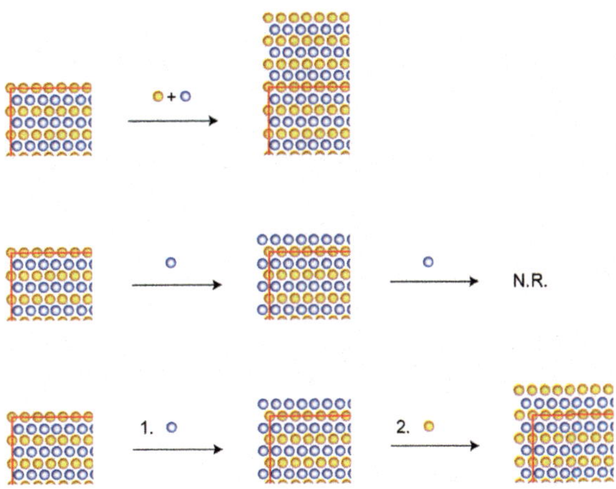

Fig. 7.2 Selective ionic layer adsorption and reaction (SILAR). A crystal may grow preferentially in one direction when all constituent elements are provided simultaneously in the appropriate oxidation state. SILAR relies on the self-limiting adsorption of an **individual** constituent of a binary compound to the crystal surface when the complementary constituent is not present. By alternating the supply between the two constituents, a conformal, layer-by-layer growth process can be achieved

capture the SILAR mechanism—enforcement of conformal growth by self-limiting surface reactions—in a process compatible with colloidal NCs (Fig. 7.3b–d).

In this segment, methods, reagents, and analysis associated with the use of alternating layer addition to achieve isotropic growth of semiconductor NCs will be presented. We focus on the formation of core/shell QD heterostructures, as this has been the principal target and has also provided an effective test case for SILAR-based elaboration of NCs based on the contrast in composition between core and shell, which facilitates analysis of growth results. Table 7.2 lists some

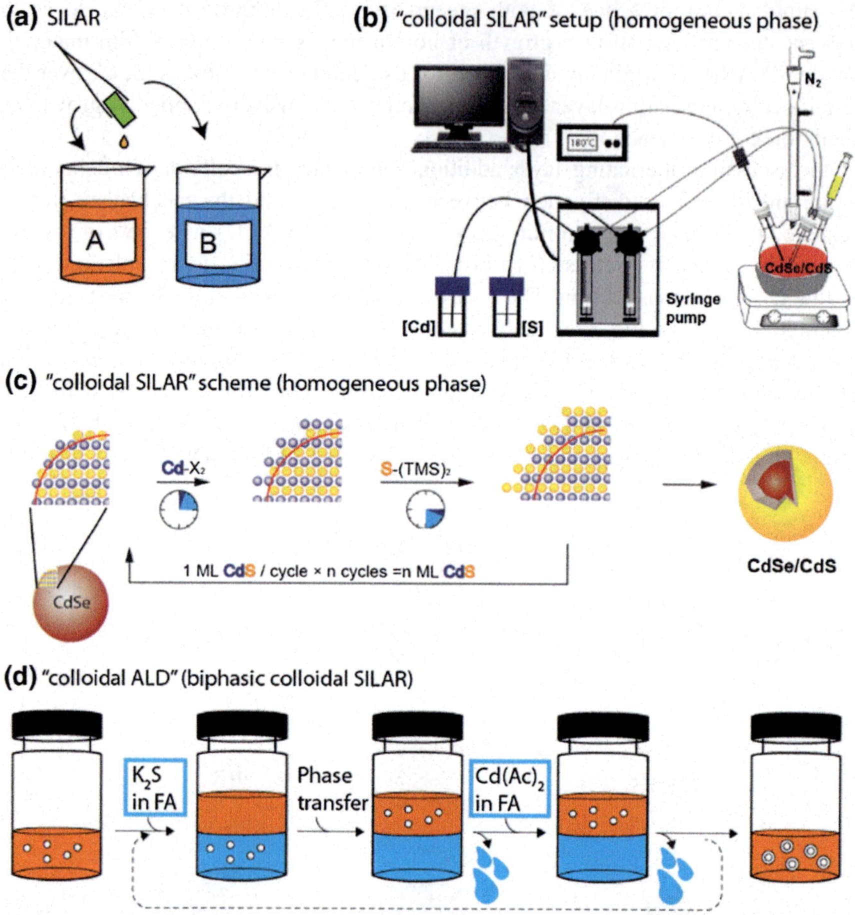

Fig. 7.3 **a** SILAR on solid-supported substrates can easily expose the sample to a large excess of each precursor, followed by easy removal. **b** An experimental configuration for colloidal SILAR (c-SILAR) permitting introduction of precise precursor doses in sequence. **c** The c-SILAR paradigm, in which saturating does are introduced in sequence to saturate available surface sites in each cycle. **d** A biphasic implementation of "colloidal ALD" in which an excess of precursors is introduced in each half-cycle, followed by removal of excess. In the example shown, precursor conversion is accompanied by a change in solubility

characteristic examples of NC shell growth via alternating layer methods, as well as representative preps using simultaneous addition for comparison.

7.5 Methods

7.5.1 Colloidal SILAR (Homogeneous Solution)

In one approach, as originally laid out by Peng and co-workers, reagents are added in doses calculated to exactly saturate the available surface sites per NC in each cycle (Fig. 7.3b, c) [51, 101]. This approach has been widely referred to as "SILAR" in the colloidal NC literature, and we will refer to it here as "colloidal SILAR" or "c-SILAR" for short. The dose necessary can be determined by considering the marginal increase in radius associated with one complete "monolayer" of the (shell) material being deposited. The change in volume associated with this increase in radius can be used to determine the molar equivalent of each reagent that is required in each cycle. To conduct a c-SILAR reaction, the NC size and the total amount of NCs must be determined. For quasi-spherical QDs, calibration curves can be established that relate the radius and the molar extinction coefficient to the LEET peak wavelength. Calibration curves have been published for Cd chalcogenides [29–31, 40] and PbS [102], and individual values are published for other materials such as InP [103]. Alternatively, the sample mass and average particle size from direct measurement (e.g. by TEM or small-angle X-ray scattering) can be used. With this information in hand, for a spherical core of radius r_0 and a nominal shell thickness t, the total dose d (in moles) can be calculated as:

$$d = n_{\mathrm{QD}} \times \frac{\frac{4\pi}{3}\left[(r_0+t)^3 - r_0^3\right] \times N_A}{V_m} \tag{7.1}$$

Here, n_{QD} is number of moles of QDs, V_{m} is the molar volume of the shell compound, and N_{A} is Avogadro's number. The incremental dose d_i for each cycle i can be calculated by considering the radius increase required in each cycle (with $d = \Sigma d_i$):

$$d_i = n_{\mathrm{QD}} \times \frac{\frac{4\pi}{3}\left(r_i^3 - r_{i-1}^3\right) \times N_A}{V_m} \tag{7.2}$$

The marginal thickness for one monolayer has frequently been taken to be the spacing between adjacent metal (or anion) atom planes in bulk structure of the shell material—for wurtzite, this corresponds to ½ of the c-axis primitive lattice vector, or about 0.337 nm in CdS. More generally, a quantity of shell precursors can then be expressed as a number of monolayer equivalents (ML eq) by considering the marginal increase in radius that would result if these reagents reacted quantitatively to form a shell of uniform thickness on all particles in the sample.

Since the initial report by Peng's group, Mews and co-workers have used c-SILAR to form multi-shell structures with gradient composition with ensemble QY of 70–85% [39]. Hollingsworth and co-workers have adapted the technique to produce thick shells [95], and Mulvaney and co-workers have used the technique to map the electronic properties of CdSe/CdS core/shell QDs over a wide range of parameter space, with reasonably good registry between nominal shell thicknesses and TEM observations [104].

Despite many successes, application of SILAR in this manner to colloidal NCs contains an inherent tradeoff that has become apparent in recent years. In thin film SILAR, a large excess of precursor is presented that drives saturation of each surface site. In the colloidal case, stoichiometric doses are added. If the precursor dose per cycle is too large, excess unreacted precursor will remain in solution and could lead to homogenous nucleation or uncontrolled growth when the complementary reagent is introduced. On the other hand, if the precursor dose is too small, surface sites will not be completely saturated, so that isotropic growth is not necessarily enforced: instead, a partial monolayer could be deposited in a regioselective manner. Importantly, even if 1 ML equivalent is added precisely, the surface reaction may not run to completion in the absence of excess reagent, instead reaching an equilibrium state with dissolved species. Indeed, in the initial description of SILAR on planar substrates, deposition of CdS and ZnS films from aqueous solutions yielded only about 0.4 ML per complete cycle [5]. The extent to which extant c-SILAR preps inhabit the mechanistic paradigm has only recently begun to be probed. As we discuss in some detail below, the choice of precursor and of reaction solvent affects precursor conversion; judicious choice of precursor dose and solvent can sustain the formation of isotropic shells with high roundness, and high ensemble QY.

7.5.2 Colloidal "Atomic Layer Deposition"

A family of biphasic alternating layer addition techniques described in the literature as "colloidal ALD" (c-ALD) [105] offers an alternative approach in which shell precursors can be applied in excess. Traditionally, it has been thought difficult to separate colloidal NCs from precursor solutions at the conclusion of each half-step as in done in thin film SILAR. NCs can be flocculated and precipitated by addition of a poor solvent to the colloidal solution, but the addition of polar and/or protic antisolvents carries the risk of re-solubilizing the surface-adsorbed ionic layer [106, 107] and/or inducing irreversible aggregation. NCs can also be separated by chromatography [108] or electrophoresis [109], but these techniques face challenges of scale and speed, allowing surfaces to re-equilibrate with a low precursor concentration. In c-ALD, SILAR is accomplished in a biphasic mixture that allows excess reagent to be added and removed rapidly; in some cases, the SILAR half-reaction is coupled to a phase transfer between the two immiscible solvents. This approach, illustrated in Fig. 7.3d, has led to core/shell QDs of high roundness, and has been adapted to deposition of conformal layers on anisotropic nanorods and nanoplatelets. Additionally, the Zamkov group has used c-ALD growth of CdS on CdS nanocluster seeds to form pure CdS NCs with narrow size distributions [110]. The temperature range may be restricted based on the boiling point or phase separation temperature of the immiscible solvents.

7.6 Precursors

NCs based on II–VI (e.g. CdSe) chalcogenide semiconductors, a few examples with Pb chalcogenides (e.g. PbSe/PbS [111]) have made up the bulk of the work done to date on the use of alternating layer approaches to enforce isotropic growth in NCs. Desirable characteristics for precursors include ease of preparation and handling, solubility, and strong reactivity toward the NC or substrate surface. Table 7.1 lists several common reagents used in the synthetic examples discussed in this chapter, and abbreviations that will be used to specify them in summaries of literature on the subject in Tables 7.2, 7.3 and 7.4.

For the metal constituents, organometallic precursors such as dimethylcadmium (Me_2Cd) and dimethyl- or diethylzinc (Et_2Zn) have historically been employed in NC synthesis, and were used in pioneering work on conformal shell growth on QDs under simultaneous addition by the Guyot-Sionnest, Alivisatos, and Bawendi groups. Additionally, Et_2Zn was used to grow conformal ZnS shells on CdSe NRs under simultaneous addition, and was also used by the Guyot-Sionnest group in vapor-source atomic layer deposition of ZnO on CdSe-based NC solids [112]. These compounds have the advantages of high solubility and high reactivity towards decomposition (to formally yield equivalents of the reduced metal) or with proton sources (to yield the 2+ ion). However, the development of c-SILAR coincided with an effort to move away from these toxic and pyrophoric organometallic precursors in favor of less hazardous alternatives.

For metals such as Cd and Zn, oleate and phosphonate salts have been widely used due to their ease of preparation from the corresponding acids and oxides, and their relative safety. Additionally, the olefin protons in oleic acid provide a unique handle for ^1H NMR, while phosphonates can be tracked with ^{31}P NMR. The preparation of these precursors from the oxides evolves water; many preps include heating and/or vacuum steps intended to remove the water after conversion of the oxides is complete. Recent work on PbS QDs indicates that at least for lead oleate, a portion of this water is retained even after heating under vacuum [113]; the possible role of retained water or hydroxide in results described in the extant c-SILAR literature has not been thoroughly explored. Anhydrous preparations exist, for example, the reaction of Me_2Cd with oleic acid to yield $Cd(oleate)_2$ and methane [114]. It has recently been suggested that Cd alkylcarboxylates may actually decarboxylate to give alkylcadmium intermediates at very high temperatures [115].

Table 7.1 Some chemical species used in nanocrystal synthesis

Abbreviation	Chemical name
Amines	
OctAm	Octylamine
HDAm	Hexadecylamine
ODAm	Octadecylamine
OLAm	Oleylamine
DOAm	Dioctylamine
THAm	Trihexylamine
Phosphines and derivatives	
TBP	Tributylphosphine
TOP	Trioctylphosphine
TOPS	Trioctylphosphine sulfide
TOPSe	Trioctylphosphine selenide
TOPTe	Trioctylphosphine telluride
Phosphonic and phosphinic acids	
PPA	n-propylphosphonic acid
HPA	*n*-hexylphosphonic acid
TDPA	n-tetradecylphosphonic acid
ODPA	n-octadecylphosphonic acid
DOPA	Dioctylphosphinic acid
TMPPA	Bis-(2,2,4-trimethylpentyl) phosphinic acid
Other reagents and solvents	
OA	Oleate (*cis*-9-octadecenoate)
Ac	Acetate
ODE	1-octadecene
NMF	N-methylformamide
TMS_2S	Bis (trimethysilyl) sulfide
DDAmBr	Didodecyldimethylammonium bromide

Considerable variation is found in the selection of chalcogenide precursors. For NC SILAR reactions in homogeneous solution, silylated chalcogenides (e.g. bis-trimethylsilyl sulfide, TMS_2S) and phosphine chalcogenides (e.g. trioctylphosphine sulfide or selenide) can serve as ready sources of these elements in the −2 oxidation state as is required for crystal growth. Many literature protocols rely on preparations of elemental S or Se in octadecene (ODE) or oleylamine (OLAm) solvents; in these cases reduction is presumably provided by reaction with the solvent [116]. An emergent approach is the preparation of substituted thio- and seleno-urea compounds [117]; these offer the ability to tune reactivity but have not yet been applied to SILAR growth. For biphasic c-ALD preps, aqueous solutions of alkali or ammonium chalcogenides are compatible with metal oleate precursors introduced in the organic phase.

An early review of shell growth in QDs including by SILAR indicated that TOP diminishes precursor conversion in sulfide shell growth [118]. However, we note that for preps in which S is supplied as the element (including ODE/S and OAm/S), if TOP is present, the relatively refractory TOPS is likely to be formed

Table 7.2 Representative growth conditions for approximately spherical core/shell nanocrystals via alternating layer methods

Core	Shell	Nominal thickness	Metal prec.	E prec.	Supporting ligands	Solvent	Temp	Remarks	Reference
c-SILAR									
CdSe, W	CdS	5 ML	$Cd(OA)_2$	ODE/S	ODAm	ODE	240 °C	Very Se-rich core prep; Cd stearate core prec.; Cd first in shell growth	Li et al. [51]
CdSe, W	CdS/ZnS, step	6 ML	$Cd(OA)_2$, $Zn(OA)_2$	ODE/S	ODAm	ODE	235	½ dose Cd to start Phosphonate prep	Xie et al. [39]
CdSe, W	CdS/ZnS, step	4 ML	$Cd(OA)_2$, $Zn(OA)_2$	TMS_2S	OLAm, TOP	ODE	180	0.8 ML/cycle; high QY, monoexp. decay	Greytak et al. [40]
CdSe, W	CdS	6 ML	$Cd(OA)_2$	TMS_2S	THAm, TOP	ODE	180	1.0 ML/cycle; note less than expected radius incr.	Tan et al. [130]
CdSe, W	(Cd,Zn)S	19 ML	$Cd(OA)_2$	ODE/S	DOAm	ODE	240	"giant" QDs	Chen et al. [95], Guo et al. [101]
CdSe, W	CdS	2.5 nm	$Cd(OA)_2$	ODE/S	OlAm	ODE	230	High crystallinity	Mahler et al. [90]
CdSe, W	CdS	3 ML	$Cd(OA)_2$	DPPS		ODE	235	Oleate cores; Low final QY and limited shape control; good control of surface enrichment	Sowers et al. [121], Wei et al. [120]
CdSe, W	CdS	3 ML	Cd $(TMPPA)_2$	ODE/S	ODAm, plus TOP or oleic acid	ODE	240	TOP interfered with shell growth (possible TOPS formation)	Van Embden et al. [118]
CdSe, W	CdS	6 ML	Cd $(TMPPA)_2$	ODE/S	ODAm, oleic acid	ODE	200–245	Used to map optical properties versus core and shell size. Uses excess Cd precursor	van Embden et al. [104]
								Raman study indicates alloying over ~3 ML thickness	Todescato et al. [132]
CdSe, W	CdS	6–8 ML	$Cd(OA)_2$	ODE/S	HDAm; oleic acid added at end	ODE	260	Compare to dot-in-plate (Table 7.4)	Cassette et al. [56]
CdSe, ZB	CdS	4 nm	$Cd(OA)_2$	ODE/S	OLAm	ODE	230	Generated stacking faults, wurtzite polytypism, platelets and rods	Mahler et al. [90]

(continued)

Table 7.2 (continued)

Core	Shell	Nominal thickness	Metal prec.	E prec.	Supporting ligands	Solvent	Temp	Remarks	Reference
CdSe, ZB	Cd(Se,Te) graded	3 ML	Cd $(MPA)_2$	NaHTe/NaHSe	MPA	Basic water	100	Graded, aqueous prep	Kaniyankandy et al. [122]
CdSe, W	(1) ZnS (2) CdSe	3 ML 3 ML	$Zn(OA)_2$ $Cd(OA)_2$	ODE/S TBPSe	OLAm	ODE	190 200	Core/barrier/shell structure	Dias et al. [133]
InAs, ZB	(1) CdSe (2) ZnSe	7 ML total	$Cd(OA)_2$ $(CH_3)_2Zn$	ODE/Se	ODAm, oleic acid	ODE	260	Measured thickness 31% of expected; good shape control	Aharoni et al. [134]
c-ALD									
CdSe, W	CdS	10 ML	$Cd(OA)_2$ or Cd $(Ac)_2$	$(NH_4)_2S$	OLAm or DDAmBr	Toluene	RT	Phosphonate-capped cores	Ithurria et al. [105]
CdSe, ZB	CdS	10 ML	$Cd(OA)_2$ or Cd $(Ac)_2$	$(NH_4)_2S$	OLAm or DDAmBr	Toluene	RT	Myristate-capped cores	Ithurria et al. [105]
Simultaneous addition									
CdSe, ZB	CdS	6 ML	Single source	Single source	OLAm, OctAm		100–140	High QY, monoexp. decay	Nan et al. [91]
CdSe, ZB	CdS	3 nm	$Cd(OA)_2$	TOPS	Trioctylamine	Oleic acid	260–280	Use of tertiary amine maintains ZB structure	Mahler et al. [90]
CdSe, W	(Cd,Zn)S	3 ML	Me_2Cd, Et_2Zn	TMS_2S (excess)	TOP, HPA	TOPO (99%)	140	Alloy shell (30% Cd) with high QY	Snee et al. [135], Liu et al. [136]
Cd (Te, Se), ZB	(Cd,Zn)S	3 ML	$Cd(OA)_2$, $Zn(OA)_2$	TOPS	Trioctylamine, TOP	(trioctylamine)	230	Primary amine led to wurtzite arms; TOPO solvent led to quenching	Pons et al. [96]

Table 7.3 Representative conditions favoring anisotropic growth of colloidal semiconductor nanocrystals

Core	Shell	Metal prec.	E prec.	Supporting ligands	Solvent	Temp	Remarks	Reference
Wurtzite—axial growth								
CdSe NR	Up to 1 ML ZnS	Me_2Cd	TBPSe	HPA	TOPO	360 °C	~6nm dia., ~20 nm length	Peng et al. [137]
CdSe NR	(Cd,Zn)S graded, conformal	NR: Me_2Cd Shell: $Me_2Cd + Et_2Zn$	NR: TBPSe Shell: TMS_2S	NR: HPA, TDPA	NR: TOPO Shell: TOPO	290 (NR) 160 (shell)	CdSe NRs coated under simultaneous addition	Manna et al. [138]
CdSe NR	–	CdO + HPA, ODPA	TBPSe		TOPO (90%)	320/290	Avoids organometallic precursors	Wang et al. [139], Tice et al. [57]
CdSe NR	–	CdO + 2-octenoic acid	TBPSe		ODE	220	Higher temp led to spherical; suggests thermodynamic product	Sapra et al. [97]
CdSe NR	–	CdO + TDPA	TOPSe	Dioctylphosphinic acid (DOPA)	TOPO (recryst.) or ODE	275/250	DOPA necessary for NR growth with Cd-TDPA in purified TOPO	Wang et al. [140]
CdS NR	–	CdO + ODPA and PPA	TOPS	TOP	TOPO	320	NRs ca. 25 nm	Ithurria et al. [105]
Cd(Se,S) NR	–	CdO + ODPA	TOPSe, TOPS	TOP	TOPO	320	Graded NRs due to greater reactivity of Se precursor; later work tunes relative reactivity	Ruberu et al. [75, 76]
CdSe QD	CdS seeded NR	Me_2Cd	TMS_2S, excess	TOP, HDAm	TOPO	130	Higher temperatures led to isotropic growth	Talapin et al. [69]
CdSe QD	CdS seeded NR	CdO + HPA, ODPA	TOPS	HPA, ODPA	TOPO	360	Very high temp; widely used	Carbone et al. [71]
CdS NR	ZnSe end-caps	$Zn(OA)_2$	TOPSe, added dropwise	ODAm, TOP	ODE	245	Selective growth of ZnSe at end caps of W CdS NRs; ZnSe has smaller lattice constant	Kirsanova et al. [78]
CdSe/CdS dot/rod	CdSe QD tip	$Cd(Ac)_2$	TOPSe	OLAm	ODE, TOPO	200	Regioselective growth of CdSe QD at (0001) tip	Deutsch et al. [141]

(continued)

Table 7.3 (continued)

Core	Shell	Metal prec.	E prec.	Supporting ligands	Solvent	Temp	Remarks	Reference
Wurtzite—planar growth								
CdSe NPL		$CdCl_2$	Octylammonium selenocarbamate	octylamine		70–100	{112′0} plane; <0001> long axis	Joo et al. [142]
CdSe NPL	–	$Cd(Ac)_2$	Selenourea	octylamine		65–100	{112′0} plane; <0001> long axis	Liu et al. [143]
Zincblende—planar growth								
CdSe NPL	–	Cd myristate + Cd acetate dihydrate	ODE/Se		ODE	195–240	{100} plane	Ithurria et al. [144, 145]
CdSe NPL	CdS “crown” (NPL extension)	Cd acetate	ODE/S	Oleic acid	ODE	240	{100} plane	Tessier et al. [146]
Zincblende—tetrapod								
CdSe, CdSe	CdSe, CdTe	Cd phosphonate (CdO + ODPA), excess	TOPSe, TOPTe	TOP	TOPO	280–300	High supersaturation leads to branches; anisotropic growth favored for shells w/larger anion	Milliron et al. [68]

Table 7.4 Formation of anisotropic structures via alternating layer methods

Core comp.	Shell comp.	Metal prec.	E prec.	Supporting ligands	Solvent	Temp	Remarks	Reference
Anisotropic substrates—supported								
TiO_2, SnO_2, Nb_2O_5, Ta_2O_5, ZnO	PbS, CdS, Ag_2S, Sb_2S_3, Bi_2S_3	Metal nitrates	Na_2S		H_2O	RT	Islanding. PbS, CdS show high IPCE	Vogel et al. [147]
TiO_2 sintered film	CdSe, 6 cycles or CdSe/CdTe	$Cd(NO_3)_2$	HSe^-, HTe^- from $NaBH_4$ reduction of SeO_2, TeO_2		EtOH	RT	Formation of QDs by islanding; some alloy	Lee et al. [123]
	CdSe, 8 cycles	$Cd(NO_3)_2$	HSe^- from $NaBH_4$ reduction of SeO_2		EtOH	RT	Starting with Se imparts SeO_x buffer layer; interconnected CdSe film after 8 cycles	Becker et al. [124]
ZnO NW	CdS	$Cd(NO_3)_2$	Na_2S		H_2O	RT	Up to 12 nm in 120 cycles; island growth during first 30 cycles	Tak et al. [148]
Si NW	CdS	$CdCl_2$ (EtOH)	Na_2S (H_2O)		EtOH/H_2O	RT, post-dep anneal 500 °C	Core/shell photodetectors	Jiang et al. [149]
PbS/CdS core/shell NC solid	CdS matrix encap.	$Cd(Ac)_2$ (MeOH)	Na_2S (H_2O)		MeOH/H_2O	RT, post-dep anneal 150 °C	CdS shell formed by ion exch. Film thickness ~300 nm	Kinder et al. [150]
CdSe colloidal NW films	Surface enrichment	$Cd(Ac)_2$ (MeOH)	Na_2S, Na_2Se (MeOH)		MeOH	RT, post-dep anneal 250 °C	Passivation of NW FETs; modulates In diffusion doping	Kim et al. [151]
PbS, PbSe NC solids	Surface enrichment	$PbCl_2$/OLAm	Na_2S, Na_2Se (MeOH)	OLAm	OLAm/MeOH	RT, post-dep anneal 65 °C	Enables electronic doping of NC films by stoichiometric excess	Oh et al. [152]

(continued)

Table 7.4 (continued)

Core comp.	Shell comp.	Metal prec.	E prec.	Supporting ligands	Solvent	Temp	Remarks	Reference
Anisotropic cores—c-SILAR								
CdSe/CdS dot/rod, W	(Cd,Zn)S, (Cd, Zn)Se	$Cd(OA)_2$, Zn $(OA)_2$	ODE/S, TOPSe	ODAm, HDAm, TOP	ODE	190–220	"Extended coverage" reported for c-SILAR shell growth	Deutsch et al. [153]
CdSe NPL, ZB	CdS, CdZnS	Cd, Zn acetate	$(TMS)_2S$, H_2S, or $(NH_4)_2S$		Hexane	RT	Oleic acid added following growth to disperse; continuous growth with thioacetamide for S	Mahler et al. [79, 154]
CdTe NW (mostly W)	CdS, 1 ML		Alkylthiol	TDPA/DOPA	Toluene/TOPO	100	CdS monolayer formed on CdTe NWs with Cd-rich surface; increased PL	Liu et al. [155]
Anisotropic cores—c-ALD								
CdS NR, W	CdS	Cd acetate	$(NH_4)_2S$	DDAmBr	Formamide/H_2O	RT	NCs undergo phase transfer with each cycle	Ithurria et al. [105]
CdSe NPL, ZB	CdS	Cd acetate	$(NH_4)_2S$		NMF	RT	NCs are precipitated by addition of toluene after each half cycle	Ithurria et al. [105]
Anisotropic growth								
c-SILAR								
CdSe QD, W	CdS (platelet)	$Cd(OA)_2$ subjected to extended heating	ODE/S	HDAm	ODE, oleic acid added at end	240	Dot in plate w/{0001} 2D facets	Cassette et al. [56]
CdSe NPL, ZB	CdS (1 ML)	–	K_2S	BF_4^-	NMF	RT	Evidence for sequential absorption to non-identical sites	Fedin et al. [131]

spontaneously. In contrast, reagents extruding the sulfide directly such as TMS_2S do not require reduction and are not intercepted by TOP. We note that TOPS has been used successfully as the sulfide source for (non-SILAR) shell growth on ZB CdSe by Dubertret and co-workers at high temperatures [90]. In this case, oleic acid was also introduced. The addition of excess free carboxylic or phosphonic acids opens a pathway whereby TOPS (or TOPSe) can extrude a formal equivalent of H_2S (or H_2Se), ultimately forming the phosphine oxide (TOPO) and an acid anhydride: this process is driven by the formation of the strong phosphorus-oxygen bond and has been studied in the context of QD nucleation and growth [119]. These intermediates, if formed, could protonate metal carboxylates or phosphonates to yield the desired shell compounds. Compared to TOPS, TOPSe is more reactive and has been widely used as a precursor for SILAR growth of selenides. Compared to the tertiary phosphine derivatives, greater reactivity is displayed by secondary phosphine chalcogenides such as diphenylphosphine sulfide, used by the Krauss group to achieve strongly sulfur-rich surfaces in the course of CdS shell growth on CdS QDs by c-SILAR [120, 121].

For c-ALD [105], aqueous c-SILAR preps [122], and traditional SILAR on heterogeneous substrates, precursors soluble in polar solvents including water are desirable. For metals these include acetate, nitrate, or chloride salts. For chalcogenides, sulfide salts Na_2S, K_2S, and $(NH_4)_2S$, and selenide salt Na_2Se have been cited as precursors. These precursors are convenient and inexpensive, though care must be taken to maintain basic conditions to avoid liberating toxic H_2S or H_2Se vapors. We note that in aqueous solution the active sulfide (selenide) species are likely to be HS^- (HSe^-) at accessible pH's. In situ formation of HSe^- from borohydride reduction of SeO_2 is reported to lead to a stable and effective selenide precursor [123, 124]. In some versions of c-ALD the metal precursor has been introduced in the organic phase as $Cd(oleate)_2$.

7.7 Analysis of the SILAR Mechanism in Colloidal NC Processes

The mechanistic details of c-SILAR and c-ALD process are important because of the possibilities that these methods holds for the control of NC structure during epitaxial growth. The essential question is whether it is possible, through manipulation of the dose sequence, to achieve nanostructures that are different than those formed by kinetic or thermodynamic control under simultaneous addition. To do so requires that nearly all of the precursor dose added in each cycle binds to the NC surface to form a self-limited monolayer—with little left in solution—and that this layer is not significantly restructured upon addition of the complementary reagent. The results for nominally isotropic growth of colloidal nanocrystals by c-SILAR and c-ALD can be analyzed to evaluate this premise.

Upon the addition of a weakly-confining shell to colloidal QDs, a large redshift in the LEET is observed due to penetration of the band-edge electron and hole state wavefunctions into the shell [36, 104]. As such, the position and shape of the LEET can be a sensitive indicator of shell growth reaction progress; however, it cannot easily be used to uniquely or quantitatively describe the extent of shell growth without evidence from supporting microscopy and analytical techniques.

More specific tools for analysis of shell growth results include TEM, which can reveal size and shape distributions. For this purpose, high-angle annular darkfield scanning TEM (HAADF-STEM) is superior to lattice-resolved high-resolution bright field TEM, because HAADF-STEM is sensitive to incoherent scattering, which exhibits high atomic number (Z) contrast and eliminates the challenge in identifying the edges of very small crystals on the basis of lattice fringes formed by coherent scattering. Powder X-ray diffraction can reveal crystal structures, and X-ray photoelectron spectroscopy (XPS) can reveal oxidation states as well as information on shell thickness on the basis of attenuation of core element signals.

In the original work by Peng's group in 200 [3, 51] CdSe QDs (cores) were treated with alternating additions of Cd oleate and ODE/S to form CdS shells. Precursors were added in amounts corresponding to 1 ML per cycle, starting with Cd. The resulting particles had a wurtzite crystal structure, and showed a red-shifted LEET as detected by UV-visible absorption and PL spectroscopy. XPS analysis was consistent with uniform coverage of the CdSe core by CdS, the PL of the resulting particles displayed reduced sensitivity to ligand exchange compared to the cores, and the prep was scalable to multi-gram samples. Although no attempt was made to directly probe precursor conversion, several features of the results indicated imperfect adherence to the SILAR reaction model. TEM images showed slightly oblate structures, the PL QY was modest at 20–40%, and photoluminescence excitation (PLE) spectroscopy indicated the formation of a considerable quantity of isolated CdS nanoparticles that contributed to absorption in the sample, but not to band-edge emission from the core/shell QDs.

Subsequent work by Mews and co-workers, also focusing on CdSe QDs, extended the colloidal SILAR technique to ZnS and CdZnS alloy and graded-composition shells [39]. Once again the crystal structure was wurtzite, and in this work, the SILAR shell growth commenced with the addition of a diminished metal oleate dose (0.5 ML eq) on the hypothesis that the initial surface was nearly equimolar and so could not sustain addition of a completed Cd monolayer. This work succeeded in raising the PL QY of the resulting particles to up to 85% in the case of CdSe/CdS core/shell QDs, and in producing total shell thicknesses that approximately correspond to the total quantity of added shell reagents after up to 6 cycles (6 ML eq of metal and sulfur reagents). The potential for lattice mismatch to lead to poor core/shell results was clearly demonstrated by the decrease in QY of CdSe QDs with pure ZnS shells when the shell thickness was extended beyond 2 ML eq; however, the graded shell, built up by a stepwise transition from CdS to CdZnS to ZnS, was able to show good results for CdSe cores with a ZnS final surface.

These two studies have served as the foundation for c-SILAR work conducted since, but neither study directly addressed whether the alternating layer process was enforcing isotropic growth or whether the reagent doses being added indeed corresponded to saturation of available surface sites. In fact, the study by Mews showed that the lattice mismatch between CdSe and ZnS leads to lattice defects and irregular shapes for CdSe/ZnS core/shell QDs with either simultaneous or SILAR addition.

Additional insight into surface saturation conditions during c-SILAR growth emerged from studies by the Mulvaney group, the Hollingsworth group, and our own investigations related to c-SILAR growth in the CdSe/CdS core/shell system. Mulvaney's group published a careful analysis of the synthetic methodology involved in the growth of core/shell nanocrystals, within an isotropic model. In particular, they emphasized the importance of the equilibrium between dissolved precursors and surface-bound forms in directing epitaxial shell growth instead of homogeneous nucleation of the shell compound [118]. They also discussed the charge-balance considerations for c-SILAR growth in solvents of low polarity. In particular, the presence of excess metal "ions" (metal atoms in the oxidation state found in the crystal) at the nanocrystal surface must be accompanied by the presence of anionic ligands to maintain near-zero overall charge for the colloidal particle. Addition of the anion (non-metal component in its reduced oxidation state) to this surface can in principle lead to formation of a nearly charge-neutral surface stabilized by neutral molecules engaging in dative bonding with available coordination sites on the surface. Colloidal SILAR preps starting with Peng and Mews had used primary amines as a component of the reaction solvent, and these amines presumably take on the role of coordinating the neutral surface (at the conclusion of the anion addition cycle) in typical chalcogenide c-SILAR preps. Mulvaney's group also investigated changes in surface enrichment and LEET wavelength under single reagent addition to CdSe QDs [125], and published a detailed article concerning the size-LEET wavelength calibration curve and molar extinction coefficient of CdSe QDs [31]. This information is essential to accurate and consistent calculations of surface ML equivalents in c-SILAR growth.

The Hollingsworth and Klimov groups sought to use c-SILAR to form CdSe-based QDs with very thick CdS shells with considerable electron delocalization, in order to limit Auger recombination in multiply-excited QDs and in an effort to suppress fluorescence intermittency (blinking) processes associated with Auger recombination and/or surface traps [92, 93, 95]. SILAR, with Cd oleate and ODE/S precursors, was used for shells with nominal thickness up to 19 ML equivalent. Notably, the authors substituted a secondary amine (dioctylamine) for the primary amine in the growth solvent—without this switch, growth of thick shells was problematic due in part to uncontrolled homogeneous nucleation that robbed the core/shell particles of shell precursors; these issues were expounded in subsequent independent work by Vela and co-workers [101].

In 2009, one of us (ABG) used a variation of the c-SILAR approach to form wurtzite CdSe/CdS core/shell QDs with near-unity QY (98% absolute QY), mono-exponential ensemble PL decay, and suppressed blinking [40]. Several features distinguished this work from the examples described above: trioctylphosphine (TOP) was used as a solvent component; $(TMS)_2S$ was used as the sulfur precursor; and the reagent dosing was only 0.8 ML eq per cycle (starting with Cd, and out to ~4 ML eq total). Part of the success of this approach can be attributed to greater precursor conversion due to the reduction in dose per cycle and the use of the silylated sulfur precursor; we have since demonstrated that the presence of TOP as a ligand on the surface of these QDs is also important to establishing the high QY of these samples at the conclusion of shell growth [126]. However, the surface saturation conditions associated with the SILAR mechanism still remained to be measured.

We have now conducted two studies that, firstly, demonstrate limited conversion of the $Cd(oleate)_2$ precursor in CdSe/CdS shell growth in primary amines at full monolayer equivalency; secondly, show that surface saturation is not necessary to achieve isotropic growth under the conditions used; and thirdly, that moving to a more highly substituted amine solvent promotes precursor conversion and the formation of high quality samples under full monolayer equivalency. These results are important details in the preparation of CdSe/CdS core/shell QDs by colloidal SILAR, but more importantly, they emphasize the importance of testing for the SILAR mechanism under the alternating layer addition process in colloidal NC growth, and provide examples of how to perform such tests.

7.7.1 Dose Dependence in c-SILAR

In order to test the hypothesis that alternating addition of stoichiometric doses of cation and anion precursors can effectively saturate surface sites, which is necessary to effectively control shape, we first examined the dose dependence of CdSe/CdS core/shell growth under c-SILAR conditions (Fig. 7.4) [127]. In the case that the reagent dose per cycle exceeds the amount necessary to saturate surface sites, or experiences limited precursor conversion at equilibrium, an increase in homogeneous nucleation is expected. At low dose per cycle, heterogeneous nucleation is expected to be diminished, but anisotropic growth is expected if precursor conversion is strongly regioselective. We used the redshift associated with CdS shell growth as a convenient process monitoring tool, as well as scanning transmission electron microscopy (STEM) and mass spectrometry measurements of free reagent concentration in these experiments.

Redshift data was first used to examine reaction rates as single reagents (Cd $(oleate)_2$ or TMS_2S) were added, and to establish that the initially-prepared surface would not react with TMS_2S to yield growth but did react with $Cd(oleate)_2$, such that $Cd(oleate)_2$ was selected as the first reagent to add in the alternating layer growth process.

For the addition of total reagent amounts equivalent to 4.0 ML, at 1.0 ML/cycle, we indeed observed the appearance of small particles in HAADF-STEM imaging (Fig. 7.4b) that are consistent with heterogeneous nucleation of CdS. When the

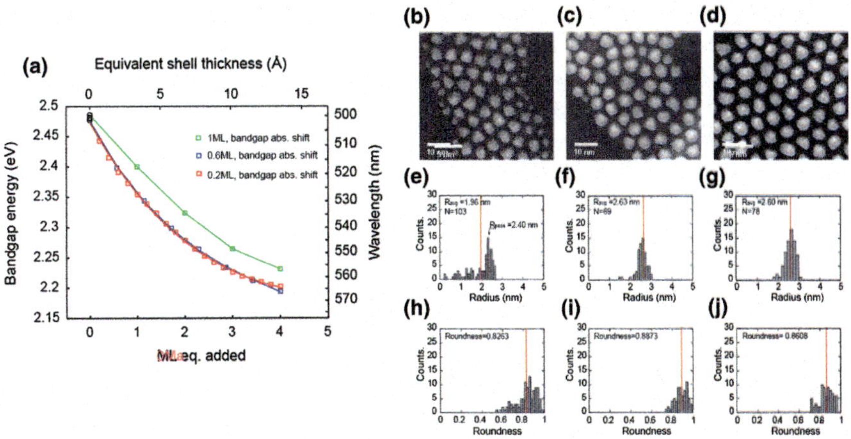

Fig. 7.4 Dependence of redshift and particle shape on reagent dose per cycle in colloidal SILAR growth of CdSe/CdS core/shell QDs in oleylamine/octadecene solvent at 180 °C. **a** Progressive redshift of lowest-energy electronic transition (LEET) with increasing nominal shell thickness ~4.0 ML total applied at 1.0, 0.6, and 0.2 ML equivalents per cycle. **b–d** Representative STEM images of samples at the conclusion of shell growth. **e–g** Distribution of total radius. **h–J** Distribution of roundness, indicating better shape control at lower dose per cycle. Adapted with permission from Ref. [127]. Copyright 2013 American Chemical Society

same total reagent amounts were added at 0.6 and 0.2 ML/cycle, the occurrence of such small particles was greatly diminished, the redshift was increased (corresponding to greater characteristic shell thickness) and the average particle radius was increased compared to 1.0 ML/cycle. These results suggested that precursor conversion was incomplete at 1.0 ML/cycle. A surprising result was found in the roundness of the particles. For a 2D projected shape with area A and major axis a, the roundness characteristic is defined as $4A/\pi a^2$. The roundness is 1 for a circle, and <1 for oblate shapes. The roundness of the core/shell QDs *increased* for low dose per cycle, whereas the SILAR mechanism posits saturating doses as being necessary to enforce isotropic growth. This initially surprising result can be understood as a combination of (1) an absence of strongly anisotropic precursor binding strength, leading to high roundness at low dose per cycle; and (2) anisotropic and non-self-limited surface growth when both precursors are present in solution at the same time due to incomplete conversion to bound forms during the SILAR cycle at high dose.

The origin of these growth results was revealed through a series of titration experiments (Fig. 7.5) in which each reagent addition was conducted in a series of submonolayer steps, permitting aliquots to be withdrawn and analyzed along the way. These steps can be extended beyond 1 ML equivalent in order to observe saturation behavior. For Cd addition (s. 5A), we found that red shifting of the LEET associated with growth extended beyond 1 ML equivalent—in other words, the surface was not completely saturated after the addition of 1 ML equivalent, but instead required an excess to drive the precursor binding equilibrium to the right.

This implies that some Cd is left in solution at the conclusion of a 1 ML addition half-cycle. Indeed, we were able to directly detect free Cd in the supernatant when QDs were removed from aliquots drawn during a titration reaction accomplishing a complete 1 ML SILAR cycle (Fig. 7.5c). In contrast, a sharp saturation signature is seen when TMS_2S is added (Fig. 7.5b); the saturation signal occurs when the amount of TMS_2S added is equal to the amount of $Cd(oleate)_2$ added previously. These results suggest that it is incomplete conversion of the Cd precursor, Cd $(oleate)_2$, to a surface-bound form, that limits SILAR performance under the reaction conditions.

We can contrast the precursor conversion reactions of the Cd and S precursors. For $Cd(oleate)_2$, precursor conversion can be thought of as simply the association of an equivalent of $Cd(oleate)_2$ to the NC surface: the $Cd(oleate)_2$ acts as an electrophilic or “Z-type” ligand. This process is reversible, and in a simple picture, can be described by an association equilibrium constant. It is known that metal carboxylate equivalents can be displaced from the surface of chalcogenide NCs in the

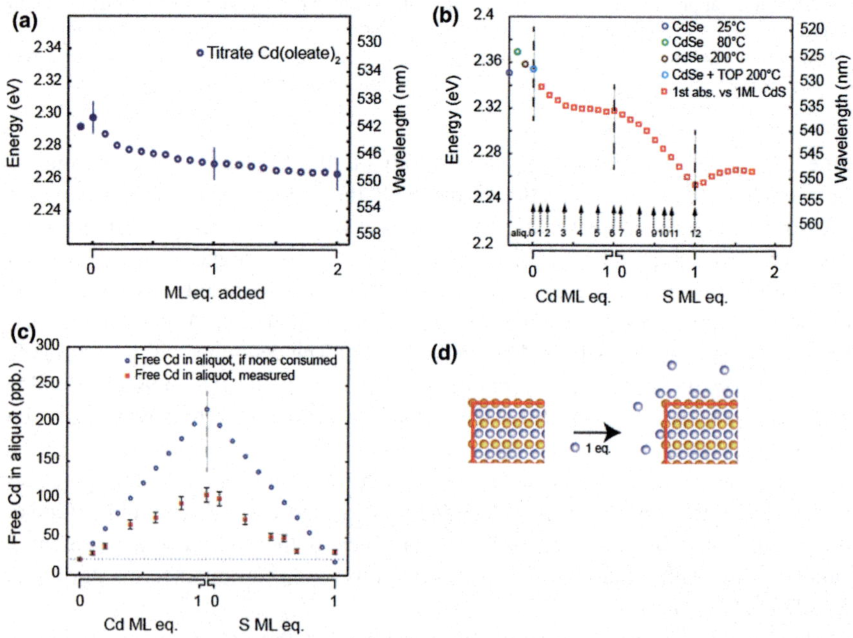

Fig. 7.5 Probing surface saturation in the colloidal SILAR process monitored by titration with a series of submonolayer reagent doses. **a** Titration of $Cd(oleate)_2$ into a solution of CdSe NCs leads to a continuous redshift over at least 2 ML equivalents. **b** The reaction of the sulfur precursor, TMS_2S, after Cd addition leads to a strong redshift and a sharp endpoint feature. **c** Plot of the concentration of Cd in aliquots taken at the points marked in **b**, after QDs are precipitated and the supernatant digested with aqua regia. **d** When 1 ML equivalent is added, a portion of the precursor remains in solution at equilibrium and fails to convert to a surface-bound form. Adapted with permission from Ref. [127]. Copyright 2013 American Chemical Society

presence of strong nucleophilic (L-type) ligands such as amines [107, 128]. In contrast, the addition of TMS_2S results in transfer of the TMS groups to oleate ligands, which deprotects the NC surface and is driven by the effectively irreversible formation of Si–O bonds; this behavior has been seen previously in the use of silylated alkylthiol ligands to drive ligand exchange reactions in high yield [129].

The reactivity of NC surfaces toward shell growth can be sensitive to the manner in which the NCs are initially prepared, due to differences in surface termination (elemental enrichment and ligand population) and/or the presence of impurities that interfere with binding of shell precursors. In our own work, we found that while purification of CdSe cores by two precipitation/redissolution cycles led to selective reactivity toward the Cd precursor as described above, purification of the cores by gel permeation chromatography (GPC), a technique that leads to a lower number of residual ligands per QD, led to (1) diminished reactivity towards the Cd precursor; and (2) aggregation and etching when TMS_2S was added first [108]. We interpret this as, in the first case, a larger amount of precursor needed to saturate the solution before surface binding could proceed, and in the second case, the rapid removal of the smaller number of residual phosphonate or oleate ligands upon silylation by TMS_2S such that colloidal stability is lost. On examination of the literature, one can see that different assessments of the optimum starting point for SILAR growth have been reached by different researchers: for example, in work by the Mulvaney group [104], a half-monolayer of Cd precursor was introduced prior to the complete c-SILAR cycles starting with S. Careful attention to the role of initial surface occupation and impurities, as well as precursor conversion and byproducts, will be important to development of increasingly sophisticated and repeatable NC growth steps, particularly in reactions targeting anisotropic shapes. Purification challenges are inherent in simultaneous addition and single-source shell growth as well: Nan et al. describe the addition of amines and phosphines to extract residual Cd carboxylate and Se species, respectively, from zincblende CdSe QD samples prior to CdS shell growth in a low-temperature, one-pot, single-source CdS shell growth reaction [91].

7.7.2 Solvent Dependence of Precursor Conversion in c-SILAR

Even for a precursor conversion reaction that is effectively a ligand association, as is the case for Cd above, it is possible to improve precursor conversion through a judicious choice of solvent such that the effective association equilibrium constant is large. In this way, high fractional occupation of surface sites can be achieved in the presence of small residual free ligand concentration.

In the case of $Cd(oleate)_2$ in CdSe/CdS core/shell growth, the oleylamine coordinating solvent used in many early SILAR preps was found to limit precursor conversion [101]. Replacing oleylamine with dioctylamine, a secondary amine of similar molecular weight, leads to a reduction in homogeneous nucleation in 1.0 ML/cycle c-SILAR growth. Our group has shown that moving to trihexylamine, a tertiary amine, leads to even greater suppression of homogeneous nucleation [130]. The effect of suppressed nucleation can be seen in STEM images of the core/shell growth products (Fig. 7.6), and in excitation spectra of the resulting products that reveal a weaker excitation energy dependence of the

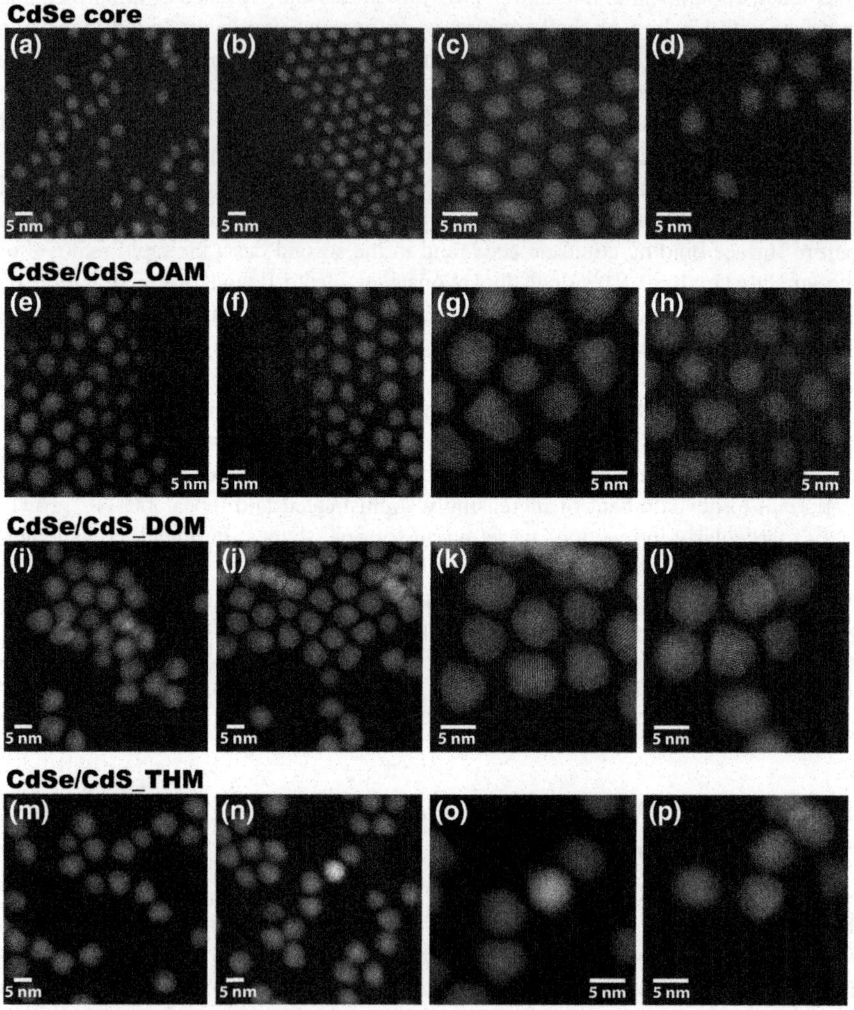

Fig. 7.6 STEM images for CdSe QDs **a–d** and for CdSe/CdS core/shell QDs formed by colloidal SILAR in solvent mixtures of octadecene with three different amines. **e–h** oleylamine (primary amine). **i–l** dioctylamine. **m–p** trihexylamine. Adapted with permission from Ref. [130]. Copyright 2015 American Chemical Society

quantum yield when growth is conducted in dioctylamine or trihexylamine instead of oleylamine.

The simplest model for precursor conversion considers a Langmuir-type mass action relationship between the fractional occupation y of surface sites and the concentration of free precursor:

$$y = \frac{K \cdot [Cd]}{1 + K \cdot [Cd]} \tag{7.3}$$

In this expression, K is the association equilibrium constant and the surface sites are assumed to be identical and non-interacting (an oversimplification, but sufficient to compare K for different precursors and/or conditions).

In a SILAR experiment, the amount of precursor introduced is known; if the amount remaining in solution can be measured, the difference can be attributed to an increase in the amount bound to the surface. In an idealized model of the colloidal SILAR process, $y = 0$ at the beginning of the reagent addition cycle (representing complete saturation of the surface by the complementary precursor and the maximum capacity for adsorption of the new precursor), and the number of sites corresponds to the estimated stoichiometry for 1 ML as described above. In practice, neither of these is necessarily true: a non-zero amount of surface occupation and non-zero free ion concentration may exist at the start of the addition cycle. Additionally, the total increase in surface occupation corresponding to self-limiting saturation may differ from the 1 ML estimate. For these reasons, it is helpful to introduce an effective surface occupation θ defined as the number of bound precursors divided by the *estimated* number of sites: when $y = 1$, θ will have its maximum value θ_{max}, with $\theta_{max} = 1$ if the estimate is perfect. At the beginning of the addition cycle, $\theta = \theta_{min}$, and the *change* in surface occupation, which is accessible experimentally, is $\Delta\theta = \theta - \theta_{min}$. We can combine this with the Langmuir expression to write

$$\Delta\theta = \theta_{max} \times y - \theta_{min} \tag{7.4}$$

$$\text{or } y = (\theta_{min} + \Delta\theta)/\theta_{max} \tag{7.5}$$

In order to experimentally characterize precursor conversion in the three solvents, we performed single reagent titrations of $Cd(oleate)_2$ into CdSe QD cores under each solvent condition. The free Cd concentration was determined by mass spectrometry analysis of aliquots drawn from the reaction over the course of the titration. Equation 7.4 above allows an experimentally determined series of $\Delta\theta$ values to be fit as a function of precursor concentration ([Cd]) with K, θ_{max}, and θ_{min} as parameters. Equation 7.5 allows the standardized y versus precursor concentration for a series of conditions to be compared. Figure 7.7 shows our results for $Cd(oleate)_2$ in each of the amine solvents. In each case, increasing solution [Cd] is associated with increased surface occupation, with an asymptotic relationship, but K is greatest in trihexylamine.

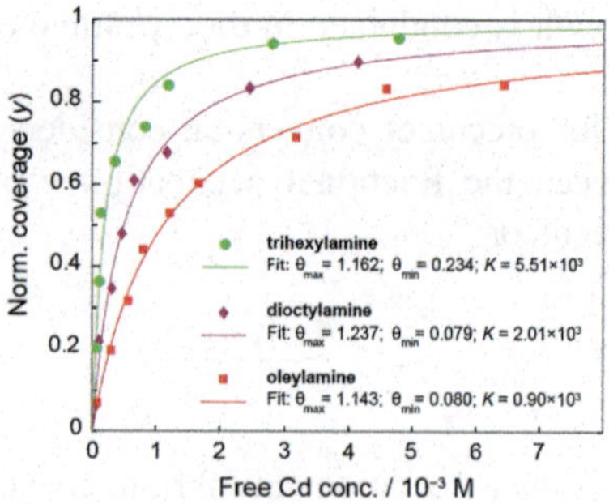

Fig. 7.7 Fractional occupation of CdSe QD surface by added Cd(oleate)$_2$ in primary, secondary, and tertiary amine-containing solvents. Free [Cd] is determined from ICP-MS analysis of aliquot supernatants following precipitation of QDs. The *K*'s are association constants, with units of M^{-1}. Adapted with permission from Ref. [130]. Copyright 2015 American Chemical Society

Amine coordination of both the QD surface and of free Cd(oleate)$_2$ in solution will tend to discourage precursor conversion and thereby decrease *K*. Increased steric hindrance in the case of secondary and tertiary amines could plausibly decreased amine binding in both cases and thereby an increase in *K* and, accordingly, higher precursor conversion (larger *y*) at 1 ML equivalent doses. We were able to show via time-resolved PL and NMR measurements that the secondary and tertiary amines bind much less strongly to the QD surface than does oleylamine. In effect, moving to more weakly coordinating solvents decreases competition for precursor binding interactions. In the presence of primary amines, high precursor conversion could only be achieved with submonolayer dose per cycle, which undermines the role of the SILAR approach in controlling NC shape. These concepts are expressed schematically in Fig. 7.8. We predict that the tertiary amine solvent will offer improved shape control for colloidal SILAR growth of anisotropic nanocrystals as well, though this has not yet been directly tested.

The approach used here to monitor the SILAR mechanism—in particular, analysis of small aliquots drawn from the reaction with mass spectrometry—was sufficient to explain the difference in growth results and offer confidence in further developing the colloidal SILAR approach to shape control, but it also entails limitations. In particular, while the reaction is conducted in a mixture of solvents at elevated temperature (here 180 °C), the aliquots are cooled to room temperature and diluted in a new solvent (hexane) when they are collected. Precipitation with a polar solvent is then used to remove QDs and allow analysis of solution species. Accordingly, it is not clear to which condition the effective association constant *K* applies: which temperature and which solvent? While meaningful comparisons can be made between different growth solvents or precursors (as seen above), an in situ method could offer greater confidence as to the actual precursor conversion and saturation conditions during growth.

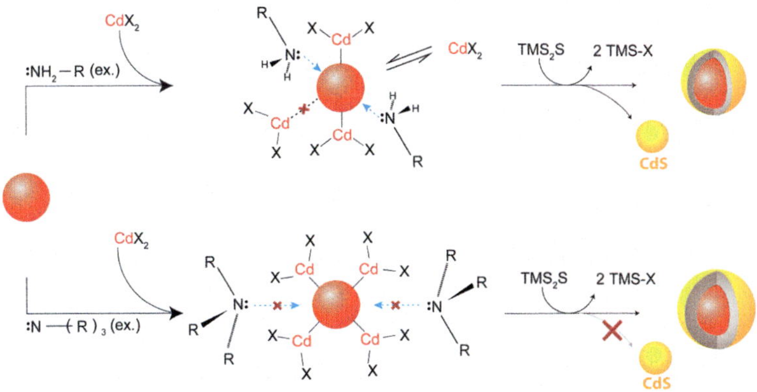

Fig. 7.8 Primary amines bind strongly to the NC surface and effectively inhibit Cd(oleate)$_2$ precursor conversion to bound species. Tertiary amines avoid this. As a result, cross-reaction with the sulfur precursor to form CdS nanoparticle side product is suppressed. Adapted with permission from Ref. [130]. Copyright 2015 American Chemical Society

7.7.3 Electrochemical In Situ Monitoring

In 2014 Fedin and Talapin described an electrochemical technique that permitted monitoring of sulfide concentration under SILAR conditions at temperatures up to 150 °C. [131] In particular, a silver (Ag) wire coated with Ag_2S was used as a non-aqueous, ion-sensitive electrode for sulfide, according to the following reaction:

$$2\,Ag + S^{2-} \rightarrow Ag_2S + 2e^-$$

The potential at this electrode, measured relative to a Ag/Ag$^+$-cryptand reference electrode, formally measures the activity of sulfide according to the Nernst equation: the potential decreases proportional to log([S^{2-}]). The sulfide potential was measured in *N*-methylformamide (NMF) solution and could be used to monitor SILAR growth either directly in NMF, or in a less polar solvent (hexane or ODE) placed in contact with it, with either K_2S or TMS_2S as sulfide sources. Using this technique, the titration of NCs with Cd^{2+}-rich surfaces with sulfide could be monitored, and showed clear endpoints that scaled with NC size and concentration (Fig. 7.9). In SILAR growth, the sulfide potential was seen to rise rapidly (corresponding to a decrease in [S^{2-}] when Cd(oleate)$_2$ precursor was introduced. As with the measurements of free Cd concentration above, equilibrium constants for sulfide addition to NC surfaces could be obtained. The biphasic SILAR system differs from colloidal ALD in that the precursors are provided stoichiometrically, rather than in excess, and separations between subsequent half-cycles are not required.

While challenging to implement, in situ measurements such as this offer highly valuable information about reaction progress and the opportunity to detect surface

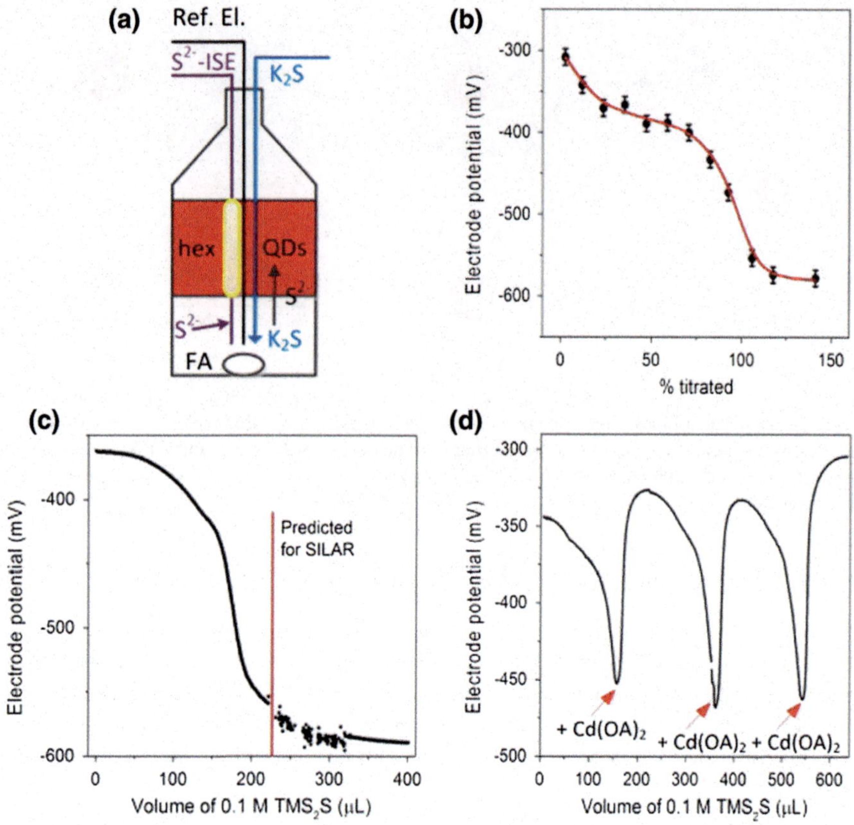

Fig. 7.9 **a** Experimental configuration for c-SILAR shell growth monitored by sulfide-sensitive electrode. FA is formamide. **b** Titration of Cd-rich QDs in hexanes with sulfide at room temperature. **c**, **d** Single-component titration and multi-cycle monitoring of c-SILAR shell growth of CdS on CdSe QDs in ODE/oleylamine at 150 °C. Adapted from Fedin and Talapin [131]. Copyright 2014 American Chemical Society

saturation on-the-fly, which could lead to improved core/shell material properties, as well as facilitating the development of shape control by SILAR.

7.7.4 XPS Monitoring

The foregoing monitoring techniques, while useful, offer only an indirect probe of the actual surface composition of the NCs during alternating layer growth. X-ray photoelectron spectroscopy (XPS) can in principle offer such information because the binding energy of each atom is highly sensitive to its local environment. XPS is often thought of as a surface-sensitive technique. However, considerable care is necessary in the interpretation of XPS signals from NC solids because the particle size is often on the order of the inelastic mean free path for photoelectrons in the core and/or shell compounds. The relative intensities of peaks for different elements, together with the appropriate sensitivity factors and mean free paths, can be used to assign elemental ratios in the NC sample (for example Cd:Se in NC core samples) and look for changes upon various treatments; our group showed that heating phosphonate-capped CdSe cores in amine solvents led to a decrease in the Cd:Se ratio in QDs that were subsequently isolated from the mixture [127].

Additional information can be obtained if the binding energy for atoms at the NC surface is sufficiently different from that of the *same* element in the NC interior that precise fits for their relative intensities can be obtained. In this way, the distribution of a single element between NC interior and surface can be obtained. Krauss and co-workers have used this technique to demonstrate control over surface enrichment in CdS QDs and during CdS shell growth on CdSe cores via c-SILAR [120, 121]. Using a secondary phosphine sulfide (diphenylphosphine sulfide) sulfur precursor, they were able to tune the surface composition as measured by XPS to >70% Cd and >70% S in the course of the SILAR cycles, and to associate these changes in surface composition with changes in PL brightness.

7.8 Rational Construction of Anisotropic Colloidal Nanocrystals with Alternating Layer Approaches

We can distinguish two different roles for alternating layer methods in the synthesis of anisotropic nanocrystals and heterostructures. The first role is the combination of conformal c-SILAR or c-ALD growth steps with anisotropic growth steps conducted under different conditions: for example, to deposit conformal shells on nanoplatelets synthesized in a previous step. There are several examples of this role and it draws heavily on experience with heterogeneous SILAR growth on solid substrates and on the quasi-spherical core/shell work described in the previous section. The second role is in directly controlling anisotropic growth by selectively saturating some sites or facets but not others. This is a newer approach, and will be aided by an understanding of the origins of anisotropic or regioselective growth in nanocrystals.

Many of the early studies in this area have focused on photophysical properties and applications of the resulting anisotropic nanostructures, for example, charge separation [156], dual emission and upconversion in double quantum dots [84], and quasi-2D electronic structures emerging in nanoplatelet systems [157]. It will be important going forward to apply advances in chemical understanding and monitoring methods such as those exemplified above to test the validity of the SILAR mechanism in these reactions and ultimately achieve even greater control of nanocrystal shape.

The core concept of SILAR—proceeding via a sequence of saturating, self-limiting ionic layers—appears well-suited to the formation of conformal shells on nanocrystals with complex shapes and anisotropic surface reactivity.

Indeed, conventional SILAR growth on planar substrates and surface-supported nanostructures can lead to complex shapes and provides instructive examples for c-SILAR and c-ALD growth on anisotropic colloidal cores; accordingly, we discuss results in this area in Sect. 7.9.

However, as described above, implementation of conformal growth via SILAR in a colloidal system poses challenges associated with achieving surface saturation while avoiding cross-reaction of excess reagents. Several studies have examined alternating layer growth to form conformal layers on pre-existing colloidal nanorods, nanoplatelets, and nanowires: we will discuss these in Sect. 7.10.

The second role for alternating layer addition methods is in deliberate regioselective extension of nanocrystal cores based on the core crystal structure. In such an approach, sub-monolayer reagent doses may be used to promote regioselective growth, while still attempting to avoid cross-reactions and maintain sample homogeneity. This is an emerging concept, and early examples and future prospects are discussed in Sect. 7.11.

Examples of template-free colloidal growth leading to anisotropic shapes are listed in Table 7.3. Efforts to achieve anisotropic nanostructures through the use of alternating layer approaches, either using anisotropic substrates or anisotropic growth, are summarized in Table 7.4.

7.9 Alternating Layer Growth on Supported Nanostructures

SILAR was initially developed for depositing conformal compound films on planar and irregularly-shaped supported substrates, and work in this area provides important background for work on alternating layer growth on anisotropic colloidal nanostructures. In particular, growth on supported substrates allows precursors and heteroepitaxial growth to be studied without having to worry about maintaining colloidal stability.

Nicolau's initial description of SILAR in 1985 demonstrated heteroepitaxial growth on planar substrates [5]. He used aqueous solutions of $CdSO_4$, $ZnSO_4$, and Na_2S to grow CdS and ZnS films on a variety of planar substrates including single-crystal semiconductors, oxides, and polycrystalline metal sheets. For ZnS deposition, optically flat polycrystalline films up to 300 nm in thickness were obtained, while epitaxial growth of both hexagonal and cubic CdS could be obtained depending on the crystalline substrate used (Difficulties were encountered with Si substrates, for which native oxide formation is difficult to avoid, and CaF_2, which is slightly soluble in the aqueous solvent). For comparison with early work on colloidal SILAR as described above, it is worth noting that the best growth rates corresponded to only about 0.4 ML/cycle despite the excess of ionic precursors available (at 5 mM concentration) in each step, and the high dielectric constant of the aqueous solution, which should act to screen surface charge. Thus, it is overly simplistic to assume that 1.0 ML/cycle can be sustained in c-SILAR or c-ALD growth; precursor conversion must be evaluated experimentally.

By 2002, planar SILAR had been extended by Park et al. to deposition of metal oxides and silicates on nitrided Si wafer substrates [6]. Alternating treatments with aqueous metal cations and hydroxide or silicate anions led to amorphous films of uniform thickness, which could be annealed into dense, polycrystalline films of MnO_2, ZrO_2, and $ZrSiO_4$ through hydrothermal treatment. These films, of several hundred nanometers thickness, retained a generally smooth and uniform appearance, even though it is common for films deposited on amorphous or lattice-mismatched substrates to experience an increase in grain size and/or islanding upon annealing. Such islanding can be desirable for photocatalysis applications as it increases the surface area. In subsequent work by Park (working with Woodward and Vogt), SILAR was used to deposit TiO_2 films on Si; hydrothermal annealing led to films with relatively uniform TiO_2 nanoparticle sizes that could be used for photoelectrochemical H_2 production [7].

Indeed, much work has focused on SILAR as a method to grow self-organized QDs on textured oxide supports for QD-sensitized solar cells [158] and photoelectrochemical devices. For example, in 1994, Weller's group demonstrated the growth of a variety of metal sulfide QDs on substrates consisting of sintered colloidal oxide particles on tin-doped indium oxide (ITO) coated glass slides [147]. Formation of metal selenide QDs has historically been more challenging due to limited stability of aqueous HSe^- solutions, but recent work by Gratzel's group has achieved good results for in situ formation of HSe^- by borohydride reduction of SeO_2 in alcohol solvent (Fig. 7.10a) [123].

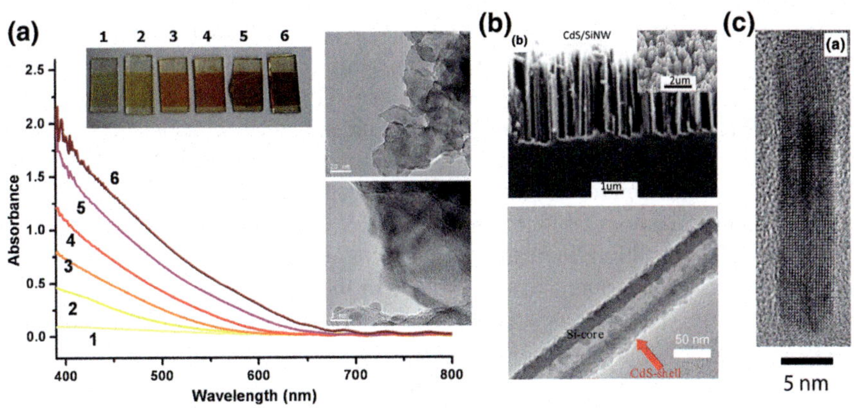

Fig. 7.10 Shell growth by alternating layer methods on anisotropic substrates. **a** Formation of CdSe nanocrystals via island growth on supported TiO_2 particle film by SILAR. Adapted with permission from Lee et al. [123]. Copyright 2009 American Chemical Society. **b** Conformal polycrystalline CdS layer formed by SILAR growth on supported Si nanowires. Adapted with permission from Jiang et al. [149]. Copyright 2015 Royal Society of Chemistry. **c** Epitaxial CdS shell growth on zincblende CdSe nanoplatelets by colloidal ALD. Adapted with permission from Ithurria et al. [105]. Copyright 2012 American Chemical Society

The QD-sensitized oxide results demonstrate the promise and limitations of SILAR as a method by which to enforce conformal growth. In particular, rather than leading to a conformal film, initial growth tends to occur as a high density of discrete islands. This presumably occurs due to dewetting of the deposited film from the oxide substrate on the basis of differences in lattice constant and surface energies. Nonetheless, nucleation at high density can be achieved on substrates with high roughness and/or aspect ratio, such as oxide nanowires. QD-sensitized solar cells can exhibit high internal quantum efficiencies for electron transfer to the reducible oxide supports, and the inorganic QDs are seen as an inexpensive alternative to dye-sensitized solar cells that can also extend the action spectrum into the infrared for increased solar energy conversion efficiency. In these approaches, as with planar substrates, excess precursor is easily removed after each step. The SILAR technique allows control of the size, the oxide-chalcogenide interface (by starting with metal or chalcogenide precursor, for example, which influences the oxide-QD interfacial electronic structure) [124], and permits the formation of core/shell structures in which the core remains in contact with the substrate. From the perspective of making QD-sensitized electrodes, this represents success, but in colloidal nanocrystals an epitaxial, single-crystal structure is often desired to minimize non-radiative recombination and inhomogeneous broadening of spectral characteristics.

Even in the absence of an epitaxial relationship, continued SILAR deposition can lead to conformal polycrystalline films on nanostructured supports, as has been demonstrated for the case of CdS shell growth by SILAR on ZnO nanowire [148] and Si nanowire [149] cores (Fig. 7.10b). This offers a low-temperature route to structures that had previously been formed by methods with a higher energy investment such as pulsed laser deposition [159].

Moving closer to the case of colloidal nanocrystals, SILAR has been employed by the Zamkov group for shell growth and matrix encapsulation in the case of substrate-supported chalcogenide nanostructures [150]. Zamkov and co-workers employed SILAR to "infill" a CdS matrix into the interstices of a close-packed film of PbS/CdS core/shell QDs; the film was supported on a glass/ITO/TiO_2 substrate. The resulting semiconductor matrix-encapsulated nanocrystal array (SMENA) material retains quantum confinement of the PbS QD cores that sets the effective bandgap, while permitting increased electronic coupling between neighboring QD sites and thereby increasing carrier mobility. In this application, SILAR is advantageous because it permits deeply recessed interstices within the QD film to be filled, while avoiding deposition of a thick CdS film on top of the QD solid.

Another application of self-limiting ionic layer addition to supported nanostructures has been the Kagan/Murray groups' use of the technique to control doping via surface enrichment in substrate-supported chalcogenide nanocrystal films. Effective control of doping is necessary in order to construct complementary logic architectures using solution-processed semiconductor nanocrystals and nanowires. A perfectly stoichiometric binary semiconductor is nominally intrinsic, but vacancies and/or impurities can create localized states that provide or trap band-edge carriers. In nanoscale forms, enrichment of the surface of by one of the component elements can also lead to excess electrons or holes, as can surface adsorption of heteroatoms. The Kagan and Murray groups tested the concept of lithographic patterning of doping type via SILAR. The first test used thin-film transistors formed with aligned CdSe colloidal NWs, because the high axial carrier mobilities permit large transconductance values [151]. Using XPS, they showed that surface enrichment of Cd, Se, and S could be achieved by exposure to Cd^{2+}, Se^{2-}, and S^{2-} salts dissolved in methanol, respectively (though limited chemical yield was found for Se^{2-} treatment). These treatments successfully modulated the threshold voltage of the TFTs, and additionally modulated the transconductance that could be achieved by diffusion doping with indium from In metal contacts. Subsequent work focused on SILAR treatment of PbSe and PbS QD solids [152], forming up to one complete ML of PbSe or PbS shell by treatment of Pb-rich QD films with chalcogenide (to achieve *p*-type films) and optionally with additional Pb^{2+} (to achieve *n*-type films). Complementary *n*-channel and *p*-channel FET performance was demonstrated following annealing treatment at less than 100 °C. It should be pointed out that this work, though described as a SILAR technique and relying on self-limiting deposition, entailed very long cycle times (up to 12 h) and the control of doping type likely includes filling of vacancies through diffusion as well as surface enrichment. Nonetheless, it indicates the stability of colloidal nanocrystals to prolonged exposure to saturation by one of the component ions, and demonstrates the possibility of achieving good electronic characteristics from SILAR growth on semiconductor nanocrystals at low temperatures.

7.10 Alternating Layer Growth on Anisotropic Colloidal Nanocrystal Cores

The chapters of this book describe a variety of strategies toward shape control in nanomaterials, and some arguments specific to shape control in colloidal semiconductor nanocrystals according to crystal structure symmetries were introduced in Sect. 7.2. The greatest amount of work has been done on those with hexagonal (e.g. wurtzite) or cubic (e.g. zincblende, rock salt) crystal structures. Anisotropic shapes arising from these structures include rod-like (quasi-1D) particles that are extended along a single crystal direction but maintain a characteristically small diameter, and plate-like (quasi-2D) structures that are extended along two directions but maintain a characteristically small thickness. One example of an alternative approach is the catalyzed growth of colloidal nanowires using metal nanoparticles (the "solution-liquid-solid" method). Selective growth steps such as these can be combined with isotropic growth and/or ion exchange reactions to arrive at sophisticated nanocrystal shapes and heterostructures.

Nucleation, isotropic growth, and anisotropic growth steps can in principle be conducted sequentially with different materials to form a wide variety of heterostructures. Additionally, many compound semiconductor nanocrystals can be modified via partial or complete ion exchange [160, 161] to alter the composition of the finished product. However, the reaction conditions available in the simultaneous presence of precursors for all component elements do not necessarily allow the extent of growth in the different crystal directions to be controlled independently, and continued nucleation events can lead to the presence of impurities or a distribution of shapes that are challenging to separate afterwards.

Conformal alternating layer growth subsequent to anisotropic growth could be used to deposit a thin layer of a higher-bandgap semiconductor to function as a barrier, leading to increased quantum yield, carrier lifetimes, and/or chemical stability. In principle, conformal growth could also be used to increase the size of a nanocrystal core prior to selective, anisotropic extension of the NC, to enable formation of NRs with independently controlled width and length.

With the colloidal SILAR method, in which calculated reagent doses are introduced and nothing is removed, reagent doses are generally calculated from the nanocrystal concentration, nanocrystal surface area, and sample volume as described above. In the case of spherical nanocrystals of known composition and crystal structure, there is typically a direct correspondence between the electronic absorption spectrum and radius, and between the radius and the molar extinction coefficient. For nanocrystals with more exotic shapes, such data is generally either not standardized or is insufficient to assign particle dimensions and concentration for c-SILAR calculations. Some work is available for zincblende nanoplatelets [162] and colloidal nanowires [163]. In principle, electron microscopy (to determine the distribution of particle dimensions) and mass or elemental analysis measurements of purified samples (to determine the total amount of the colloidal compound) can be used to accurately determine the concentration and total surface area of a nanoparticle sample, allowing c-SILAR doses to be calculated. Precursor conversion measurements can be used to detect surface saturation during SILAR growth. However, the c-ALD method is able to sidestep these requirements through the use of excess precursors in a biphasic mixture, and indeed shell growth on CdSe nanoplatelets and CdS nanorods were among the first targets for c-ALD. Below, we describe some of the goals and challenges specific to shape control in NRs, NPLs, and NWs and the ways c-SILAR and c-ALD are being used to address them.

7.10.1 Wurtzite Nanorods

Nanorod structures can be achieved by selective growth on one or both hexagonal facets of nanocrystals with the wurtzite structure. For CdSe and CdS nanocrystals, the wurtzite and zincblende bulk crystal structures are similar in energy [164], but nucleation and growth of NCs with the wurtzite structure tends to be favored for syntheses conducted at very high temperatures and/or in the presence of primary amines [90].

Work on chalcogenide colloidal NRs was principally launched by Peng and Alivisatos [137], who reported in 2000 the growth of wurtzite CdSe "quantum rods" with roughly hexagonal cross-sections of ~ 6 nm and lengths along the *c* axis of ~ 20 nm. The shape control was achieved by addition of *n*-hexylphosphonic acid (HPA) to the TOPO solvent during nucleation and growth at high temperature from Me_2Cd and TBPSe precursors. Phosphonic acids are believed to bind strongly to the sidewalls, such that growth from precursors occurs selectively at the less protected *c* planes. Later NR preps often use a mixture of short- and long-chain phosphonic acids (e.g., HPA and tetradecylphosphonic acid, TDPA) to control the NR dimensions. These CdSe NRs could be coated with thin layers of CdS or ZnS under simultaneous addition. Under the conditions used for NR growth, it is likely that Me_2Cd reacts with phosphonic acids to form the corresponding Cd phosphonate (through elimination of methane), and indeed, later work by other groups showed that comparable results for wurtzite CdSe NR growth could be achieved with a CdO starting material (through elimination of water). These preps used a TOPO solvent. While TOPO was frequently described as a coordinating solvent in early QD work, it is now increasingly accepted that it binds only weakly to QD surfaces [126, 165]. However, technical grades of TOPO contain phosphonic and phosphinic acid impurities that can influence growth. The deliberate addition of alkylphosphonic acids that led to CdSe NR growth represented an effort to gain control of the NC coordination chemistry during growth. However, recent work from the Buhro group using a CdO + TDPA precursor in solvents of recrystallized TOPO or ODE found that NR growth was only replicated when a phosphinic acid, DOPA, was present [140]. Even so, the phosphonate-based prep has been extended to growth of wurtzite CdS NRs in a TOPO/TOP solvent mixture [105]. A nominally phosphonate-free route to CdSe NRs was reported by Sapra et al., in which CdO was combined with 2-octenoic acid to form the Cd precursor, which was combined with TBPSe in ODE solvent; once again selective ligand binding is believed to promote axial growth [97].

The quasi-1D growth characteristic of wurtzite NRs lends itself to axial heterostructures. A common motif is the dot-rod structure in which a wurtzite QD is used to nucleate the extension of a "shell" material in one or both of the two dissimilar c-axis directions. The product is essentially a NR with a QD embedded within it. CdSe/CdS dot/rod structures have received continued attention because the electronic structure provides a large valence band offset leading to confinement of the hole, but minimal conduction band offset leading to a delocalized electron wavefunction [69, 81]. The structures can achieve high QY, high per-particle brightness due to light harvesting by the CdS NR, and offer a route to responsive nanostructures with dual emission and a large quantum-confined Stark shift (see also Sect. 7.5) [59, 84]. The two $\{0001\}$ facets present different reactivity, as seen by differing lengths of extension of CdS nanorod arms from CdSe QD cores, and this property has been used to achieve selective modification of just one or both tips by ion exchange, additional growth, or metal deposition.

In the first such preparation, by Talapin et al. in 2003, a rod-like CdS shell was nucleated on CdSe cores, using Me_2Cd and TMS_2S precursors at low temperature (130 °C) [69]. A large excess of the S precursor and the hexadecylamine solvent were found to promote selective axial growth. Higher temperatures led to isotropic growth, leading the authors to conclude that the rod growth is a kinetic effect, occurring when the temperature is not high enough to overcome the activation barrier for addition to facets orthogonal to the {0001} plane. More recent work has largely relied on a newer prep by Carbone et al. that uses rapid injection of CdSe QDs and TOPS to seed the growth of CdS NR shells [71]. A CdO starting material is used to generate Cd phosphonates for the Cd precursor. The growth proceeds at 320–360 °C, but the guiding principle is thought to be similar to that of the Talapin prep, with the higher temperature being required to achieve the necessary reactivity from the TOPS precursor. Alternatively, NRs with axially varying Cd(S,Se) composition have been achieved by performing NR growth with a mixture of S and Se precursors: the more reactive precursor is preferentially incorporated first [75, 76].

A family of NR heterostructures can be formed by similar methods. The anisotropic surface reactivity of wurtzite NCs in the presence of phosphonate ligands, as well as strain arising from lattice mismatch, play important roles in determining the shapes that result. Several notable examples include the selective deposition of ZnSe tips on CdS NRs to form heterostructures displaying evidence for photoinduced charge separation [78], and the selective deposition of a CdSe QD at just one tip of a CdSe/CdS dot/rod, leading to a "double QD" bridged by the CdS NR [141]. Additionally, nanoscale tetrapods and related structures can be formed from the nucleation of wurtzite arms on the {111} facets of ZB NC cores [68].

Despite all of these achievements, there are few examples of conformal shell growth on NRs, even though an encapsulating shell would likely prolong charge-separation lifetimes and QY in dot-rod nanostructures. One example, by Manna et al. and employing organometallic metal precursors, succeeded in raising the QY of CdSe NRs through deposition of a graded CdS/ZnS shell [138]. In this prep the reagents were added dropwise over ~10 min at 160 °C. This can be contrasted with the dot-rod "shell" growth conditions, which when applied to CdSe NR cores led to a CdS shell that was axially extended [166].

Recently, Oron's group have adopted c-SILAR as a means to control the shape of NR core/shell heterostructures. In particular, c-SILAR was used to achieve "extended coverage" of CdZnS and CdZnSe shells on CdSe/CdS dot/rod cores, whereas simultaneous addition was used for selective deposition at the nanorod tip [59, 153]. The group's review of double QDs emphasizes the role for c-SILAR in shape control in NR heterostructures [84]. The initial experimental results, conducted with metal oleates, ODE/S, and TOPSe as precursors and in primary amine-containing solvents, achieve results qualitatively different than simultaneous addition but do not yet include evidence for completely conformal coverage, making this a key area for future investigation with conditions that promote high precursor conversion.

Shell growth of CdS on pre-existing wurtzite CdS NRs had been reported via c-ALD [105]. Homoepitaxial shell growth was detected by an increase in NR diameter, indicating that the c-ALD process is able to displace the native ligand coordination that had promoted anisotropic growth in the initial synthesis. Specifically, excess $(NH_4)_2S$ was found to draw the initially alkylphosphonate-capped NRs into the polar phase, by forming a sulfide-terminated surface and releasing ammonium alkylphosphonate. As outlined in Fig. 7.3d, sulfide-terminated NRs could be brought back into the non-polar phase with a phase-transfer agent, didodecyldimethylammonium bromide, whereupon, after rinsing, the NRs could be brought into contact with a polar phase containing Cd acetate precursor. The change in the distribution of NR lengths and widths was consistent with conformal growth, and could offer a route to a precise and independent specification of the length and diameter of chalcogenide NRs, but it will be important to establish whether defect-free NR heterostructures with good electronic properties can be achieved via this method.

7.10.2 Colloidal Nanoplatelets with Wurtzite Structure

In the past several years, nanoplatelets, colloidal particles with two relatively extended dimensions and the third strongly confined, have attracted considerable attention due to the narrow emission linewidths, giant oscillator strength, and low optical gain thresholds that arise from their quasi-2D electronic structures. Since the modern discovery of graphene as a stable 2D allotrope of carbon, a variety of 2D materials derived from intrinsically layered bulk crystal structures has been explored, including transition metal dichalcogenides such as MoS_2. These 2D crystals have a "closed shell" within a single layer, requiring no covalent bonds between atoms in neighboring layers. As such the 2D faces do not have dangling bonds. The formation of heterostructures with these materials is an active area of research, but will generally require a different approach than SILAR. The binary chalcogenides and related compounds considered in this chapter do not fall in this category, having primarily wurtzite and cubic structures. Thus, strong ligand interactions are required to form and stabilize the large surface to volume ratios found in 2D shapes.

In the case of wurtzite, one route to 2D shapes that is consistent with the crystal symmetry is the formation of platelets with large and parallel {0001} facets (orthogonal to *c* axis). Under many solvent conditions used for nanocrystal growth, the polar {0001} facets are less stable than the neutral sidewall facets, and so nanorod structures extended along the *c* axis are favored instead. Nonetheless, in 2012 the Dubertret group reported formation of CdSe/CdS core/shell structures in which the CdS shell grew laterally hexagonal platelet surrounding the wurtzite CdSe core with flat {0001} planes. Remarkably, these nanocrystals were grown under c-SILAR conditions. Strain associated with the lattice mismatch between CdSe and CdS was cited as a possible reason for this growth habit. The hexagonal dot-in-plate structures appeared when the Cd precursor, formed from CdO and oleic acid, was subjected to extended heating above the temperature required for consumption of the CdO, while conventional isotropic growth occurred when this was eliminated. This suggests that a $Cd(oleate)_2$ decomposition product could be responsible for controlling the resulting shape.

A more extensively studied example of 2D wurtzite NC growth are nanoplatelets in which the *c* axis lies in-plane, with the {112′0} planes making up the 2D facets. Extension along the mutually orthogonal <11′00> direction leads to platelets with rectangular symmetry; examples with much greater length along the *c* axis are referred to as "nanobelts." These structures have been developed by the Hyeon and Buhro groups, and have been recently reviewed. In the case of CdSe, nanoplatelets or nanobelts of just a few nm thickness and smooth surfaces are formed when cadmium salts are dissolved in high concentrations of a primary amine (octylamine) and exposed to a source of selenide at moderate temperature. This growth is currently believed to proceed by templating and oriented attachment of CdSe clusters having a common structure ("magic size" clusters). The clusters are organized to a lamellar structure by primary amine ligands that strongly chelate the {1120} facets. Once formed, the nanoplatelets/nanobelts can be dispersed in suitable solvents. They can display intrinsically high quantum yields owing to high crystallinity and smooth surfaces, but core/shell growth on these 2D nanostructures has not yet been reported. Effective shell growth techniques must displace the ligand coating on the {1120} facets while maintaining colloidal stability. Such shell growth could lead to high performance electronic devices with remote doping.

7.10.3 Colloidal Nanoplatelets with Zincblende Structure

Growth of colloidal nanoplatelets (NPLs) with the zincblende structure has been achieved for CdSe, CdS, and CdTe. In each of these systems the reduction in symmetry from the cubic structure occurs due to self-assembly of a strongly-bound, anionic ligand layer to Cd-rich facets on the top and bottom of the platelet. In the most common synthesis, described by Ithurria and Dubertret [144], the addition of Cd acetate to a mixture of a longer-chain Cd carboxylate (e.g. oleate) and a chalcogenide precursor causes a shift from radial QD growth to lateral extension into NPLs with flat {001} facets. The thickness can be controlled by adjusting the time at which the Cd acetate is added. The acetate is evidently not a stringent requirement as Peng's group has reported CdSe NPL synthesis with {001} and with {111} facets using a range of carboxylate chain lengths in the absence of acetate [167]. However, the acetate appears to grant a substantial degree of control. Compared to the wurtzite NPLs, the zincblende NPL growth is performed at higher temperatures and in the absence of amines. Since the initial report, zincblende colloidal NPLs have extensively studied due to their 2D electronic properties [145] and apparent ease of preparation. The synthesis and electronic properties of zincblende NPLs have recently been reviewed [157, 160].

Heterostructures emerging from zincblende NPLs include "core/shell" structures featuring epitaxial growth of a shell compound on all sides including the 2D facets, and "core/crown" structures formed by continuing lateral growth with a different semiconductor compound (typically once again with a mixture of acetate and longer carboxylate metal precursors) [146]. Formation of core/shell NPLs requires displacement of the ligand coating that previously dictated 2D growth. This has been achieved through c-ALD of a CdS shell on CdSe NPLs using $(NH_4)_2S$ as the S precursor (Fig. 7.10c), and was hailed as a major success for the technique at the time, as some earlier c-SILAR attempts (unpublished) had evidently been unsuccessful [105]. A c-ALD process using TMS_2S as the sulfur precursor also succeeded in completing a monolayer of CdS growth on CdSe NPLs [79]. Eventually a simultaneous addition process using thioacetamide and octylamine to extrude H_2S in situ was discovered and found to yield good quality core/shell structures as well [79, 154]. Common to all of these successful shell depositions is the use of a precursor that supplies S in the proper −2 oxidation state.

Published work on c-SILAR growth on zincblende NPL cores is more limited. One of the challenges for c-SILAR, in which the reaction is performed in a single phase, is to accurately determine the dose required in each step. We describe in some detail below the use of in situ electrochemical monitoring to detect surface saturation and this technique is being applied to NPL surface reactions [131]. Additionally, a study on the spectra and extinction coefficients of zincblende CdSe NPLs as a function of size [162] has could serve as a valuable point of reference in dose calculations.

7.10.4 Colloidal Nanowires

Semiconductor nanowires, with nanoscale diameters and lengths that extend to microns or more, have attracted considerable interest among nanostructures in part because they can easily be contacted by lithographically defined electrodes [168–170]. The diameters can be smaller than characteristic length scales for minority carrier diffusion, depletion width, and in some cases quantum confinement. Because of these characteristics, semiconductor NWs have been studied for high density logic and memory arrays [171, 172], solar cells [173, 174], and as the basis for chemical and biological sensors and neural interfaces [175].

To achieve free-standing semiconductor crystals with such high aspect ratios, catalyzed growth is employed. In the earliest examples of this approach, gold droplets were used to catalyze the physical deposition of Si vapor on the surfaces of Si bulk crystals [176]. At the growth temperature, Si dissolves in the Au particles to form a liquid alloy; continued supply of Si leads to saturation of the mixture, whereupon it deposits as a solid crystal of pure Si according to the Au–Si phase diagram. This leads to the growth of Si wires or "whiskers" because at modest supersaturation, the Si vapor readily adheres to and dissolves in the liquid alloy droplet, but deposition onto the Si surface on the wire sidewalls occurs slowly. The catalyst is not consumed because it is insoluble in the solid NW crystal. This "vapor-liquid-solid" (VLS) process allowed growth of single-crystal wires with arbitrary length and diameter defined by the catalyst particle size and surface tension. Later on the technique was extended to compound semiconductors by Hiruma et al. [177]. A major shift in emphasis occurred when Lieber and co-workers extended the method to growth of NWs of many elemental and compound semiconductors with small diameters and on non-crystalline substrates [168, 178, 179]. Importantly, the Lieber group and many others showed that VLS-derived NWs could be suspended in solvents and transferred to foreign substrates, including glass and plastic, to form heterostructures and integrated devices. VLS syntheses typically employ gas-phase precursors and high temperatures, with NW growth occurring on the surface of a solid substrate. Though this leads to high crystallinity, there are some limitations. It is challenging to produce large quantities of NW material in this way, it can be challenging to achieve single-digit nano diameters [180], and the native surface that emerges is sensitive to oxidation.

A colloidal analogue to VLS is "solution-liquid-solid" NW growth, in which a colloidal solution of a low-melting-point metal (typically bismuth) is added to solution reaction conditions similar to those used in typical colloidal quantum dot syntheses to catalyse the growth of NWs [170]. NWs synthesized in this way are protected by a coordinating solvent or ligand that gives a measure of colloidal stability. SLS growth has been used to prepare colloidal NWs with compositions including Ge [181], CdS, CdSe, CdTe, and InAs.

For both VLS and SLS NWs, as with other semiconductor nanostructures, shell growth has been investigated as a means to control properties. As with QDs, confining shells can potentially be used to protect the core from oxidation and improve PL QY by suppressing surface recombination. Additionally, control of carrier concentrations via electronic doping is particularly important for NW device applications, but incorporation of substitutional impurities during catalyzed axial growth has been problematic [182]. Shell growth methods can be used for surface doping, epitaxial growth of a doped shell, or formation of a core/shell heterostructure that yields accomplishes remote doping of the core by charge transfer or surface Fermi level pinning [183–187]. Each of these shell growth goals has been accomplished in the case of VLS NWs, often by conducting shell deposition at an elevated temperature and/or precursor partial pressure such that surface growth, normally suppressed at axial growth temperatures, can proceed. However, shell growth on colloidal NWs has been challenging.

Challenges in controlling shell growth on colloidal NWs include lattice mismatch between core and shell materials, which can tend to lead to islanding; that ligand coordination can kinetically or thermodynamically stabilize rough surfaces; and the limited temperature range accessible in solution. Shell growth for colloidal NWs has been particularly sought as the PL QY of the initially formed NWs is very low in comparison to QDs of the same compositions. One reason for this is that exciton diffusion along the NW axis amplifies the effect of any defects that can act as recombination centers; additionally, wurtzite-zincblende polytypism through stacking faults along the NW axis is common in colloidal NWs. In an early report, the Kuno group formed a variety of II–VI core/shell heterostructures with colloidal NWs using organometallic precursors in a manner similar to early shell growth work on QDs [188]. The Kuno group also published extinction coefficient data that can be used to select shell precursor amounts [189]. However, the QY was found to decrease rather than increase and the growth conditions (250 °C) revealed some thermal instability of the core NWs. In 2010, Li and co-coworkers reported growth of CdS shells on CdSe NWs using a single-source precursor, cadmium hexadecylxanthate [190]. This succeeded in increasing the PL quantum yield, but led to a very rough shell: the shell initially deposited as a rough forest of small-diameter CdS rods, which could be annealed into somewhat larger and more uniform domains, but still containing numerous defects.

The self-limiting aspect of SILAR was successfully adopted in a unique method of forming a monolayer CdS shell on CdTe colloidal NWs, reported by the Buhro group in 2012 [155]. CdTe NWs with a principally wurtzite structure ({0001} growth axis) were treated with ethanethiol and then heated in air. This is believed to result in installation of sulfide and displacement of (likely oxidized) Te species. Subsequently, introduction of TOP, which can act as a reducing agent and is known to improve the PL QY of CdS-terminated QDs [126], along with photoannealing, led to a large increase in PL QY.

Surface supported colloidal NWs have been treated with conventional SILAR methods to control doping as described above [151], which validates alternating layer addition as a strategy for shell growth on NWs while retaining good structural characteristics, but additional work is necessary to develop a technique to achieve good results in solution.

7.11 Regioselective Growth Under SILAR Conditions

Though alternating layer addition methods have traditionally been developed as a means to enforce conformal growth, regioselective growth is achieved under certain conditions, and with suitable control, can offer a strategy for rational design of nanocrystals with anisotropic shapes. To date only a few examples exist. Below, we discuss two strategies by which anisotropic and/or regioselective growth might be achieved. Both strategies rely on the core nanocrystal having significantly different reaction rates and/or equilibrium constants for precursor binding on different regions of the nanocrystal surface, under the reaction conditions. These differences are tunable based on solvent, ligand competition, temperature, and lattice mismatch.

In the first strategy, growth on some surfaces is strongly suppressed under the reaction conditions, such that crystal growth on such surfaces will not occur even when an excess of the precursor is available in each cycle. Here, the resulting shape is not strongly dependent on the dosage per cycle, but the alternating layer approach may help to suppress nucleation or inhomogeneous growth rates among the nanocrystals in the ensemble.

In the second strategy, precursor conversion proceeds with a large association constant and sufficient rate on all surfaces, but some surfaces exhibit a significantly greater association constant or rate, such that when a sub-monolayer equivalent dose is introduced, preferential growth is achieved. In this case, the dose per cycle is an important control on the resulting nanocrystal shape.

7.11.1 Regioselective Growth Under Saturating Conditions

One example of this approach in c-SILAR growth was reported by Cassette et al. in 2012 [56], in a study in which dot-in-plate nanostructures were formed by extension of wurtzite CdSe cores with CdS. The authors found that a change in the preparation of the Cd shell precursor led to quasi-2D growth of the shell, forming hexagonal platelets with {0001} faces, centered on the CdSe core. In particular, under fairly standard SILAR conditions with $Cd(oleate)_2$ and ODE/S as precursors for CdS growth by SILAR in hexadecylamine/ODE solvent, spherical core/shell nanocrystals were obtained. In this prep, the $Cd(oleate)_2$ was prepared by heating CdO in excess oleic acid at 160 °C, followed by degassing to remove H_2O. When the Cd precursor was instead prepared by heating the CdO and oleic acid at 300 °C, anisotropic growth was observed. A portion of the Cd precursor could be replaced with similarly prepared Zn to achieve dot-in-plate structures with CdZnS alloy or gradient shells. The resulting platelets displayed 2D-polarized band-edge emission. The chemical differences in the Cd precursor that resulted from the high temperature preparation and led to the change in growth habit are not yet clear.

The initial description of c-ALD also includes evidence for regioselective and anisotropic growth under saturating conditions [105]. In particular, CdS shell growth on CdSe wurtzite cores proceeded preferentially on one of the {0001} directions, with growth on the opposing hexagonal facet largely suppressed. Growth in the presence of oleylamine led to slightly oblong core/shell nanocrystals while more pronounced hexagonal-pyramid-like faceting was observed for growth in toluene with ionic stabilization only. For zincblende CdSe cores, c-ALD growth of CdS in the presence of oleylamine led to octahedral shapes while in its absence, tetrahedral shapes with more pronounced faceting was observed. These observations are consistent with strong primary amine coordination of the surface, which may stabilize undercoordinated surface atoms and minimize differences in the thermodynamics of precursor binding on the various crystal surfaces.

7.11.2 Shape Control Via Reagent Dosing

With sufficient understanding of nanocrystal stability, surface chemistry, and precursor conversion reactions, it may become possible to use the reagent dosing sequence in c-SILAR to select between regioselective and isotropic growth modes. This would allow nanocrystal chemists to program the growth of nanocrystals and heterostructures in a variety of shapes that are derived from the underlying crystal symmetries, but that are not limited by the relative growth rates available through simultaneous addition or c-ALD growth results. To do so requires that nearly all of the precursor dose added in each cycle reacts at the NC surface to form a self-limited monolayer – with little left in solution – and that this layer is not significantly restructured upon addition of the complementary reagent.

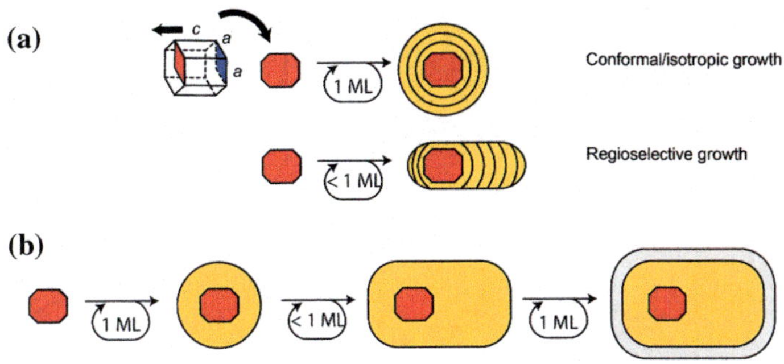

Fig. 7.11 Under appropriate conditions, c-SILAR could afford the ability to select between conformal and regioselective growth. **a** In one proposed scheme, shown for a hexagonal (e.g. wurtzite) crystal structure, c-SILAR enforces isotropic growth when the dose is high enough to saturate all surface sites, while regioselective growth occurs when only the most reactive sites are populated. **b** The ability to select among isotropic and anisotropic growth could permit the design of complex single-crystal heterostructures with independently programmed structural dimensions

A case for morphological control of wurtzite NCs via SILAR is outlined in Fig. 7.11. Under simultaneous addition, surfaces on which growth is kinetically or thermodynamically favorable can be selectively elaborated and nucleation is not blocked. Under the SILAR paradigm, growth under alternating-layer conditions should enforce isotropic growth with full monolayer equivalents, even in cases where anisotopic growth is favored under simultaneous addition. At sub-monolayer equivalency, the result would appear to depend on the distribution of binding free energies among the possible sites; in cases where the quasi-equilibrium reached after sub-monolayer addition is described by partial occupancy on all surfaces, isotropic growth will be preserved, while in cases where the surface enrichment displays a strong thermodynamic preference for one facet over another, anisotropic growth may emerge.

To date, this level of control in shell growth via SILAR has not been demonstrated. However, we can point to several design considerations and encouraging results for future development. Success is likely to require growth at relatively low temperatures (favoring strong fractionation of added material towards the sites with most negative free energy of binding), high precursor conversion, precise determination of precursor equivalencies (perhaps assisted by in situ monitoring), and highly uniform and repeatable core nanocrystal preparations.

The in situ monitoring result by Fedin et al. [131]. included an important observation that illustrates the promise and challenge of this approach (Fig. 7.12). Using an electrode sensitive to the sulfide (S^{2-}) concentration, the authors titrated S^{2-} (as K_2S) into solutions of Cd chalcogenide nanocrystals that had been stripped of covalently-bound ligands by treatment with Meerwein's salt (triethyloxonium tetrafluoroborate, a strong alkylating agent) leaving a Cd^{2+}-rich surface stabilized by BF_4^- counterions. The electrode potential, which is proportional to $-\log [S^{2-}]$, decreases slowly as S^{2-} is titrated in, since most binds to the nanocrystal surface,

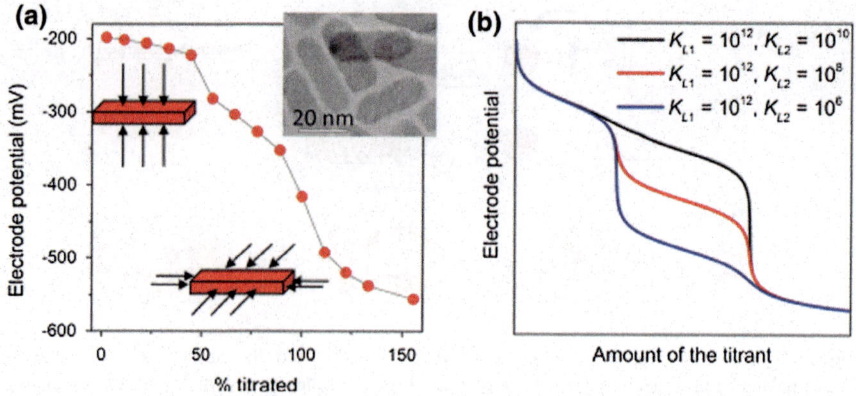

Fig. 7.12 Experimental (**a**) and calculated (**b**) electrochemical titration curves for sulfide addition to a sample of nanoplatelets; two distinct saturation features are seen for sufficiently different values of the association constants. Adapted from Fedin and Talapin [131]. Copyright 2014 American Chemical Society

and then drops sharply when an endpoint is reached corresponding to saturation of surface sites. In the case of quasi-2D CdSe/CdS core/shell nanoplatelets, two distinct titration endpoints are observed. The authors attributed this to distinct equilibrium constants for S^{2-} binding to different sets of crystal facets, with binding to the large, Cd^{2+} rich {100} facets offered as a plausible candidate for the higher association constant. If this explanation is correct, one could predict regioselective growth for a continued SILAR process with doses that are only sufficient to saturate the strongest-binding sites, versus isotropic growth with doses large enough to fill all sites. The authors indicate, based on a multiple-site Langmuir absorption model, that a difference in K_a of several orders of magnitude, and weaker binding sites having K_a of at least $\sim 10^6$ M^{-1}, is necessary to cleanly discriminate between the two sets of sites at room temperature. If this can be achieved, shape control via reagent dosing could be used to form nanorod and nanoplate structures with precisely programmed dimensions, and core/shell structures with regionally variable shell thickness, which would be advantageous for charge and energy transfer applications.

7.12 Applications

Colloidal alternating layer addition methods have already proven to be a valuable approach in controlling the shape and shell thickness of quasi-spherical QDs, particularly those based on CdSe, and this experience has aided in development of the methods and associated analytical techniques. There is currently a limited number of examples in which c-SILAR and c-ALD have been used to direct the formation of anisotropic shapes, but we anticipate that this will change in the near future. This chapter highlights three general areas in which such shape control is anticipated to be of high value.

7.12.1 Double Quantum Dots and Related Dual-Emission Structures for Temperature Measurement and Upconversion

Most luminescent molecules emit light from only one electronic state—excitation into a higher excited state results in rapid thermalization into the emissive state. This roughly characterizes the typical behavior of colloidal QDs as well: excitation at high energies still results in light emission from the lowest-energy exciton state because of rapid cooling.

Double quantum dots are inorganic nanostructures that include two localized emitters that are coupled through wavefunction overlap and/or long-range energy transfer (FRET), but can both emit light when the system is excited (rather than from the lowest energy excited state alone). For example, in concentric core/barrier/shell heterostructures, the core can form one quantum dot, while the outer shell forms a quantum well that can also emit. Double quantum dots can also be formed in nanorod heterostructures in which two quantum dots are separated by a barrier. In these systems, excitation at high energies (above the bandgap of the barrier) can result in stochastic cooling process in which energy is funneled into one or the other of the emitting states, and light emission subsequently occurs before the emitters can reach equilibrium. If the two coupled QDs have different lowest-energy exciton energies, their emission can be resolved spectroscopically. Dual emission can also result if the lower energy emitter has a low radiative rate (long lifetime) so that the two emitters can reach thermal equilibrium and yet emit light at comparable intensity. This is achieved in some types of doped QDs in which a localized excited state on an impurity atom is close in energy to the LEET of the QD.

Double quantum dots are highly sought after for several applications, as described in a review by Oron and co-workers [84]. Dual emission can be used to intentionally broaden the emission spectrum of QDs to be used in solid state lighting. Double QDs can also be used as the basis for ratiometric optical sensors, in which local conditions can be read out from the ratio of the intensities at the two emission wavelengths. For example, doped QDs have already been demonstrated as temperature sensors based on thermal equilibrium among band-edge and localized excited states [191]. Additionally, double QDs are being studied as optical gain media and photoluminescence upconversion materials (fluorophores that emit at shorter wavelengths than the excitation light, a rare property that permits low-background fluorescence imaging and microscopy in biomedicine).

Good structural control including control of shell growth is critical to the preparation of double QDs. Colloidal SILAR has been used to form concentric

CdSe/ZnS/CdSe/ZnS core/barrier/shell nanocrystals for white light emission [192], and the Oron and Weiss groups have shown the ability to form double QDs in nanorods by growing a PbS or CdZnSe tip at the far end of a CdSe/CdS dot-in-rod heterostructure [141]. The Oron group has been one of the first to show, in 2012, that c-SILAR could be used to impose "extended" shell growth on nanorod cores: material is deposited on the sidewalls as well as at the more reactive tips of the wurtzite nanorod [153].

Looking forward, many proposed uses of double QDs will require not only dual emission, but a high overall quantum yield. Encapsulation of concentric double QDs has been demonstrated and continued improvement in c-SILAR and c-ALD methods can be expected to yield even better results. On the other hand, the nanorod-based double QDs could likely benefit from complete encapsulation within a higher-bandgap shell, and this has not yet been conclusively demonstrated. The nanorod-based double QDs show considerable promise for upconversion, optical gain, and sensing based on the ability to tune the energetic landscape within, and so the achievement of conformal shell growth on anisotropic nanorod heterostructures is likely to be an important application area for c-SILAR and c-ALD methods in the future.

7.12.2 Cell Membrane Voltage Sensing

Considerable interest has arisen in the possibility of using colloidal NCs to detect and image cell membrane potentials in neuroscience research. Nearly all biological cells maintain a nonzero membrane potential (voltage) across the lipid bilayer that forms the plasma membrane. From a rest potential around −70 mV with respect to the outside, firing results in a jump to the action potential around +40 mV [193]. For a typical lipid bilayer membrane thickness of 5 nm, this is a change in the electric field of >200 kV/cm.

The importance of membrane potentials to respiration, neuron function, contractile function in heart muscle, and sensory perception including vision has spurred the development of techniques that permit membrane potentials to be measured and applied with greater throughput and spatial registration than is possible with traditional patch-clamp techniques. Fluorescent voltage probes offer a field of view encompassing many cells in culture or in a tissue [194]. These fluorophores include genetically-encoded transmembrane proteins [195] and synthetic dyes with hydrophilic and hydrophobic parts [196] that experience a change in absorption or emission upon a change in the membrane potential. Limitations associated with low water solubility, low dynamic range, and the speed of the response continue to fuel the search for improved sensors. QDs are also known to exhibit a strong quantum confined Stark effect (shift of electronic spectrum in response to electric field) [197] that could make them useful as sensors of

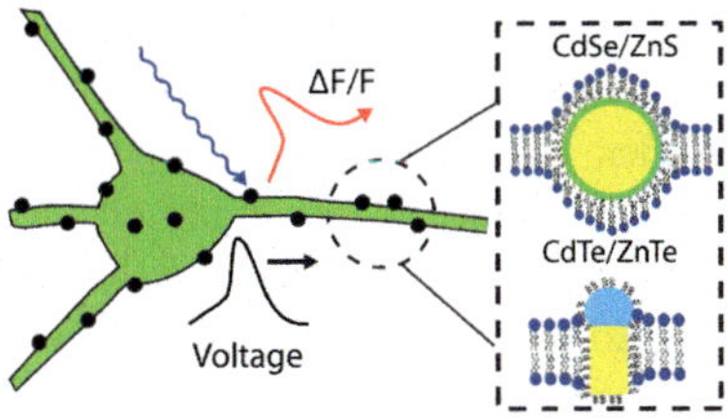

Fig. 7.13 Anisotropic NCs positioned within the lipid bilayer could provide rapid detection of the magnitude and sign of membrane voltage changes for neuroscience studies, through the quantum confined Stark effect. Adapted with permission from Marshall and Schnitzer [62]. Copyright 2013 American Chemical Society

membrane potential if they can be localized within the cell membrane. It is known from cryo-TEM experiments that QDs with a hydrophobic ligand coating can be captured within the lipid bilayer of synthetic liposomes [60].

Marshall and Schnitzer analyzed the case for NC-based cell membrane voltage sensing theoretically in 2013 [62]. They concluded that heterostructured nanorods with a type II band alignment, such as CdTe/ZnTe, would give the best membrane voltage sensors (Fig. 7.13). This is because while for spherical QDs the Stark shift nominally depends on the square of the electric field, an anisotropic heterostructure with a highly polarized excited state will generate a more linear response. Yet, the lipid bilayer is only a few nm in thickness. Successful sensors will likely be small and anisotropic, but must also maintain high QY in a live cell environment. Colloidal SILAR and/or c-ALD could assist in programming regioselective growth to form the voltage-sensing heterostructure, followed by growth of a thin, encapsulating shell.

7.12.3 Fluorescence Anisotropy in 1D and 2D Nanocrystals

Semiconductor nanocrystals can exhibit polarization anisotropy with respect to PL emission and electronic excitation. Such anisotropy is strongly dependent on the crystal structure and the overall shape. The prospect of NCs as bright and stable sources of polarized light has generated considerable interest for uses in photonics, and in bio-imaging, where polarized fluorescence can reveal orientation and dynamics of protein complexes and other intracellular structures in live cells [198, 199]. Fluorescence polarization in wurtzite CdSe NRs was observed in the initial report by Peng and Alivisatos [137]. In recent years increased attention has been devoted to the influence of nanocrystal shape on polarization anisotropy in the case of 1D and 2D nanocrystals [56–58, 166]. Applications will demand large anisotropy, high quantum yield, and homogeneous behavior among NCs in the ensemble; accordingly, this is an area in which techniques capable of growing and encapsulating anisotropic nanocrystals with a high degree of control can be anticipated to have a significant impact.

Polarization anisotropy in NCs has two essential origins: anisotropy in the electronic transition dipoles, and a classical effect from dielectric contrast between the nanocrystal and the surrounding ligands and solvent that modulates the external electric field. An anisotropic crystal structure can give rise to intrinsically polarized electronic transitions. In the case of wurtzite nanocrystals, this is manifested in the fine structure of the lowest-energy exciton state. A total of eight possible excited states are split according to their total angular momentum (electron plus hole) [58, 200]. For spherical QDs in the strong confinement limit, these are labeled ± 2, $\pm 1^{L}$, 0^{L}, $\pm 1^{U}$, and 0^{U} in order of increasing energy. Of these, ± 2 and 0^{L} are optically forbidden. Of the remaining five, the $\pm 1^{L}$ and $\pm 1^{U}$ states give rise to plane polarized transition dipoles oriented orthogonal to the *c* axis, while 0^{U} gives rise to a linearly polarized transition dipole parallel to the *c* axis. In spherical QDs all of these transitions are close in energy at room temperature. Little absorption polarization or emission polarization is observed in solution samples, and single-QD measurements show elliptical polarization. However, measurements of wurtzite NRs have shown strong emission polarization oriented parallel to the elongated *c* axis [57, 58, 166], and measurements of core/shell nanoplatelets extended in the *ab* plane have shown emission polarization oriented perpendicular to the *c* axis [56].

Emission polarization (polarization anisotropy in PL emitted by the sample) nominally probes the lowest-energy emitting state and both of these results have been rationalized on the basis of the ordering and energy splitting of the exciton fine structure. Excitation polarization (dependence of total PL intensity on polarization of excitation light) can be measured as a function of wavelength to build a more complete picture of the origin of PL anisotropy. In the case that anisotropy in the electronic transition dipoles is responsible, excitation polarization will be emphasized near the band edge, and can help to assign the exciton fine structure. On the other hand, anisotropy due to dielectric contrast will show excitation polarization that does not depend strongly on energy. Recent investigations of excitation anisotropy in wurtzite CdSe and CdSe/$Cd_{1-x}Zn_xS$ NRs found that near-band-edge transition dipoles are chiefly responsible for PL anisotropy in ensemble NR samples. Surprisingly, though, the excitation dipoles were widely distributed in angle relative to the NR axis when examined on a single-particle basis [58]. This is problematic for potential applications of such materials in single-particle tracking and orientation, where consistent excitation polarization is desirable. A possible roles of surface defects was discussed as a reason for the distribution of transition dipole angles. Improved control of NR and NPL dimensions and shell growth on anisotropic nanocrystals through c-SILAR or c-ALD approaches could lead to better understanding of PL anisotropy in colloidal nanocrystals and thereby pave the way for applications.

7.13 Concluding Remarks

Improved nanocrystal shape control will enable a range of new applications and opportunities for basic research on charge transfer, energy transfer, and ligand dynamics in nanoscale systems. In the time since c-SILAR was introduced, successive ionic layer addition methods have been applied by a host of researchers around the globe, demonstrating the accessibility and repeatability of these approaches. Much of the work so far has centered on quasi-spherical core/shell heterostructures, where suppression of nucleation and (at least nominal) enforcement of self-limiting and isotropic growth contribute to high quantum yields in core/shell QDs, and control of excited state wavefunctions. Fewer examples exist of using successive ionic layer addition methods to design and synthesize anisotropic nanostructures, yet we have described several salient results that lead us to anticipate rapid development in this area.

One of the characteristic challenges of SILAR-like approaches applied to nanoparticles in solution is to saturate the desired reaction sites while avoiding undesired cross reactions. Technical methods devised to accomplish the alternating reagent additions include colloidal SILAR, in which calculated reagent doses are added to a homogeneous liquid phase; and colloidal ALD, in which reagents are added in excess, followed by separation and removal of unreacted material. We can point to three key areas of understanding that researchers should have in mind when pursuing either of these methods. Firstly, it is important to select appropriate precursors and helpful to identify the reactants and products in the precursor conversion reactions. Secondly, careful monitoring of the extent of precursor conversion in each cycle through ex situ or in situ chemical analysis can help to identify conditions compatible with shape control by detecting saturation events. For example, conversion of Cd oleate to bound forms was found to depend on the presence of nucleophilic solvent constituents. Thirdly, the charge and ligand coordination environment at the nanocrystal surface is important to controlling solubility and reactivity, and the initial character of the surface can be affected by purification methods.

Due to advances in each of these three areas, our understanding of nanocrystal growth via c-SILAR and c-ALD has been greatly improved, and we expect rapid progress in extending successive ionic layer methods to the controlled formation of anisotropic nanocrystals and heterostructures by selecting between conformal and regioselective growth modes. In addition to improved shape control, we anticipate an expanded material scope for growth of a wider variety of compounds, and the application of homoepitaxial c-SILAR and c-ALD methods to small nuclei in order to achieve a fine degree of size and crystal structure control in nanocrystals of homogeneous composition. The principles and examples described in this chapter will continue to serve as a guide for such endeavors.

References

1. George, S.M., A.W. Ott, and J.W. Klaus. 1996. Surface chemistry for atomic layer growth. *Journal of Physical Chemistry* 100 (31): 13121–13131.
2. Hausmann, D., J. Becker, S.L. Wang, and R.G. Gordon. 2002. Rapid vapor deposition of highly conformal silica nanolaminates. *Science* 298 (5592): 402–406.
3. George, S.M. 2010. Atomic layer deposition: An overview. *Chemical Reviews* 110 (1): 111–131.
4. Knez, M., K. Nielsch, and L. Niinistö. 2007. Synthesis and surface engineering of complex nanostructures by atomic layer deposition. *Advanced Materials* 19 (21): 3425–3438.
5. Nicolau, Y.F. 1985. Solution deposition of thin solid compound films by a successive ionic-layer adsorption and reaction process. *Applications of Surface Science* 22: 1061–1074.
6. Park, S., B.L. Clark, D.A. Keszler, J.P. Bender, J.F. Wager, T.A. Reynolds, and G.S. Herman. 2002. Low-temperature thin-film deposition and crystallization. *Science* 297 (5578): 65.
7. Park, S., E. DiMasi, Y.-I. Kim, W. Han, P.M. Woodward, and T. Vogt. 2006. The preparation and characterization of photocatalytically active TiO_2 thin films and nanoparticles using successive-ionic-layer-adsorption-and-reaction. *Thin Solid Films* 515 (4): 1250–1254.
8. Lindroos, S., and M. Leskelä. 2008. Successive ionic layer adsorption and reaction (SILAR) and related sequential solution-phase deposition techniques. In *Solution processing of inorganic materials*, Ed. Mitzi, D.B., 239–282. Wiley.
9. Brus, L.E. 1984. Electron electron and electron-hole interactions in small semiconductor crystallites—the size dependence of the lowest excited electronic state. *Journal of Chemical Physics* 80 (9): 4403–4409.
10. Efros, A., and M. Rosen. 2000. The electronic structure of semiconductor nanocrystals. *Annual Review of Materials Science* 30: 475–521.
11. Smith, A.M., and S. Nie. 2010. Semiconductor nanocrystals: structure, properties, and band gap engineering. *Accounts of Chemical Research* 43 (2): 190–200.
12. Talapin, D.V., J.-S. Lee, M.V. Kovalenko, and E.V. Shevchenko. 2010. Prospects of colloidal nanocrystals for electronic and optoelectronic applications. *Chemical Reviews* 110 (1): 389–458.
13. Erathodiyil, N., and J.Y. Ying. 2011. Functionalization of inorganic nanoparticles for bioimaging applications. *Accounts of Chemical Research* 44 (10): 925–935.
14. Han, Z., F. Qiu, R. Eisenberg, P.L. Holland, and T.D. Krauss. 2012. Robust photogeneration of H2 in water using semiconductor nanocrystals and a nickel catalyst. *Science* 338 (6112): 1321–1324.
15. Lan, X., O. Voznyy, F.P. García de Arquer, M. Liu, J. Xu, A.H. Proppe, G. Walters, F. Fan, H. Tan, M. Liu, Z. Yang, S. Hoogland, and E.H. Sargent. 2016. 10.6% certified colloidal quantum dot solar cells via solvent-polarity-engineered halide passivation. *Nano Letters* 16 (7): 4630–4634.
16. Geyer, S.M., J.M. Scherer, N. Moloto, F.B. Jaworski, and M.G. Bawendi. 2011. Efficient luminescent down-shifting detectors based on colloidal quantum dots for dual-band detection applications. *ACS Nano* 5 (7): 5566–5571.
17. Anikeeva, P.O., J.E. Halpert, M.G. Bawendi, and V. Bulovic. 2007. Electroluminescence from a mixed red-green-blue colloidal quantum dot monolayer. *Nano Letters* 7 (8): 2196–2200.
18. Krause, M.M., J. Mooney, and P. Kambhampati. 2013. Chemical and thermodynamic control of the surface of semiconductor nanocrystals for designer white light emitters. *ACS Nano* 7 (7): 5922–5929.
19. Stein, J.L., E.A. Mader, and B.M. Cossairt. 2016. Luminescent InP quantum dots with tunable emission by post-synthetic modification with lewis acids. *Journal of Physical Chemistry Letters* 7 (7): 1315–1320.

20. Bruchez, M., M. Moronne, P. Gin, S. Weiss, and A.P. Alivisatos. 1998. Semiconductor nanocrystals as fluorescent biological labels. *Science* 281 (5385): 2013–2016.
21. Larson, D.R., W.R. Zipfel, R.M. Williams, S.W. Clark, M.P. Bruchez, F.W. Wise, and W. W. Webb. 2003. Water-soluble quantum dots for multiphoton fluorescence imaging in vivo. *Science* 300 (5624): 1434–1436.
22. Howarth, M., W.H. Liu, S. Puthenveetil, Y. Zheng, L.F. Marshall, M.M. Schmidt, K.D. Wittrup, M.G. Bawendi, and A.Y. Ting. 2008. Monovalent, reduced-size quantum dots for imaging receptors on living cells. *Nature Methods* 5 (5): 397–399.
23. Somers, R.C., R.M. Lanning, P.T. Snee, A.B. Greytak, R.K. Jain, M.G. Bawendi, and D.G. Nocera. 2012. A Nanocrystal-based ratiometric pH sensor for natural pH ranges. *Chemical Science* 3 (10): 2980–2985.
24. Somers, R.C., M.G. Bawendi, and D.G. Nocera. 2007. CdSe nanocrystal based chem-/bio-sensors. *Chemical Society Reviews* 36 (4): 579–591.
25. Freeman, R., and I. Willner. 2012. Optical molecular sensing with semiconductor quantum dots (QDs). *Chemical Society Reviews* 41 (10): 4067.
26. Medintz, I.L., H.T. Uyeda, E.R. Goldman, and H. Mattoussi. 2005. Quantum dot bioconjugates for imaging, labelling and sensing. *Nature Materials* 4 (6): 435–446.
27. Petryayeva, E., W.R. Algar, and I.L. Medintz. 2013. Quantum dots in bioanalysis: A review of applications across various platforms for fluorescence spectroscopy and imaging. *Applied Spectroscopy* 67 (3): 215–252.
28. Chauhan, V.P., Z. Popović, O. Chen, J. Cui, D. Fukumura, M.G. Bawendi, and R.K. Jain. 2011. Fluorescent nanorods and nanospheres for real-time in vivo probing of nanoparticle shape-dependent tumor penetration. *Angewandte Chemie (International Edition in English)* 50 (48): 11417–11420.
29. Leatherdale, C.A., W.-K. Woo, F.V. Mikulec, and M.G. Bawendi. 2002. On the absorption cross section of CdSe nanocrystal quantum dots. *Journal of Physical Chemistry B* 106 (31): 7619–7622.
30. Yu, W.W., L. Qu, W. Guo, and X. Peng. 2003. Experimental determination of the extinction coefficient of CdTe, CdSe, and CdS nanocrystals. *Chemistry of Materials* 15 (14): 2854–2860.
31. Jasieniak, J., L. Smith, J. van Embden, P. Mulvaney, and M. Califano. 2009. Re-examination of the size-dependent absorption properties of CdSe quantum dots. *The Journal of Physical Chemistry C* 113 (45): 19468–19474.
32. Blanton, S.A., A. Dehestani, P.C. Lin, and P. Guyot-Sionnest. 1994. Photoluminescence of single semiconductor nanocrystallites by two-photon excitation microscopy. *Chemical Physics Letters* 229 (3): 317–322.
33. Wu, X., H. Liu, J. Liu, K.N. Haley, J.A. Treadway, J.P. Larson, N. Ge, F. Peale, and M. P. Bruchez. 2003. Immunofluorescent labeling of cancer marker Her2 and other cellular targets with semiconductor quantum dots. *Nature Biotechnology* 21 (1): 41–46.
34. Murray, C., D. Norris, and M. Bawendi. 1993. Synthesis and characterization of nearly monodisperse CdE (E = sulfur, selenium, tellurium) semiconductor nanocrystallites. *Journal of the American Chemical Society* 115 (19): 8706–8715.
35. Clark, M.D., S.K. Kumar, J.S. Owen, and E.M. Chan. 2011. Focusing nanocrystal size distributions via production control. *Nano Letters* 11 (5): 1976–1980.
36. Peng, X., M.C. Schlamp, A.V. Kadavanich, and A.P. Alivisatos. 1997. Epitaxial growth of highly luminescent CdSe/CdS core/shell nanocrystals with photostability and electronic accessibility. *Journal of the American Chemical Society* 119 (30): 7019–7029.
37. Hines, M.A., and P. Guyot-Sionnest. 1996. Synthesis and characterization of strongly luminescing ZnS-capped CdSe nanocrystals. *Journal of Physical Chemistry* 100 (2): 468–471.
38. Dabbousi, B.O., J. Rodriguez-Viejo, F.V. Mikulec, J.R. Heine, H. Mattoussi, R. Ober, K.F. Jensen, and M.G. Bawendi. 1997. (CdSe)ZnS core—shell quantum dots: synthesis and characterization of a size series of highly luminescent nanocrystallites. *The Journal of Physical Chemistry B* 101 (46): 9463–9475.

39. Xie, R., U. Kolb, J. Li, T. Basche, and A. Mews. 2005. Synthesis and characterization of highly luminescent CdSe-core CdS/Zn0.5Cd0.5S/ZnS multishell nanocrystals. *Journal of the American Chemical Society* 127 (20): 7480–7488.
40. Greytak, A.B., P.M. Allen, W. Liu, J. Zhao, E.R. Young, Z. Popović, B.J. Walker, D.G. Nocera, and M.G. Bawendi. 2012. Alternating layer addition approach to CdSe/CdS core/shell quantum dots with near-unity quantum yield and high on-time fractions. *Chemical Science* 3 (6): 2028–2034.
41. Chen, O., J. Zhao, V.P. Chauhan, J. Cui, C. Wong, D.K. Harris, H. Wei, H.-S. Han, D. Fukumura, R.K. Jain, and M.G. Bawendi. 2013. Compact high-quality CdSe–CdS core–shell nanocrystals with narrow emission linewidths and suppressed blinking. *Nature Materials* 12 (5): 445–451.
42. Cui, J., A.P. Beyler, L.F. Marshall, O. Chen, D.K. Harris, D.D. Wanger, X. Brokmann, and M.G. Bawendi. 2013. Direct probe of spectral inhomogeneity reveals synthetic tunability of single-nanocrystal spectral linewidths. *Nature Chemistry* 5: 602–606.
43. Dubertret, B., P. Skourides, D.J. Norris, V. Noireaux, A.H. Brivanlou, and A. Libchaber. 2002. In vivo imaging of quantum dots encapsulated in phospholipid micelles. *Science* 298 (5599): 1759–1762.
44. Susumu, K., E. Oh, J.B. Delehanty, J.B. Blanco-Canosa, B.J. Johnson, V. Jain, W.J. Hervey, W.R. Algar, K. Boeneman, P.E. Dawson, and I.L. Medintz. 2011. Multifunctional compact zwitterionic ligands for preparing robust biocompatible semiconductor quantum dots and gold nanoparticles. *Journal of the American Chemical Society* 133 (24): 9480–9496.
45. Liu, W., M. Howarth, A.B. Greytak, Y. Zheng, D.G. Nocera, A.Y. Ting, and M.G. Bawendi. 2008. Compact biocompatible quantum dots functionalized for cellular imaging. *Journal of the American Chemical Society* 130 (4): 1274–1284.
46. Liu, W., A.B. Greytak, J. Lee, C.R. Wong, J. Park, L.F. Marshall, W. Jiang, P.N. Curtin, A. Y. Ting, D.G. Nocera, D. Fukumura, R.K. Jain, and M.G. Bawendi. 2010. Compact biocompatible quantum dots via RAFT-mediated synthesis of imidazole-based random copolymer ligand. *Journal of the American Chemical Society* 132 (2): 472–483.
47. Palui, G., H.B. Na, and H. Mattoussi. 2012. Poly(ethylene glycol)-based multidentate oligomers for biocompatible semiconductor and gold nanocrystals. *Langmuir* 28 (5): 2761–2772.
48. Viswanath, A., Y. Shen, A.N. Green, R. Tan, A.B. Greytak, and B.C. Benicewicz. 2014. Copolymerization and synthesis of multiply binding histamine ligands for the robust functionalization of quantum dots. *Macromolecules* 47 (23): 8137–8144.
49. McBride, J., J. Treadway, L.C. Feldman, S.J. Pennycook, and S.J. Rosenthal. 2006. Structural basis for near unity quantum yield core/shell nanostructures. *Nano Letters* 6 (7): 1496–1501.
50. Orfield, N.J., J.R. McBride, J.D. Keene, L.M. Davis, and S.J. Rosenthal. 2015. Correlation of atomic structure and photoluminescence of the same quantum dot: Pinpointing surface and internal defects that inhibit photoluminescence. *ACS Nano* 9 (1): 831–839.
51. Li, J.J., Y.A. Wang, W.Z. Guo, J.C. Keay, T.D. Mishima, M.B. Johnson, and X.G. Peng. 2003. Large-scale synthesis of nearly monodisperse CdSe/CdS core/shell nanocrystals using air-stable reagents via successive ion layer adsorption and reaction. *Journal of the American Chemical Society* 125 (41): 12567–12575.
52. Arnspang, E.C., J.R. Brewer, and B.C. Lagerholm. 2012. Multi-color single particle tracking with quantum dots. *PLoS ONE* 7 (11): e48521.
53. Choi, H.S., W. Liu, P. Misra, E. Tanaka, J.P. Zimmer, B.I. Ipe, M.G. Bawendi, and J.V. Frangioni. 2007. Renal clearance of quantum dots. *Nature Biotechnology* 25 (10): 1165–1170.
54. Burns, A.A., J. Vider, H. Ow, E. Herz, O. Penate-Medina, M. Baumgart, S.M. Larson, U. Wiesner, and M. Bradbury. 2009. Fluorescent silica nanoparticles with efficient urinary excretion for nanomedicine. *Nano Letters* 9 (1): 442–448.

55. Koberling, F., U. Kolb, G. Philipp, I. Potapova, T. Basché, and A. Mews. 2003. Fluorescence anisotropy and crystal structure of individual semiconductor nanocrystals. *Journal of Physical Chemistry B* 107 (30): 7463–7471.
56. Cassette, E., B. Mahler, J.-M. Guigner, G. Patriarche, B. Dubertret, and T. Pons. 2012. Colloidal CdSe/CdS dot-in-plate nanocrystals with 2D-polarized emission. *ACS Nano* 6 (8): 6741–6750.
57. Tice, D.B., D.J. Weinberg, N. Mathew, R.P.H. Chang, and E.A. Weiss. 2013. Measurement of wavelength-dependent polarization character in the absorption anisotropies of ensembles of CdSe nanorods. *Journal of Physical Chemistry C* 117 (25): 13289–13296.
58. Diroll, B.T., T. Dadosh, A. Koschitzky, Y.E. Goldman, and C.B. Murray. 2013. Interpreting the energy-dependent anisotropy of colloidal nanorods using ensemble and single-particle spectroscopy. *Journal of Physical Chemistry C* 117 (45): 23928–23937.
59. Park, K., Z. Deutsch, J.J. Li, D. Oron, and S. Weiss. 2012. Single molecule quantum-confined stark effect measurements of semiconductor nanoparticles at room temperature. *ACS Nano* 6 (11): 10013–10023.
60. Kloepfer, J.A., N. Cohen, and J.L. Nadeau. 2004. FRET between CdSe quantum dots in lipid vesicles and water- and lipid-soluble dyes. *Journal of Physical Chemistry B* 108 (44): 17042–17049.
61. Lo, S.S., T. Mirkovic, C.-H. Chuang, C. Burda, and G.D. Scholes. 2011. Emergent properties resulting from type-II band alignment in semiconductor nanoheterostructures. *Advanced Materials* 23 (2): 180–197.
62. Marshall, J.D., and M.J. Schnitzer. 2013. Optical strategies for sensing neuronal voltage using quantum dots and other semiconductor nanocrystals. *ACS Nano* 7 (5): 4601–4609.
63. Tomaselli, M., J.L. Yarger, M. Bruchez, R.H. Havlin, D. DeGraw, A. Pines, and A. P. Alivisatos. 1999. NMR study of InP quantum dots: Surface structure and size effects. *The Journal of Chemical Physics* 110 (18): 8861–8864.
64. Rosenthal, S.J., J. McBride, S.J. Pennycook, and L.C. Feldman. 2007. Synthesis, surface studies, composition and structural characterization of CdSe, core/shell, and biologically active nanocrystals. *Surface Science Reports* 62 (4): 111–157.
65. Rempel, J.Y., B.L. Trout, M.G. Bawendi, and K.F. Jensen. 2005. Properties of the CdSe (0001), (0001), and (1120) single crystal surfaces: Relaxation, reconstruction, and adatom and admolecule adsorption. *Journal of Physical Chemistry B* 109 (41): 19320–19328.
66. Rempel, J.Y., B.L. Trout, M.G. Bawendi, and K.F. Jensen. 2006. Density functional theory study of ligand binding on CdSe (0001), (0001), and (1120) single crystal relaxed and reconstructed surfaces: Implications for nanocrystalline growth. *Journal of Physical Chemistry B* 110 (36): 18007–18016.
67. Gong, K., and D.F. Kelley. 2014. A predictive model of shell morphology in CdSe/CdS core/shell quantum dots. *The Journal of Chemical Physics* 141 (19): 194704.
68. Milliron, D.J., S.M. Hughes, Y. Cui, L. Manna, J.B. Li, L.W. Wang, and A.P. Alivisatos. 2004. Colloidal nanocrystal heterostructures with linear and branched topology. *Nature* 430 (6996): 190–195.
69. Talapin, D.V., R. Koeppe, S. Götzinger, A. Kornowski, J.M. Lupton, A.L. Rogach, O. Benson, J. Feldmann, and H. Weller. 2003. Highly emissive colloidal CdSe/CdS heterostructures of mixed dimensionality. *Nano Letters* 3 (12): 1677–1681.
70. Talapin, D.V., J.H. Nelson, E.V. Shevchenko, S. Aloni, B. Sadtler, and A.P. Alivisatos. 2007. Seeded growth of highly luminescent CdSe/CdS nanoheterostructures with rod and tetrapod morphologies. *Nano Letters* 7 (10): 2951–2959.
71. Carbone, L., C. Nobile, M. De Giorgi, F.D. Sala, G. Morello, P. Pompa, M. Hytch, E. Snoeck, A. Fiore, I.R. Franchini, M. Nadasan, A.F. Silvestre, L. Chiodo, S. Kudera, R. Cingolani, R. Krahne, and L. Manna. 2007. Synthesis and micrometer-scale assembly of colloidal CdSe/CdS nanorods prepared by a seeded growth approach. *Nano Letters* 7 (10): 2942–2950.

72. Choi, C.L., K.J. Koski, S. Sivasankar, and A.P. Alivisatos. 2009. Strain-dependent photoluminescence behavior of CdSe/CdS nanocrystals with spherical, linear, and branched topologies. *Nano Letters* 9 (10): 3544–3549.
73. Borys, N.J., M.J. Walter, J. Huang, D.V. Talapin, and J.M. Lupton. 2010. The role of particle morphology in interfacial energy transfer in CdSe/CdS heterostructure nanocrystals. *Science* 330 (6009): 1371–1374.
74. Guo, Y., S.R. Alvarado, J.D. Barclay, and J. Vela. 2013. Shape-programmed nanofabrication: Understanding the reactivity of dichalcogenide precursors. *ACS Nano* 7 (4): 3616–3626.
75. Ruberu, T.P.A., and J. Vela. 2011. Expanding the one-dimensional CdS–CdSe composition landscape: Axially anisotropic $CdS_{1-x}Se_x$ nanorods. *ACS Nano* 5 (7): 5775–5784.
76. Ruberu, T.P.A., H.R. Albright, B. Callis, B. Ward, J. Cisneros, H.-J. Fan, and J. Vela. 2012. Molecular control of the nanoscale: Effect of phosphine-chalcogenide reactivity on CdS–CdSe nanocrystal composition and morphology. *ACS Nano* 6 (6): 5348–5359.
77. Vela, J. 2013. Molecular chemistry to the fore: New insights into the fascinating world of photoactive colloidal semiconductor nanocrystals. *Journal of Physical Chemistry Letters* 4 (4): 653–668.
78. Kirsanova, M., A. Nemchinov, N.N. Hewa-Kasakarage, N. Schmall, and M. Zamkov. 2009. Synthesis of ZnSe/CdS/ZnSe nanobarbells showing photoinduced charge separation. *Chemistry of Materials* 21 (18): 4305–4309.
79. Mahler, B., B. Nadal, C. Bouet, G. Patriarche, and B. Dubertret. 2012. Core/shell colloidal semiconductor nanoplatelets. *Journal of the American Chemical Society* 134 (45): 18591–18598.
80. Müller, J., J.M. Lupton, A.L. Rogach, J. Feldmann, D.V. Talapin, and H. Weller. 2005. Monitoring surface charge migration in the spectral dynamics of single CdSe/CdS nanodot/nanorod heterostructures. *Physical Review B* 72 (20): 205339.
81. Müller, J., J.M. Lupton, P.G. Lagoudakis, F. Schindler, R. Koeppe, A.L. Rogach, J. Feldmann, D.V. Talapin, and H. Weller. 2005. Wave function engineering in elongated semiconductor nanocrystals with heterogeneous carrier confinement. *Nano Letters* 5 (10): 2044–2049.
82. Mauser, C., T. Limmer, E. Da Como, K. Becker, A.L. Rogach, J. Feldmann, and D.V. Talapin. 2008. Anisotropic optical emission of single CdSe/CdS tetrapod heterostructures: Evidence for a wavefunction symmetry breaking. *Physical Review B* 77 (15): 153303.
83. Lutich, A.A., C. Mauser, E. Da Como, J. Huang, A. Vaneski, D.V. Talapin, A.L. Rogach, and J. Feldmann. 2010. Multiexcitonic dual emission in CdSe/CdS tetrapods and nanorods. *Nano Letters* 10 (11): 4646–4650.
84. Teitelboim, A., N. Meir, M. Kazes, and D. Oron. 2016. Colloidal double quantum dots. *Accounts of Chemical Research* 49 (5): 902–910.
85. Saba, M., S. Minniberger, F. Quochi, J. Roither, M. Marceddu, A. Gocalinska, M.V. Kovalenko, D.V. Talapin, W. Heiss, A. Mura, and G. Bongiovanni. 2009. Exciton-exciton interaction and optical gain in colloidal CdSe/CdS dot/rod nanocrystals. *Advanced Materials* 21 (48): 4942–4946.
86. Khon, E., K. Lambright, R.S. Khnayzer, P. Moroz, D. Perera, E. Butaeva, S. Lambright, F. N. Castellano, and M. Zamkov. 2013. Improving the catalytic activity of semiconductor nanocrystals through selective domain etching. *Nano Letters* 13 (5): 2016–2023.
87. Wu, W.-Y., S. Chakrabortty, C.K.L. Chang, A. Guchhait, M. Lin, and Y. Chan. 2014. Promoting 2D growth in colloidal transition metal sulfide semiconductor nanostructures via halide ions. *Chemistry of Materials* 26 (21): 6120–6126.
88. Kortan, A.R., R. Hull, R.L. Opila, M.G. Bawendi, M.L. Steigerwald, P.J. Carroll, and L.E. Brus. 1990. Nucleation and growth of cadmium selenide on zinc sulfide quantum crystallite seeds, and vice versa, in inverse micelle media. *Journal of the American Chemical Society* 112 (4): 1327–1332.

89. Galland, C., Y. Ghosh, A. Steinbrück, M. Sykora, J.A. Hollingsworth, V.I. Klimov, and H. Htoon. 2011. Two types of luminescence blinking revealed by spectroelectrochemistry of single quantum dots. *Nature* 479 (7372): 203–207.
90. Mahler, B., N. Lequeux, and B. Dubertret. 2010. Ligand-controlled polytypism of thick-shell CdSe/CdS nanocrystals. *Journal of the American Chemical Society* 132 (3): 953–959.
91. Nan, W., Y. Niu, H. Qin, F. Cui, Y. Yang, R. Lai, W. Lin, and X. Peng. 2012. Crystal structure control of zinc-blende CdSe/CdS core/shell nanocrystals: Synthesis and structure-dependent optical properties. *Journal of the American Chemical Society* 134 (48): 19685–19693.
92. Park, Y.-S., A.V. Malko, J. Vela, Y. Chen, Y. Ghosh, F. García-Santamaría, J.A. Hollingsworth, V.I. Klimov, and H. Htoon. 2011. Near-Unity quantum yields of biexciton emission from CdSe/CdS nanocrystals measured using single-particle spectroscopy. *Physical Review Letters* 106 (18): 187401.
93. Htoon, H., A.V. Malko, D. Bussian, J. Vela, Y. Chen, J.A. Hollingsworth, and V.I. Klimov. 2010. Highly emissive multiexcitons in steady-state photoluminescence of individual "giant" CdSe/CdS core/shell nanocrystals. *Nano Letters* 10 (7): 2401–2407.
94. Pal, B.N., Y. Ghosh, S. Brovelli, R. Laocharoensuk, V.I. Klimov, J.A. Hollingsworth, and H. Htoon. 2012. "Giant" CdSe/CdS core/shell nanocrystal quantum dots as efficient electroluminescent materials: Strong influence of shell thickness on light-emitting diode performance. *Nano Letters* 12 (1): 331–336.
95. Chen, Y., J. Vela, H. Htoon, J.L. Casson, D.J. Werder, D.A. Bussian, V.I. Klimov, and J.A. Hollingsworth. 2008. Giant multishell CdSe nanocrystal quantum dots with suppressed blinking. *Journal of the American Chemical Society* 130 (15): 5026–5027.
96. Pons, T., N. Lequeux, B. Mahler, S. Sasnouski, A. Fragola, and B. Dubertret. 2009. Synthesis of near-infrared-emitting, water-soluble CdTeSe/CdZnS core/shell quantum dots. *Chemistry of Materials* 21 (8): 1418–1424.
97. Sapra, S., J. Poppe, and A. Eychmüller. 2007. CdSe nanorod synthesis: A new approach. *Small* 3 (11): 1886–1888.
98. Boles, M.A., D. Ling, T. Hyeon, and D.V. Talapin. 2016. The surface science of nanocrystals. *Nature Materials* 15 (2): 141–153.
99. Peterson, M.D., L.C. Cass, R.D. Harris, K. Edme, K. Sung, and E.A. Weiss. 2014. The role of ligands in determining the exciton relaxation dynamics in semiconductor quantum dots. *Annual Review of Physical Chemistry* 65 (1): 317–339.
100. Green, M. 2010. The nature of quantum dot capping ligands. *Journal of Materials Chemistry* 20 (28): 5797–5809.
101. Guo, Y., K. Marchuk, S. Sampat, R. Abraham, N. Fang, A.V. Malko, and J. Vela. 2012. Unique challenges accompany thick-shell CdSe/nCdS (N > 10) nanocrystal synthesis. *Journal of Physical Chemistry C* 116 (4): 2791–2800.
102. Moreels, I., K. Lambert, D. Smeets, D. De Muynck, T. Nollet, J.C. Martins, F. Vanhaecke, A. Vantomme, C. Delerue, G. Allan, and Z. Hens. 2009. Size-dependent optical properties of colloidal PbS quantum dots. *ACS Nano* 3 (10): 3023–3030.
103. Talapin, D.V., N. Gaponik, H. Borchert, A.L. Rogach, M. Haase, and H. Weller. 2002. Etching of colloidal InP nanocrystals with fluorides: Photochemical nature of the process resulting in high photoluminescence efficiency. *Journal of Physical Chemistry B* 106 (49): 12659–12663.
104. van Embden, J., J. Jasieniak, and P. Mulvaney. 2009. Mapping the optical properties of CdSe/CdS heterostructure nanocrystals: The effects of core size and shell thickness. *Journal of the American Chemical Society* 131 (40): 14299–14309.
105. Ithurria, S., and D.V. Talapin. 2012. Colloidal atomic layer deposition (c-ALD) using self-limiting reactions at nanocrystal surface coupled to phase transfer between polar and nonpolar media. *Journal of the American Chemical Society* 134 (45): 18585–18590.
106. Hassinen, A., I. Moreels, K. De Nolf, P.F. Smet, J.C. Martins, and Z. Hens. 2012. Short-chain alcohols strip X-type ligands and quench the luminescence of PbSe and CdSe

quantum dots, acetonitrile does not. *Journal of the American Chemical Society* 134 (51): 20705–20712.
107. Anderson, N.C., M.P. Hendricks, J.J. Choi, and J.S. Owen. 2013. Ligand exchange and the stoichiometry of metal chalcogenide nanocrystals: Spectroscopic observation of facile metal-carboxylate displacement and binding. *Journal of the American Chemical Society* 135 (49): 18536–18548.
108. Shen, Y., M.Y. Gee, R. Tan, P.J. Pellechia, and A.B. Greytak. 2013. Purification of quantum dots by gel permeation chromatography and the effect of excess ligands on shell growth and ligand exchange. *Chemistry of Materials* 25 (14): 2838–2848.
109. Lhuillier, E., P. Hease, S. Ithurria, and B. Dubertret. 2014. Selective electrophoretic deposition of CdSe nanoplatelets. *Chemistry of Materials* 26 (15): 4514–4520.
110. Razgoniaeva, N., L. Carrillo, D. Burchfield, P. Moroz, P. Adhikari, P. Yadav, D. Khon, and M. Zamkov. 2016. Colloidal synthesis of monodisperse semiconductor nanocrystals through saturated ionic layer adsorption. *Chemistry of Materials* 28 (8): 2823–2833.
111. Xu, J., D. Cui, T. Zhu, G. Paradee, Z. Liang, Q. Wang, S. Xu, and A.Y. Wang. 2006. Synthesis and surface modification of PbSe/PbS core-shell nanocrystals for potential device applications. *Nanotechnology* 17 (21): 5428–5434.
112. Pourret, A., P. Guyot-Sionnest, and J.W. Elam. 2009. Atomic layer deposition of ZnO in quantum dot thin films. *Advanced Materials* 21 (2): 232–235.
113. Zherebetskyy, D., M. Scheele, Y. Zhang, N. Bronstein, C. Thompson, D. Britt, M. Salmeron, P. Alivisatos, and L.-W. Wang. 2014. Hydroxylation of the surface of PbS nanocrystals passivated with oleic acid. *Science* 344 (6190): 1380–1384.
114. García-Rodríguez, R., M.P. Hendricks, B.M. Cossairt, H. Liu, and J.S. Owen. 2013. Conversion reactions of cadmium chalcogenide nanocrystal precursors. *Chemistry of Materials* 25 (8): 1233–1249.
115. Anderson, N.C., and J.S. Owen. 2013. Soluble, chloride-terminated CdSe nanocrystals: Ligand exchange monitored by 1H and 31P NMR spectroscopy. *Chemistry of Materials* 25 (1): 69–76.
116. Thomson, J.W., K. Nagashima, P.M. Macdonald, and G.A. Ozin. 2011. From sulfur—amine solutions to metal sulfide nanocrystals: Peering into the oleylamine—sulfur black box. *Journal of the American Chemical Society* 133 (13): 5036–5041.
117. Hendricks, M.P., M.P. Campos, G.T. Cleveland, I. Jen-La Plante, and J.S. Owen. 2015. A tunable library of substituted thiourea precursors to metal sulfide nanocrystals. *Science* 348 (6240): 1226–1230.
118. van Embden, J., J. Jasieniak, D.E. Gomez, P. Mulvaney, and M. Giersig. 2007. Review of the synthetic chemistry involved in the production of core/shell semiconductor nanocrystals. *Australian Journal of Chemistry* 60 (7): 457–471.
119. Liu, H.T., J.S. Owen, and A.P. Alivisatos. 2007. Mechanistic study of precursor evolution in colloidal group II-VI semiconductor nanocrystal synthesis. *Journal of the American Chemical Society* 129 (2): 305–312.
120. Wei, H.H.-Y., C.M. Evans, B.D. Swartz, A.J. Neukirch, J. Young, O.V. Prezhdo, and T.D. Krauss. 2012. Colloidal semiconductor quantum dots with tunable surface composition. *Nano Letters* 12 (9): 4465–4471.
121. Sowers, K.L., Z. Hou, J.J. Peterson, B. Swartz, S. Pal, O. Prezhdo, and T.D. Krauss. 2016. Photophysical properties of CdSe/CdS core/shell quantum dots with tunable surface composition. *Chemical Physics* 471: 24–31.
122. Kaniyankandy, S., S. Rawalekar, and H.N. Ghosh. 2013. Charge carrier cascade in Type II CdSe–CdTe graded core–shell interface. *Journal of Materials Chemistry C* 1 (15): 2755–2763.
123. Lee, H., M. Wang, P. Chen, D.R. Gamelin, S.M. Zakeeruddin, M. Graetzel, and M.K. Nazeeruddin. 2009. Efficient CdSe quantum dot-sensitized solar cells prepared by an improved successive ionic layer adsorption and reaction process. *Nano Letters* 9 (12): 4221–4227.

124. Becker, M.A., J.G. Radich, B.A. Bunker, and P.V. Kamat. 2014. How does a SILAR CdSe film grow? Tuning the deposition steps to suppress interfacial charge recombination in solar cells. *Journal of Physical Chemistry Letters* 5 (9): 1575–1582.
125. Jasieniak, J., and P. Mulvaney. 2007. From Cd-rich to Se-rich—the manipulation of CdSe nanocrystal surface stoichiometry. *Journal of the American Chemical Society* 129 (10): 2841–2848.
126. Shen, Y., R. Tan, M.Y. Gee, and A.B. Greytak. 2015. Quantum yield regeneration: Influence of neutral ligand binding on photophysical properties in colloidal core/shell quantum dots. *ACS Nano* 9 (3): 3345–3359.
127. Tan, R., D.A. Blom, S. Ma, and A.B. Greytak. 2013. Probing surface saturation conditions in alternating layer growth of CdSe/CdS core/shell quantum dots. *Chemistry of Materials* 25 (18): 3724–3736.
128. Morris-Cohen, A.J., M. Malicki, M.D. Peterson, J.W.J. Slavin, and E.A. Weiss. 2013. Chemical, structural, and quantitative analysis of the ligand shells of colloidal quantum dots. *Chemistry of Materials* 25 (8): 1155–1165.
129. Caldwell, M.A., A.E. Albers, S.C. Levy, T.E. Pick, B.E. Cohen, B.A. Helms, and D. J. Milliron. 2011. Driving oxygen coordinated ligand exchange at nanocrystal surfaces using trialkylsilylated chalcogenides. *Chemical Communications* 47 (1): 556–558.
130. Tan, R., Y. Shen, S.K. Roberts, M.Y. Gee, D.A. Blom, and A.B. Greytak. 2015. Reducing competition by coordinating solvent promotes morphological control in alternating layer growth of CdSe/CdS core/shell quantum dots. *Chemistry of Materials* 27 (21): 7468–7480.
131. Fedin, I., and D.V. Talapin. 2014. Probing the surface of colloidal nanomaterials with potentiometry in situ. *Journal of the American Chemical Society* 136 (32): 11228–11231.
132. Todescato, F., A. Minotto, R. Signorini, J.J. Jasieniak, and R. Bozio. 2013. Investigation into the heterostructure interface of CdSe-based core-shell quantum dots using surface-enhanced raman spectroscopy. *ACS Nano* 7 (8): 6649–6657.
133. Dias, E.A., S.L. Sewall, and P. Kambhampati. 2007. Light harvesting and carrier transport in core/barrier/shell semiconductor nanocrystals. *Journal of Physical Chemistry C* 111 (2): 708–713.
134. Aharoni, A., T. Mokari, I. Popov, and U. Banin. 2006. Synthesis of InAs/CdSe/ZnSe core/shell1/shell2 structures with bright and stable near-infrared fluorescence. *Journal of the American Chemical Society* 128 (1): 257–264.
135. Snee, P.T., Y.H. Chan, D.G. Nocera, and M.G. Bawendi. 2005. Whispering-gallery-mode lasing from a semiconductor nanocrystal/microsphere resonator composite. *Advanced Materials* 17 (9): 1131–1136.
136. Liu, W., H.S. Choi, J.P. Zimmer, E. Tanaka, J.V. Frangioni, and M. Bawendi. 2007. Compact cysteine-coated CdSe(ZnCdS) quantum dots for in vivo applications. *Journal of the American Chemical Society* 129 (47): 14530–14531.
137. Peng, X., L. Manna, W. Yang, J. Wickham, E. Scher, A. Kadavanich, and A.P. Alivisatos. 2000. Shape control of CdSe nanocrystals. *Nature* 404 (6773): 59–61.
138. Manna, L., E.C. Scher, L.-S. Li, and A.P. Alivisatos. 2002. Epitaxial growth and photochemical annealing of graded CdS/ZnS shells on colloidal CdSe nanorods. *Journal of the American Chemical Society* 124 (24): 7136–7145.
139. Wang, W., S. Banerjee, S. Jia, M.L. Steigerwald, and I.P. Herman. 2007. Ligand control of growth, morphology, and capping structure of colloidal CdSe nanorods. *Chemistry of Materials* 19 (10): 2573–2580.
140. Wang, F., and W.E. Buhro. 2012. Morphology control of cadmium selenide nanocrystals: Insights into the roles of di-n-octylphosphine oxide (DOPO) and di-n-octylphosphinic acid (DOPA). *Journal of the American Chemical Society* 134 (11): 5369–5380.
141. Deutsch, Z., L. Neeman, and D. Oron. 2013. Luminescence upconversion in colloidal double quantum dots. *Nature Nanotechnology* 8 (9): 649–653.
142. Joo, J., J.S. Son, S.G. Kwon, J.H. Yu, and T. Hyeon. 2006. Low-temperature solution-phase synthesis of quantum well structured CdSe nanoribbons. *Journal of the American Chemical Society* 128 (17): 5632–5633.

143. Liu, Y.-H., V.L. Wayman, P.C. Gibbons, R.A. Loomis, and W.E. Buhro. 2010. Origin of high photoluminescence efficiencies in CdSe quantum belts. *Nano Letters* 10 (1): 352–357.
144. Ithurria, S., and B. Dubertret. 2008. Quasi 2D colloidal CdSe platelets with thicknesses controlled at the atomic level. *Journal of the American Chemical Society* 130 (49): 16504–16505.
145. Ithurria, S., M.D. Tessier, B. Mahler, R.P.S.M. Lobo, B. Dubertret, and A.L. Efros. 2011. Colloidal nanoplatelets with two-dimensional electronic structure. *Nature Materials* 10 (12): 936–941.
146. Tessier, M.D., P. Spinicelli, D. Dupont, G. Patriarche, S. Ithurria, and B. Dubertret. 2014. Efficient exciton concentrators built from colloidal core/crown CdSe/CdS semiconductor nanoplatelets. *Nano Letters* 14 (1): 207–213.
147. Vogel, R., P. Hoyer, and H. Weller. 1994. Quantum-sized PbS, CdS, Ag2S, Sb2S3, and Bi2S3 particles as sensitizers for various nanoporous wide-bandgap semiconductors. *Journal of Physical Chemistry* 98 (12): 3183–3188.
148. Tak, Y., S.J. Hong, J.S. Lee, and K. Yong. 2009. Fabrication of ZnO/CdS core/shell nanowire arrays for efficient solar energy conversion. *Journal of Materials Chemistry* 19 (33): 5945–5951.
149. Jiang, Y., C. Li, W. Cao, Y. Jiang, S. Shang, and C. Xia. 2015. Large scale fabrication of well-aligned CdS/P-Si shell/core nanowire arrays for photodetectors using solution methods. *Physical Chemistry Chemical Physics: PCCP* 17 (26): 16784–16790.
150. Kinder, E., P. Moroz, G. Diederich, A. Johnson, M. Kirsanova, A. Nemchinov, T. O'Connor, D. Roth, and M. Zamkov. 2011. Fabrication of all-inorganic nanocrystal solids through matrix encapsulation of nanocrystal arrays. *Journal of the American Chemical Society* 133 (50): 20488–20499.
151. Kim, D.K., A.T. Fafarman, B.T. Diroll, S.H. Chan, T.R. Gordon, C.B. Murray, and C.R. Kagan. 2013. Solution-based stoichiometric control over charge transport in nanocrystalline CdSe devices. *ACS Nano* 7 (10): 8760–8770.
152. Oh, S.J., N.E. Berry, J.-H. Choi, E.A. Gaulding, H. Lin, T. Paik, B.T. Diroll, S. Muramoto, C.B. Murray, and C.R. Kagan. 2014. Designing high-performance PbS and PbSe nanocrystal electronic devices through stepwise, post-synthesis, colloidal atomic layer deposition. *Nano Letters* 14 (3): 1559–1566.
153. Deutsch, Z., O. Schwartz, R. Tenne, R. Popovitz-Biro, and D. Oron. 2012. Two-color antibunching from band-gap engineered colloidal semiconductor nanocrystals. *Nano Letters* 12 (6): 2948–2952.
154. Tessier, M.D., B. Mahler, B. Nadal, H. Heuclin, S. Pedetti, and B. Dubertret. 2013. Spectroscopy of colloidal semiconductor core/shell nanoplatelets with high quantum yield. *Nano Letters* 13 (7): 3321–3328.
155. Liu, Y.-H., F. Wang, J. Hoy, V.L. Wayman, L.K. Steinberg, R.A. Loomis, and W.E. Buhro. 2012. Bright core-shell semiconductor quantum wires. *Journal of the American Chemical Society* 134 (45): 18797–18803.
156. Wu, K., H. Zhu, and T. Lian. 2015. Ultrafast exciton dynamics and light-driven H2 evolution in colloidal semiconductor nanorods and Pt-tipped nanorods. *Accounts of Chemical Research* 48 (3): 851–859.
157. Lhuillier, E., S. Pedetti, S. Ithurria, B. Nadal, H. Heuclin, and B. Dubertret. 2015. Two-dimensional colloidal metal chalcogenides semiconductors: Synthesis, spectroscopy, and applications. *Accounts of Chemical Research* 48 (1): 22–30.
158. Selinsky, R.S., Q. Ding, M.S. Faber, J.C. Wright, and S. Jin. 2013. Quantum dot nanoscale heterostructures for solar energy conversion. *Chemical Society Reviews* 42 (7): 2963–2985.
159. Hayden, O., A.B. Greytak, and D.C. Bell. 2005. Core-shell nanowire light-emitting diodes. *Advanced Materials* 17 (6): 701–704.
160. Bouet, C., D. Laufer, B. Mahler, B. Nadal, H. Heuclin, S. Pedetti, G. Patriarche, and B. Dubertret. 2014. Synthesis of zinc and lead chalcogenide core and core/shell nanoplatelets using sequential cation exchange reactions. *Chemistry of Materials* 26 (9): 3002–3008.

161. Li, H., M. Zanella, A. Genovese, M. Povia, A. Falqui, C. Giannini, and L. Manna. 2011. Sequential cation exchange in nanocrystals: Preservation of crystal phase and formation of metastable phases. *Nano Letters* 11 (11): 4964–4970.
162. Yeltik, A., S. Delikanli, M. Olutas, Y. Kelestemur, B. Guzelturk, and H.V. Demir. 2015. Experimental determination of the absorption cross-section and molar extinction coefficient of colloidal cdse nanoplatelets. *Journal of Physical Chemistry C* 119 (47): 26768–26775.
163. Vietmeyer, F., M.P. McDonald, and M. Kuno. 2012. Single nanowire microscopy and spectroscopy. *Journal of Physical Chemistry C* 116 (23): 12379–12396.
164. Yeh, C.-Y., Z.W. Lu, S. Froyen, and A. Zunger. 1992. Zincblende-wurtzite polytypism in semiconductors. *Physical Review B* 46 (16): 10086–10097.
165. Williams, E.S., K.J. Major, A. Tobias, D. Woodall, V. Morales, C. Lippincott, P.J. Moyer, and M. Jones. 2013. Characterizing the influence of TOPO on exciton recombination dynamics in colloidal CdSe quantum dots. *Journal of Physical Chemistry C* 117 (8): 4227–4237.
166. Sitt, A., A. Salant, G. Menagen, and U. Banin. 2011. Highly emissive nano rod-in-rod heterostructures with strong linear polarization. *Nano Letters* 11 (5): 2054–2060.
167. Li, Z., and X. Peng. 2011. Size/shape-controlled synthesis of colloidal CdSe quantum disks: Ligand and temperature effects. *Journal of the American Chemical Society* 133 (17): 6578–6586.
168. Lieber, C.M. 2011. Semiconductor nanowires: A platform for nanoscience and nanotechnology. *MRS Bulletin* 36 (12): 1052–1063.
169. Yang, P., R. Yan, and M. Fardy. 2010. Semiconductor nanowire: What's next? *Nano Letters* 10 (5): 1529–1536.
170. Wang, F., A. Dong, J. Sun, R. Tang, H. Yu, and W.E. Buhro. 2006. Solution—liquid—solid growth of semiconductor nanowires. *Inorganic Chemistry* 45 (19): 7511–7521.
171. Whang, D., S. Jin, Y. Wu, and C.M. Lieber. 2003. Large-scale hierarchical organization of nanowire arrays for integrated nanosystems. *Nano Letters* 3 (9): 1255–1259.
172. Dong, Y, G. Yu, M.C. McAlpine, W. Lu, and C.M. Lieber. 2008. Si/a-Si core/shell nanowires as nonvolatile crossbar switches. *Nano Letters*.
173. Boettcher, S.W., J.M. Spurgeon, M.C. Putnam, E.L. Warren, D.B. Turner-Evans, M.D. Kelzenberg, J.R. Maiolo, H.A. Atwater, and N.S. Lewis. 2010. Energy-conversion properties of vapor-liquid-solid-grown silicon wire-array photocathodes. *Science* 327 (5962): 185–187.
174. Kempa, T.J., R.W. Day, S.-K. Kim, H.-G. Park, and C.M. Lieber. 2013. Semiconductor nanowires: A platform for exploring limits and concepts for nano-enabled solar cells. *Energy & Environmental Science* 6 (3): 719–733.
175. Patolsky, F., G. Zheng, and C.M. Lieber. 2006. Nanowire sensors for medicine and the life sciences. *Nanomedicine* 1 (1): 51–65.
176. Wagner, R.S., and W.C. Ellis. 1053. Vapor-liquid-solid mechanism of crystal growth and its application to silicon. *Transactions of the Metallurgical Society of AIME* 1965: 233.
177. Hiruma, K., M. Yazawa, K. Haraguchi, K. Ogawa, T. Katsuyama, M. Koguchi, and H. Kakibayashi. 1993. GaAs freestanding quantum-size wires. *Journal of Applied Physics* 74 (5): 3162–3171.
178. Morales, A.M., and C.M. Lieber. 1998. A laser ablation method for the synthesis of crystalline semiconductor nanowires. *Science* 279 (5348): 208–211.
179. Duan, X.F., and C.M. Lieber. 2000. General synthesis of compound semiconductor nanowires. *Advanced Materials* 12 (4): 298–302.
180. Wu, Y., Y. Cui, L. Huynh, C.J. Barrelet, D.C. Bell, and C.M. Lieber. 2004. Controlled growth and structures of molecular-scale silicon nanowires. *Nano Letters* 4 (3): 433–436.
181. Heath, J.R., and F.K. Legoues. 1993. A liquid solution synthesis of single-crystal germanium quantum wires. *Chemical Physics Letters* 208 (3–4): 263–268.
182. Koren, E., J.K. Hyun, U. Givan, E.R. Hemesath, L.J. Lauhon, and Y. Rosenwaks. 2011. Obtaining uniform dopant distributions in VLS-grown Si nanowires. *Nano Letters* 11 (1): 183–187.

183. Barrelet, C.J., Y. Wu, D.C. Bell, and C.M. Lieber. 2003. Synthesis of CdS and ZnS nanowires using single-source molecular precursors. *Journal of the American Chemical Society* 125 (38): 11498–11499.
184. Greytak, A.B., L.J. Lauhon, M.S. Gudiksen, and C.M. Lieber. 2004. Growth and transport properties of complementary germanium nanowire field-effect transistors. *Applied Physics Letters* 84 (21): 4176–4178.
185. Lauhon, L.J., M.S. Gudiksen, and C.M. Lieber. 1819. Semiconductor nanowire heterostructures. *Philos Transact A Math Phys Eng Sci* 2004 (362): 1247–1260.
186. Lu, W., J. Xiang, B.P. Timko, Y. Wu, and C.M. Lieber. 2005. One-dimensional hole gas in germanium/silicon nanowire heterostructures. *Proceedings of the National Academy of Sciences of the United States of America* 102 (29): 10046–10051.
187. Li, Y., J. Xiang, F. Qian, S. Gradecak, Y. Wu, H. Yan, D.A. Blom, and C.M. Lieber. 2006. Dopant-free GaN/AlN/AlGaN radial nanowire heterostructures as high electron mobility transistors. *Nano Letters* 6 (7): 1468–1473.
188. Goebl, J.A., R.W. Black, J. Puthussery, J. Giblin, T.H. Kosel, and M. Kuno. 2008. Solution-based II-VI core/shell nanowire heterostructures. *Journal of the American Chemical Society* 130 (44): 14822–14833.
189. Protasenko, V., S. Gordeyev, and M. Kuno. 2007. Spatial and intensity modulation of nanowire emission induced by mobile charges. *Journal of American Chemical Society.*
190. Li, Z., X. Ma, Q. Sun, Z. Wang, J. Liu, Z. Zhu, S.Z. Qiao, S.C. Smith, G.M. Lu, and A. Mews. 2010. Synthesis and characterization of colloidal core-shell semiconductor nanowires. *European Journal of Inorganic Chemistry* 2010 (27): 4325–4331.
191. McLaurin, E.J., L.R. Bradshaw, and D.R. Gamelin. 2013. Dual-emitting nanoscale temperature sensors. *Chemistry of Materials* 25 (8): 1283–1292.
192. Sapra, S., S. Mayilo, T.A. Klar, A.L. Rogach, and J. Feldmann. 2007. Bright white-light emission from semiconductor nanocrystals: By chance and by design. *Advanced Materials* 19 (4): 569–572.
193. Purves, W.K., G.H. Orians, and H.C. Heller. 1995. *Life: The science of biology*, 4th ed.; W. H. Freeman & Co.
194. Zhang, F., A.M. Aravanis, A. Adamantidis, L. de Lecea, and K. Deisseroth. 2007. Circuit-breakers: Optical technologies for probing neural signals and systems. *Nature Reviews Neuroscience* 8 (8): 577–581.
195. Villalba-Galea, C., W. Sandtner, D. Dimitrov, H. Mutoh, T. Knöpfel, and F. Bezanilla. 2009. Charge movement of a voltage-sensitiive fluorescent protein. *Biophysical Journal* 96 (2): L19.
196. Patrick, M.J., L.A. Ernst, A.S. Waggoner, D. Thai, D. Tai, and G. Salama. 2007. Enhanced aqueous solubility of long wavelength voltage-sensitive dyes by covalent attachment of polyethylene glycol. *Organic & Biomolecular Chemistry* 5 (20): 3347.
197. Empedocles, S.A., and M.G. Bawendi. 1997. Quantum-confined stark effect in single cdse nanocrystallite quantum dots. *Science* 278 (5346): 2114–2117.
198. Mattheyses, A.L., M. Kampmann, C.E. Atkinson, and S.M. Simon. 2010. Fluorescence anisotropy reveals order and disorder of protein domains in the nuclear pore complex. *Biophysical Journal* 99 (6): 1706–1717.
199. DeMay, B.S., N. Noda, A.S. Gladfelter, and R. Oldenbourg. 2011. Rapid and quantitative imaging of excitation polarized fluorescence reveals ordered septin dynamics in live yeast. *Biophysical Journal* 101 (4): 985–994.
200. Norris, D.J., A.L. Efros, M. Rosen, and M.G. Bawendi. 1996. Size dependence of exciton fine structure in CdSe quantum dots. *Physical Review B* 53 (24): 16347–16354.

Chapter 8
Plasmon Drag Effect. Theory and Experiment

M. Durach and N. Noginova

Abstract The plasmon drag effect was discovered about a decade ago in 2005. Since then, it has attracted considerable attention from the nanotechnology and photonics community due to fundamental physics of this effect as manifestation of electron plasmon coupling and myriad of possible applications in nanoelectronics, photonics and sensing. In this chapter, we review the recent advances in the plasmon drag effect studies.

Keywords Plasmonics · Plasmon drag effect · Photoinduced responses · Optoelectronics

8.1 Introduction

Continuous progress in the information technology plays a crucial role for the development of our society in XXI century. Progressing in electronic and optoelectronic devices towards the smaller, faster, and more efficient ones has also brought many challenges. Currently, the most significant problem preventing significant increase in processor speed is delay issues associated with electronic interconnections. Increase in data transport and data processing operating speeds requires fundamentally new approaches such as implementation of photonic technologies. However, use of optical elements in electronics has many challenges, including a diffraction-limited size of the optical components and a large size mismatch between electronic and dielectric photonic components. The field of nanoplasmonics is capable to modernize the information technology by combining advantages of compact size and optical speed of operations [1–5]. Plasmonic components of different functionality [6–16] are proposed (examples are shown in

M. Durach (✉)
Department of Physics, Georgia Southern University, Statesboro, GA 30460, USA
e-mail: mdurach@georgiasouthern.edu

N. Noginova
Center for Materials Research, Norfolk State University, Norfolk, VA 23504, USA

S.E. Hunyadi Murph et al. (eds.), *Anisotropic and Shape-Selective Nanomaterials*, Nanostructure Science and Technology, DOI 10.1007/978-3-319-59662-4_8

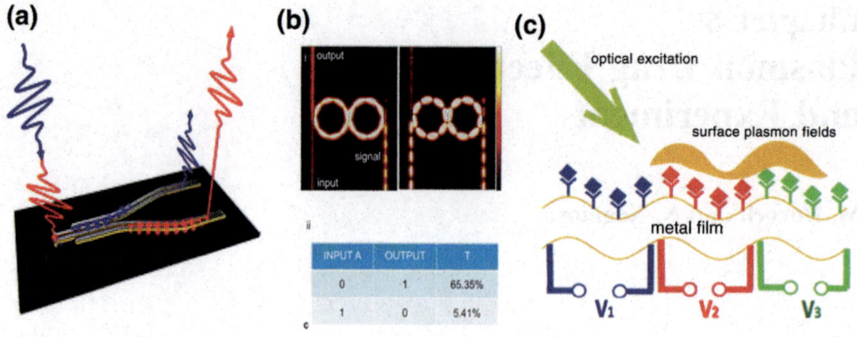

Fig. 8.1 **a** Plasmonic transmission line with ultra fast control [13]; **b** Plasmonic C-NOT gates [16, reproduced courtesy of The Electromagnetics Academy]; **c** Possible plasmonic-based chemical sensors with compact electric reading

Fig. 8.1a, b) which can enable the information processing at the nanoscale with optical frequencies and bridge the gap between the world of microscale photonics and nanoscale electronics.

Coupling of electric and optical effects in plasmonic circuit elements is of imminent importance for photonic nanocircuitry applications as it may provide an opportunity to include plasmonic elements in electronic circuits and/or control surface plasmon effects electrically. It would be also advantageous for biomedical applications of plasmonics [17–20] such as plasmonic-based sensors with electrical detection, which would allow one to substitute bulky optical detectors with compact electronic reading, Fig. 8.1c.

Plasmon drag effect (PLDE) is a giant enhancement of the photon drag effect taking place in metal films and nanostructures under conditions of surface plasmon resonance (SPR). PLDE presents interest for various applications as it may provide the direct way for incorporation of plasmonic elements in electronic and optoelectronic devices. From fundamental point of view, it provides information on plasmon-assisted momentum transfer process in light-matter interaction in metal.

Let us first discuss photon drag effect (PDE), well known for semiconductors, which is generation of dc electric currents or electromotive force under optical illumination. It has been first observed in 1970 [21, 22]. In first approximation, PDE is commonly discussed in terms of radiation pressure [23], i.e. momentum (or quasi-momentum) transfer from photons to charge carriers. The efficiency of this classical process is quite low due to the large disparity in masses of interacting particles. In semiconductors where photon drag is significant, other mechanisms enabling this effect were shown to play a major role, such as momentum-selective interband or intraband transitions and different mobilities in corresponding bands [24–32]. The sign and the magnitude of PDE can be different from those solely determined by a direct photon-to-electron momentum transfer, depending on the energy band structure, electronic transitions, and relaxation processes involved.

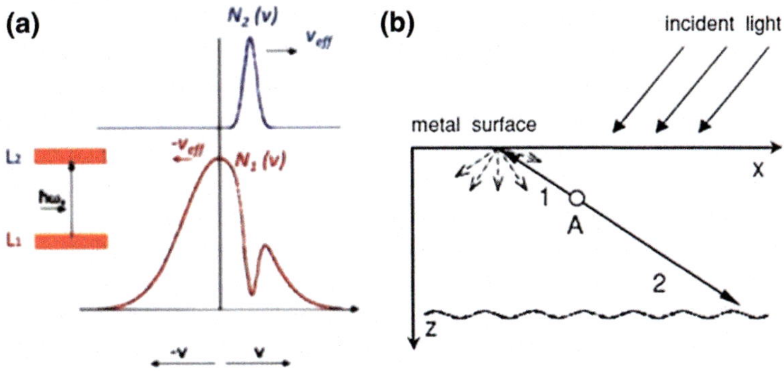

Fig. 8.2 **a** Schematics of momentum selective transitions in solids [27, 28]. **b** Diffused scattering of electrons at the metal surface, contributing to photogalvanic effect [56]

One of the mechanisms of enhanced PDE in semiconductors and metals originates from a combination of directional transition selectivity associated with the Doppler effect and different carriers mobilities in ground and excited states [27, 28], see Fig. 8.2a.

Semiconductors such as Ge and Te have already found applications as broad-band "photon drag" detectors of high energy pulsed laser and microwave radiation [33, 34].

In bulk metals, the photon drag effect is weak and commonly discussed in the terms of light pressure mechanism. The photogalvanic model [29–31] also takes into account diffusive scattering of electrons off the metal surface that can result in the net flow of electrons away from the surface and against the in-plane component of the photon wave-vector, Fig. 8.2b. The theory predicts the photoinduced current as

$$j = r\frac{e\tau P}{2mc}\sin 2\theta, \tag{8.1}$$

where r is the absorbance, e and m are the electron charge and an effective mass, θ is the angle of incidence, P is the power density of the incident light, and τ is the electron scattering time.

Similar considerations were used in the hydrodynamic model of PDE, [35] based on a combination of the Maxwell's equations and the hydrodynamic equations for a jellium surface obliquely illuminated by a monochromatic light.

The PDE in bulk metal has been observed experimentally [36] at low temperatures. The detected signal demonstrated reasonable agreement with the photogalvanic model [30, 31].

Photoinduced electric effects in nanoscale metal systems include generation of hot electrons. Under conditions of plasmon resonance, a metal particle acts as a source of electrons, which are strongly accelerated by plasmonic fields and escape along the trajectories governed by the particle geometry [37]. Novel effects associated with catalytic activity of hot electrons are reported [38–43]. Possible applications include solar energy harvesting as well [40].

The plasmon drag effect discussed in this work takes place within the metal, in contrast to the effects related to hot electrons escaping from the metal. However, as it will be discussed below, these effects might be closely related.

Before discussing the PLDE in details, let us briefly review basics of surface plasmons and their excitation methods. Plasmons are collective oscillations of free electron density in metal at optical frequency. Two classes of this phenomenon are commonly discussed: localized surface resonances (LSPs) which can be excited in 0-D metal nanoparticles, and surface plasmon polaritons (SPPs) propagating at the interface between metal and dielectric (Fig. 8.3a) [44, 45]. The resonance frequency of LSP can be adjusted by the size, the shape of the particle, and dielectric properties of the surrounding medium. LSPs have been observed on rough surfaces [46, 47], in engineered nanostructures [48–50], as well as in clusters of nanoparticles [51], where they can lead to extremely large enhancement of local field amplitudes [52].

In flat films direct optical excitation of surface plasmon polaritons is forbidden because the SPP dispersion curve lies entirely below that of photon dispersion curve in the dielectric (Fig. 8.3b), and an additional Δk is needed for the plasmon excitation. One of common methods for SPP excitation in thin films is the Kretschmann

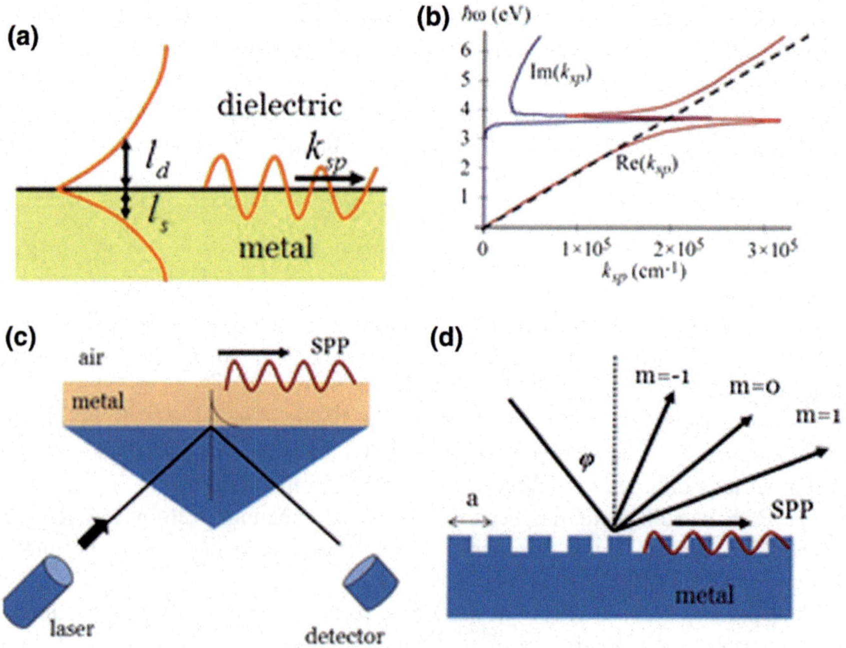

Fig. 8.3 **a** Surface plasmon polariton schemaics. **b** SPP dispersion curve; **c** Kretschmann geometry of SPP excitation; **d** diffraction grating assisted SPP excitation [44]

configuration [44, 45], where the metal film is bounded by dielectric (air) on one side and by a high refractive index glass prism on the other side (Fig. 8.3c) and is illuminated through the prism. The SPP is excited at the metal-dielectric interface excitation at [44]

$$\frac{\omega}{c}\cdot n\cdot \sin\theta_{spp} = k_{SPP} = \frac{\omega}{c}\cdot\sqrt{\frac{\varepsilon_1\cdot\varepsilon_2}{\varepsilon_1\cdot\varepsilon_2}}\cdots. \tag{8.2}$$

Under these resonant conditions, a sharp minimum is observed in the reflectivity profile, manifesting energy transfer from light to the SPP. The profile of the reflection dip in a p-polarized light is described by the formula [44]:

$$R = \left|\frac{r^p_{01} + r^p_{01}e^{2ik_{x1}d}}{1 + r^p_{01}r^p_{12}e^{2ik_{x1}d}}\right|^2. \tag{8.3}$$

Here R—reflectivity, r^p_{ik} is the amplitude reflection coefficient for p polarized light at the interface between media i and k (i, k = 0, 1, 2), εi, k is the dielectric constant, d is the thickness of the metallic film, k_{zi} is the wave vector in the direction perpendicular to the surface of the metal, kx is the wave vector in the direction of the SPP propagation, $kphot$ is the wave vector of a photon in the glass prism, and θ is the angle of incidence; the subscripts 0, 1, and 2 correspond to glass, silver, and air, respectively.

Another way to produce SPP excitation is to use a diffraction grating [44] such as a periodic structure on the top of the metal film (Fig. 8.3d). The SPP will be excited at

$$k_{\mathrm{SPP}} = k_{\mathrm{x}} + \mathrm{n}\,2\pi/a, \tag{8.4}$$

where n = 0, ±1, ±2,... is a diffraction orders and a is the periodicity of the grating. As with prism coupling, excitation of SPPs is detected as a minimum in the reflected light. The SPP modes can be generated at non-periodic structures such as sharp edges, rough surfaces of surfaces with particles or holes [44, 53, 54]. If the dimensions of the defects are smaller than the wavelength of the incident light, a broad spectrum of k_x vectors can be generated (stemming from the spatial Fourier spectrum of the particular defect), in which a mode satisfying the coupling condition (Eq. 8.4) can be found. The plasmon excitation efficiency depends on the particular profile of the defect. As an example, a single subwavelength hole in gold film can generate SPPs with efficiency up to 28% depending on the hole size and shape [54].

8.2 Experiment

8.2.1 Photoinduced Electric Effects in Flat Metal Films

In the first experiments on the photoinduced electric effects in thin metal films, SPPs were excited using Kretschmann configuration. Strips of thin gold or silver films were deposited on the glass prism with the electric contacts placed on the opposite sides of the metal strips. The photoinduced electric voltage was recorded as the function of the light incidence angle. In [55] the enhancement of photoinduced electric signal was reported in gold films for p-polarized light at infrared range ($\lambda = 930$ nm) at the angles of incidence corresponding to the SPP resonance angles (Fig. 8.4). This enhancement was initially explained with the photon pressure mechanism taking into account the resonant absorption.

The effect was studied in more detail in silver films in Ref. [56], see Fig. 8.5. The film was illuminated with *p* polarized laser light with approximately 5 ns

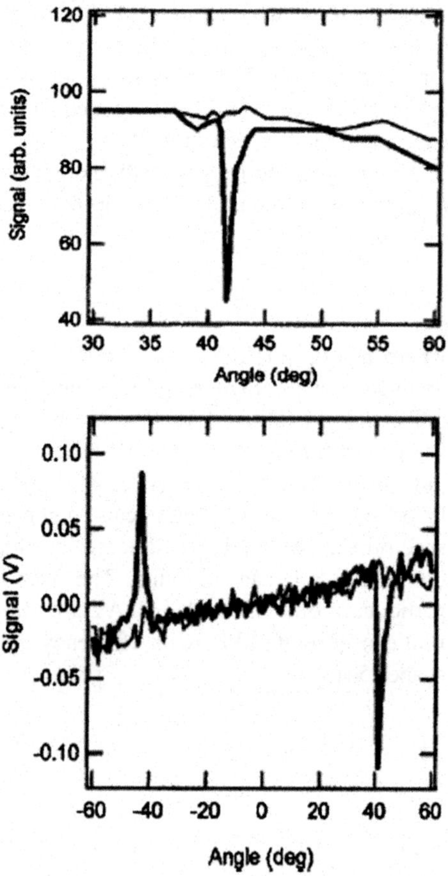

Fig. 8.4 Reflectivity (*top*) and the electric signal (*bottom*) in flat gold films for p-polarized (*thick curves*) and s-polarized(*thin curves*) laser light illumination. Reprinted from Ref. [55], with the permission of AIP Publishing

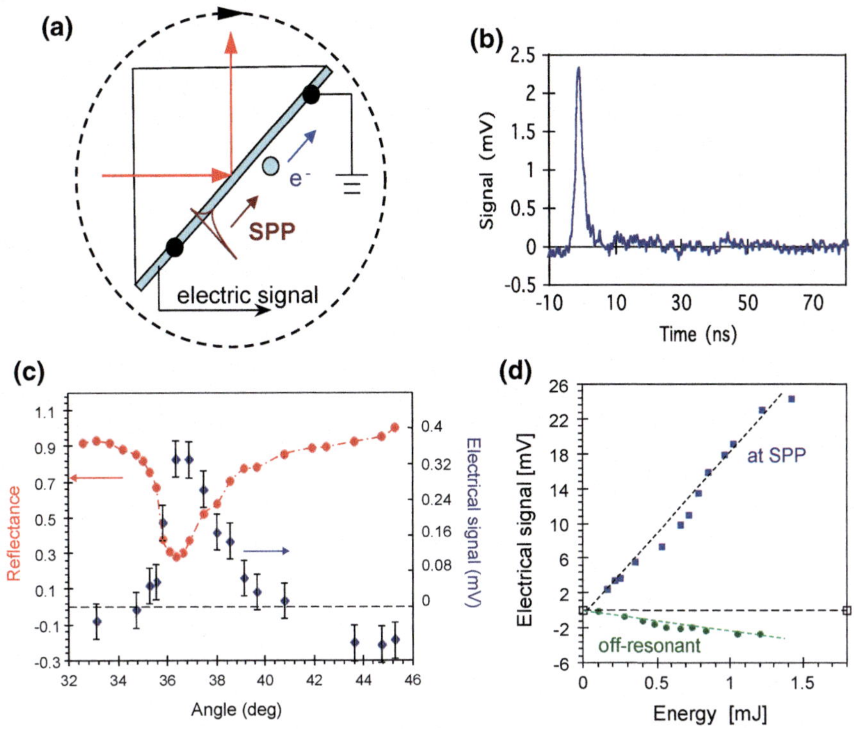

Fig. 8.5 PLDE in flat silver films. **a** Experimental setup. **b** Temporal profile of the signal. **c** SPP reflectance angular profile (*red circles*) and angular profile of the electric signal (*blue diamonds*). **d** PLDE magnitude versus pulse power [56]

pulses of an optical parametric oscillator, which was tunable between 430 and 700 nm. The photoinduced electrical signal, was measured using 1 GHz oscilloscope and 50 Ω input impedance. The temporal profile of the photoinduced electric signal approximately corresponded to that of the laser pulse. The effect has been observed in the broad range of incidence angles. However, there was a clear correlation between angular reflection and electrical signal, as it is seen from Fig. 8.5c, where the minimum of the SPP reflectance profile is nearly coincided with the maximum of the electric signal. The polarity of the photoinduced electric signal at the SPP *resonance* angle was different from that at *off-resonance* angles. When the film was illuminated from the other cathetus face of the prism, with the reversed direction of the in-plane component of the optical k-vector, the signs of both on-resonance and off-resonance electric signals flipped to opposite. The polarity of the photoinduced electric signal corresponded to the drift of electrons along the projection of the photon k-vector at the resonant SPP excitation conditions and against the wavevector's projection at the angles at which SPPs were not resonantly excited.

The magnitude of the electric signal at resonant excitation significantly exceeded that of the off-resonant signal. Both resonant and off-resonant electric signals were nearly linearly proportional to the laser light intensity at the small excitation energies and showed some saturation at the higher light intensity, Fig. 8.5d. Despite of the fact, that relatively small voltages measured, the current density peak in the experiments was very high, up to 10–20 A/mm^2, comparable to the maximum of the working current density in the insulated copper wires, 6 A/mm^2 [57]. Note that the photogalvanic model (Eq. 8.1) cannot properly describe the effect. As an example, for the experimental values $P = 6$ kW/mm^2, $\theta = 30°$, currents predicted by Eq. (8.1) were in the order of J = ~ 1 µA (with account for increase of resonant absorption), while the values of 20 µA and higher were observed in the experiment. It was suggested that not the incident photons but propagating SPPs are the main source of the photoinduced electric currents under the SPP resonance conditions. The off-resonant signal can be associated with the plasmons as well, which are excited due to a small roughness.

The direct relationship between SPPs and photoinduced electric effects was clearly demonstrated in Ref. [58], where SPPs in a thin gold film were excited using spatially modulated illumination. In the experiment, a periodic structure composed of 200 nm-thick resist with the periodicity 700 nm was deposited on top 50 nm-thick Au film. The structure was illuminated with the pulsed laser light in the wavelength range of 700–1400 nm. The photoinduced electric signal well correlated with the SPP excitation. The polarity of the signal corresponded to the drift of electrons in the SPP propagation direction regardless of the direction of the photon pressure force, see Fig. 8.6a. The estimations made in Ref. [58] were more than the order of magnitude lower than the experimental values Fig. 8.6b.

The fact that the direction of the SPP propagation determines the polarity of the photoinduced voltage was also confirmed in the experiments with asymmetric gratings. [59]

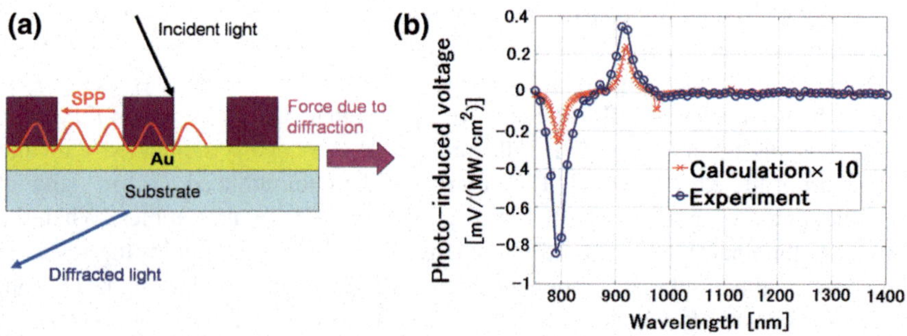

Fig. 8.6 **a** Schematic of the structure and forces at $\lambda = 925$ nm. **b** Comparison of experiment and theoretical prodictions. Adapted from Ref. [58]

Thus, in the experiments with flat metal films, it was clearly demonstrated that:

- **Maximum of photoinduced electric signal is observed at SPP resonance conditions;**
- **Electrons are dragged in the direction of SPP propagation regardless of the direction of light pressure force;**
- **Magnitude of the effect was significantly higher than predicted by light pressure.**

This is a new phenomenon of a *plasmon drag effect (PLDE)*. More experimental and theoretical studies are required for understanding and a proper theoretical description of the PLDE. Results of recent works in this direction are presented in the next paragraphs.

8.2.2 Experiment. PLDE in Nanostructured Films

It has been predicted in Ref. [60] that in the conditions of extreme confinement the optical forces acting on electrons should increase dramatically resulting in a giant surface plasmon induced drag effect (SPIDEr). In order to better understand the mechanism of the *plasmon*-related optical drag and to explore the possibility that this effect can be directly controlled through the engineering design of the nanoscale geometry, PLDE was studied in several nanostructured systems as a function of (i) surface roughness and topology, (ii) angle of incidence and (iii) wavelength of illumination [61].

The systems under study (see Fig. 8.7) included (1) gold or silver films with random roughness (introduced through moderate laser damage or with gold nanoparticles spread on the gold film surface); (2) gold and silver quasi-periodic nanomeshes (produced by thermal deposition of metal onto porous anodic-alumina membranes (AAO) of different periodicities (30–250 nm)), and (3) periodic arrays of gold-coated silicon nanopillars, (fabricated following [62]).

Effect of roughness and nanostructuring. The optically induced electric signals were very weak in nominally flat films and were dramatically enhanced in rough and nanostructured gold samples. Practically no signal was observed in smooth silver films. After the films were exposed to high intensity laser light above the damage threshold, and received a moderate laser damage, the photoinduced electrical signal of a significant magnitude was observed [61]. In silver and gold nanomesh samples, magnitudes of the signals were much higher than in flat films. The role of the nanostructuring in providing the amplified effect is clearly seen in Fig. 8.8, where the photo-induced voltage is recorded for various laser spot positions on the Au-coated silicon nanopillar arrays: with the laser spot focused directly on the array (green curve), on the flat gold film to the left (purple curve) and to the right (blue curve) of the array. When the nanopillar array was illuminated, the signal was drastically enhanced in comparison to the surrounding flat gold films, and was

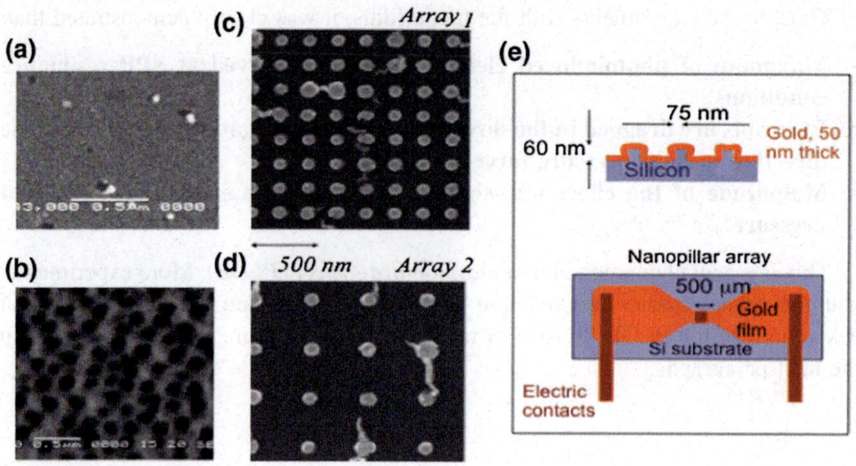

Fig. 8.7 **a** SEM of a gold film after exposure to intense laser light; **b** SEM of a gold nanomesh on AAO; **c, d** SEM of the arrays with the periods of 250 and 500 nm; **e** schematic of the nanopillar array sample [61]

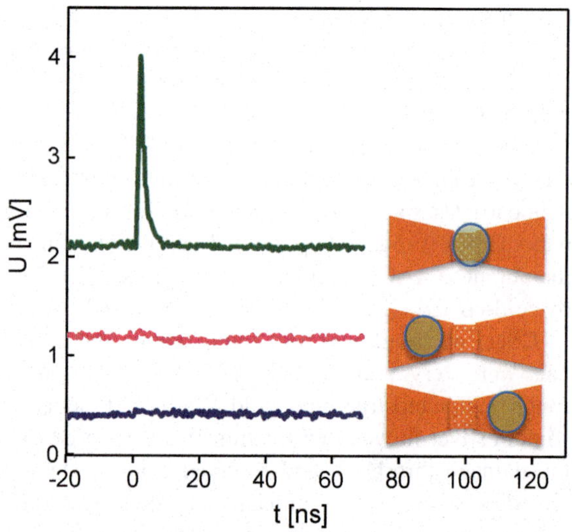

Fig. 8.8 Electric signal in the nanopillar Array 2 as observed at different positions of the illumination spot as indicated. The incident power (upon the array) was 0.03 mJ, $\lambda = 600\,\text{nm}$, $\theta = 45°$ [61]

much higher than the signals observed in other nanostructured samples under similar experimental conditions. Thus, rough and nanostructured surfaces clearly not only exhibit the PLDE effect but also provide an enhancement of the photo-induced voltage in comparison with that observed at flat metal surfaces.

Dependence of PLDE on the angle of incidence. Typical dependences of the photo-induced voltage on the angle of incidence, θ, in rough, but nominally flat films are shown in Fig. 8.9a, b. The dependences can be approximately described with the function sinθ. The signals in the nanopillar arrays grew with increasing angle at small angles, reaching a maximum near ~50°, and sharply decreased with the further increase in the incidence angle.

As a rule, in flat films, the polarity of the photo-induced current corresponded to the electron drag in the direction of $\boldsymbol{k}_x$. In some cases (in particular, when a film with small roughness was illuminated from the other side through the glass substrate), a small signal of the opposite polarity was observed, corresponding to electron drift anti-parallel to the *k*-vector projection on the film plane.

Unidirectional PLDE was observed in nanomesh samples with asymmetric profiles fabricated through the oblique deposition of metal on AAO. In that case, the electron drag was only observed in one direction regardless of the direction of the optical *k*-vector Fig. 8.9d, e.

Dependence on the wavelength of illumination. Figure 8.10 shows the spectral dependence of PLDE in various samples, including a rough silver film (with the laser-induced nanoscale surface damage), gold and silver nanomesh samples, and the two gold nanopillar arrays with different periodicity of nanopillars. The amplitude of the signal is normalized to the pulse energy incident on the sample. The spectral dependence is non-monotonous, with a maximum in the range of ~440 nm in silver samples, ~550 nm in gold nanomesh, and around 610 nm

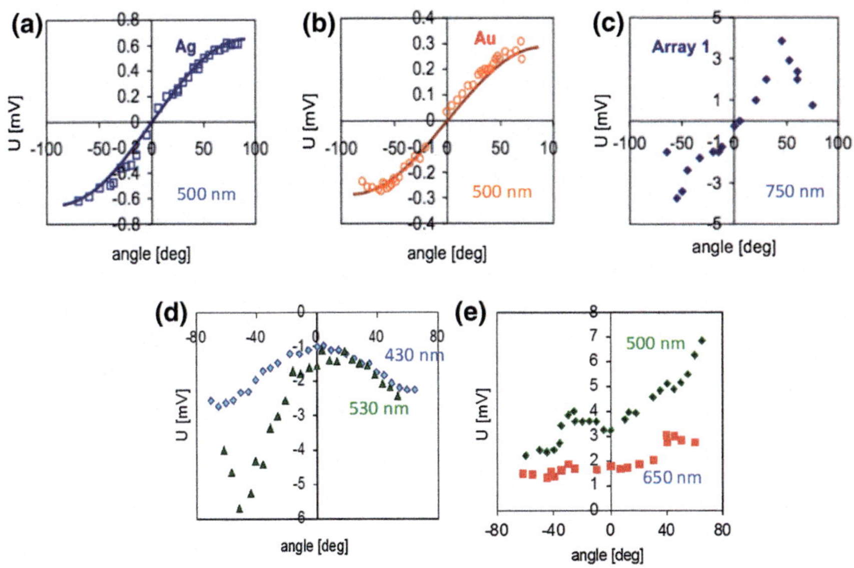

Fig. 8.9 Angular dependences of the PLDE magnitude in rough silver (**a**), rough gold (**b**), nanopillar array (**c**). *Solid lines* are sinθ fits; Same in gold nanomeshes (**d**, **e**) with asymmetric profiles. The wavelength of illumination is indicated [61]

in nanopillar arrays. In all the cases, the peak of PLDE was in the range of the localized plasmon resonance as measured via transmission or reflection spectra. The reflection spectra of the gold nanopillar array, shown in Fig. 8.10c, had several features due to its periodic structure. The feature around 600 nm corresponds to the localized plasmon resonance of an individual gold nanopillar [62, 63]. Another interesting result was that in the nanopillar arrays, the magnitude of PLDE is almost twice as large in the array with the 250 nm pitch in comparison to the 500 nm pitch, which is roughly consistent with the increase in the number of nanopillars in a row in these arrays.

Effect of light polarization. In flat films where SPPs are excited only with p-polarized light, no PLDE is observed for s-polarized light illumination [55], Fig. 8.4. However, in rough and nanostructured samples, s-polarized illumination can excite surface plasmons and produce the electrical signal [64]. As was discussed in Refs. [64–68], in metasurfaces with structures designed for the optical magnetic resonance, the polarity of the photoinduced voltage can be manipulated by the wavelength and handedness of circularly polarized light. Experimentally, control of photoinduced voltages with light polarization has been reported in arrays of gold holes [66], in light-bending metasurfaces [67], nanoporous gold films [64], and chiral metamaterials [68]. As example, Fig. 8.11 depicts the experiments with porous gold [64]. Both longitudinal and transverse electrical signals were observed

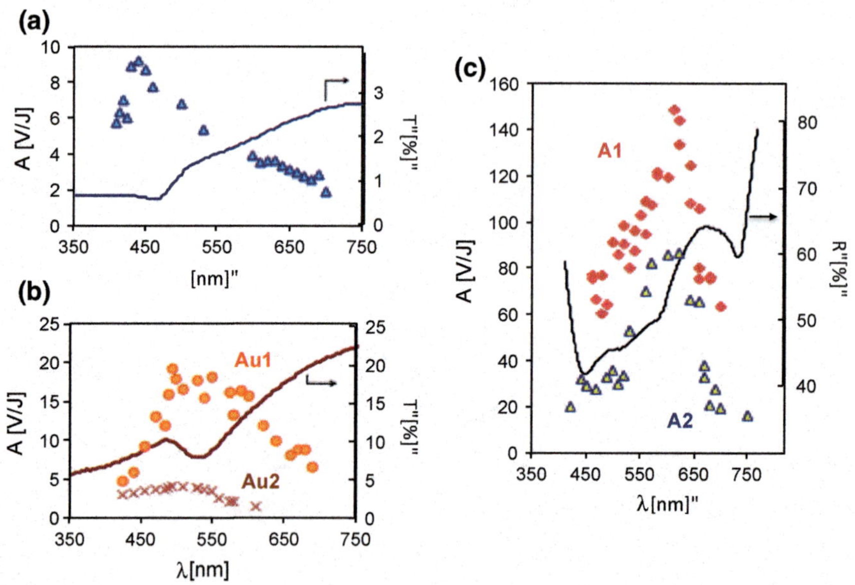

Fig. 8.10 Spectral dependence of PLDE in **a** silver nanomesh with periodicity of 30 nm (Ag1), and a roughened silver film (Ag2); **b** gold on AAO with periodicity of ~250 nm (Au1) and 150 nm (Au2); **c** nanopillar Array 1 (A1) and Array 2 (A2). *Solid traces* are the corresponding transmission (T) and reflection (R) optical spectra [61]

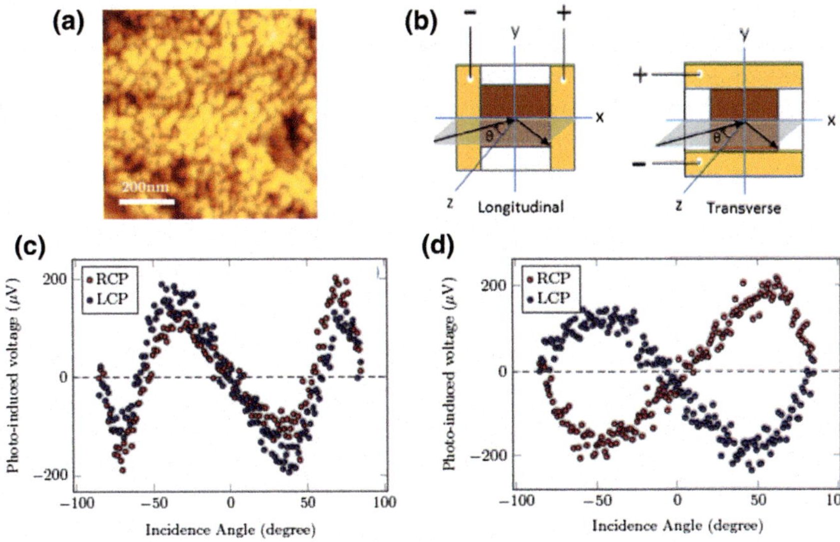

Fig. 8.11 Adapted from Ref. [64]. **a** AFM image of gold films; **b** schematics of electric measurements; **c** Longitudinal voltage, λ = 580 nm; **d** transverse voltage, λ = 550 nm

at illumination with the circular light polarization. The transverse signal was shown to change sign at the change of the polarization rotation.

The main conclusions of the experimental studies in nanostructured samples are:

- **The effect sharply increases with increase in roughness and degree of nanostructuring;**
- **Photoinduced voltages are orders of magnitude higher than predicted by the radiation pressure model;**
- **Polarity of PLDE voltage can be controlled with the direction of illumination, light polarization and nanoscale geometry;**
- **Maximum of the effect is observed close to the conditions of localized plasmon resonance.**

8.2.3 *Effect of Highly Nonhomogeneous Illumination*

Two possible models (see the PLDE theory section below) can explain a sharp increase in the PLDE with increase in roughness. According to the SPIDEr model [60], the PLDE is created by SPPs and LSPs, which create rectified pressure and striction forces. According to Ref. [61], a source of photoinduced voltage comes from an internal nonlinearity of metals driven by field enhancements at the surface of nanoparticles or nanoholes. In order to better understand the role of roughness and nanostructuring, photoinduced electric signals were studied in rough silver films in comparison with the samples under highly inhomogeneous illumination. In similarity to previous experiments [61], no signal was resolved at a smooth silver film. To introduce conditions for plasmon excitation, two different methods were used. One series of the samples was exposed to intense laser light, which induced film damage such as nanosize holes and roughness, see Fig. 8.12a. In the second method, barium titanate glass microspheres with the ~5μm diameter were randomly placed on the surface of a smooth film (Fig. 8.12b). Such microspheres are known to focus light into subwavelength "photonic jet" [70], which can provide conditions for SPP excitation.

For both types of samples, photoinduced electric signals were observed under the laser light illumination. In rough films the signal was stronger than that in the films with microspheres. However, for small incidence angles, the signals were of comparable magnitude, see Fig. 8.12c. In the films with laser damage, the angular dependence of the photoinduced signal follows $\sim\sin(\theta)$, with the polarity of the signal corresponding to the electron drift in the direction of the optical k-vector. In contrast, in the sample with the microspheres, the signal has the opposite polarity. In this sample, under the laser light illumination, electrons move against the k-vector, with the maximum of the effect being observed at the angle of incidence 20–30°.

This experiment points to the propagating plasmons as the most probable source of PLDE. In films with random roughness, the majority of SPPs were excited in the direction of the optical k-vector, dragging electrons in the direction of the SPP

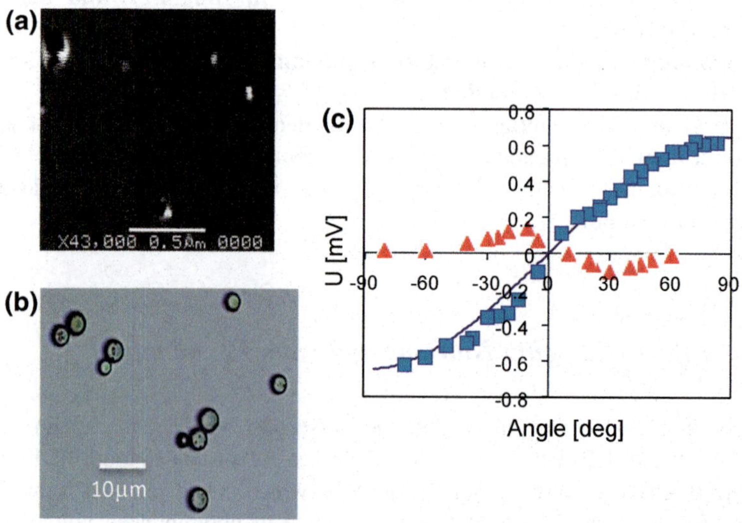

Fig. 8.12 **a** Silver film surface after laser damage; **b** microspheres on the film. **c** Signal versus incidence angle in the films with laser damage (*squares*) and microsphere on the surface (*triangles*) [69]

propagation. In order to explain the opposite behavior in the film with microspheres, one can consider the photonic jets formed on the back of microspheres in the tight air gap near the contact of the microsphere with metal. Such illumination can excite the SPPs inside the effective channel of the near field contact between a sphere and metal in backward direction through the end-fire mechanism [71]. According to geometrical estimations, the most efficient excitation corresponds to $\sim 30^\circ$ angle of incidence, which is reasonably close to experimental observations. At higher angles, the photonic jet is formed on the microsphere-air surface far from the microsphere metal contact, which is unfavorable for excitation of SPPs.

8.3 PLDE Theory

8.3.1 Macroscopic Forces Acting on Polarized Matter

The general theory of optical forces in dielectric media was presented in Ref. [60], where a general formula has been derived describing momentum transfer between electromagnetic fields and extended polarizable structures. Classically, light-matter interaction is well described by macroscopic Maxwell equations. The response of matter is represented by the polarization vector $\boldsymbol{P} = \chi \boldsymbol{E}$, where χ is susceptibility of the material, the induced polarization charge density $\rho = -\nabla \cdot \boldsymbol{P}$, the current density $\boldsymbol{j} = \partial \boldsymbol{P}/\partial t$, and surface charge density $\sigma = (\boldsymbol{P} \cdot \boldsymbol{n})$ at the surface of the metal, where $\boldsymbol{n}$ is the normal to the surface pointing outward. We do not consider systems with optical magnetism, whose operation at optical frequency is problematic [72, 73]. Therefore $\boldsymbol{B} = \boldsymbol{H}$, which precludes the existence of surface currents. Correspondingly, the Lorentz force density represents the rate of momentum transfer from the field to matter per unit volume as

$$\boldsymbol{f}_{Lv} = -\boldsymbol{E}(\nabla \cdot \boldsymbol{P}) + \frac{1}{c}\frac{\partial \boldsymbol{P}}{\partial t} \times \boldsymbol{B}. \tag{8.5}$$

The surface polarization charges experience the Lorentz force with a surface density

$$\boldsymbol{f}_{Ls} = \sigma \boldsymbol{E}. \tag{8.6}$$

The volume force density can be represented in the following way

$$\boldsymbol{f}_{Lv} = -\mathrm{div}(\boldsymbol{P} \otimes \boldsymbol{E}) + \mathrm{grad}(\boldsymbol{P}^c \cdot \boldsymbol{E}) + \frac{1}{c}\frac{\partial(\boldsymbol{P} \times \boldsymbol{B})}{\partial t}, \tag{8.7}$$

where superscript "c" exempts a vector from differentiation. The total force acting on the polarization charges in the SPP field is

$$\boldsymbol{F}_L = \int_V \boldsymbol{f}_{Lv} dV + \oint_S \boldsymbol{f}_{Ls} ds, \qquad (8.8)$$

where the first integration runs over volume V of the structure, and the second is over its surface S. Noticing that the volume integral of the first term on right hand side of Eq. (8.7) is equal to

$$-\int_V \mathrm{div}(\boldsymbol{P} \otimes \boldsymbol{E}) dV = -\oint_S (\boldsymbol{P} \cdot \boldsymbol{n}) \boldsymbol{E} ds = -\oint_S \boldsymbol{f}_{Ls} ds, \qquad (8.9)$$

we cancel terms in Eq. (8.4) to simplify it and obtain the following general expression for the time-averaged total force acting upon the object [60]

$$\bar{\boldsymbol{F}} = \frac{1}{2} \mathrm{Re} \int_V \left(\mathrm{grad}(\boldsymbol{P}^c \cdot \boldsymbol{E}^*) + \frac{1}{c} \frac{\partial (\boldsymbol{P} \times \boldsymbol{B}^*)}{\partial t} \right) dV. \qquad (8.10)$$

The result (8.10) is of fundamental importance for processes involving interaction of nanoplasmonic fields with metal electrons. Equation (8.10) is similar to the force acting on a point dipole moment (see Ref. [74]), but is more general. It is valid for a wide range of problems with a general material equation $\boldsymbol{P} = \hat{\chi} \boldsymbol{E}$, including those where operator $\hat{\chi}$ describes anisotropic or non-local media.

The second term in Eq. (8.10) is known as the Abraham force. In a monochromatic field, this force averaged over the period of oscillations is zero, but in the field of a pulsed excitation it has a finite, but small magnitude. The expression in the first term can be represented in components using $\mathrm{grad}(\boldsymbol{P}^c \cdot \boldsymbol{E}^*)_i = P_\alpha \partial_i E_\alpha$, where $i, \alpha = x, y, z$ and summation over α is implied. After averaging over an oscillation period the rectified momentum transfer rate per unit volume is obtained as [60]

$$\bar{f}_{Li} = \frac{1}{2} \mathrm{Re} \{ P_\alpha \partial_i E_\alpha^* \}. \qquad (8.11)$$

Equation (8.11) can be decomposed as

$$\bar{f}_{Li} = \frac{1}{2} \mathrm{Re} \chi \cdot \mathrm{Re} \{ E_\alpha \partial_i E_\alpha^* \} - \frac{1}{2} \mathrm{Im} \{ \chi \} \, \mathrm{Im} \{ E_\alpha \partial_i E_\alpha^* \}. \qquad (8.12)$$

The first term corresponds to the striction force, also known as the gradient or ponderomotive force. The first term is curl-free, it produces no overall work on electrons travelling through a circuit, but can result in rectified polarization and serve as a source for intrinsic nonlinearities in metals [75]. The second term is the electromagnetic pressure force and the total PLDE electromotive force (emf) in a closed circuit can only stem from this pressure force.

8.3.2 The Quantum Aspect of Relationship Between PLDE Emf and Absorption

Quantization of plasmonic oscillators or objects that they interact with has become a hot topic known as *quantum plasmonics*. Several major results in this research area have been obtained in recent years including the prediction and demonstration of coherent stimulated emission of plasmons [76, 77], control over spontaneous emission of single quantum emitters by plasmonic nanostructures [78–80], non-classical quantum optics states of plasmons [81, 82], and influence of quantum wave properties of metal plasma on plasmons [83, 84]. Note that quantum features of these effects are retained in classical or semi-classical considerations.

In Sect. 10.2.1 it was noted that there is strong correlation between PLDE emf and absorption at SPR. We demonstrated [85] that this is an inherent property of the force given by Eq. (8.10). Energy transfer rate (absorption) per unit volume is given by $Q = \frac{\partial \boldsymbol{P}}{\partial t} \cdot \boldsymbol{E}$ and for monochromatic fields after averaging over the period of oscillations,

$$\bar{Q} = -\frac{\omega}{2}\mathrm{Im}\{P_\alpha E_\alpha^*\}. \tag{8.13}$$

According to Eq. (8.12), due to high fields and high field gradients in plasmonic conditions, optical forces acting on charge carriers can strongly exceed predictions from the traditional radiation pressure mechanism, as shown in Ref. [60]. They can be very different depending on local positions, but note that for a single-mode field, the time-averaged force is related to the time-averaged power. This has a certain similarity to the momentum transfer from photons however with the plasmon wave vector $\boldsymbol{k}_{SPP}$ and plasmon-enhanced absorption. Indeed, comparing the pressure force in Eq. (8.12) with Eq. (8.13), one can see that the rectified force density is directly related to the energy transfer as $\bar{\boldsymbol{f}}_L = \hbar \boldsymbol{k}_{SPP} Q_{SPP}/(\hbar\omega_{SPP})$ which can be interpreted from the quantum point of view as the number of quanta absorbed from the plasmonic field multiplied by their momentum, unambiguously signifying the quantum aspect of the problem. The quantum aspect is retained in our classical consideration since the ratio $\hbar k_{SPP}/\hbar\omega$ stays constant in the classical limit $\hbar \to 0$. We demonstrate below that this quantum aspect is retained in multimode fields in metal films with modulated profile.

8.3.3 Kinetic Renormalization of PLDE

Both electromagnetic field and free electrons are many-body entities, which should be viewed in the scope of statistical mechanics. The approach gives rise, in particular, to the emerging field of *hot-electron plasmonics* [38–43, 86–95], which includes studies of plasmon-induced non-equilibrium electron distributions and thermalization processes in metal [86–88, 92–95] and the interaction of plasmon-generated hot electrons with materials and molecules outside the metal [42, 43, 89–91].

The momentum transferred from the SPP to electrons can be found as $P_{tr} = \int_{V_{il},t_p} \bar{f}_{L_x}(z,t)dVdt$ where integration is carried out over the illuminated volume V_{il} of the metal film and the duration of the pulse t_p. This momentum is distributed over the free electrons with the steady-state value of

$$P_{tr} = n_e V_{il} \frac{t_p}{\tau_m} m^* v_d. \tag{8.14}$$

Here n_e is the free electron density, m^* is their effective mass, v_d is the resulting drift velocity and τ_m is the time constant which describes the relaxation of the momentum. We assume that τ_m is on the order of energy relaxation time, the thermalization time t_{therm}. The assumption $\tau_m = t_{therm}$ is confirmed by a very good agreement of the theoretical predictions of Eq. (8.14) for the PLDE magnitude and its experimental values, as will be shown below. The applicability of the thermalization concept to the relaxation of momentum in excited metal plasma indicates that PLDE is essentially a *hot-electron* plasmonics effect. From this we estimate the PLDE current density j as

$$j = env_d = \frac{e}{m^*} t_{therm} \bar{f}_{L_x}, \tag{8.15}$$

This result can be also obtained from considering the effect of the plasmonic pressure on the kinetic distribution of electrons, $f = f(v_x, v_\perp)$. For simplicity, here we consider the steady state case, and conditions of closed circuit (in which the dc polarization field $E_{dcx} \approx 0$) and homogeneous illumination. The distribution function f should depart from the equilibrium distribution f_0 and satisfy the steady-state Boltzmann equation

$$\left(\frac{\bar{f}_{L_x}/n - eE_{dcx}}{m^*}\right)\frac{df}{dv_x} = -\frac{(f - f_0)}{t_{therm}}, \tag{8.16}$$

where E_{dcx} is the dc polarization field. There are two opposite limits in which Eq. (8.16) permits analytical solutions. The first one, in which action of the optical forces upon open-circuit metal nanostructure leads purely to induction of dc polarization electric fields, corresponds to $enE_{dcx} = \bar{f}_{L_x}$ and was considered in Ref. [60]. Depending on the dimensions and the geometry of the nanostructure and illumination spot, the edge optical effects in the induced polarization charge distribution may lead to an additional diffusion term in Eq. (8.16) in the open-circuit problem. Here we consider another scenario of a closed-circuit structure, where $E_{dcx} \approx 0$ and diffusion is insignificant.

In this case the strongly non-equilibrium hot-electron distribution and electric currents are formed. Denoting $v_d = \frac{\bar{f}_{L_x}}{m^* n} t_{\text{therm}}$ Eq. (8.16) can be solved analytically as

$$f = \int_0^{+\infty} e^{-u} f_0(v_x + u v_p, v_\perp) du = \sum_{n=0}^{\infty} v_d^n \frac{d^n f_0}{d v_x^n} \tag{8.17}$$

which gives Eq. (8.15) for electric current $\boldsymbol{j} = e \int \boldsymbol{v} f(\boldsymbol{v}) d^3 v$. Equation (8.17) shows direct relation between PLDE and the energy distribution of non-equilibrium electrons in presence of SPP excitation and allows an additional tool to study thermalization of plasmon-induced hot-electrons in metal nanostructures.

8.3.4 PLDE in Flat Metal Films in Kretschmann Geometry

Consider the electric field of a wave propagating in metal films as

$$\boldsymbol{E}(x, z) = \boldsymbol{E}_0(z) e^{i(k_x x - \omega t)} \tag{8.18}$$

Following Eq. (8.12) and taking Eq. (8.13) into account, the PLDE force density acting in the direction along the film can be calculated as

$$\bar{f}_{L_x}(z) = -\frac{k_x}{2} \operatorname{Im}\{\chi\} |E_0(z)|^2 = \hbar k_x \frac{\bar{Q}}{\hbar \omega}. \tag{8.19}$$

The corresponding PLDE emf, U, normalized by incident intensity I, after averaging over the film thickness h can be found as

$$\frac{U}{I} = \frac{j}{\sigma} \frac{L}{I} = \frac{t_{therm}}{\tau} \frac{L/I}{e n_e} \frac{1}{h} \int_h \bar{f}_{L_x}(z) dz = \frac{FL}{eI}. \tag{8.20}$$

Here L is the diameter of the illuminated spot, F is the effective force acting on each electron. We use the experimentally measured conductivity of silver $\sigma = 6.3 \times 10^7\ \text{S/m}$ and express our result using the Drude collision time τ [96]. Using Eqs. (8.19) and (8.20) F can be expressed as

$$F = \hbar k_x \frac{t_{therm}}{\tau} \frac{1}{n_e} \frac{dn_{pl}}{dt}, \tag{8.21}$$

where $\frac{dn_{pl}}{dt} = \frac{1}{h} \int_h \frac{\bar{Q}(z)}{\hbar \omega} dz$ is the rate of plasmonic quanta absorption per unit volume. Below we use $t_{therm} = 1\,\text{ps}$ [97] and $\tau = 31\,\text{fs}$ [96], requiring approximately 30 collisions for electrons to thermalize.

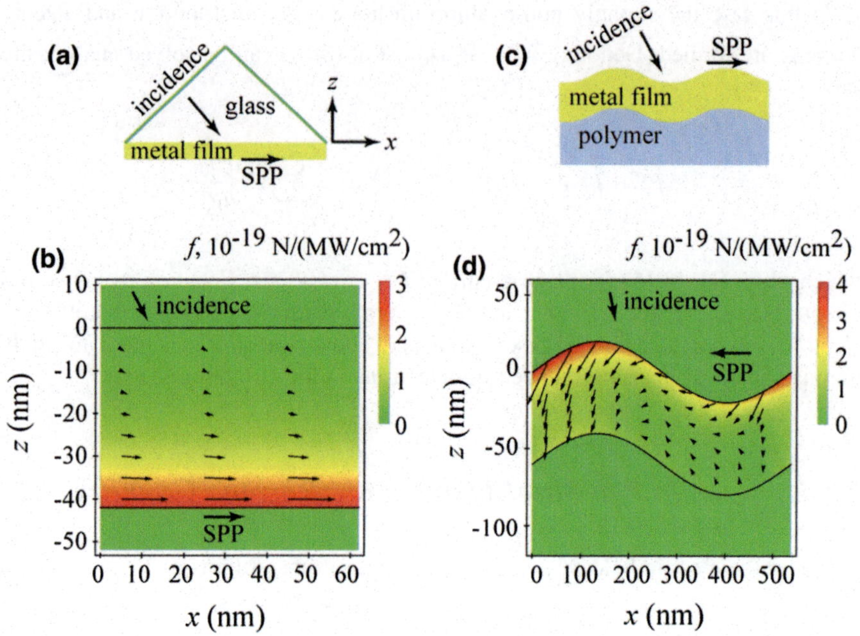

Fig. 8.13 Plasmonic pressure force in metal films. Flat films: **a** schematic of SPP excitation, **b** plasmonic pressure force distribution (Eq. 8.2) per electron at SPR resonance at **λ = 480 nm**. Films with modulated profile: **c** schematic, **d** plasmonic pressure force distribution at SPR resonance with "backward" propagating plasmon at **λ = 608 nm** > d = 538 nm [85]

As a numerical example we consider the PLDE observed in a silver film deposited upon a glass prism, as reported in Ref. [56]. Illumination through the prism at a certain angle of incidence (Kretschmann geometry) excites surface plasmon resonance (SPR), see Fig. 8.13a. The corresponding PLDE pressure force (shown in Fig. 8.13b) is in the direction of SPP propagation and strongest at the back of the film, reaching $3 \times 10^{-19} \frac{\mathrm{N}}{\mathrm{MW/cm^2}}$ per electron.

Let us compare the theoretical predictions of Eq. (8.20) for PLDE with the experimental data previously obtained in the first PLDE experiments in silver films [56] for two different experimental samples, S1 and S2, see Fig. 8.14a, b. Our theoretical estimations (shown in red) were calculated for the following parameters, corresponding those used in the experiment: film thickness $h = 42$ nm, size of the spot $L = 3$ mm, refractive index of the prism $n_{pr} = 1.78$. One can see a good agreement between theory and experiment in the magnitude of the photoinduced voltage. To our knowledge, this agreement between theory and experiment for PLDE is achieved for the first time.

There are two major differences between the theoretical and experimental data. First, the experimental curves at SPR are significantly broader than simulation results, which can be expected since the calculations assume perfectly flat films.

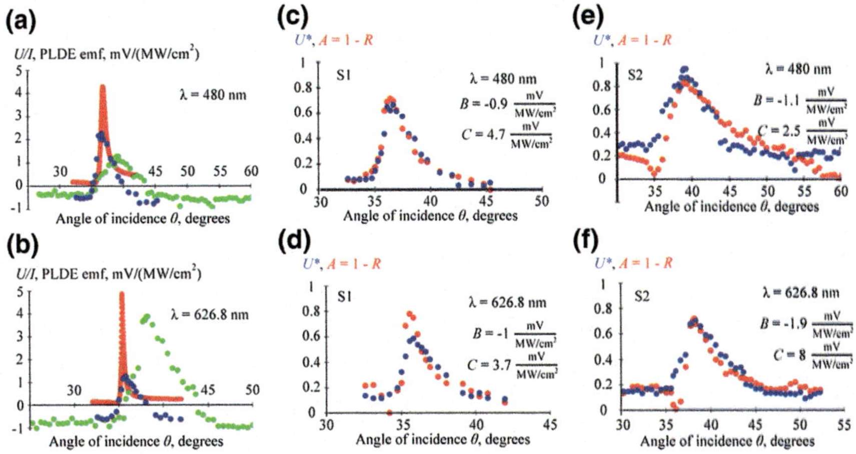

Fig. 8.14 **a**, **b** PLDE emf in flat silver films, theory (*red*) and experiment in the samples S1 (*blue*), S2 (*green*); **c–e** comparison of $U^* = (U/I - B)/C$ (*blue*, *green*) and $A = 1 - R$ (*red*), the sample, wavelength and parameters B and C are indicated [85]

Nevertheless, the angular position of SPR is reproduced in the theory. Theoretically, the enhanced PLDE emf at SPR is due to enhanced fields that enter Eq. (8.20) via force given by Eq. (8.19), which also coincides with enhanced absorption rate density Q. Second, the off-resonant signal in the experiment has the polarity opposite to the main effect and theoretical predictions, as well as being greater in magnitude than the off-resonance signal in calculations. Nevertheless, this experimental result is completely in line with the general picture of the plasmon drag in the "plasmonic pressure" model. In Fig. 8.14c–e, we compare the angular profiles of the experimental reflectivity $R(\theta)$ and the corresponding profiles of photoinduced electrical signals U/I for both samples at two different wavelengths. As one can see, in all the cases these profiles practically coincide if a proper offset and scaling are introduced.

The close relationship between U and $A = 1 - R$, and the need for an offset B can be explained by taking into account the fact that contributions to both absorbance and PLDE emf come from two kinds of sources: (i) propagating SPPs excited at the resonance conditions with the wave vector $k_x = k_{SPP}$, and (ii) other plasmonic modes excited in the experimental samples due to small scale surface roughness, which include plasmons with high values of k.

Consider the ratio,

$$C = \frac{U - U_{rough}}{A - A_{rough}}. \tag{8.22}$$

Here we offset the full values of A and U by the contributions A_{rough} and U_{rough} from the small-scale roughness. In ideally flat films, $U_{rough} = B_{th} = 0$, $A_{rough} = 0$, and PLDE is associated only with SPP excited in Kretschmann geometry. The constant C can be found using Eqs. (8.20–8.21), $\frac{dn_{pl}}{dt} = \frac{I}{h} A \cdot \cos\theta/\hbar\omega$ and $I = \frac{c}{8\pi} n_{pr} |E_0|^2$, so that

$$C = \frac{t_{therm}}{\tau} \frac{1}{n_e ec} \frac{Ln_{pr}}{h} \sin\theta\cos\theta \approx 6 \frac{\mathrm{mV}}{\frac{\mathrm{MW}}{\mathrm{cm}^2}} \text{(for our experimental conditions)}. \quad (8.23)$$

In the presence of other plasmon excitations, contributions of plasmonic modes to emf are proportional to their contributions to absorption $U \propto \frac{\hbar k}{\hbar\omega} \cdot A$, according to Eqs. (8.12–8.13). Our results can be fitted as $U_{spp} \approx U_{rough}$ and $A_{spp} \gg A_{rough}$. This is only possible if $k_{rough} \gg k_{spp}$, which confirms that the offsets come from the small-scale roughness. Note that $A_{rough} \approx 0$ is not quite true for sample 2 as seen in Fig. 8.14d. This fact correlates with the broader SPR in this sample signifying a larger-scale roughness.

One can show that in films with periodically modulated profile, SPPs propagating in the direction opposite to excitation can be excited in the case of small periods of modulation, d, satisfying $\lambda > d > \lambda/(1 + k_{SPP}/k_0)$. The opposite polarity of the off-resonance signal observed in the experiments (negative values of B) can be ascribed to a predominant contribution of such "backward" propagating plasmonic modes generated at the rough surface. The sensitivity of PLDE to such modes can provide interesting opportunities for applications such as electrical detection of dark modes, inaccessible via far-field optical response.

8.3.5 PLDE in Metal Films of Modulated Profile

We would like to extend our model of PLDE to a more complicated geometry including films with surface modulation and investigate if the relationship between PLDE and absorption still holds for multi-mode fields. Consider a metal film with thickness h whose interfaces are given by $z = a(x)$ and $z = a(x) - h$, and $a(x)$ is a periodic function with period d. Such structures support SPP excitations, provided that the excitation wave vector satisfies the quasi-momentum conservation $k_x + \frac{2\pi m}{d} = k_{SPP}$, where m is an integer number. Electromagnetic field distribution in such structure can be found using the Chandezon's method [98] which is based on solving Maxwell's equations in a transformed coordinate system with new coordinates $u = x$ and $v = z - a(x)$, in which the field interfaces become flat.

The resulting electric fields in the metal can be represented as [98]

$$E_l = b_j v^l_{jm} e^{ir_j v} e^{i\alpha_m u} = b_j v^l_{jm} e^{ir_j (z - a(x))} e^{i\alpha_m x}. \quad (8.24)$$

Here index $l = u, v$ characterizes the projection of the field, v^l_{jm} is m-th element of j-th eigenvector of the Chandezon's method with eigenvalue r_j, corresponding to

m-th diffraction wave with wave number $\alpha_m = k_x + \frac{2\pi m}{d}$. The amplitude of the j-th eigenvector, found from the boundary conditions, is b_j. In Eq. (8.24) summation over j and m is implied.

Consider electron drift along a periodically modulated film, characterized by the position-dependent angle $\theta(x, y)$ between electric current and the x-axis. Using Eqs. (8.12), (8.13) and (8.24), the work done on electrons by the PLDE pressure force over a period can be found as

$$\begin{aligned}&\overline{f_{Lx} + \tan\theta \cdot f_{Lz} \cdot d}\\&= \sum_m \frac{\hbar\alpha_m}{\hbar\omega}\overline{Q_m}d + \overline{(\tan\theta - a')f_{Lz}d} - \frac{1}{2}\mathrm{Im}\{\chi\}\overline{a''\mathrm{Im}\{E_x^* E_z\}d},\end{aligned} \tag{8.25}$$

where bars denote averaging over a period, film thickness, and time and Q_m is absorption of fields in m-th diffraction wave. In order to clarify the physical meaning of Eq. (8.25), let us consider the drift of electrons along trajectories parallel to the film profile $a(x)$. For such electrons the second term on the right-hand side of Eq. (8.25) vanishes and, if the last term can be neglected (at certain conditions discussed below), the momentum transfer from light to electrons is fully determined by the energy transfer [the first term on the right-hand side of Eq. (8.25)].

Assume that the direction of electric current in the film is parallel to the film profile $a(x)$, such that $\tan\theta = a'$ and the electrons travel along the film following laminae parallel to each other. With the assumption of such laminar electron current, Eq. (8.25) allows us to extend Eq. (8.20) for the PLDE emf on multi-mode plasmonic fields as

$$\frac{U}{I} = \frac{1}{I}\frac{t_{therm}}{\tau}\frac{L}{n_e e}\overline{f_{Lx} + a' f_{Lz}} = \frac{1}{I}\frac{t_{therm}}{\tau}\frac{L}{n_e e}\sum_m \frac{\hbar\alpha_m}{\hbar\omega}\overline{Q_m} = \frac{1}{I}\sum_m U_m. \tag{8.26}$$

As an example, we consider a sine-wave gold film (Fig. 8.1c) with the profile $a(x) = \Omega\sin(2\pi x/d)$ with the period $d = 538\,\mathrm{nm}$ and amplitude $\Omega = 20\,\mathrm{nm}$. Substituting fields (8.24) into Eq. (8.12) and normalizing per incident intensity we obtain the distribution of the forces inside the metal at SPR conditions, as illustrated at Fig. 8.13d.

Our numerical calculations of the PLDE (Eq. 8.25) in this structure confirm the direct relationship (8.26) between the PLDE emf and the energy transfer for both single mode and multimode plasmonic fields [99]. This can be expected in the conditions of predominantly laminar current flow and a relatively small modulation amplitude $\Omega \ll d$, when derivatives of the profile $a(x)$ can be omitted and the emf is determined primarily by the first term on the right hand side of Eq. (8.25). However, in a general case, in particular for nanostructured surfaces with a steep height profile, the last two terms in Eq. (8.25) cannot be excluded and the simple relationship between momentum and energy transfer does not hold. The strict result given by Eq. (8.25) would allow one to describe or predict PLDE in small-scale,

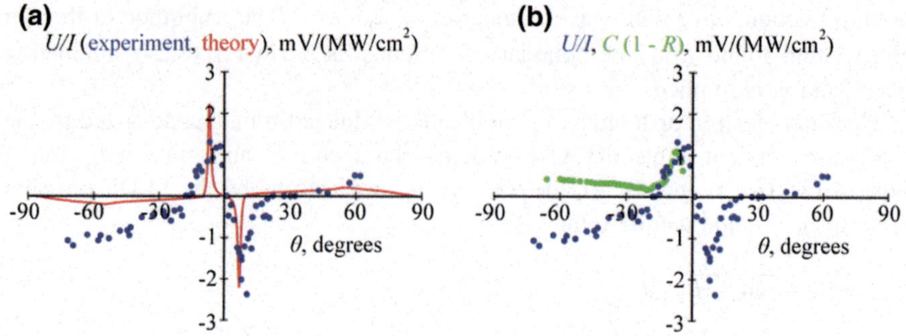

Fig. 8.15 **a**, **b** PLDE emf in sine wave gold films (*green circles* and *stars* two different measurements on the same sample) compared with theory and absorption scaled by C = $3.5\,\mathrm{mV}/(\mathrm{MW}/\mathrm{cm}^2)$ (*red* in **a**, **b** respectively) [85]

irregular nanostructures, where energy absorption is not directly tied to momentum transfer in the same sense as in this paper, paving a way for additional shape-dependent control and engineering of PLDE.

We compare the calculations to the PLDE measurements performed on sine-wave films in Fig. 8.15a. The experimental data was obtained in a sine-wave gold film with the period of 538 nm and peak to peak modulation height of $\sim$50 nm at pulsed laser light illumination with wavelength of 630 nm and pulse duration of 5 ns. Details of the structure fabrication, experiment and calculations are presented in Ref. [99]. As one can see the theory and experiment correspond well to each other similarly to flat films. One can see the direct correlation of experimental PLDE emf and absorption at SPR in Fig. 8.15b with fitting constant $C = 3.5\,\mathrm{mV}/(\mathrm{MW}/\mathrm{cm}^2)$, which coincides with the value for C, needed to fit theoretical emf and absorption, assuming the spot size $L = 2\,\mathrm{mm}$, and kinetic coefficient $t_{therm}/\tau = 30$.

8.3.6 *PLDE in Nanostructures*

Metal nanostructures larger than the skin depth ($\sim$25 nm) support surface plasmon polaritons (SPPs) that exert forces on electrons causing an SPP-enhanced PDE or PLDE [55, 56, 85]. As was discussed in the previous sections the plasmonic pressure force in such structures provides the induced potential differences in the mV range with the response time in the 100 fs–1 ps range which is suitable for conjugation with electronic circuits. In this section we describe the drag effect under the conditions of strong nanoplasmonic confinement, when the SPP localization radius is less than the skin depth ($\sim$25 nm) [60, 61]. In this case the illumination leads to a giant surface plasmon induced drag effect (SPIDEr) [60] with a response on the femtosecond time scale. Ultrashort, nanolocalized SPP pulses exert forces on

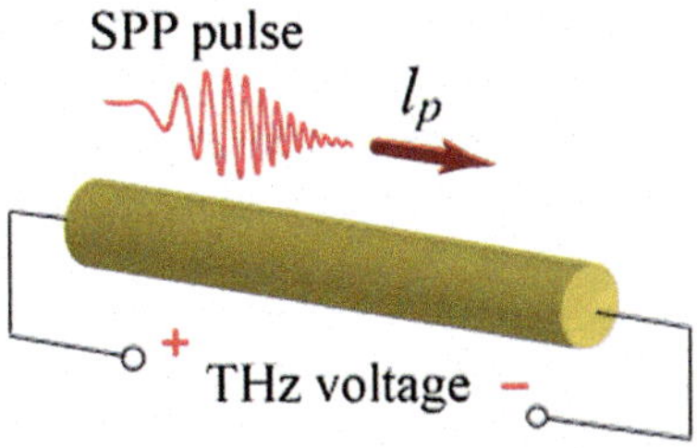

Fig. 8.16 Schematic of a SPIDEr in metal nanowire. Propagating SPPs create forces acting on carriers in the nanowire, which leads to THz-band voltage (emf) between the ends of the wire. Picosecond or femtosecond pulses can be used to manipulate the time dependence of the created emf [60]

electrons in the nanowires, inducing giant THz electromotive force along the SPP propagation direction (see schematic in Fig. 8.16). Additionally, internal nonlinearity of metals driven by field enhancements at the very thin layer on the surface of nanoparticles or nanoholes would contribute to the photoinduced voltage [61].

According to our estimations, in thin (~ 5 nm radius) wires the photoinduced emf can reach ~ 10 V over the SPP decay length, with nanolocalized THz fields as high as ~ 1 MV/cm. Such THz field have previously been generated in the far zone [100], where they produce non-perturbative effects [101], but not on the nanoscale. In contrast, the plasmonic metal nanowires can serve as nanolocalized sources of high THz fields [60]. Thin metal nanowires can be used as broadband nanoscopic photodetectors with extremely fast response due to the femtosecond momentum relaxation times in metals. The nature of the giant enhancement of the SPIDEr is novel in nanoplasmonics: it is not the enhancement of the optical fields per se (the maximum magnitude of the local fields is limited by the breakdown at the metal surface that occurs for fields ~ 1 V/Å) but the effect of very high gradients of these fields. The SPIDEr is ultrafast because it is a non-resonant effect whose bandwidth is comparable to that of the entire optical spectrum.

We apply the general Eq. (8.10) to describe the SPIDEr. To be specific, consider a metal nanowire with radius r and dielectric susceptibility χ, which is oriented along the z-axis and embedded in a dielectric with a dielectric permittivity of ε_d. This wire propagates an SPP pulse, which can be excited by external sources using, e.g., the effect of adiabatic compression [102, 103]. In the case of extreme nanoplasmonic confinement ($r \ll l_s$, where l_s is the skin depth), r becomes the only relevant quantity of the dimensionality of length [102]. Therefore there is a scaling of all magnitudes in r. The SPP wave power P scales as $P \sim E^2 r^2 v_g$, where $v_g \sim \omega r$ is the SPP group velocity. The SPIDEr-induced potential difference [electromotive force (emf)] U is proportional to the pressure produced by the force from (8.10), $U \sim F/E^2$. The propagation length of the SPP, $l_p \sim rQ$, where Q is the SPP figure of merit independent of r [102]. These arguments allow one to predict a scaling of the SPIDEr force F, emf U, rectified electric field E_r (which for femtosecond SPP pulses possesses THz frequencies), and the maximum rectified field E_{mr} (at the maximum tolerable power P), $F \propto Pr^{-1}$, $U \propto Pr^{-3}$, $E_r \propto Pr^{-4}$, $E_{mr} \propto r^{-1}$. The scaling implies that all the effects caused by the SPIDEr increase with decrease in the wire radius as its powers. This enhancement is not resonant and therefore has the bandwidth comparable to that of the entire optical spectrum. The scaling describes only the dependence on r. There are also prefactors describing the dependence on dielectric permittivities, frequency, etc., accounting for an additional enhancement close to the SP resonant frequency, which is multiplicative in magnitude. In the next session, we show that the strict theory confirm such a scaling.

8.3.6.1 SPIDEr in Metal Nanowires

The SPPs are transverse magnetic (TM) modes, and their complex fields have the form

$$\boldsymbol{E} = A(t')\left(\tilde{E}_z\hat{\boldsymbol{z}} + \tilde{E}_\rho\hat{\boldsymbol{\rho}}\right)e^{i(kz-\omega t)}, \quad H = A\left(t'\right)\tilde{H}_\varphi\hat{\boldsymbol{\varphi}}e^{i(kz-\omega t)} \tag{8.27}$$

where $t' = t - z/v_g$, v_g is the SPP group velocity at the pulse carrier frequency ω, and k is the SPP wave number (see Fig. 8.16). The total power flowing through the plane $z = 0$ at the moment t is $P(t) = 2\pi\int_0^\infty \rho d\rho \bar{S}_z(\boldsymbol{r}, t)|_{z=0}$, where $\bar{S}_z(r,t) = \left(\frac{c}{8\pi}\right)\mathrm{Re}[\boldsymbol{E}\times\boldsymbol{H}^*]$ is the Poynting vector averaged over the SPP period. Considering the azimuthally symmetric (TM0) modes, the functions $\tilde{E}_z$, $\tilde{E}_\rho$ and $\tilde{H}_\varphi$ depend only on radius ρ. We normalize them for the real amplitude $A(t)$ to satisfy a relation $A^2(t) = P(t)$. Disregarding the group velocity dispersion in Eq. (8.27) is valid for pulses with duration of tens of femtoseconds and greater, and frequencies not too close to the SP resonance [104].

Illumination of a nanostructure with femtosecond SPP pulses results in a rectified force exerted on electrons which varies with time in THz scale. Since the electron momentum-relaxation time is on the scale of femtoseconds, electrons come to a local equilibrium in the process of this rectification. Therefore, the hydrodynamic approximation can be applied, for which the pressure p and electrostatic potential ϕ satisfy an equation $p + ne\phi = const$. From this, we can find the emf $U = \Delta\phi$, which is the total change of potential in the direction of SPP propagation, $U = \Delta p/(ne)$, where $\Delta p = \bar{F}_z/(\pi r^2)$ is the full change of the pressure. Here, $\bar{F}_z$ is z-component of the force (8.10) averaged over the period of SPP oscillations.

Now let us calculate the force $\bar{F}_z$ in the fields given by (8.27)

$$\begin{aligned}\bar{F}_z &= \frac{1}{2}Re\int_V \left(\tilde{P}_z\tilde{E}_z^* + \tilde{P}_\rho\tilde{E}_\rho^*\right)\cdot\left\{-ikP(t') + \frac{1}{2}\frac{\partial P(t')}{\partial z}\right\}e^{-2k''z}\cdot dV \\ &= \left(\frac{\pi\chi''Q}{2}\Phi(t) - \frac{\pi\chi''}{2}P(t)\right)\cdot\int_0^R \left|\tilde{E}\right|^2\rho d\rho\end{aligned} \tag{8.28}$$

where a new function was defined via

$$\Phi(t) = \frac{1}{l_p}\int_0^{\infty} dz \cdot P(t'')e^{-z/l_p} = \frac{e^{-\frac{t}{t_p}}}{t_p}\int_{-\infty}^{t} dt'' \cdot P(t')e^{-t''/t_p} \tag{8.29}$$

The total force $\bar{F}_z = f_{pr} + f_{st} + f_A$ is composed of three forces: the SPP pressure, striction, and a small-magnitude Abraham force, whose z-components are, correspondingly,

$$f_{pr} = \frac{\pi r^2}{cA_{pr}}\Phi(t), \quad f_{st} = \frac{\pi r^2}{cA_{st}}P(t), \quad f_A = \frac{\pi r^2}{c^2 L_A}\Phi'(t) \tag{8.30}$$

In Eq. (8.30) coefficients A_{pr} and A_{st} have dimensionality of area, and L_A has dimensionality of length. They are defined as

$$A_{pr} = \left(\frac{c\chi'' Q}{2r^2}\int_0^r \tilde{E}^2 \rho d\rho\right)^{-1}, \quad A_{st} = -\left(\frac{c\chi'}{2r^2}\int_0^r \tilde{E}^2 \rho d\rho\right)^{-1},$$
$$L_A = \left(\frac{cl_p}{r^2}\int_0^r \mathrm{Re}\left(\chi \tilde{E}_\rho \tilde{H}_\varphi^*\right)\rho d\rho\right)^{-1}, \tag{8.31}$$

where $Q = \mathrm{Re}k/\mathrm{Im}k$ is the SPP figure of merit, $|\tilde{E}|^2 = |\tilde{E}_z|^2 + |\tilde{E}_\rho|^2$, $l_p = 1/(2\mathrm{Im}\ k)$ is the SPP propagation length, and $t_p = l_p/v_g$ is the SPP pulse lifetime.

These three forces result in three terms of the SPIDEr emf [60]

$$U(t) = R_H\left(\frac{\Phi(t)}{A_{pr}} + \frac{P(t)}{A_{st}} + \frac{P'(t)}{cL_A}\right) \tag{8.32}$$

where $R_H = -1/(nec)$ is the Hall constant.

The Abraham force contribution [the third term in Eq. (8.32)] is small under the condition $k_0 l_p/(kc\tau) \ll 1$, where $k_0 = \omega/c$, and τ is the pulse duration.

Let us consider two limiting cases pertaining to Eq. (8.32): a quasi-monochromatic regime of long pulses $\tau \gg t_p$ and a regime of short pulses $\tau \ll t_p$, where t_p is the SPP dissipation time.

SPIDEr as a THz Source

Consider first the quasi-monochromatic regime $(\tau \gg t_p)$, where $\Phi(t) \approx P(t)$ (see Eq. (8.29)). In such a case, the total emf $U(t)$ follows the pulse-envelope time dependence $P(t)$. The computations will be made for a silver [96] nanowire embedded in vacuum ($\varepsilon_d = 1$).

For the quasi-monochromatic case, the total emf U as a function of frequency ω and wire radius r is displayed in Fig. 8.17a. In contrast to the case of dielectric media, in the present plasmonic case the SPP pressure and striction contributions to the emf have the same sign since $\chi' < 0$. These two contributions are equal if a condition $\chi'' Q < \chi'$ is satisfied. The black solid line represents this condition; to the left of this line the pressure dominates, and to the right the striction force gives the major contribution to the SPIDEr. This is understandable because close to the SP resonance of the wire (at ~3.7 eV), the gradient of the SPP intensity increases due to the high loss: the striction force is of a gradient nature, therefore dominating. Similarly, with the decrease of r, the intensity gradient increases due to the increased confinement, which also leads to the relative increase of the striction with respect to the pressure force, as we clearly can see from this and other panels of Fig. 8.17. General increase of the SPIDEr at the SP resonant frequency can be seen as a broad red peak.

Magnitude of the SPIDEr emf relative to the SPP wave power, U/P, is illustrated in Fig. 8.17b as a function of the wire radius r for a frequency of $\omega = 2.9\,\text{eV}$. The SPIDEr effect is gigantically enhanced for strong nanoplasmonic confinement: by four orders of magnitude when r decreases from 100 to 5 nm at the same SPP power. There is a pronounced scaling $U/P \propto r^{-3}$ at $r \ll l_s$, in accord with the discussion in the beginning of Sect. 8.3.5.

The spectral dependence of the relative SPIDEr emf, U/P, for a wire of the smallest radius considered, $r = 5\,\text{nm}$, is depicted in Fig. 8.17c [60]. For significantly thinner wires, nonlocal-response effects may become significant, cf. Refs. [105, 106]. The large magnitude shows that the SPIDEr effect can be used for the photodetection on the nanoscale, i.e., in the role that previously was deemed only available for semiconductors. In this sense, it belongs to the area of active nanoplasmonics [104].

By nonlinear optics classification, the SPIDEr is an optical second-order nonlinear effect: the magnitude of the SPIDEr emf U is proportional to the power P of the SPP field. Therefore, the maximum achievable magnitude of the emf is determined by the maximum P that the wire can tolerate. This we estimate setting the optical field E at the surface of the wire equal to 1 V/Å [107, 108]. For fields significantly higher than this, there will be massive ionization and damage of the metal surface. We plot in Fig. 8.17d this maximum intensity as a function of the wire radius for three SPP frequencies. Note a very good scaling, $P \propto r^3$, in the region $r < l_s$ of strong nanoplasmonic confinement. These values of the P_m in comparison to the data of Fig. 8.17a–c show that the large values of the SPIDEr emf are realistically achievable.

One of the most important for applications properties of the giant SPIDEr is the high local electric field E_R generated due to the SPIDEr optical rectification near the metal nanowire. Such field (averaged over the SPP decay length l_p) can be found as $E_r = U/l_p$. We display the maximum achievable rectified field E_{mr} (at the

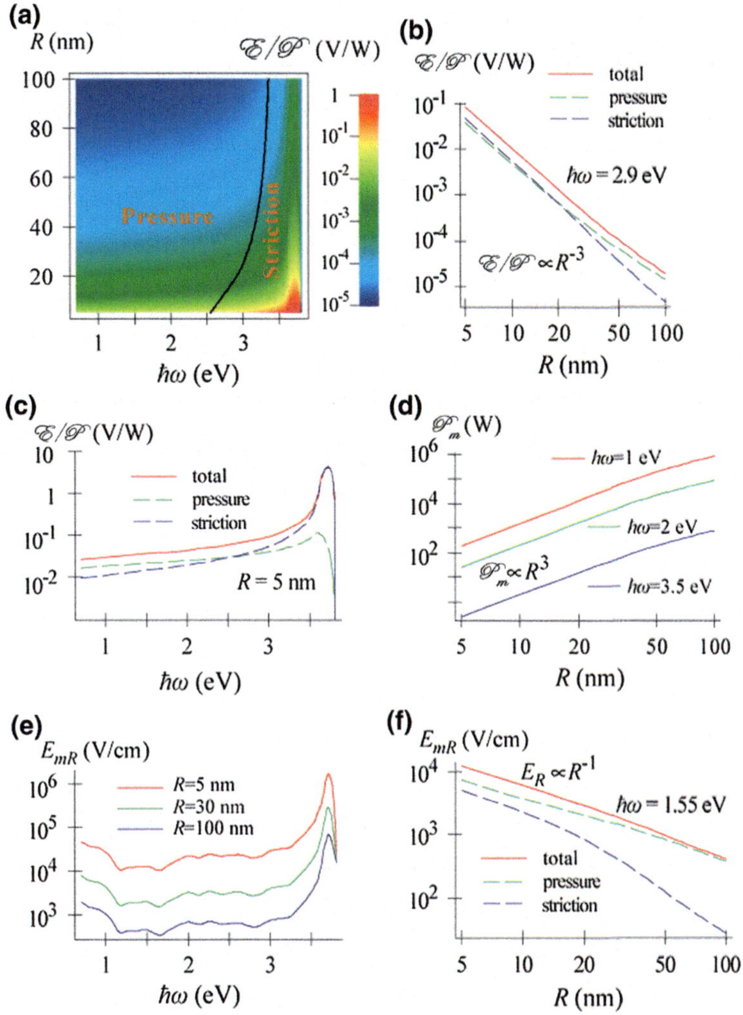

Fig. 8.17 SPIDEr for quasi-monochromatic SPP pulses: emf and rectified field dependence on the frequency $\hbar\omega$ and wire radius R. Note the logarithmic scale for the magnitude of the effect. **a** Dependence of the SPIDEr emf per unit SPP power U/P on wire radius and frequency. The *black broken curve* indicates the parameters at which SPP pressure is equal to striction. The magnitude of the effect is denoted by the *color-coding bar*. **b** Dependence of SPIDEr magnitude U on wire radius R per unit power of the SPP wave (*solid red curve*). The contributions of the pressure and striction to the total magnitude of SPIDEr are shown by the *dashed curves*. **c** Dependence of SPIDEr magnitude per unit power U/P on frequency ω for $R = 5$nm. **d** Maximum power that a wire can tolerate P_m as a function of wire radius R for different frequencies ω. **e** The maximum SPIDEr rectified field E_{mR} (for the maximum tolerable power P_m) as a function of frequency for three wire radii R = 5, 30, and 100 nm. **f** The maximum SPIDEr rectified field E_{mR} (for the maximum tolerable power P_m) as a function of the wire radius R for frequency $\hbar\omega = 1.55\,\text{eV}$ [60]

propagating SPP power of P_m) in Fig. 8.17e, f. As we see from panel (e), the spectral dependence of the SPIDEr rectified THz field is very similar for all wire sizes, but the magnitude of this field is much greater for the 5 nm wire: $E_{mr} \sim 10^5 - 10^6$ V/cm. The nanolocalized THz fields of such a magnitude will excite a wealth of nonlinear THz responses at the nanoscale.

SPIDEr as a Femtosecond Detector

Now let us consider dynamics of the emf response to SPP pulses that differ in duration τ with respect to the SPP pulse dissipation time t_p. The latter is displayed in Fig. 8.18a as a function of the frequency ω. As we can see time t_p is in the range from 10 to 150 fs. The temporal dependencies of the emf in comparison with the power P of the SPP pulses for various pulse durations are illustrated in Fig. 8.18b–d. For a relatively long pulse ($\tau = 1\text{ps} \gg t_p$) shown in Fig. 8.18b, the shape of the emf $U(t)$ repeats that of the

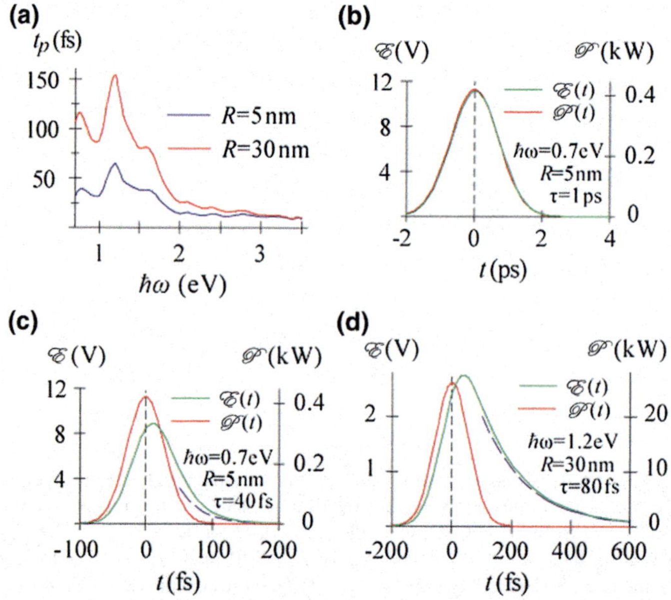

Fig. 8.18 SPIDEr created by ultrashort SPP pulses: fast femtosecond emf response. **a** The dependence of the SPP lifetime $t_p = l_p/v_g$ on the frequency $\hbar\omega$ for $R = 5$ nm and $R = 30$ nm. **b** The time dependence of the emf $U(t)$ (*green line, left scale*) and input power $P(t)$ (*red line, right scale*). The pulse duration is $\tau = 1$ ps $\gg t_p \approx 30$ fs and the emf closely follows the SPP pulse dynamics. **c** The same for much shorter pulse with $\tau = 40$ fs. The pressure-induced emf leads to a small broadening in the emf dynamics (*green line*). The limiting exponential decay is outlined by the broken blue line. **d** Emf induced by the short pulse in nanowire with $R = 30$ nm with frequency $\hbar\omega = 1.2$ eV. The emf response is broadened, since $\tau = 80$ fs, while $t_p = 150$ fs [60]

power $P(t)$. This relatively long, picosecond response, nevertheless, corresponds to a 1 THz bandwidth for this nanowire used as a nanoscale photodetector.

For a much shorter, $\tau = 40\,\text{fs}$, SPP pulse and the same 5 nm wire, as shown in Fig. 8.18c, there is a small broadening and delay of the voltaic response (emf) $U(t)$ with respect to the excitation SPP pulse $P(t)$. This broadening is due to the pressure force that decays exponentially for long times, as Eq. (8.29) suggests, and the broken blue line in the figure indicates. However, under the conditions considered, this delay and broadening are not large. The frequency-response bandwidth of this wire as an SPP photodetector on the nanoscale is very large, ~20 THz, which is characteristic of the extreme nanoplasmonic confinement.

For a much thicker nanowire of $R = 30\,\text{nm}$ (weak plasmonic confinement case) and $\hbar\omega = 1.2\,\text{eV}$, illustrated in Fig. 8.18d, the SPP decay time becomes much longer $\left(t_p = 150\,\text{fs}\right)$. This leads to a very significant delay and temporal broadening of the emf response with a pronounced exponential part due to the pressure forces shown by the broken blue line. This behavior is due to the much longer SPP lifetimes for the weak confinement where a significant fraction of the SPP energy propagates in the dielectric (vacuum). Nevertheless, the emf response bandwidth is still very large, on the order of 5 THz, and its amplitude is $U > 1\,\text{V}$.

8.3.6.2 "Batteries" Model Based on Nonlinearity of Metal and Asymmetric Boundary Conditions

This model was proposed in Ref. [61] for explaining the experimental findings on spectral dependences of the photoinduced voltage in metal nanostructures, where the electrical signal reached maximum at an incident wavelength close to the localized surface plasmon resonance of an individual nanostructure. This experimental fact indicates the predominant role of nanoscale features. In the experiments with nanopillar arrays, the nanopillar array with the twice higher number of pillars per row provided approximately twice the magnitude of the photo-induced voltage in comparison to the second nanopillar array with the smaller number of pillars [61]. This difference clearly illustrated that the individual plasmonic resonators played a major role in the photoinduced effect. These experimental observations suggest the following scenario: each nanopillar when excited with light, works as a source of the electromotive force (see Fig. 8.19). Using the data from Fig. 8.10 and taking into account the total size of the array, (500 μm), estimation of the *emf* of an individual nanopillar is in the microvolt range for a light intensity of 2 MW/cm^2.

This scenario can be explained with intrinsic nonlinearity of nanostructured metal. As was discussed in [109, 110] sharp boundaries of nanostructures, nanopillars or nanoparticles can lead to nonlinear effects, including harmonics generation. Nonlinearity of metal nanoparticles was also discussed in [111] where electrons were considered as a negatively charged compressive fluid in the frames of Drude's model. Following Ref. [111], let us assume that under the periodic field, $E_0 \cos(\omega t)$ negative charges can be compressed indefinitely on one side of the

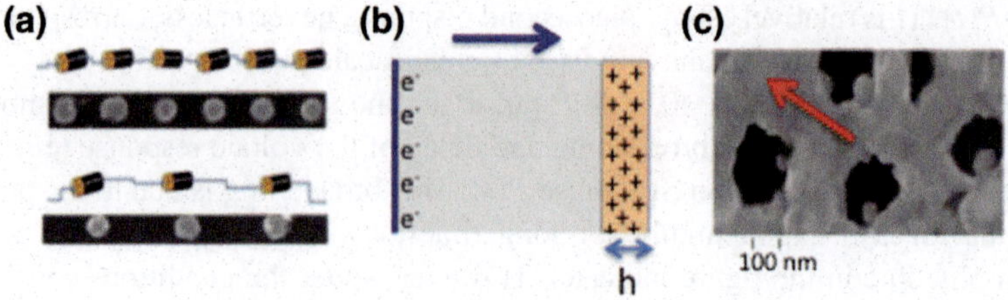

Fig. 8.19 **a** Equivalent circuit of PLDE in nanopillar arrays. **b** Asymmetry of positive and negative charges motion under driving field in nanostructure according to Ref. [25]. **c** SEM image of the gold nanomesh on AAO fabricated under oblique deposition. *Arrow* shows the direction of the photoinduced electron drag [61]

particle, while a positively charged layer with the width, h, is formed on the other side. Thus, the range of the oscillations of the positive charge is different from that of the negative charge. According to Ref. [111], such a difference can result in the generation of high harmonics.

For a slab (Fig. 8.19b), the width of the positive layer, h, can be estimated as $h = |\sigma|/\rho$, where $\sigma = \varepsilon_0 \Lambda(\omega) E_0 \cos(\omega t + \varphi)$ is the charge induced by the optical field, $\rho = Ze/l^3$ is the background positive charge density, Z is the number of electrons per atom, e is the electron charge, l is the lattice constant, φ is the phase shift and $\Lambda(\omega)$ is the resonant factor describing the field enhancement. Let us now consider a case when the motion of the charges in a nanoparticle under the driving field is highly asymmetric and assume that a positively charged layer with the width $h(t)$ is formed only on one side of the nanoparticle for half of the period of the oscillating field. Asymmetric shapes of the particle, boundary conditions or illumination conditions can be possible sources for such asymmetry. This asymmetric motion would result in the generation of a *dc* electric field. The voltage can be found with averaging over the period of oscillation, as follows:

$$U = \langle E(t) \cdot h(t)/2 \rangle = \frac{1}{2\pi} \int_0^{\pi} \Lambda E_0 \cos(\omega t) \frac{\varepsilon_0 \Lambda E_0 \cos(\omega t)}{2\rho} dt = \frac{I \Lambda^2 l^3}{4cZe} \tag{8.33}$$

where I is the light intensity.

At $Z = 1$, $l \sim 0.5$ nm, and the experimental value of light power, $P = 2$ MW/cm^2, Eq. (8.33) yields, $U \sim 10^{-8} \Lambda^2 V$. The factor Λ is on the order of ~ 10 at the resonance conditions [112]. Equation (8.33) predicts the induced voltage to fall within the microvolt range, which is in line with the experimental observations.

8.4 Conclusions

Plasmonics is a major branch of nanotechnology which encompasses a broad range of phenomena on scales from 1 nm to 1 micron, covering the whole visible frequency band and beyond. Plasmon drag effect is relevant at all scales in three different regimes. In large structures, it takes a form of a quantum momentum transfer, directly related with energy transfer, between plasmons and electron-hole pairs, resulting in strong non-equilibrium distribution of hot-electrons and enhanced plasmon-induced electrical currents. In smaller plasmonic structures PLDE takes the form of SPIDEr, i.e. induction of rectified plasmon-induced electric polarization, which is dramatically increased in conditions of strong plasmonic confinement. At the nanometric scale the plasmon-induced metal nonlinearities in thin surface layers of metal nanostructures lead to the large rectified fields in the nano-batteries regime. In all cases PLDE provides an invaluable for nanotechnology coupling between plasmonic and electronic degrees of freedom promising a myriad of possible applications ranging from electro-plasmonic integrated circuits to biomedical sensors with nanoscopic selectivity.

References

1. Rattner, J. "Plenary: The Future of Silicon Photonics," In *Integrated Photonics Research, Silicon and Nanophotonics and Photonics in Switching,* OSA Technical Digest, Optical Society of America, paper JTuA1.
2. Kirchain, R., and L. Kimerling. 2007. A roadmap for nanophotonics. *Nature Photonics* 1: 303.
3. Brongersma, M.L., and V.M. Shalaev. 2010. The case for plasmonics. *Science* 328: 440–441.
4. Zia, R., J.A. Schuller, A. Chandran, and M.L. Brongersma. 2006. Plasmonics: The next chip-scale technology. *Materials Today* 9: 20–27.
5. Sorger, V.J., R.F. Oulton, R.-M. Ma, and X. Zhang. 2012. Toward integrated plasmonic circuits. *MRS Bulletin* 37: 728.
6. Engheta, N. 2007. Circuits with light at nanoscales: Optical nanocircuits inspired by metamaterials. *Science* 317: 1698.
7. MacDonald, K.F., and N.I. Zheludev. 2010. Active plasmonics: Current status. *Laser & Photonics Reviews* 4: 562.
8. Dionne, J.A., K. Diest, L.A. Sweatlock, and H.A. Atwater. 2009. PlasMOStor: A metal-oxide-Si field effect plasmonic modulator. *Nano Letters* 9: 897.
9. Sorger, V.J., N. Pholchai, E. Cubukcu, R.F. Oulton, P. Kolchin, C. Borschel, M. Gnauck, C. Ronning, and X. Zhang. 2011. Strongly enhanced molecular fluorescence inside a nanoscale waveguide gap. *Nano Letters* 11: 4907.
10. Ebbesen, T.W., C. Genet, and S. Bozhevolnyi. 2008. Surface plasmon circuity. *Physics Today* (May): 44–50.
11. deLeon, N.P., M.D. Lukin, and H.D. Park. 2012. Quantum plasmonic circuits. *IEEE Journal of Selected Topics in Quantum Electronics* 18: 1781.
12. Kriesch, A., S.P. Burgos, D. Ploss, H. Pfeifer, H.A. Atwater, and U. Peschel. 2013. Functional plasmonic nanocircuits with low insertion and propagation losses. *Nano Letters* 13 (9): 4539–4545.

13. Rewitz, C., G. Razinskas, P. Geisler, E. Krauss, S. Goetz, M. Pawłowska, B. Hecht, and T. Brixner. 2014. Coherent control of plasmon propagation in a nanocircuit. *Physical Review Applied* 1: 014007. doi:https://doi.org/10.1103/PhysRevApplied.1.014007.
14. Fang, Yu, and M. Sun. 2015. Nanoplasmonic waveguides: Towards applications in integrated nanophotonic circuits. *Light: Science and Applications* 4: e294. doi:10.1038/lsa.2015.67.
15. Wei, H., Z.X. Wang, X.R. Tian, M. Kall, and H.X. Xu. 2011. Cascaded logic gates in nanophotonic plasmon networks. *Nature Communications* 12 (2): 387. doi:10.1038/ncomms1388.
16. Wu, Yaw-Dong, Yung-Ta, Hsueh, and Tien-Tsorng, Shih. 2013. Novel all-optical logic gates based on microring metal-insulator-metal plasmonic waveguides, 169–172, Stockholm, August 12–15, 2013.
17. Brolo, A.G. 2012. Plasmonics for future biosensors. *Nature Photonics* 6: 709–713.
18. Anker, J.N., W.P. Hall, O. Lyandres, N.C. Shah, J. Zhao, and R.P. Van Duyne. 2008. Biosensing with plasmonic nanosensors. *Nature Materials* 7: 442–453.
19. Hill, Ryan T. 2015. Plasmonic biosensors. *WIREs Nanomedicine Nanobiotechnolgy* 7: 152–168. doi:10.1002/wnan.1314.
20. Nguyen, H.H., J. Park, S. Kang, and M. Kim. 2015. Surface plasmon resonance: A versatile technique for biosensor applications. *Sensors* 15: 10481–10510. doi:10.3390/s150510481.
21. Gibson, A.F., M.f. Kimmitt, A.C. Walker. 1970. Photon drag in germanium. *Applied Physics Letters* 17: 75.
22. Danishevskii, A.M., A.A. Kastal'skii, S.M. Ryvkin, and I.D. Yaroshetskii. 1970. Dragging of free carriers by photons in direct interband transitions in semiconductors. *Soviet Physics JETP* 31: 292.
23. Serafetinides, A.A., and M.F. Kimmitt. 1978. Photon-drag detection in p-type silicon. *Journal of Physics D. Applied Physics* 11: L97.
24. Gibson, A.F., and S. Montasser. 1975. A theoretical description of the photon-drag spectrum of p-type germanium. *Journal of Physics C* 8: 3147.
25. Luryi, S. 1987. Photon-drag effect in intersubband absorption by a two-dimensional electron gas. *Physical Review Letters* 58: 2263–2266.
26. Grinberg, A.A., and S. Luryi. 1991. Light-induced drift of quantum-confined electrons in semiconductor heterostructures. *Physical Review Letters* 67: 156.
27. Shalaev, V.M., C. Douketis, and M. Moskovits. 1992. Light-induced drift of electrons in metals. *Physics Letters A* 169: 205.
28. Shalaev, V.M., C. Douketis, J.T. Stuckless, and M. Moskovits. 1996. Light-induced kinetic effects in solids. *Physical Review B* 53: 11388.
29. Gurevich, V.L., R. Laiho, and A.V. Lashkul. 1992. Photomagnetism of metals. *Physical Review Letters* 69: 180.
30. Gurevich, V.L., and R. Laiho. 1993. Photomagnetism of metals: Microscopic theory of the photoinduced surface current. *Physical Review B* 48: 8307.
31. Gurevich, V.L., and R. Laiho. 2000. Photomagnetism of metals. First observation of dependence on polarization of light. *Physics of the Solid State* 42: 1807.
32. Stockman, M.L., L.N. Pandey, and T.F. George. 1990. *Physical Review Letters* 65: 3433.
33. Garate, E., R. Cook, C. Shaughnessy, G. Boudreaux, and J. Walsh. 1986. Boradband photon drag detector for pulsed, high-power radiation detection. *International Journal of Infrared and Microwave Waves* 7: 1827.
34. Rogalski, A. 2010. *Infrared detectors*, 2nd ed, 898 p. Boca Raton: CRC Press.
35. Goff, J.E., and W.L. Schaich. 1997. Hydrodynamic model of photon drag. *Physical Review B* 56: 15421.
36. Laiho, R. 1995. Observation of photinduced bulk current in metal. *Physical Review B* 52: 15054.
37. Linic, S., P. Christopher, and D.B. Ingram. 2011. Plasmonic-metal nanostructures for efficient conversion of solar to chemical energy. *Nature Materials* 10: 911–921.

38. Govorov, A.O., H. Zhang, H.V. Demir, Yu. K. Gun'ko. 2014. Photogeneration of hot plasmonic electrons with metal nanocrystals: Quantum description and potential applications. *Nanotoday* 9: 85–101.
39. Brongersma, M.L., N.J. Halas, and P. Nordlander. 2015. Plasmon-induced hot carrier science and technology. *Nature Nanotechnology* 10: 25–34.
40. Clavero, C. 2014. Plasmon-induced hot-electron generation at nanoparticle/metal-oxide interfaces for photovoltaic and photocatalytic devices. *Nature Photonics* 8: 95–103.
41. Dombi, P., A. Hörl, P. Rácz, I. Márton, A. Trügler, J.R. Krenn, and U. Hohenester. 2013. Ultrafast strong-field photoemission from plasmonic nanoparticles. *Nano Letters* 13: 674–678.
42. Mukherjee, S., F. Libisch, and N. Large. 2012. Hot electrons do the impossible: Plasmon-induced dissociation of H_2 on Au. *Nano Letters* 13: 240–247.
43. Lee, S.J., B.D. Piorek, C.D. Meinhart, and M. Moskovits. 2010. Photoreduction at a distance: Facile, nonlocal photoreduction of Ag ions in solution by plasmon-mediated photoemitted electrons. *Nano Letters* 10: 1329–1334.
44. H. Raether. 1988. Surface plasmons on smooth and rough surfaces and on gratings. In *Springer tracts in modern physics*, 111. New York: Springer.
45. Kretschmann, E. 1972. *Optics Communications* 5: 331.
46. Ritchie, R.H. 1973. Surface plasmons in solids. *Surface Science* 34: 1–19.
47. Moskovits, M. 1985. Surface-enhanced spectroscopy. *Reviews of Modern Physics* 57: 783–826.
48. Quinten, M., A. Leitner, J.R. Krenn, and F.R. Aussenegg. 1998. Electromagnetic energy transport via linear chains of silver nanoparticles. *Optics Letters* 23: 1331–1333.
49. Averitt, R.D., S.L. Westcott, and N.J. Halas. 1999. Linear optical properties of gold nanoshells. *Journal of the Optical Society of America* B 1 (6): 1824–1832.
50. Mock, J.J., M. Barbic, D.R. Smith, D.A. Schultz, and S. Schultz. 2002. Shape effects in plasmon resonance of individual colloidal silver nanoparticles. *Chemical Physics* 116 (116): 6755–6759.
51. Kreibig, U., and M. Vollmer. 1995. *Optical properties of metal clusters*, vol. 25. New York: Springer.
52. Su, K.-H., Q.-H. Wei, X. Zhang, J.J. Mock, D.R. Smith, and S. Schultz. 2003. Interparticle coupling effects on plasmon resonances of nanogold particles. *Nano Letters* 3: 1087–1090.
53. Noginov, M.A., G. Zhu, M. Mayy, B.A. Ritzo, N. Noginova, and V.A. Podolskiy. 2008. Stimulated emission of surface plasmon polaritons. *Physical Review Letters* 101: 226806.
54. Baudrion, A.-L., F. de León-Pérez, O. Mahboub, A. Hohenau, H. Ditlbacher, F. J. García-Vidal, J. Dintinger, T.W. Ebbesen, L. Martín-Moreno, and J.R. Krenn. 2008. Coupling efficiency of light to surface plasmon polariton for single subwavelength holes in a gold film. *Optics Express* 16: 3420.
55. Vengurlekar, A., and T. Ishiara. 2005. Surface plasmon enhanced photon drag in metal films. *Applied Physics Letters* 87: 091118.
56. Noginova, N., A.V. Yakim, J. Soimo, L. Gu, and M.A. Noginov. 2011. Light-to-current and current-to-light coupling in plasmonic systems. *Physical Review B* 84: 035447.
57. MWS Wire Industries Specifications, http://www.mwswire.com/.
58. Kurosawa, H., and T. Ishihara. 2012. Surface plasmon drag effect in a dielectrically modulated metallic thin film. *Optics Express* 20 (2): 1561–1574. doi:10.1364/OE.20.001561.
59. Kurosawa, H., T. Ishihara, N. Ikeda, D. Tsuya, M. Ochiai, and Y. Sugimoto. 2012. Optical rectification effect due to surface plasmon polaritons at normal incidence in a nondiffraction regime. *Optics Letters* 37 (14): 2793–2795. doi:10.1364/OL.37.002793.
60. Durach, M., A. Rusina, and M.I. Stockman. 2009. Giant surface-plasmon-induced drag effect in metal nanowires. *Physical Review Letters* 103: 186801–1–4.
61. Noginova, N., V. Rono, F.J. Bezares, and J.D. Caldwell. 2013. Plasmon drag effect in metal nanostructures. *New Journal of Physics* 15: 113061. doi: http://dx.doi.org/10.1088/1367–2630/15/11/113061.

62. Caldwell, J.D., O. Glembocki, F.J. Bezares, N.D. Bassim, R.W. Rendell, M. Feygelson, M. Ukaegbu, R. Kasica, L. Shirey, and C. Hosten. 2011. Plasmonic nanopillar arrays for large-area, high-enhancement surface-enhanced Raman scattering sensors. *ACS Nano* 5: 4046.
63. Simpkins, B.S., J.P. Long, O.J. Glembocki, J. Guo, J.D. Caldwell, and J.C. Owrutsky. 2012. Pitch-dependent resonances and near-field coupling in infrared nanoantenna arrays. *Optics Express* 20: 27725.
64. Akbari, M., M. Onoda, and T. Ishihara. 2015. Photo-induced voltage in nano-porous gold thin film. *Optics Express* 23 (2): 823–832.
65. Ishihara, T., T. Hatano, H. Kurosawa, Y. Kurami, N. Nishimura. 2012. Transverse voltage induced by circularly polarized obliquely incident light in plasmonic crystals. In *Proceedings of SPIE 8461, Spintronics V*, 846117 (October 1, 2012); doi:10.1117/12.933279.
66. Ni, X., S. Xiao, Yu. Wang, Yang, and X. Zhang, Photon spin induced collective electron motion on a metasurface. CLEO:2015, OSA 2015, # FW4E.1.
67. Bai, Q. 2015. Manipulating photoinduced voltage in metasurface with circularly polarized light. 23 (4): 5348–5356. doi:10.1364/OE.23.005348.
68. Kang, L., S. Lan, Y. Cui, S.P. Rodrigues, Y. Liu, D.H. Werner, and W. Cai. 2015. An Active Metamaterial Platform for Chiral Responsive Optoelectronics. *Advanced Materials* 27: 4377–4383. doi:10.1002/adma.201501930.
69. Noginova, N., V. Rono, A. Jackson, and M. Durach. 2015. Controlling plasmon drag with illumination and surface geometry. OSA Technical Digest (online) (Optical Society of America, 2015), paper FTh3E.7. doi:10.1364/CLEO_QELS.2015.FTh3E.7.
70. Allen, K.W., N. Farahi, Y. Li, N.I. Limberopoulos, D.E. Walker, A.M. Urbas, and V.N. Astratov. 2015. Overcoming the diffraction limit of imaging nanoplasmonic arrays by microspheres and microfibers. *Optics Express* 23: 24484–24496.
71. Stegeman, G., R. Wallis, and A. Maradudin. 1983. Excitation of surface polaritons by end-fire coupling. *Optics Letters* 8: 386–388.
72. Landau, L.D., and E.M. Lifshitz. 2005. *Theoretical physics. Electrodynamics of continuous media*, vol. 8. Moscow: Nauka.
73. Merlin, R. 2009. Metamaterials and the Landau-Lifshitz permeability argument: Large permittivity begets high-frequency magnetism. In *Proceedings of the National Academy of Sciences of the United States of America*, vol. 106, no. (6), 1693–1698.
74. Shimizu, Y., and H. Sasada. 1998. Mechanical Force in Laser Cooling and Trapping. *American Journal of Physics* 66 (11): 960–967.
75. Ginzburg, P., A. Hayat, N. Berkovitch, and M. Orenstein. 2010. Nonlocal ponderomotive nonlinearity in plasmonics. *Optics Letters* 35 (10): 1551–1553.
76. Bergman, D.J., and M.I. Stockman. 2003. Surface plasmon amplification by stimulated emission of radiation: Quantum generation of coherent surface plasmons in nanosystems. *Physical Review Letters* 90 (2): 027402.
77. Noginov, M.A., G. Zhu, A.M. Belgrave, R. Bakker, V.M. Shalaev, E.E. Narimanov, S. Stout, E. Herz, T. Suteewong, and U. Wiesner. 2009. Demonstration of a spaser-based nanolaser. *Nature* 460 (7259): 1110–1112.
78. Chang, D.E., A.S. Sørensen, P.R. Hemmer, and M.D. Lukin. 2006. Quantum optics with surface plasmons. *Physical Review Letters* 97 (5): 053002.
79. Iorsh, I., A. Poddubny, A. Orlov, P. Belov, and Y.S. Kivshar. 2012. Spontaneous emission enhancement in metal–dielectric metamaterials. *Physics Letters A* 376 (3): 185–187.
80. Hussain, R., D. Keene, N. Noginova, and M. Durach. 2014. Spontaneous emission of electric and magnetic dipoles in the vicinity of thin and thick metal. *Optics Express* 22 (7): 7744–7755.
81. Huck, A., S. Smolka, P. Lodahl, A.S. Sørensen, A. Boltasseva, J. Janousek, and U.L. Andersen. 2009. Demonstration of quadrature-squeezed surface plasmons in a gold waveguide. *Physical Review Letters* 102 (24): 246802.

82. Di Martino, G., Y. Sonnefraud, S. Kéna-Cohen, M. Tame, S.K. Özdemir, M.S. Kim, and S. A. Maier. 2012. Quantum statistics of surface plasmon polaritons in metallic stripe waveguides. *Nano Letters* 12 (5): 2504–2508.
83. Zuloaga, J., E. Prodan, and P. Nordlander. 2009. Quantum description of the plasmon resonances of a nanoparticle dimer. *Nano Letters* 9 (2): 887–891.
84. Zuloaga, J., E. Prodan, and P. Nordlander. 2010. Quantum plasmonics: Optical properties and tunability of metallic nanorods. *ACS Nano* 4 (9): 5269–5276.
85. Durach, M., and N. Noginova. 2016. On nature of plasmonic drag effect. *Physical Review B (Rapid Communication)* 93: 161406(R).
86. Fann, W.S., R. Storz, H.W.K. Tom, and J. Bokor. 1992. Electron thermalization in gold. *Physical Review B* 46 (20): 13592.
87. Sun, C.K., F. Vallée, L.H. Acioli, E.P. Ippen, and J.G. Fujimoto. 1994. Femtosecond-tunable measurement of electron thermalization in gold. *Physical Review B* 50 (20): 15337.
88. Link, S., C. Burda, Z.L. Wang, and M.A. El-Sayed. 1999. Electron dynamics in gold and gold–silver alloy nanoparticles: The influence of a nonequilibrium electron distribution and the size dependence of the electron–phonon relaxation. *The Journal of Chemical Physics* 111 (3): 1255–1264.
89. Knight, M.W., H. Sobhani, P. Nordlander, and N.J. Halas. 2011. Photodetection with active optical antennas. *Science* 332 (6030): 702–704.
90. Mukherjee, F., F. Libisch, N. Large, O. Neumann, L.V. Brown, J. Cheng, J.B. Lassiter, E.A. Carter, P. Nordlander, and N.J. Halas. 2012. Hot electrons do the impossible: Plasmon-induced dissociation of H2 on Au. *Nano Letters* 13 (1): 240–247.
91. Clavero, C. 2014. Plasmon-induced hot-electron generation at nanoparticle/metal-oxide interfaces for photovoltaic and photocatalytic devices. *Nature Photonics* 8 (2): 95–103.
92. Khurgin, J.B. 2015. How to deal with the loss in plasmonics and metamaterials. *Nature Nanotechnology* 10 (1): 2–6.
93. Khurgin, J.B. 2015. Ultimate limit of field confinement by surface plasmon polaritons. *Faraday Discussions* 178: 109–122.
94. Brown, A.M., R. Sundararaman, P. Narang, W.A. Goddard III, and H.A. Atwater. 2015. Non-radiative plasmon decay and hot carrier dynamics: Effects of phonons, surfaces and geometry. *ACS Nano* 10 (1): 957–966.
95. Brown, A.M., R. Sundararaman, P. Narang, W. A. Goddard III, and H. A. Atwater. 2016. Ab initio phonon coupling and optical response of hot electrons in plasmonic metals. *Physical Review B* 94: 075120.
96. Johnson, P.B., and R.W. Christy. 1972. Optical constants of the noble metals. *Physical Review B* 6 (12): 4370.
97. Fann, W.S., R. Storz, H.W.K. Tom, and J. Bokor. 1992. Electron thermalization in gold. *Physical Review B* 46 (20): 13592.
98. Chandezon, J., G. Raoult, and D. Maystre. 1980. A new theoretical method for diffraction gratings and its numerical application. *Journal of Optics* 11 (4): 235.
99. Noginova, N., M. LePain, V. Rono, S. Masshadi, R. Hussain, and M. Durach. 2016. Plasmon drag in profile-modulated gold film. Theory and experiment. *New Journal of Physics* 18: 093036.
100. Sell, A., A. Leitenstorfer, and R. Huber. 2008. Phase-locked generation and field-resolved detection of widely tunable terahertz pulses with amplitudes exceeding 100 MV/cm. *Optics Letters* 33 (23): 2767–2769.
101. Leinss, S., T. Kampfrath, K. Volkmann, M. Wolf, J.T. Steiner, M. Kira, S.W. Koch, A. Leitenstorfer, and R. Huber. 2008. Terahertz coherent control of optically dark paraexcitons in Cu_2O. *Physical Review Letters* 101 (24): 246401-4.
102. Stockman, M.I. 2004. Nanofocusing of optical energy in tapered plasmonic waveguides. *Physical Review Letters* 93 (13): 137404.
103. Verhagen, E., M. Spasenovic, A. Polman, and L. Kuipers. 2009. Nanowire plasmon excitation by adiabatic mode transformation. *Physical Review Letters* 102(20): 203904–4.

104. MacDonald, K.F., Z.L. Samson, M.I. Stockman, and N.I. Zheludev. 2009. Ultrafast active plasmonics. *Nature Photonics* 3 (1): 55–58.
105. Larkin, I.A., and M.I. Stockman. 2005. Imperfect perfect lens. *Nano Letters* 5 (2): 339–343.
106. Aizpurua, J., and A. Rivacoba. 2008. Nonlocal effects in the plasmons of nanowires and nanocavities excited by fast electron beams. *Physical Review B* 78 (3): 035404–035414.
107. Gault, B., F. Vurpillot, A. Bostel, A. Menand, and B. Deconihout. 2005. Estimation of the tip field enhancement on a field emitter under laser illumination. *Applied Physics Letters* 86 (9): 094101–094103.
108. Sha, G., A. Cerezo, and G.D.W. Smith. 2008. Field evaporation behavior during irradiation with picosecond laser pulses. *Applied Physics Letters* 92(4): 043503-3.
109. Ginzburg, P., A. Krasavin, S. Sonnefraud, A. Murphy, R. J. Pollard, S. A. Maier. A. V. Zayats. 2012. Nonlinearly coupled localized plasmon resonances: Resonant second-harmonic generation. *Physical Review B* 86: 085422 (2012).
110. Ginzburg, P., A. Krasavin, and A.V. Zayats. 2013. Cascaded second-order surface plasmon solitons due to intrinsic metal nonlinearity. *New Journal of Physics* 15: 013031.
111. Panasyuk, G.Y., J.C. Schotland, and V.A. Markel. 2008. Classical theory of optical nonlinearity in conducting nanoparticles. *Physical Review Letters* 100: 047402.
112. Masuhara, H,, S. Kawata. *Nanoplasmonics: from fundamentals to applications*, 334. Amsterdam: Elsevier.

Chapter 9
Dimensional Variations in Nanohybrids: Property Alterations, Applications, and Considerations for Toxicological Implications

Nirupam Aich, Arvid Masud, Tara Sabo-Attwood, Jaime Plazas-Tuttle and Navid B. Saleh

Abstract Hybridization of nano-scale entities lead to higher dimensional ensemble materials with multifunctionality. Such hierarchical complex materials though are engineered with output properties in mind, these evolved nanostructures possess unique shapes and physico-chemical attributes. Nanotoxicological considerations hinge on physical size and shape factors; thus, dramatic alterations to shape and dimensionality of ensemble nanohybrids (NHs) necessitate careful evaluation of this 'horizon' material class. This chapter reviews size/shape/dimensionality variations of nanomaterials due to hybridization and discusses property alteration of these NHs, relevant to applications and nanotoxicology. The chapter also discusses nano-bio interactions of novel nanohybrids in relation to their size, shape, and dimensionality, and outlines future research needs and strategies.

Keywords Nanotoxicity · Dimensionality · Inflammation · Asbestos · Graphene

N. Aich (✉) · A. Masud
Department of Civil, Structural and Environmental Engineering, University at Buffalo, the State University of New York, 14260 Buffalo, NY, USA
e-mail: nirupama@buffalo.edu

T. Sabo-Attwood
Department of Environmental and Global Health, College of Public Health and Health Professions, Center for Human and Environmental Toxicology, University of Florida, 32611 Gainesville, FL, USA

J. Plazas-Tuttle · N.B. Saleh
Department of Civil, Architectural and Environmental Engineering, University of Texas at Austin, 78712 Austin, TX, USA

S.E. Hunyadi Murph et al. (eds.), *Anisotropic and Shape-Selective Nanomaterials*, Nanostructure Science and Technology, DOI 10.1007/978-3-319-59662-4_9

9.1 Introduction

Progress in nanoscale material development has paved the way for synthesis and use of nanohybrids (NHs) by conjugating multiple nanomaterials (NMs) to form nano heterostructures. In recent years, there has been an exponential growth in the research and development of these complex multi-component NHs [1]. The increasing trend of NHs development and usage for a wide range of applications can be attributed to their enhanced–or multi-functionality compared to their component materials [2]. This has been possible due to the manipulation of component NMs' fundamental attributes including physico-chemical, electrical, mechanical, and optical properties. The extensive array of applications of NHs includes prominent uses in electronics [3–5], energy [6–8], sensors [9], biomedicine [10, 11], optical imaging [12, 13], catalysis [14, 15], environmental remediation [16–18], among others. The enhanced production and use of these novel nano-assemblies warrant fundamental understanding of environmental and biological safety during manufacture, usage, and end-of-life stages of these materials [2].

The working definition of the NHs proposed in recent literature is as follows: *"when more than one NM of unique chemical origin or differing dimensionality are conjugated by molecular or macromolecular links or physico-chemical forces or when one nanomaterial overcoats another possessing a unique chemical identity or when complex soft molecules are engineered to chemically bind to NM surfaces, all to enhance the existing functionality or achieve multifunctional usage, can be defined as NHs"* [2]. These "horizon materials" show enhanced performance by exhibiting desired multiple functional properties in comparison to the component NMs. For example, nanoscale zero-valent iron (nZVI) can be very effective in removing heavy metals from water due to their high reducing capability. Practical application of the nZVI is limited due to their high tendency to agglomerate via enhanced van der Waals and magnetic attractions. Upon hybridization with graphene, the tendency of nZVI agglomeration is reduced, while other functional properties including graphene's large reactive surface area, high adsorption properties, and excellent electron transport properties improve the contaminant degradation abilities of the nZVI by many folds [19]. However, formation of such heterostructures via hybridization brings about physical and chemical property alteration in the NH itself, which can lead to uncertainties related to their potential environmental health and safety (EHS) [20, 21]. Among these altered properties, dimensional attributes (e.g., size, shape, morphology etc.) need to be considered carefully for adequate assessment of their environmental and ecological impact [2, 22].

Hybridization often causes emergent dimensional modification in the NHs compared to the component materials. For example, when zero-dimensional (0-D) fullerenes are covalently bonded with one-dimensional (1-D) carbon nanotubes (CNTs), the resulting NHs (often termed as nanobud) acquire three dimensional (3-D) physical attribute [23]. These dimensional modifications can cause altered toxic responses when interacting with microorganisms or higher order biological species. Nanotoxicity from passive and singular nanostructures has been correlated with sizes and shapes; e.g.., asbestos-like pulmonary effects that is demonstrated by 1-D CNTs. Heterostructures of CNTs and other carbonaceous NMs thus will likely present uncertainty on demonstrated nanotoxicity from the singular carbon nanotubes. Furthermore, when metal nanocrystals are deposited onto CNTs, the resulting heterostructures can achieve unique dissolution (e.g., for silver) or photoactive properties (e.g., reactive oxygen species or ROS generation from TiO_2)–two major mechanisms dictating most toxicological manifestations from metal nanocrystals.

This book chapter presents a review of expanding classes of hybridized nano-ensembles based on variations in their size and shape, and dimensionality and the potential implications of these altered attributes. Finally, strategies for unraveling complex nano-bio interactions and reducing uncertainty of nano EHS prediction is outlined.

9.2 Dimensional Variations in Nanohybrids: Altered Properties and Applications

Wide ranges of NHs are being prepared and studied using various combinations of primary NM constituents for multidirectional applications. The most commonly pursued combinations include: carbon-carbon (e.g., graphene-CNT [24], CNT-fullerene [23]), carbon-metal (e.g., graphene-nZVI [19], CNT-TiO_2 [25]), metal-metal (e.g., Ag–Au [26], Ag–TiO_2 [27]) etc. [2]. Each of these classes of NHs has application in multiple fields including electronics [27–29], biomedical [30, 31], environmental [32–34], etc. Alteration in dimensional features from parent NMs is one of the distinct consequences of hybridization. Changes in dimensional attributes like size and shape result in modification of the surface area and morphological characteristics of NHs, which is responsible for altered physico-chemical properties. This section attempts to categorize NHs into different classes based on their dimensionality (Fig. 9.1) and discusses relevant physico-chemical property alterations intended for applications.

(a) Zero dimensional (0-D) heterostructures commonly are core-shell bimetallic NHs (e.g., Cu–ZnO [35], Ag–TiO_2 [38], Cu–Ag [39]), where the parent NMs are also nanoscale 0-D entities (Fig. 9.1a). Au–Ag bimetallic core-shell NH is such an example, where an Au nanoparticle (Au NP) core is surrounded by a nanometer thick coating of Ag. Individual spherical Au and Ag NPs have single surface plasmon resonance (SPR) peaks around 520 and 400 nm, respectively, and have been investigated for SPR based chemical or bio-sensing [40]. For Au–Ag core-shell NH, two distinct SPR peaks are observed in the spectrum corresponding to both Au and Ag NPs. However, the SPR peak positions shift due to the layered structure of the bimetallics. The peak corresponding to Au displays blue shift from

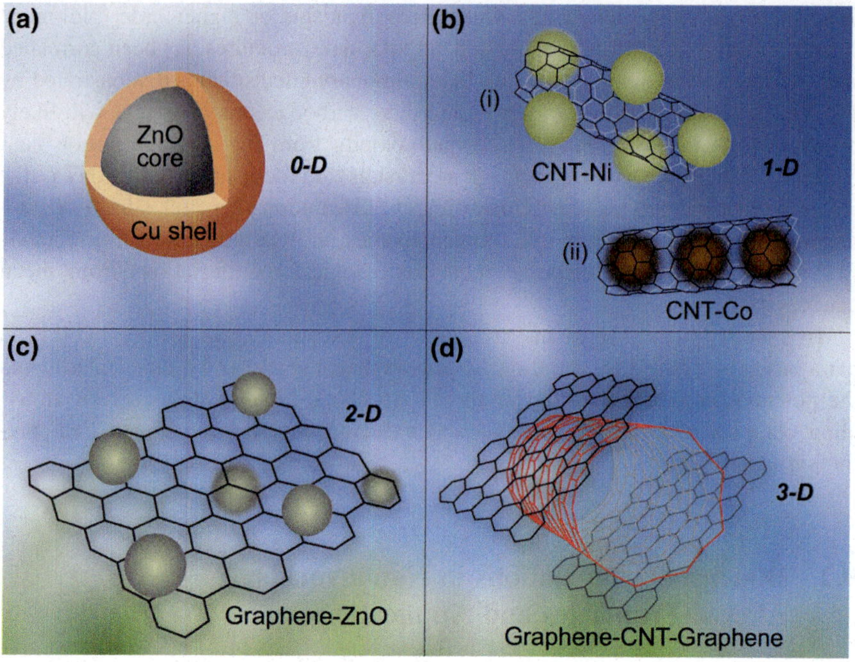

Fig. 9.1 Classification of NHs based on dimensionality. Examples of **a** zero dimensional (0-D) core-shell ZnO–Cu NH [35], **b** one dimensional (1-D) CNT-metal NH [36], **c** two-dimensional (2-D) planar graphene–ZnO NH [33], **d** three dimensional (3-D) graphene-CNT-graphene NH [37]

520–480 nm whereas that of Ag shows red shift from 400–450 nm. Such plasmon shift is found to be a function of Ag layer thickness (as it varies from 0–15 nm), as the core radius of Au is kept constant at 10 nm [41]. Thus, controlling the shell thickness of the Au–Ag NH can allow for tuning of SPR spectrum and enable efficient bio-sensing [40]. In the case of Ag–Au core-shell NH, similar shifts in SPR spectrum can be observed due to the thickness changes in Au-shell. However, in this case, both Au and Ag SPR peaks red shift towards higher wavelength (Fig. 9.2).

Another example of plasmonic core-shell NHs is Au–TiO_2, developed for usage in dye-sensitized solar cells (DSSCs). Similar to the previous one, the functional enhancement of this plasmonic NH depends on the tuning of the shell thickness. Thinner TiO_2 shells enhance the short-circuiting in a solar cell. On the contrary, thicker TiO_2 shells allow for generating open-circuit voltage activated by the semiconducting property of the TiO_2. Thicker shell also resists corrosion of the Au core [42]. Considering both of these aspects, an optimum shell thickness of 5 nm of TiO_2 was found to give the highest performance with 23% power conversion efficiency in DSSCs (Fig. 9.3). Several other similar examples of core-shell NHs are listed in Table 9.1 alongside with their size-dependent physico-chemical property alterations and potential applications.

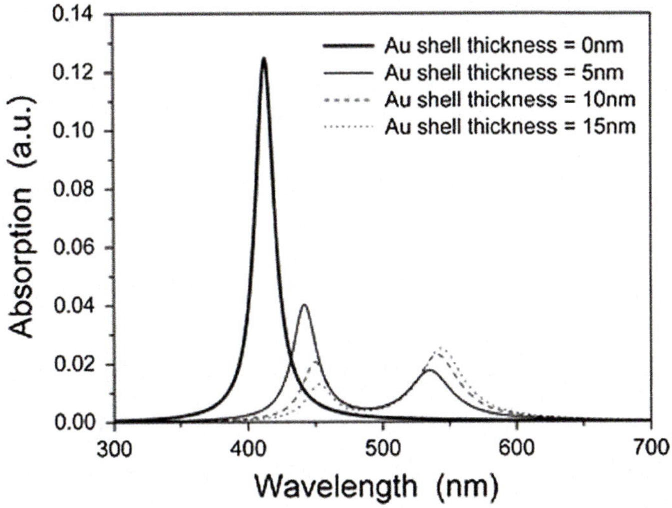

Fig. 9.2 Light absorption spectra of Ag–Au core-shell NH with different thicknesses of Au shell exhibiting different SPR peak shifts, when the core radius of Ag is kept constant at 10 nm. [Reprinted from Zhu [41]. Copyright 2009 by Jian Zhu. Reprinted with permission.]

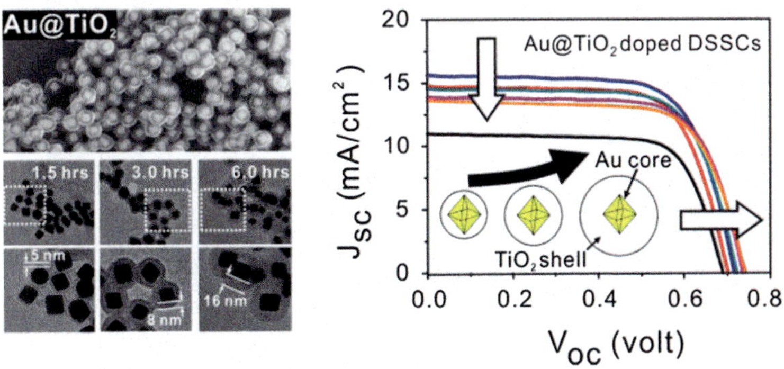

Fig. 9.3 *Left* Transmission electron microscopy (TEM) images of Au–TiO_2 NHs with increasing thickness of TiO_2 shell, i.e., 5, 8, and 16 nm which is a function of reaction time involving stirring of Au NP suspensions in titanium isopropoxide (TTIP); *Right* Short circuit current (J_{sc}) versus open circuit voltage (V_{oc}) characteristic of DSSCs for different Au–TiO_2 NHs with TiO_2 shell thickness of 0 nm (*red*), 5 nm (*blue*), 8 nm (*cyan*), 12 nm (*magenta*), and 16 nm (*orange*). TiO_2 shell thickness of 5 nm showed the optimum current-voltage characteristic in DSSCs. [Reprinted from Liu [42]. Copyright 2013 by The Royal Society of Chemistry. Reprinted with permission.]

(b) One dimensional (1-D) structures are mostly composed of tubular, rod-like, and ribbon shaped NHs that are predominantly stretched in one direction (Fig. 9.1b). CNT-metal NHs are the most common 1-D NHs [36]. The metal NPs can either be encapsulated within the CNT's hollow structure (endohedrally) or conjugated at the exterior of the CNT walls (exohedrally). When metal NP is encapsulated within CNT

Table 9.1 Specific examples of nanohybrid class, enhanced property, and application

NH Class	Specific NH	Enhanced property	Application	Reference
Zero-dimensional (0-D)	Au–Ag	Change of Plasmon band and enhanced stability based on the diameter of the core and thickness of the shell	Biosensor/Biomedical	Widayanti et al. [40]
	Au–TiO_2	Electrical conductivity and photovoltaic property depends on the relative thickness of Au and TiO_2	Solar cell	Liu et al. [42]
	Au–Co	Enhanced catalytic performance due to modified electronic structure; optimum performance at certain thickness of core (2.5 nm) and shell (4.5 nm)	Renewable energy	Yan et al. [77]
	Au–Pt	Enhanced catalytic property due to higher catalytically active surface area, which depends on the shape of the core. Most enhanced performance for octahedron shape of Au core	Catalysis	Kim et al. [78]
	Pd–Pt	Enhanced electrocatalytic performance due to interaction between Pd and Pt, dependent on the Pt shell thickness	Electro-catalysis	Kim et al. [79]
One-dimensional (1-D)	Poly (3-hexylthiophene) (P3HT)-CdS nanowire	Enhanced charge separation and charge transport depending on the increased interfacial area and interpenetrating morphology	Solar cell	Ren et al. [80]
	PbS–CdS (core-shell)-MWCNT	Enhanced photo- and thermal-stability dependent on confinement of the thickness of shell within $\leq$ 0.7 nm	Photo-voltaic device	Zhao et al. [81]
	CNT-Cu	Enhanced photoresponse property due to the interaction of CNT with Cu. Photoresponse depends on the size of Cu nanoparticle. Enhanced property diminishes with the increase of Cu size beyond 0.2 nm	Photo-electro-chemical devices	Scarselli et al. [47]
	SnO_2 nanowire-Ag	Enhanced gas sensing property depending on coverage of SnO_2 surface by Ag. The property diminishes in case of continuous Ag film compared to dispersed particles	Gas Sensor	Hwang et al. [82]

(continued)

Table 9.1 (continued)

NH Class	Specific NH	Enhanced property	Application	Reference
Two-dimensional (2-D)	GO–Au	Surface enhanced Raman scattering, which is dependent on the shape of the Au nanomaterial (enhanced performance is for rod-shaped Au)	Sensor and imaging device	Lee et al. [55]
	GO–ZnO	Enhanced photoluminescence property is a function of ZnO size, exhibiting higher performance with lower size of ZnO	Photo-catalyst	Thuan et al. [57]
	MoS_2–Au	Enhanced gas sensing property for high surface area and catalytic activity. The property depends on the rough surface morphology of MoS_2, which ensures even distribution of Au	Gas sensor	Yan et al. [63]
Three-Dimensional (3-D)	PbS–CdS (core-shell)-mesoporous TiO_2	Enhanced electron lifetime and transit time, the trade-off between enhanced property and decrease of light absorption and charge injection depends on thickness of CdS shell	Solar cell	Lai et al. [83]
	Pb–Pt (core-shell)-C	Enhanced electrolytic property which is a function of shell thickness of Pt with an optimum thickness 0.94 nm	Catalysis for oxygen/reduction reaction	Choi et al. [84]
	CNT-graphene	Enhanced conductive network and surface area due to interconnected chamber-shaped morphology	Li ion battery	Shi et al. [37]
	MoS_2–Graphene	Potential barrier of MoS_2/Graphene can be adjusted by tuning the thickness of MoS_2	Semi-conductor	Su et al. [85]
	BN/graphene/BN	Variable energy band gap, which is dependent on the thickness of the layer	Radio frequency device	Kim et al. [86]

(e.g., CNT-Ni, CNT-Co), the NH masks the zero-dimensionality of the metal NP, making the hetero-structure one-dimensional [43]. Magnetic NPs, in many cases, are encapsulated within CNTs to minimize their interaction with the surrounding environment. This enables protection of the encapsulated magnetic NPs from oxidation or other chemical reactions mediated by various atmospheric agents [36]. For example, Fe–Co NPs are encapsulated within CNTs, which are used in arsenic removal from water [44]. The enclosed Fe-Co NPs achieve higher chemical stability in environment while the CNT backbone works as a closed nano-reactor system within which the Fe-Co can efficiently adsorb arsenic from water. Even if the metal NP is conjugated on the outer surface of CNT, the NH still can be considered 1-D as long as the size of the spherical metal NP is negligible compared to the length of CNT. For example, CNT-TiO_2 exohedral NH has length in the range of few microns, whereas the diameter of TiO_2 is in the range of 5–10 nm [45]. Similarly, multiwalled CNT (MWCNT)-ZnO NHs have length more than a micron while the ZnO NPs are below 10 nm [46]. These exohedrally conjugated CNT-metal NHs show various preferred multifunctionality. For example, MWCNT-Cu NH shows enhanced photo-response property combining excellent charge transfer abilities of 1-D MWCNT and catalytic capability of Cu NP for usage in photo-electrochemical devices [47]. The performance of these MWCNT-Cu NHs also depends on the size of Cu NPs. The photoresponse measured in terms of incident photon-to-charge carrier generation efficiency (IPCE) across the visible and near ultraviolet range varies depending on the size of the Cu NPs on the MWCNT surfaces. For an optimum thickness of 0.2 nm of Cu NP on MWCNT, the IPCE value has shown to reach 15%, which is 2.5 time higher than that of MWCNT only. However, the IPCE value decreases as the thickness of the Cu coverage increases beyond 0.2 nm and the materials lose photoresponsive abilities when the thickness increases above 1 nm (Fig. 9.4). Carbon-carbon NHs like nanopeapods (e.g., fullerenes encapsulated in CNT [48]) and tubular metal oxides including TiO_2 and ZnO can also form 1-D NHs (e.g., TiO_2 nanotube-Pt, ZnO nanotube-Cu_2O NHs [49, 50]). Similar examples of 1-D NHs are provided in Table 9.1.

(c) Two-dimensional (2-D) NHs are formed with parent materials possessing 2-D sheet-like structures (Fig. 9.1c). The size of this class of NHs may vary along the plane, however, the thickness is confined within the range of few nanometers. Recent advances in graphene and similar 2-D nanostructures have accelerated the synthesis and application of 2-D NHs. Graphene based 2-D NHs are the most common, where metal/metal oxide NP are hybridized with graphene (G), graphene oxides (GO), and reduced graphene oxides (rGO) (e.g., G–Pd [51], rGO–Ag [52], G–ZnO [53], rGO–CuO [54], etc.). Atomically thin 2-D planar surfaces of graphene provide extraordinary properties including enhanced conductivity, zero-band gap, high mechanical strength, enhanced surface area, and excellent adsorption properties [2]. These unique properties improve the functional characteristics of the supported metal NPs. For example, GO–Au NH exhibits better optical property than pure Au NP, which can be utilized in sensing and imaging applications [55, 56]. The enhanced optical property of GO–Au NH is also dependent on the shape (i.e., sphere, octahedral, rod, etc.) of the Au NP on the GO surface [55]. Figure 9.5 shows extinction spectra, corresponding TEM images, and surface-enhanced

Raman scattering (SERS) spectra of different GO–Au NHs with octahedral, spherical, and rod shaped Au NPs. The extinction spectra with different GO loading ratios for each shape of the NH suggest that the shape of Au NP as well as the loading ratios of GO have an effect on the peaks of the extinction curve. Figure 9.5g indicates that the change in SERS signal for the GO–Au NHs is a function of the shape of Au NPs. The SERS signal of the NHs is shown to have enhanced by factors of 9, 16, and 20, respectively, for spherical, octahedral, and rod shaped Au NPs in comparison to that for GO only cases [55].

In the case of GO–ZnO NH, 6 nm ZnOs tend to show better photodegradation rate of methylene blue compared to 12 nm ZnOs [57]. The size decrease and corresponding increase of surface area have been reported to have caused such enhancement. However, the same NH decorated with 3 nm ZnOs exhibits reduced photocatalytic ability compared to the 6 nm cases. This has been explained by the impediment of free movement of electron due to electron confinement, which result in the decrease in photocatalytic ability.

In recent years, other two dimensional (2-D) structures, e.g., transition metal dichalcogenide (TMD) NMs (e.g. MoS_2 [58], TiS_2 [59], WS_2 [60], etc.) and NHs based on TMDs (e.g. MoS_2–Au [61], WS_2–Cu [62] etc.) are being pursued extensively for various applications. In the case of MoS_2–Au NH, Au NPs enhance the gas sensing ability of MoS_2 nanosheets due to increased surface area and catalytic activity. The enhanced gas sensing property depends on the homogeneity of dispersion and isolation of Au NPs from each other on the MoS_2 surfaces [63]. In another case, Cadmium Selenide (CdSe) quantum dots are hybridized on the surface of MoS_2 layers for non-radiative energy transfer, which enables the NH to be used in optoelectronic devices [64]. The CdSe quantum dots enhance the energy absorptive property of the MoS_2 sheets whereas, the efficiency of energy transfer depends on the number of MoS_2 layers. Single layer MoS_2-CdSe NH exhibits the highest energy transfer efficiency. With the increase in the number of the MoS_2 layers in the NH, the energy transfer rate reduces [64].

(d) Three-dimensional (3-D) NHs or heterostructures are formed when blocks of synthesized NHs expand their dimensional features in all three spatial coordinates (Fig. 9.1d). For example, in CNT-graphene NH, CNTs serve as the backbone and graphene sheets as walls and when conjugated these form 3-D structures with large surface area and inter connected chambers [65]. Such structure and shape resist clustering of constituent NMs and maintain gap between the consecutive graphene layers. The CNT-graphene NHs possess a better conductive network, which makes them potential candidates for lithium ion battery application. In such structures, the CNTs bridge the defects for electron transfer of graphene [37]. When CNTs are vertically mounted between graphene layers, the CNT-graphene NH show high specific capacitance value, which makes these NHs attractive for use as supercapacitors [66]. On the other hand, when the CNT and graphene are hybridized in parallel alignment, the planar CNT-graphene NHs show better photovoltaic conversion efficiency in solar cells due to enhancement in conductivity. However, in the case of planar CNT-graphene NH films, the graphene layers tend to stack together [67]. Thus, the directional arrangements of 1-D CNT and 2-D graphene can have significant impact on the physico-chemical property alterations. Metal-metal NHs can also form 3-D structures. Metal-semiconductor composites like Ag–TiO_2 are an example of this class of NH, where the TiO_2 forms 3-D mesoporous morphology, which enables to evenly disperse Ag NPs (Fig. 9.6) [68]. This NH shows enhanced optoelectronic and catalytic properties due to the well dispersed Ag NP on TiO_2 mesostructure [69].

3-D heterostructures or NHs, formed by vertically stacked alternate layers of recently developed 2-D NMs, are being extensively studied for use in modern semiconductor and optoelectronic devices due to the enhanced photocurrent generation property of these heterostructures [70, 71]. Examples of these NHs include WS_2–MoS_2 [72], graphene-WS_2 [73], and graphene–MoS_2 [74] heterostructures, where the complementary physico-chemical properties from each of the constituent NMs can provide a synergistic effect on the desired functionality. The alternate nanosheets are linked with weak van der Waals interaction [72] and have edges surrounding the layered sheets. In the case of solar cell, when a layer of MoS_2 is added between the graphene-silicone layers, the photovoltaic property of the tri-layer NH enhances. This is due to the effective passivation and electron-blocking/hole transport function of the MoS_2 layer. The photovoltaic property of the NH increases with decrease in the thickness of MoS_2 layer and with the increase in the number of graphene layers [75]. While used in photovoltaic devices, the graphene–MoS_2 heterostructures can have very high photon adsorption and electron-hole creation property due to the presence of MoS_2 and high resistance against radiation damage due to the mechanical strength of graphene [76].

9.3 Nano-Bio Interactions of Nanohybrids: Importance of Dimensionality

The mechanistic understanding of environmental health and safety of individual NMs is still emerging, while NHs are considered as 'horizon materials'. However, significant progress has been made in understanding the toxic responses from singular NM insults on a wide range of organisms, which include prokaryotic microorganisms like bacteria, eukaryotic human cell lines, and hierarchical organisms like rodents and fish [87]. The key mechanisms attributed to the toxicological effects of NMs include: dissolution and associated cell damage [88], oxidative stress through reactive oxygen species (ROS) generation [89, 90], asbestos-like effects and inflammation [90, 91], cell rupture via physical interaction and direct contact [92]. These toxic manifestations are controlled by physico-chemical properties of NMs including size [93], shape [94], surface area [93], dimensionality [95], etc. These physico-chemical attributes will also likely influence NH toxicity; however, the emergent alteration in dimensionality through hybridization and the presence of multiple nano-scale entities with unique chemical composition, will undergo a delicate interplay and can result in unexpected toxic stresses. In the following paragraphs, several toxicity mechanisms will be discussed in light of such potential uncertainty in manifested toxicity.

Morphological aspects have contributed significantly towards nano-bio interaction and manifested toxicity as shown by carbon based NMs. 1-D CNTs are reported to cause asbestos-like inflammation in lungs and penetration through cell membranes attributed to their needle-like tubular shape [96]. The long fibrous CNTs are considered to get retained in the parietal pleura region of lungs when inhaled [97]. These fibrous CNT structures initiate injury to the pleural mesothelium (membrane of pleura), which causes pulmonary inflammation. Over time, this may cause fibrosis and mesothelioma. However, if the CNTs are short in length and tangled together, these no longer have the fibrous needle-like shape. As a result, the short or tangled CNTs are not retained in pleura causing no significant inflammation. Another important process affecting length and shape dependent inflammation response from CNTs is the inability of the macrophage to enclose longer tubes (generally more than 15 μm). Figure 9.7 shows that macrophage can enclose short or tangled fibers of CNTs whereas it is unable to enclose the long and needle-shaped CNT as compared to asbestos-induced inflammation pattern. Similar physical interaction based toxicity has also been observed for other high aspect ratio metallic NMs such as TiO_2 [98] and ZnO [99]. Moreover, planar size variation of 2-D graphene nanosheets can play an important role in cytotoxicity depending on their surface chemistry. While smaller sizes ($\sim$33 nm) of graphene oxides (GO) show highest bio-uptake and biocompatibility in HeLa cells [100], the reduced graphene oxides (rGO) with smaller sizes ($\sim$11 nm) show significantly higher cyto–and geno-toxicity when tested with human stem cells [101]. The mechanisms of such toxicity variations for different sizes can be attributed to a complex interplay of ROS generation and physical cell membrane damage [101]. As the dimensional characteristics are altered for 1-D CNTs (or other 1-D tube/rod-shaped NMs) by hybridization with 2-D graphene (or other similar 2-D NMs), these heterostructures are likely exhibit unanticipated toxic responses. For example, it is quite uncertain whether and how CNT-graphene 3-D heterostructures will modify the asbestos-like inflammation response of CNTs. It can be argued that the 3-D shape (Fig. 9.1d) may inhibit the CNT penetration through the cell membrane. Moreover, the hierarchical 3-D expansion may result in a much larger structure and thus modulate their interactions with the cells and macrophages. These ambiguities in understanding such interactions also will depend on the effects of other variables including number of stacked layers, vertical or horizontal alignments of the components, chirality of the CNTs, rigidity or mechanical properties of the NMs, co-valent versus non-covalent bonding between the component NMs, types of surfactants/polymers as coupling agents, among others.

One other key mechanism for NMs to cause cellular toxicity is the generation of oxygen-containing radicals, collectively termed as ROS [102]. Through ROS generation, NMs impart oxidative stress to the affected cells causing lipid peroxidation and cell organelle damage. For metal oxide NPs, electronic property i.e., position of electron conduction band (E_c) has been correlated with ROS generation abilities and potential nanotoxicity—demonstrating that only a handful of metal oxide NPs are

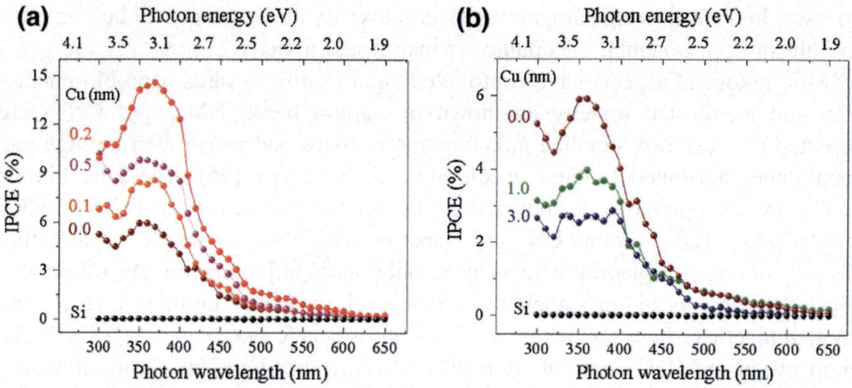

Fig. 9.4 IPCE spectra of MWCNT-Cu NH with different thicknesses of Cu NP with respect to bare MWCNT (*brown*) and silicon substrate (*grey*). **a** MWCNT-Cu with 0.1 nm Cu (*orange*), 0.5 nm Cu (*pink*), and 0.2 nm Cu (*red*); **b** MWCNT-Cu with 3 nm Cu (*blue*) and 1 nm Cu (*green*). [Reprinted from Scarselli [47]. Copyright 2011 the IOP Publishing Ltd. Reprinted with permission.]

ROS-active [103]. However, conjugation of the apparently non-ROS-active metal oxides with 1-D CNTs or 2-D graphene can alter their band architecture to modulate their ROS-generation ability. The excellent electron shuttling capability of CNTs or graphene on top of the enhanced available reactive surface area for supporting non-aggregating metal oxide NPs can play a strong role in such improvements of ROS production. For example, semiconducting ZnO NP has a conduction band (−3.9 eV) outside the range of the ROS-producing cellular redox potential (−4.12 to −4.84 eV); however, ZnO-graphene hybridization causes band-bending to move the conduction band within the cellular redox potential range [20]. Similarly, deposition of UV-active TiO_2 NPs onto CNTs or graphene make the heterostructures photocatalytic, also in the visible spectrum via band gap reduction [104].

Dissolution of metal (e.g., Ag) and metal oxide (e.g., ZnO) NPs are considered to be a major mechanism to incur nanotoxicity. Dissolved metal ions damage cell functions through direct cellular interaction and ROS generation [20]. For example, Ag NPs when undergo oxidative dissolution, Ag^+ ions bind with thiol groups of cellular protein and enzymes, impeding ion transport across cell membranes and affecting cellular respiration [105]. Dissolved Ag^+ ions also contribute to microbial toxicity by exerting ROS within cells [106]. The dissolution of Ag NPs is also size dependent due to Kelvin effect (highest dissolution occurs for NM sized < 10 nm) [107]. The dissolution also depends on the shape of the Ag NPs; platelet-like and spherical Ag NPs have shown higher dissolution rate compared to rod-like or cubic NPs [108]. Hybridization of Ag NPs with other nanomaterials can alter their dissolution properties. In case of bimetallic core-shell NHs, Ag can both be placed as a core or on the shell with corresponding inert Au. The choice of the arrangement will dictate their morphology as well as subsequent dissolution characteristics. For example, 20 nm thick Ag NP shell surrounding an Au core has higher dissolution

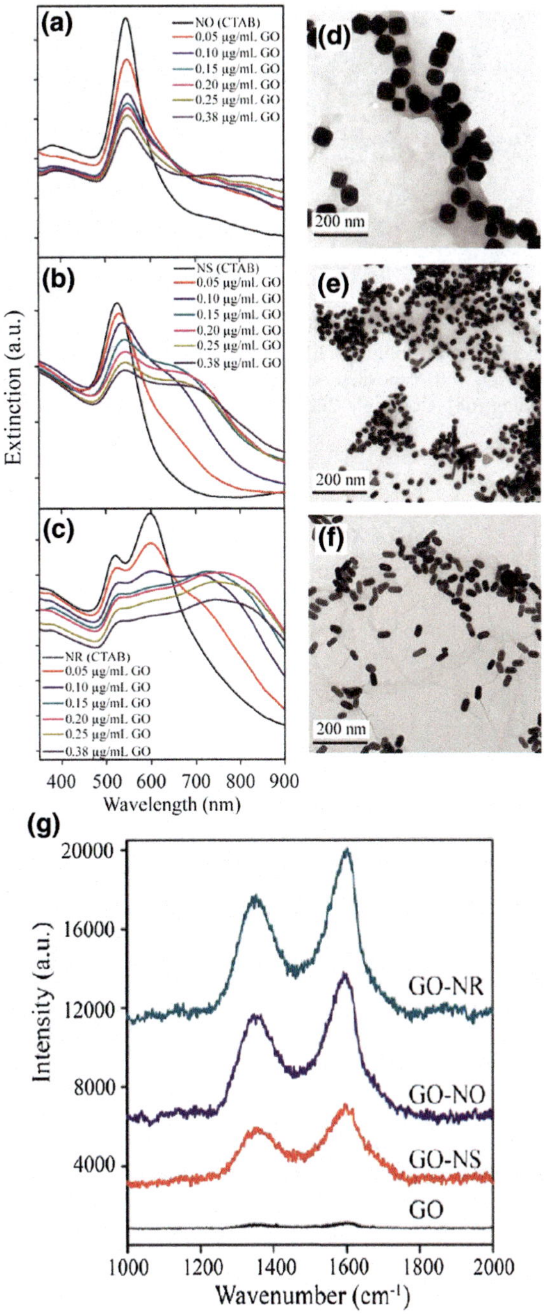

Fig. 9.5 Different extinction spectra of cetyltrimethylammonium bromide (CTAB) capped **a** GO–Au nanooctahedra (GO–NO), **b** GO–Au nanosphere (GO–NS), **c** GO–Au nanorod (GO–NR) with different ratios of GO loading; **d–f** corresponding TEM images of Au/GO NHs; **g** SERS signals of GO-Au NHs for different shapes of Au NP, i.e., GO–NR, GO–NO and GO–NS. [Reprinted from Lee [55]. Copyright 2011 by The American Chemical Society. Reprinted with permission.]

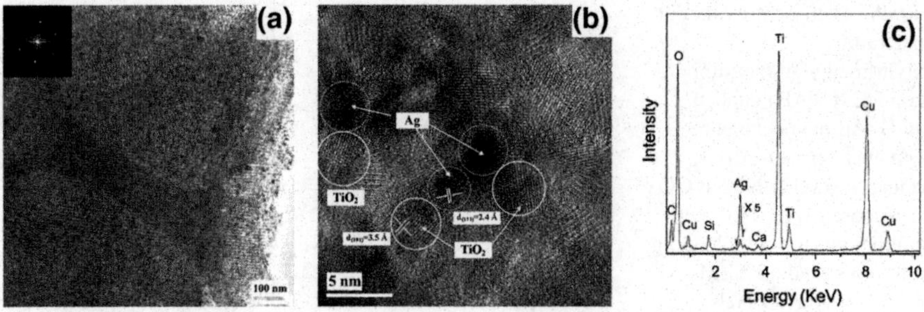

Fig. 9.6 **a** TEM image of Ag–TiO_2 NH, in the inset Fourier transformed pattern of the selected area; **b** high-resolution TEM image of Ag–TiO_2 NH showing the mesoporous surface morphology of TiO_2 with deposited Ag NP; **c** energy dispersive X-ray of the Ag-TiO_2 NH. [Reprinted from Wang [68]. Copyright 2005 by The Royal Society of Chemistry. Reprinted with permission.]

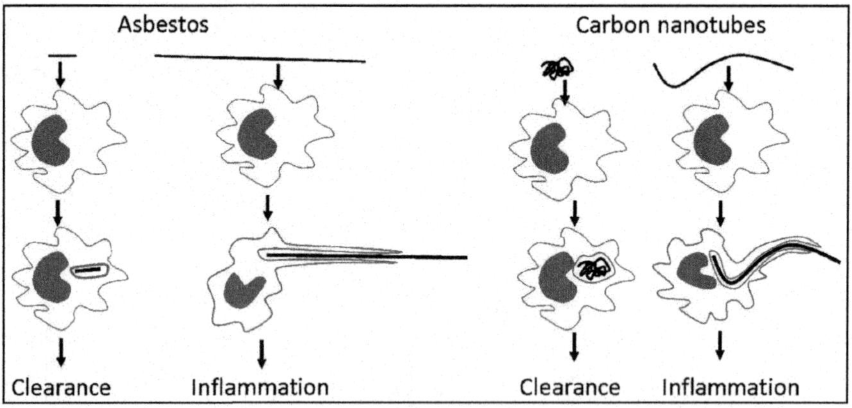

Fig. 9.7 Enclosing mechanism of macrophage for asbestos (*left side*) and CNTs (*right side*). Macrophage can enclose and clear asbestos and CNTs when these are present in short and tangled form, respectively. However, macrophages are unable to enclose both asbestos and CNTs when present in long fibrous form. [Reprinted from Donaldson [97]. Copyright 2010 by Ken Donaldson. Reprinted with permission.]

characteristics than that of the singular Ag NP of the same size [109]. Dissolution is exacerbated due to higher crystalline disorder on Ag NP shell structure. On the contrary, when Ag core is covered with a relatively inert Au shell, the stability of the NH increases substantially as observed via lowering of dissolution [110]. Thus, depending on the core-shell structure as well as shape and size of the materials, Ag^+ mediated toxicity of Ag NPs will vary. Similarly, Ag NP deposition on carbon-based NMs can alter their dissolution properties. 1-D CNT or 2-D graphene can provide high surface area to reduce agglomeration of Ag NPs and provide sustained ion release, thus influence manifested antimicrobiality. It is important to note that these dimensional effects on toxicological consequences of metallic NPs will be further exemplified by the physical perturbation of the cell-membranes by 1-D CNTs or 2-D graphenes as previously discussed or simply by providing a larger surface presence for attachment of cells and subsequent interactions with metal/metal oxide nanoparticles. On the contrary, the presence of metallic NPs on the surface can alter the mechanical properties of CNT or graphene (e.g., physical stiffness) resulting in modification of the physical interactions and bio-uptake [111]. These effects can be further influenced by the sizes and shapes of deposited metallic NPs on CNT or graphene—however, no systematic studies have thus far investigated such structure-activity relationships.

9.4 Environmental and Toxicological Significance

The above discussion summarizes the importance of dimensional considerations for the NHs' toxicological impacts. With the rapidly increasing numbers of these hierarchical ensembles and their applications, NHs' potential environmental and biological exposure and subsequent toxicological manifestation should be systematically evaluated. Change in dimensionality via conjugation, passivation of surfaces in core-shell particles, modulation of band architecture, and introduction of multiple chemical entities to bio species in tandem present emergent uncertainties in predicting toxic behavior of these heterostructures. A singular NM though may be benign to environmental and biological systems, can become toxic when hybridized with a secondary NM. Characterization tools to evaluate heterostructure emergent properties and techniques to detect these structures within cellular and hierarchical biological entities are necessitated. Advanced methods, such as hyper-spectral imaging, Raman-imaging, in situ aberration corrected transmission electron microscopy, single particle inductively coupled plasma mass spectrometry may provide new insights into the complex nano-bio processes with these heterostructures. Since there is an infinite combination of materials possible to form hybrid structures, material selection for extensive toxicological studies should be carefully made. Reports of demonstrated toxicity from component materials can serve as a guide for such material selection.

9.5 Conclusions

With the increasing complexity of material systems at nanoscale, there is extended need to understand the hybridized nanomaterials and their interactions at the environmental and biological interfaces. Hybridization of nanomaterials leads to different sizes, shapes, and dimensionalities—which may significantly alter the structure-property relationships for toxicological consequences. Future research will need to distinguish emerging properties of novel nanohybrids including dimensional variations and identify their relevance in nanotoxicity.

Acknowledgements This work is partially supported by a United States National Science Foundation award, bearing award#1602273 (Navid B. Saleh and Tara Sabo-Attwood).

References

1. Roco, M.C. 2011. The long view of nanotechnology development: the National Nanotechnology Initiative at 10 years. *Journal of Nanoparticle Research* 13 (2): 427–445.
2. Aich, N., et al. 2014. A critical review of nanohybrids: synthesis, applications and environmental implications. *Environmental Chemistry* 11 (6): 609–623.
3. Ipe, B.I., and C.M. Niemeyer. 2006. Nanohybrids composed of quantum dots and cytochrome P450 as photocatalysts. *Angewandte Chemie. International Edition* 45 (3): 504–507.
4. Wang, K., et al. 2011. TiO2-decorated graphene nanohybrids for fabricating an amperometric acetylcholinesterase biosensor. *Analyst* 136 (16): 3349–3354.
5. Ding, S., et al. 2011. Graphene-supported anatase TiO2 nanosheets for fast lithium storage. *Chemical Communications* 47 (20): 5780–5782.
6. Jafri, R.I., et al. 2010. Nanostructured Pt dispersed on graphene-multiwalled carbon nanotube hybrid nanomaterials as electrocatalyst for PEMFC. *Journal of the Electrochemical Society* 157 (6): B874–B879.
7. Tokarev, I., and S. Minko. 2012. Tunable plasmonic nanostructures from noble metal nanoparticles and stimuli-responsive polymers. *Soft Matter* 8 (22): 5980–5987.
8. Luo, Y., et al. 2014. Gold nanoparticles embedded in Ta2O5/Ta3N5 as active visible-light plasmonic photocatalysts for solar hydrogen evolution. *Journal of Materials Chemistry A* 2 (36): 14927–14939.
9. Yogeswaran, U., S. Thiagarajan, and S.-M. Chen. 2008. Recent updates of dna incorporated in carbon nanotubes and nanoparticles for electrochemical sensors and biosensors. *Sensors* 8 (11): 7191–7212.
10. Prakash, S., et al. 2011. Polymeric nanohybrids and functionalized carbon nanotubes as drug delivery carriers for cancer therapy. *Advanced Drug Delivery Reviews* 63 (14–15): 1340–1351.
11. Wei, G., et al. 2011. Novel 1-D biophotonic nanohybrids: protein nanofibers meet quantum dots. *Soft Matter* 7 (5): 2011–2018.
12. Ren, D.M., et al. 2008. A novel soluble Tin(IV) porphyrin modified single-walled carbon nanotube nanohybrid with light harvesting properties. *International Journal of Molecular Sciences* 9 (1): 45–55.
13. Si, H.-Y., et al. 2009. Shell-controlled photoluminescence in CdSe/CNT nanohybrids. *Nanoscale Research Letters* 4 (10): 1146–1152.
14. Karousis, N., et al. 2008. Carbon nanotubes decorated with palladium nanoparticles: Synthesis, characterization, and catalytic activity. *Journal of Physical Chemistry C* 112 (35): 13463–13469.
15. Li, Y., et al. 2010. Gold nanoparticles-graphene hybrids as active catalysts for Suzuki reaction. *Materials Research Bulletin* 45 (10): 1413–1418.
16. Ruiz-Hitzky, E., et al. 2010. Hybrid materials based on clays for environmental and biomedical applications. *Journal of Materials Chemistry* 20 (42): 9306–9321.
17. Ghosh, S., et al. 2012. ZnO/Ag nanohybrid: synthesis, characterization, synergistic antibacterial activity and its mechanism. *RSC Advances* 2 (3): 930–940.

18. Chen, B., et al. 2014. One-pot, solid-phase synthesis of magnetic multiwalled carbon nanotube/iron oxide composites and their application in arsenic removal. *Journal of Colloid and Interface Science* 434: 9–17.
19. Li, J., et al. 2015. Nanoscale zero-valent iron particles supported on reduced graphene oxides by using a plasma technique and their application for removal of heavy-metal ions. *Chemistry–An Asian Journal* 10 (6): 1410–1417.
20. Saleh, N.B., et al. 2014. Emergent properties and toxicological considerations for nanohybrid materials in aquatic systems. *Nanomaterials* 4 (2): 372–407.
21. Saleh, N.B., et al. 2015. Research strategy to determine when novel nanohybrids pose unique environmental risks. *Environmental Science Nano* 2 (1): 11–18.
22. Plazas-Tuttle, J., et al. 2015. Dynamism of stimuli-responsive nanohybrids: Environmental implications. *Nanomaterials* 5 (2): 1102–1123.
23. Nasibulin, A.G., et al. 2007. A novel hybrid carbon material. *Nature Nanotechnology* 2 (3): 156–161.
24. Jiang, L.L., et al. 2015. Densely packed graphene nanomesh-carbon nanotube hybrid film for ultra-high volumetric performance supercapacitors. *Nano Energy* 11: 471–480.
25. Zhang, H.B., et al. 2011. Enhanced catalytic activity of sub-nanometer titania clusters confined inside double-wall carbon nanotubes. *Chemsuschem* 4 (7): 975–980.
26. Major, K.J., C. De, and S.O. Obare. 2009. Recent advances in the synthesis of plasmonic bimetallic nanoparticles. *Plasmonics* 4 (1): 61–78.
27. Sun, M.L., et al. 2011. One-step synthesis of coaxial Ag/TiO2 nanowire arrays on transparent conducting substrates: Enhanced electron collection in dye-sensitized solar cells. *Electrochemistry Communications* 13 (12): 1324–1327.
28. Alley, N.J., et al. 2012. Effect of carbon nanotube-fullerene hybrid additive on P3HT:PCBM bulk-heterojunction organic photovoltaics. *Synthetic Metals* 162 (1–2): 95–101.
29. Kim, H.I., et al. 2012. Solar photoconversion using Graphene/TiO2 composites: Nanographene shell on TiO2 core versus TiO2 nanoparticles on graphene sheet. *Journal of Physical Chemistry C* 116 (1): 1535–1543.
30. Chen, M.L., et al. 2012. Quantum dots conjugated with Fe3O4-filled carbon nanotubes for cancer-targeted imaging and magnetically guided drug delivery. *Langmuir* 28 (47): 16469–16476.
31. Larson, T.A. et al. (2007). Hybrid plasmonic magnetic nanoparticles as molecular specific agents for MRI/optical imaging and photothermal therapy of cancer cells. *Nanotechnology* 18(32).
32. Liu, X.J., et al. 2011. Microwave-assisted synthesis of TiO2-reduced graphene oxide composites for the photocatalytic reduction of Cr(VI). *RSC Advances* 1 (7): 1245–1249.
33. Kavitha, T., et al. 2012. Glucose sensing, photocatalytic and antibacterial properties of graphene-ZnO nanoparticle hybrids. *Carbon* 50 (8): 2994–3000.
34. Li, M.H., et al. 2011. Synergistic bactericidal activity of Ag–TiO2 nanoparticles in both light and dark conditions. *Environmental Science and Technology* 45 (20): 8989–8995.
35. Kalidindi, S.B., and B.R. Jagirdar. 2008. Synthesis of Cu@ZnO core-shell nanocomposite through digestive ripening of Cu and Zn nanoparticles. *Journal of Physical Chemistry C* 112 (11): 4042–4048.
36. Eder, D., and A.H. Windle. 2008. Carbon-inorganic hybrid materials: The carbon-nanotube/TiO2 interface. *Advanced Materials* 20 (9): 1787–1793.
37. Shi, W.L., et al. 2016. Novel three-dimensional carbon nanotube-graphene architecture with abundant chambers and its application in lithium-silicon batteries. *Journal of Physical Chemistry C* 120 (25): 13807–13814.
38. Zhang, D., et al. 2005. Preparation and characterization of Ag@TiO2 core-shell nanoparticles in water-in-oil emulsions. *European Journal of Inorganic Chemistry* 9: 1643–1648.
39. Xue, W.P., et al. 2016. Catalytic oxidation of 1,2-Propanediol over bimetallic Cu@Au core/shell nanoparticles. *Catalysis Letters* 146 (6): 1139–1152.

40. Widayanti and K. Abraha. (2016). Study on the effect of nanoparticle bimetallic coreshell Au-Ag for sensitivity enhancement of biosensor based on surface plasmon resonance. In *13th South-East Asian Congress of Medical Physics 2015*. F. Haryanto, et al. ed.
41. Zhu, J. 2009. Surface plasmon resonance from bimetallic interface in Au–Ag core-shell structure nanowires. *Nanoscale Research Letters* 4 (9): 977–981.
42. Liu, W.-L., et al. 2013. The influence of shell thickness of Au@TiO2 core-shell nanoparticles on the plasmonic enhancement effect in dye-sensitized solar cells. *Nanoscale* 5 (17): 7953–7962.
43. Yao, Y.J., et al. 2016. Fe Co, Ni nanocrystals encapsulated in nitrogen-doped carbon nanotubes as Fenton-like catalysts for organic pollutant removal. *Journal of Hazardous Materials* 314: 129–139.
44. Mahalingam, P., et al. 2014. Characterization of magnetic metal encapsulated in multi-walled carbon nanotubes synthesized from methyl ester of pongamia pinnata oil and its application for removal of arsenic ions from aqueous solution. *Asian Journal of Chemistry* 26 (14): 4167–4171.
45. Wen, Z., et al. 2013. TiO2 nanoparticles-decorated carbon nanotubes for significantly improved bioelectricity generation in microbial fuel cells. *Journal of Power Sources* 234: 100–106.
46. Köse, H., et al. 2015. A facile synthesis of zinc oxide/multiwalled carbon nanotube nanocomposite lithium ion battery anodes by sol–gel method. *Journal of Power Sources* 295: 235–245.
47. Scarselli, M., et al. (2011). Influence of Cu nanoparticle size on the photo-electrochemical response from Cu-multiwall carbon nanotube composites. *Nanotechnology* 22(3).
48. Smith, B.W., and D.E. Luzzi. 2000. Formation mechanism of fullerene peapods and coaxial tubes: a path to large scale synthesis. *Chemical Physics Letters* 321 (1–2): 169–174.
49. Qin, Y., et al. 2016. Efficiently visible-light driven photoelectrocatalytic oxidation of as(III) at low positive biasing using Pt/TiO2 nanotube electrode. *Nanoscale Research Letters* 11 (1): 1–13.
50. Abd-Ellah, M., et al. 2016. Enhancement of solar cell performance of p-Cu2O/n-ZnO-nanotube and nanorod heterojunction devices. *Solar Energy Materials and Solar Cells* 152: 87–93.
51. Esrafili, M.D., P. Nematollahi, and R. Nurazar. 2016. Pd-embedded graphene: An efficient and highly active catalyst for oxidation of CO. *Superlattices and Microstructures* 92: 60–67.
52. Khan, S., et al. 2016. Synthesis of reduced graphene oxide and enhancement of its electrical and optical properties by attaching Ag nanoparticles. *Physica E: Low-dimensional Systems and Nanostructures* 81: 320–325.
53. Yu, M., et al. 2014. An alumina stabilized ZnO-graphene anode for lithium ion batteries via atomic layer deposition. *Nanoscale* 6 (19): 11419–11424.
54. Rai, A.K., et al. 2013. Facile approach to synthesize CuO/reduced graphene oxide nanocomposite as anode materials for lithium-ion battery. *Journal of Power Sources* 244: 435–441.
55. Lee, Y.H., et al. 2012. Enhanced optical properties of graphene oxide-Au nanocrystal composites. *Langmuir* 28 (1): 321–326.
56. Jasuja, K., and V. Berry. 2009. Implantation and growth of dendritic gold nanostructures on graphene derivatives: Electrical property tailoring and raman enhancement. *ACS Nano* 3 (8): 2358–2366.
57. Thuan, D.V., et al. 2015. Photoluminescence properties and photocatalytic performance of different size ZnO nanoparticles decorated graphene oxide. *Journal of Nanoscience and Nanotechnology* 15 (11): 8896–8900.
58. Ishihara, S., et al. (2016). Properties of single-layer MoS2 film fabricated by combination of sputtering deposition and post deposition sulfurization annealing using (t-C4H9)(2)S-2. *Japanese Journal of Applied Physics* 55(6).
59. Wan, C.L., et al. 2015. Flexible n-type thermoelectric materials by organic intercalation of layered transition metal dichalcogenide TiS2. *Nature Materials* 14 (6): 622–627.

60. Wu, K., et al. 2015. WS2 as a saturable absorber for ultrafast photonic applications of mode-locked and Q-switched lasers. *Optics Express* 23 (9): 11453–11461.
61. Jin, K., et al. 2016. Au-modified monolayer MoS2 sensor for DNA detection. *Journal of Physical Chemistry C* 120 (20): 11204–11209.
62. Zhao, X., X.Q. Dai, and C.X. Xia. 2015. Magnetic properties of two nearest Cu-doped monolayer WS2: A first-principles study. *Solid State Communications* 217: 66–69.
63. Yan, H.H., et al. 2016. A low temperature gas sensor based on Au-loaded MoS2 hierarchical nanostructures for detecting ammonia. *Ceramic International* 42 (7): 9327–9331.
64. Prins, F., A.J. Goodman, and W.A. Tisdale. 2014. Reduced dielectric screening and enhanced energy transfer in single–and few-layer MoS2. *Nano Letters* 14 (11): 6087–6091.
65. Zhu, Y., et al. 2012. A seamless three-dimensional carbon nanotube graphene hybrid material. *Nature Communication* 3: 1225.
66. Du, F., et al. 2011. Preparation of tunable 3D pillared carbon nanotube-graphene networks for high-performance capacitance. *Chemistry of Materials* 23 (21): 4810–4816.
67. Chen, L., et al. 2015. Harnessing light energy with a planar transparent hybrid of graphene/single wall carbon nanotube/n-type silicon heterojunction solar cell. *Electrochimica Acta* 178: 732–738.
68. Wang, X., et al. 2005. A robust three-dimensional mesoporous Ag/TiO2 nanohybrid film. *Chemical Communications* 17: 2262–2264.
69. Ohko, Y., et al. 2003. Multicolour photochromism of TiO2 films loaded with silver nanoparticles. *Nature Materials* 2 (1): 29–31.
70. Gannett, W., et al. 2011. Boron nitride substrates for high mobility chemical vapor deposited graphene. *Applied Physics Letters* 98 (24): 242105.
71. Britnell, L., et al. 2012. Field-effect tunneling transistor based on vertical graphene heterostructures. *Science* 335 (6071): 947–950.
72. Gong, Y., et al. 2014. Vertical and in-plane heterostructures from WS2/MoS2 monolayers. *Nature Materials* 13 (12): 1135–1142.
73. Georgiou, T., et al. 2013. Vertical field-effect transistor based on graphene-WS2 heterostructures for flexible and transparent electronics. *Nature Nano* 8 (2): 100–103.
74. Yu, W.J., et al. 2013. Highly efficient gate-tunable photocurrent generation in vertical heterostructures of layered materials. *Nature Nanotechnology* 8 (12): 952–958.
75. Tsuboi, Y., et al. 2015. Enhanced photovoltaic performances of graphene/Si solar cells by insertion of a MoS2 thin film. *Nanoscale* 7 (34): 14476–14482.
76. Jiang, J.-W. 2015. Graphene versus MoS2: A short review. *Frontiers of Physics* 10 (3): 287–302.
77. Yan, J.M., et al. (2010). One-step seeding growth of magnetically recyclable Au@Co core-shell nanoparticles: Highly efficient catalyst for hydrolytic dehydrogenation of ammonia borane. *Journal of the American Chemical Society* 132(15): 5326–5327.
78. Kim, Y., et al. 2010. Synthesis of AuPt Heteronanostructures with Enhanced Electrocatalytic Activity toward Oxygen Reduction. *Angewandte Chemie. International Edition* 49 (52): 10197–10201.
79. Kim, Y., et al. 2014. One-pot synthesis and electrocatalytic properties of Pd@Pt Core-shell nanocrystals with tailored morphologies. *Chemistry–A European Journal* 20 (26): 7901–7905.
80. Ren, S., et al. 2011. Inorganic-organic hybrid solar cell: Bridging quantum dots to conjugated polymer nanowires. *Nano Letters* 11 (9): 3998–4002.
81. Zhao, H., et al. 2014. Investigating photoinduced charge transfer in double–and single-emission PbS@CdS core@shell quantum dots. *Nanoscale* 6 (1): 215–225.
82. Hwang, I.-S., et al. 2011. Facile control of C2H5OH sensing characteristics by decorating discrete ag nanoclusters on SnO2 nanowire networks. *ACS Applied Materials and Interfaces* 3 (8): 3140–3145.
83. Lai, L.-H., et al. 2014. Sensitized solar cells with colloidal PbS-CdS core-shell quantum dots. *Physical Chemistry Chemical Physics: PCCP* 16 (2): 736–742.

84. Choi, R., et al. 2013. Designed synthesis of well-defined Pd@Pt core-shell nanoparticles with controlled shell thickness as efficient oxygen reduction electrocatalysts. *Chemistry–A European Journal* 19 (25): 8190–8198.
85. Su, W.-J., et al. 2016. Two dimensional MoS2/graphene p-n heterojunction diode: Fabrication and electronic characteristics. *Journal of Alloys and Compounds* 671: 276–282.
86. Kim, D., et al. 2013. Thickness dependent band gap and effective mass of BN/graphene/BN and graphene/BN/graphene heterostructures. *Surface Science* 610: 27–32.
87. Jones, C.F., and D.W. Grainger. 2009. In vitro assessments of nanomaterial toxicity. *Advanced Drug Delivery Reviews* 61 (6): 438–456.
88. Levard, C., et al. 2012. Environmental transformations of silver nanoparticles: Impact on stability and toxicity. *Environmental Science and Technology* 46 (13): 6900–6914.
89. Choi, O., and Z.Q. Hu. 2008. Size dependent and reactive oxygen species related nanosilver toxicity to nitrifying bacteria. *Environmental Science and Technology* 42 (12): 4583–4588.
90. Zhu, X., et al. 2009. Acute toxicities of six manufactured nanomaterial suspensions to Daphnia magna. *Journal of Nanoparticle Research* 11 (1): 67–75.
91. Poland, C.A., et al. 2008. Carbon nanotubes introduced into the abdominal cavity of mice show asbestos-like pathogenicity in a pilot study. *Nature Nanotechnology* 3 (7): 423–428.
92. Akhavan, O., and E. Ghaderi. 2010. Toxicity of Graphene and Graphene Oxide Nanowalls Against Bacteria. *ACS Nano* 4 (10): 5731–5736.
93. Scown, T.M., et al. 2010. Effects of aqueous exposure to silver nanoparticles of different sizes in rainbow trout. *Toxicological Sciences* 115 (2): 521–534.
94. Pal, S., Y.K. Tak, and J.M. Song. 2007. Does the antibacterial activity of silver nanoparticles depend on the shape of the nanoparticle? A study of the gram-negative bacterium Escherichia coli. *Applied and Environmental Microbiology* 73 (6): 1712–1720.
95. Ma-Hock, L., et al. (2013). Comparative inhalation toxicity of multi-wall carbon nanotubes, graphene, graphite nanoplatelets and low surface carbon black. *Particle and Fibre Toxicology* 10 (23).
96. Lam, C.W., et al. 2006. A review of carbon nanotube toxicity and assessment of potential occupational and environmental health risks. *Critical Reviews in Toxicology* 36 (3): 189–217.
97. Donaldson, K., et al. (2010). Asbestos, carbon nanotubes and the pleural mesothelium: A review of the hypothesis regarding the role of long fibre retention in the parietal pleura, inflammation and mesothelioma. *Particle and Fibre Toxicology* 7.
98. Allegri, M., et al. 2016. Shape-Related Toxicity of Titanium Dioxide Nanofibres. *PLoS ONE* 11 (3): e0151365.
99. Hua, J., et al. 2014. Particle-specific toxic effects of differently shaped zinc oxide nanoparticles to zebrafish embryos (Danio rerio). *Environmental Toxicology and Chemistry* 33 (12): 2859–2868.
100. Zhang, H., et al. 2013. Uniform ultrasmall graphene oxide nanosheets with low cytotoxicity and high cellular uptake. *ACS Applied Materials and Interfaces* 5 (5): 1761–1767.
101. Akhavan, O., E. Ghaderi, and A. Akhavan. 2012. Size-dependent genotoxicity of graphene nanoplatelets in human stem cells. *Biomaterials* 33 (32): 8017–8025.
102. Fu, P.P., et al. 2014. Mechanisms of nanotoxicity: Generation of reactive oxygen species. *Journal of Food and Drug Analysis* 22 (1): 64–75.
103. Zhang, H., et al. 2012. Use of metal oxide nanoparticle band gap to develop a predictive paradigm for oxidative stress and acute pulmonary inflammation. *ACS Nano* 6 (5): 4349–4368.
104. Wu, C.-H., C.-Y. Kuo, and S.-T. Chen. 2013. Synergistic effects between TiO2 and carbon nanotubes (CNTs) in a TiO2/CNTs system under visible light irradiation. *Environmental Technology* 34 (17): 2513–2519.
105. Fabrega, J., et al. 2011. Silver nanoparticles: Behaviour and effects in the aquatic environment. *Environment International* 37 (2): 517–531.

106. Peretyazhko, T.S., Q. Zhang, and V.L. Colvin. 2014. Size-controlled dissolution of silver nanoparticles at neutral and acidic pH conditions: Kinetics and size changes. *Environmental Science and Technology* 48 (20): 11954–11961.
107. Sotiriou, G.A., and S.E. Pratsinis. 2011. Engineering nanosilver as an antibacterial, biosensor and bioimaging material. *Current Opinion in Chemical Engineering* 1 (1): 3–10.
108. Helmlinger, J., et al. 2016. Silver nanoparticles with different size and shape: equal cytotoxicity, but different antibacterial effects. *RSC Advances* 6 (22): 18490–18501.
109. Munusamy, P., et al. 2015. Comparison of 20 nm silver nanoparticles synthesized with and without a gold core: Structure, dissolution in cell culture media, and biological impact on macrophages. *Biointerphases* 10 (3): 031003.
110. Mott, D.M., et al. 2012. Electronic transfer as a route to increase the chemical stability in gold and silver core–shell nanoparticles. *Advances in Colloid and Interfaces* 185–186: 14–33.
111. Subbiah, R., et al. (2013). Evaluation of cytotoxicity, biophysics and biomechanics of cells treated with functionalized hybrid nanomaterials. *Journal of the Royal Society Interface* 10(88).

Chapter 10
Assemblies and Superstructures of Inorganic Colloidal Nanocrystals

Swati Naik and Gabriel Caruntu

Abstract Inorganic colloidal nanoparticles (NPs) possess the ability to self-assemble into complex hierarchical structures with unique properties that are different from their individual counterparts. The assembly of particles into compact 2D and 3D structures occurs only when they have a narrow size distribution and uniform shape. In this chapter, we review how various types of forces and fundamental interactions at the nanoscale govern the assembly of colloidal nanocrystals, the available methods for assembling colloidal nanocrystals with discrete geometries and the application of such assembled structures in various fields ranging from catalysis, biological diagnosis, plasmonics and electronics.

Keywords Colloidal nanocrystals · Solution-based synthesis · Shape anisotropy · Self-assembly · Superstructures

10.1 Introduction

Although the concept of nanotechnology has been coined relatively recently, materials at the nanometer-length scale have been empirically known for centuries, being used in cosmetic preparations, fermentation processes, advanced glass manufacturing and other technological applications. For example, in ancient Egypt the hair dyeing was performed by a procedure involving the formation of galenite (PbS) nanoparticles, which allowed for an even and steady dyeing [1]. In the medieval period, the European knights have experienced during crusades the extraordinary strength, sharp edges and shatter resistances of blades made of ultra-strong Damascus steel which incorporated nanowires and tube-like Fe_3C (cementite) nanostructures with sizes of 40–50 nm [2]. Moreover, metallic nanoparticles have been also used at large scale in ceramics and lustre decorations

S. Naik · G. Caruntu (✉)
Science of Advanced Materials Program, Central Michigan University,
1200 S. Franklin St., Mount Pleasant, MI 48858, USA
e-mail: carun1g@cmich.edu

S.E. Hunyadi Murph et al. (eds.), *Anisotropic and Shape-Selective Nanomaterials*,
Nanostructure Science and Technology, DOI 10.1007/978-3-319-59662-4_10

in many parts of Asia, Middle East and Europe. As such, Celtic art used copper (Cu) and copper oxide (Cu_2O) nanoparticles for the fabrication of red enamels [3] whereas colloidal gold and silver nanoparticles were used in ancient Rome (4th century AD) in the manufacture of dichroic glasses which exhibit spectacular color changing properties depending on the light source (the Lycurgus cup) or in the decoration of majolica during the Renaissance period [4].

With the increasing tendency towards miniaturization of electronics, the range of technological applications of nanomaterials diversified enormously in the past decade and nowadays they permeate all aspects of the human life, being used in electronics, energy storage and conversion, biosciences, medicine [5], catalysis and data storage [6], respectively. Nanomaterials can be defined as materials made of submicroscopic grains which possess at least one dimension smaller than 100 nm. The small size of the constituents in such materials is associated with a large surface/volume ratio which, in turn, will lead to a very important contribution of surface effects to the physical properties of nanomaterials, thereby making them substantially different from those of their bulk counterparts. As an example, barium titanate ($BaTiO_3$), an archetypal ferroelectric perovskite, is characterized by multiple phase transitions as the material is cooled down from elevated temperatures. Among these, technological relevance is a paraelectric to ferroelectric phase transition which occurs around 120 °C in the bulk material. This transition is microscopically described by the off-center shift of the Ti^{4+} ions occupying the centers of corner-sharing TiO_6 octahedra which leads to the formation of electrical dipoles. The cooperative interactions between these electrical dipoles eventually results into a sizeable intrinsic lattice polarization associated with a ferroelectric behavior. However, until now there is no clear consensus as to where ferroelectricity is suppressed in nanoscale titanium-containing perovskites. To fully understand the polar ordering in nanoscale perovskites, it is crucial to rationally design non-aggregated, free-standing monodisperse colloidal nanocrystals with well-defined morphology and tunable surface composition and chemical functionality [7].

The approaches conventionally applied in the synthesis of nanostructured materials can be classified into physical and chemical methods. Physical approach relates to the fabrication of nanometer length-scale structures from larger ones, and chemical methods employ a bottom-up approach wherein small particles result from nucleation and growth processes. Both the methods have their set of advantages and disadvantages, the physical method finds application in an electronic industry but have the resolution limited to few tens of nanometers. On the other hand, a chemical approach is carried out on a small scale with a very high precision [8]. Wet-chemistry/solution-phase methods can be utilized to achieve the best control of particle sizes, shapes, and composition. The molecular precursors along with the structure directing agents manipulate the structural growth of the nanocrystals. The range of materials that can be synthesized includes the magnetic oxide such as Fe_3O_4, MFe_2O_4 (M = Fe, Co, Mn) and Mn_3O_4, noble metals such as Pt and Pd, plasmonic metals such as Au, Ag, and semiconductors consisting of CdS and CdSe [9].

Controlling the morphology of a nanocrystal and thereby stabilizing its structure is critical in nanoparticle synthesis. The focus of nanoscience and nanotechnology is gradually shifting from individual components to larger assemblies since the first report published in 1989 by Benton and coworkers [10]. Using the bottom-up approach is characteristic for nanofabrication and to further assemble particles into larger ordered architectures by employing specific interaction forces. The process of formation of the long-range ordered structure from individual nanocrystals serves as a connecting bridge between the nano to micro worlds. Nanoparticles can assemble in various dimensionalities (1D, 2D, and 3D) which thereby induces collective properties. To understand how specific property is related to ordering of atoms in nanocrystals, consider the case of opals which consist of sub-micrometer sized silicate particles. The colorless appearance in opals is due to disordered silicate particles while size segregation of ordered particles creates reflectivity in the material. The change in optical behavior shows intrinsic property dependence on particle ordering at large scale [11]. During self-assembly, the building blocks arrange itself in an ordered pattern caused by direct interactions or indirectly using external stimuli. The stacking of nanoparticles (1–100 nm diameter) in 2D and 3D arrangement results in superlattices which now exhibit new physical, chemical and mechanical properties due to near field coupling. Superparticles are colloidal particles in nano dimension which assemble leading to superlattice structures [12]. When nanocrystals assemble themselves in a certain crystallographic order such as *fcc*, *bcc*, and *hcp*, they are known as supracrystals. Understanding these nanocrystal assemblies and giving rise to new superlattice phases will establish our ability to create new structure-property relationship towards the emerging class of complex nanomaterials.

10.2 Forces at Nanoscale

Monodisperse nanoparticles have been synthesized in a wide variety of sizes and shapes, such as spheres [13], cubes [14], rods [15], ellipsoids [16], plates [17], tetrapods [18], core/shell [19], nanocages [20], dumbbells [21] and so on. A span of morphologies stabilized among the most commonly studied class of materials that includes metals [22], semiconductors [23], oxides [24] and polymers [25]. These new materials with its unique properties can find a variety of applications only when they organize in purposeful ways and can interact with each other. The classification of self-assembly processes into three categories comprises of one-dimensional wires, two-dimensional sheets and closed packed three-dimensional lattice structures. The spontaneity of particles to organize itself into superlattices/supralattices combines both attractive as well as non-covalent forces. Novel properties such as charge transport, conductivity is enhanced due to synergistic effects caused by the self-assembly [26]. The forces that interplay between the nanoparticle building blocks are similar to those seen in the assembly of small molecules or polymers [27]. Before understanding the contribution of these complex forces, it is crucial to consider the overall interaction potential of nanoparticles for the assembly process. If the obtained assembly is generated or caused randomly will depend on the magnitude and length scale of nanoparticle interaction [28].

When nanoparticles interact at short-range, gels are formed, whereas long-range interaction result in ordered structures. The equilibrium theory of physical clusters suggests that the formation energy for dimers from spherical particles can be expressed as shown in Eq. 10.1.

$$\Delta F_2 = -kT \ln \frac{Q_2}{Q_1} \tag{10.1}$$

where Q represents the canonical partition functions [29]. In a dilute solution, the cluster formation will result only when particle-particle interaction reaches several kT to compensate for entropy [30] while, for a concentrated solution, entropy will be a driving force to cause ordered assemblies thereby maximizing interparticle free volume and overall entropy. The shape of the particles plays a crucial role in deciding the nature of the formation. For example, with spherical particles; close-packed structures are favorable as it has nothing to do with orientational entropy while, for anisotropic structures (cubes, rods) the assembly will vary depending on the chosen shapes. Maximization of the entropy of the system is a key factor that results in any formation thereby restricting both translational and rotational motion [31]. We can now develop our understanding towards a complex scenario that consists of several interactions that results from solvent molecules, binding agents etc. The various internal forces that play a deciding factor in causing the assembly are as discussed below.

10.2.1 Van der Waals Interactions

The most dominating force in self-assembly processes is non-covalent by nature. van der Waals interactions have been recognized as the fundamental forces present in a non-covalent interaction between molecules resulting in the assembly. It arises due to the electromagnetic fluctuations caused by the constant movement of charges in all materials [31]. van der Waals force can be widely classified into three types namely Keesom forces, Debye forces and London dispersion forces of which the dispersion forces play a crucial role [31]. All these forces depend on the inverse of the separation between the dipoles, as they scale with $1/r^6$ and their strength is comparable to that of other types of long-range interactions characteristic to nanoparticles, such as electrostatic interactions. An exact calculation of these types of forces should necessarily take into consideration intrinsic characteristics of the nanoparticles, such as the chemical composition, the size and shape, as well as the separation between nanoparticles and can be performed by using the Dzyaloshinskii-Lifshitz-Pitaevskii (DLP) theory using the Derjaguin approximation. Though Keesom and dispersion forces are attractive interactions, Keesom forces appear between permanent dipoles regardless of their spatial orientation and dispersion forces result from the fluctuation of the polarization associated with the electron distribution within atoms. Since dispersion forces are associated with molecules, they play an important role in the assembly of capping ligand-passivated nanoparticles. Equation 10.2 is used to calculate the strength of these dispersion interactions.

$$E_{AB}^{disp} \approx \frac{3I_A I_B}{2I_A + I_B} \frac{\alpha^A \cdot \alpha^B}{r^6} \tag{10.2}$$

where, I_A and I_B denote first ionization potentials of atoms, r is the intermolecular distance, and α is dipole polarizability. The dispersion forces are proportional to $1/r^6$ and hence at nanoscale the strength of electrostatic interaction is comparable to that of van der Waals forces.

10.2.1.1 Examples of Nanoparticle Self-assemblies

Spherical monodisperse nanoparticles use van der Waals forces resulting in 2-D close packing which arises due to the 6-fold symmetry (Fig. 10.1a) [32] and in three dimensions, it results in the formation of the superstructures (Fig. 10.1b) [33]. In case of anisotropic particles such as nanorods, van der Waals forces result in highly directional interactions. For nanorods with large aspect ratio side-by-side assembly is most common as illustrated in Fig. 10.1c [34] as compared to end-to-end assembly as shown in Fig. 10.1d [34]. The potential in side-by-side assembly is given by U_{SbS} whereas; end-to-end potential is given by U_{EtE}. When the ratio of these interactions is close to 1, then the assembly becomes complicated while in case the ratio is greater than unity than side-by-side assembly is preferred over end-to-end.

10.2.2 *Induced Self-assembly*

Solvent evaporation is one of the many ways used to assemble colloidal particles on any given substrate. On solvent evaporation, isotropic particles with nearly same sizes result in a closed packed array. The capping ligands present on the nanoparticle surface contribute weak electrostatic repulsion while the most dominating force being the van der Waals attraction [35]. Temperature plays an important role in this assembly process as it controls the evaporation rate of the solvent. This concept was tested by Talapin et al. who performed a systematic study on formation of superlattice structures of $Fe_xO/CoFe_2O_4$. They observed that at highest temperature of 45 °C, *fcc* (face centered cubic) was formed. When the temperature further decreases, the hcp (hexagonal close packed) structure begins to crop in. At temperatures as low as −20 °C, all of the *fcc* is converted completely into *hcp* structure [36]. The formation of *fcc* and *hcp* is supported by both entropic

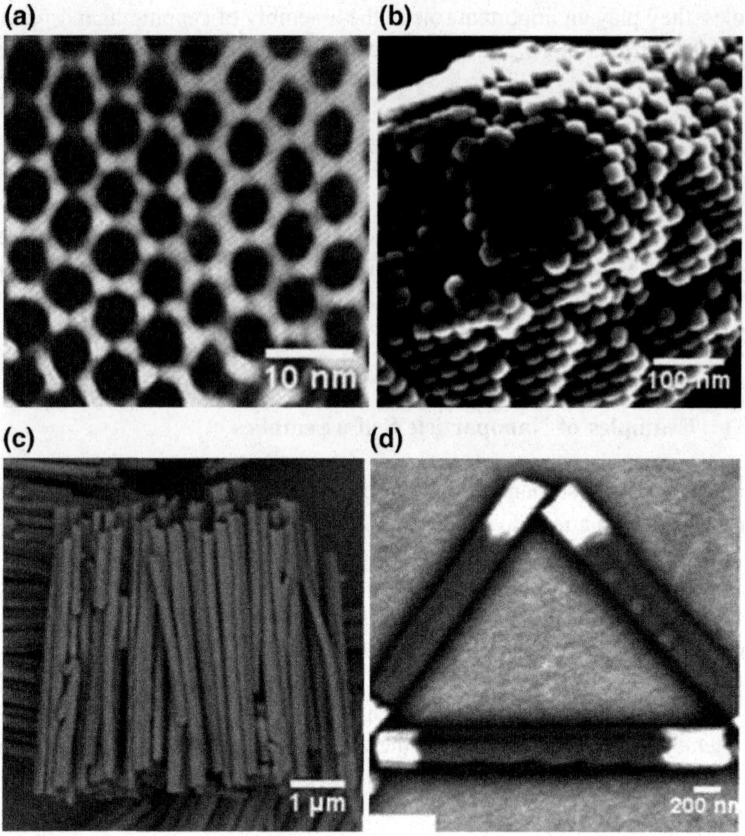

Fig. 10.1 **a** TEM image of 2D close packing of Ag nanocrystal array (Reprinted with permission from Ref. [32]. Copyright 1996, American Chemical Society), **b** SEM image of superstructures of Ag_2S sphere-topview (Reprinted with permission from Ref. [33]. Copyright 2013, American Chemical Society), **c** SEM image of 3D bundles of Au rods and **d** SEM image of 2D assembly formed using Au-Ni-Au rods. Reprinted with permission from Ref. [34]. Copyright 2004, The American Chemical Society

and isotropic van der Waals interactions. Along with temperature, the solvent plays an important role in stabilizing of *fcc* structure [36].

Another parameter that regulates van der Waals interactions in the self-assembly process is the solvent. To understand the solvent effect Quan et al. assembled Pt nanocubes dispersed in two different solvents such as aliphatic hexane and aromatic toluene. The superstructure obtained in toluene is simple cubic (*sc*) as ligand-ligand interactions are stronger in comparison to ligand-solvent interactions. However, ligands are completely miscible in hexane due to which the ligand-solvent interactions prevails and results in body-centered tetragonal (*bct*) lattice as can be seen in Fig. 10.2 [37].

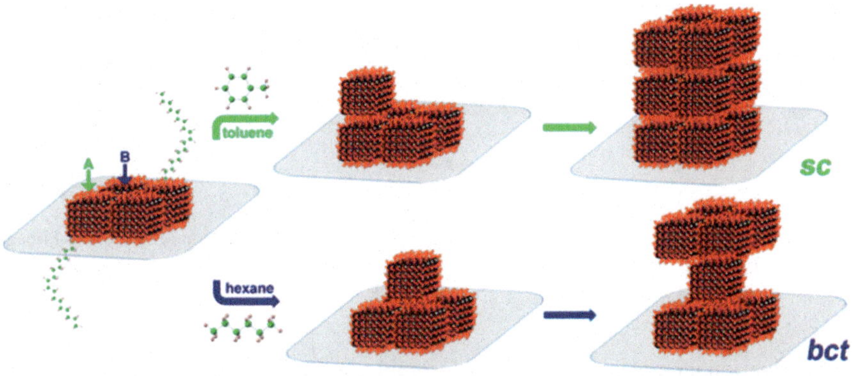

Fig. 10.2 Schematic illustration of self-assembled Pt nanocubes in sc and bct superstructures in presence of solvents like toluene and hexane. Reprinted with permission from Ref. [37]. Copyright 2014, The American Chemical Society

10.2.3 Electrostatic Interactions

Bearing similarity to the van der Waals forces, electrostatic interactions are known to play a pivotal role on the assembly process at various scales. Such interactions can be both attractive as well as repulsive in nature and in particles with asymmetric surface charge distribution they can be directional [38]. Additionally, the extent of the electrostatic interaction can be varied by the choice of the solvent, solution pH, and electrolyte concentration. The length of the electrostatic interaction between nanoparticles is calculated by the Debye equation and is shown in Eq. 10.3:

$$k^{-1} = \sqrt{\frac{\varepsilon_o \varepsilon_r k_B T}{e^2 \sum c_i z_i^2}} k^{-1} \cdot \sqrt{\frac{\varepsilon_o \varepsilon_r k_B T}{e^2 \sum c_i z_i^2}} \tag{10.3}$$

where, ε_o is the vacuum permittivity, ε_r is the dielectric constant of the solvent, k_B is the Boltzmann constant, TT is the absolute temperature, e is the elementary charge, c_i is number densities and z_i is the valence of the electrolyte ions.

10.2.3.1 Examples of Self-assembly of Nanoparticles

Electrostatic interactions have been investigated to facilitate the formation of structures that are used for various applications. Caruntu et al. successfully attached 2–3 nm gold particles onto the surface of ~10 nm Fe_3O_4 nanocrystals. This nanostructure opens up a new possibility of using magnetic particles for in vivo biomedical applications [39]. Similar technique was applied for assembly of bimetallic nanocrystals that display excellent electrocatalytic activity and durability for fuel cells applications. Zhang et al. used electrostatic assembly to synthesize

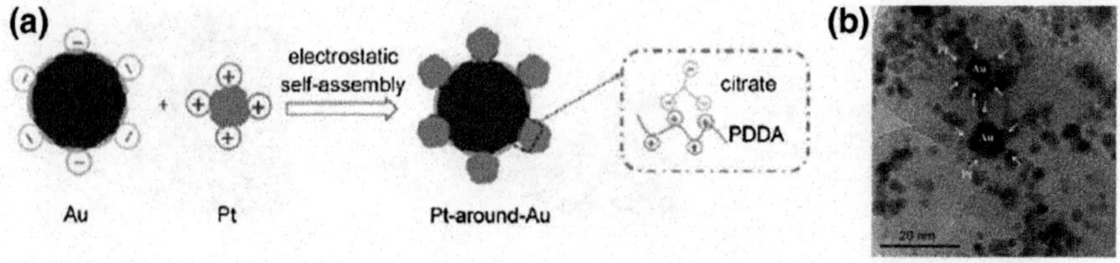

Fig. 10.3 **a** Schematic illustration of electrostatic self-assembly of Pt decorated Au, **b** TEM image of Au nanoparticle being surrounded by Pt nanoparticles. Reprinted with permission from Ref. [40]. Copyright 2010, Wiley

Pt/Au binary nanocrystals. The whole assembly process is shown as a schematic in Fig. 10.3a [40]. Pt nanoparticles are prepared in poly-(diallyldimethylammonium) (PDDA) solution while negatively charged Au nanoparticles are synthesized in sodium citrate solution. The attachment of Pt to the surface of Au nanoparticles results in a complex structure which is held together by the electrostatic forces between the two charged particles as evidenced in the TEM image in Fig. 10.3b. If comparable amounts of positively and negatively charged particles are mixed, it would lead to aggregation of nanoparticles as there are not enough net charges to form discrete particles. In the case mentioned above, there is a large excess of positively charged Pt nanoparticles which generate electrostatic repulsion interaction. This repulsion stabilizes the nanocomposite and prevents precipitation in solution [41, 42]. Nanoparticles can be also self-assembled into various structures with different complexities in the presence of external magnetic or electric fields. For example, Singh and coauthors recently demonstrated that a strict control of the interplay between the van der Waals interactions and the magnetic forces, Zeeman coupling and entropic forces between monodisperse cubic Fe_3O_4 colloidal nanocrystals allow them to be assembled into complex arrays with a helicoidal geometry [43].

10.2.4 Magnetic Interactions

When magnetic particles are arranged in an orderly fashion, such as a self-assembly, they tend to align their magnetic moments in the direction of the magnetic field. The alignment gives certain directionality to the magnetic interactions, thereby enabling magnetic nanoparticles to form 1D, 2D and 3D assemblies. All fundamental particles possess a spin, associated with which is a corresponding magnetic moment. When magnetic particles approach each other, the dipole-dipole interactions can significantly drive the particle assembly. A dipole-induced assembly causes the domains of magnetic particles to align resulting in a ring or chain-like structures. The above behavior has been observed by Dong et al. [44] wherein, they synthesized magnetic Ni-Co alloy using microwave irradiation method. Each nanoparticle can be considered as a mini magnet due to its own magnetic moment. Now depending upon the strength of the alloy, the nanoparticle assembly will be decided. Figure 10.4a shows a schematic depiction of how the alloy nanoparticles behave in both presence and absence of a magnetic field. In solution, single domain nanoparticles can move freely.

As evidenced in Fig. 10.4c, when the magnetic field is applied, nanoparticles are attracted forming short nanochains. In the absence of the field, the direction of the magnetic moment is random resulting in nanorings formation as seen in Fig. 10.4b. The formation of ring structure implies that the magnetic moment is strong enough such that the magnetostatic field surpasses the external magnetic field leading to the formation of nanorings. In a recent work by Taheri et al. [45] a controlled way to assemble nanoparticles in solution has been proposed. Cubic-shaped iron-oxide nanoparticles with sub-15 nm diameter are synthesized using oleic acid as a stabilizing agent. The assembly of cubic nanoparticles in 1D and 3D structures is induced by the magnetic field as evidenced in Fig. 10.5a, b. The external field aligns the dipole moment of the nanoparticles such that the particles come in close contact with each other and the van der Waals interactions can further cause a face-to-face arrangement.

It is hard to predict the spatial arrangement of the superstructures by knowing the constituents of the building blocks as there exists a delicate balance between competing forces that includes van der Waals, magnetic, electrostatic and others [31].

The role of the interplay between the van der Waals interactions and the magnetic forces, Zeeman coupling and entropic forces on the assembly of colloidal nanocrystals has been demonstrated recently by Singh and coworkers. To illustrate this, they assembled superparamagnetic, ~15 nm monodisperse cubic Fe_3O_4

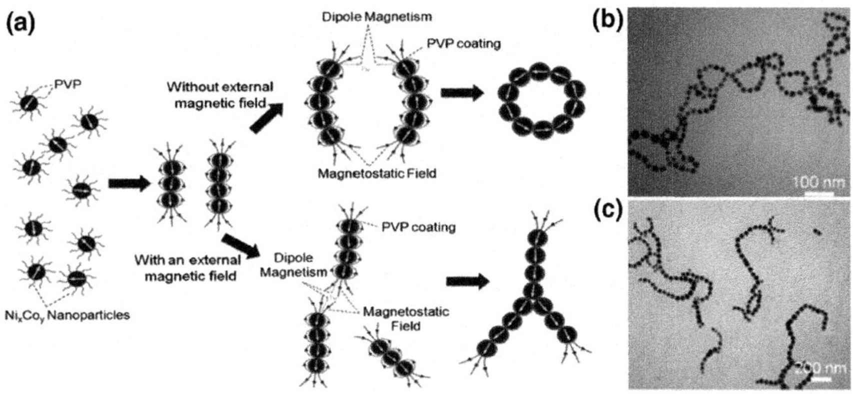

Fig. 10.4 **a** Schematic depiction of self-assembly of Ni-Co alloy; the nanoparticles are encapsulated by surfactant. In absence of an external magnetic field, nanoparticles connect with each other forming short nanochains and random magnetic alignment thereby resulting in nanorings. In presence of external magnetic field long nanochains are formed which have consistent magnetic alignment; **b** TEM image of Ni-Co nanorings; **c** TEM image of Ni-Co nanochains. Reprinted with permission from Ref. [43]. Copyright 2013, Wiley

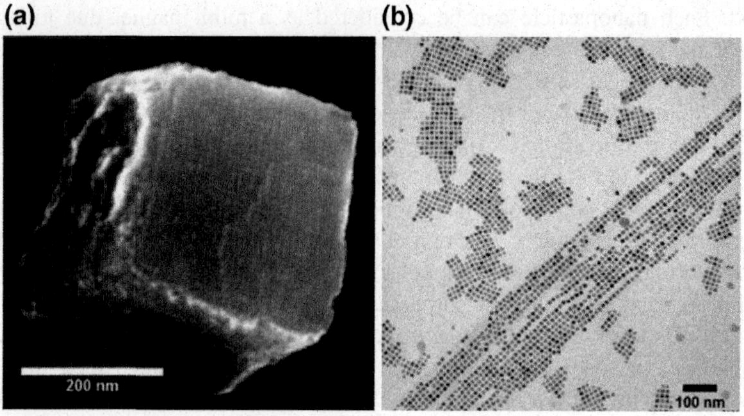

Fig. 10.5 **a** SEM image of 3D cuboid, **b** cryo-TEM image of cubic nanoparticles in toluene. Reprinted with permission from Ref. [44]. Copyright 2015, National Academy of Sciences, U.S.A.

colloidal nanocrystals in the presence of an external magnetic field. The colloidal nanocrystals possess a single-domain structure and the alignment of dipoles is the result of the interplay between the dipole-dipole interactions and the coupling with an external magnetic field. As such, when a horizontal magnetic field is applied and its strength is progressively increased the magnetic dipoles, which initially possess a zig-zag orientation, end up being oriented parallel to the lines of the external magnetic field leading to the formation of uniform belts of magnetite nanocubes. Conversely, when the direction of the applied magnetic field is perpendicular to the liquid-air interface, the arrays will possess a hexagonal close packing, as a result of the mutual repulsion between the individual pillars. By increasing the monolayer coverage the magnetite nanocubes will orient their [111] axes thereby leading to the formation of single-stranded helices [43]. The nanocubes have been in hexane solution and assembled onto diethylene glycol/air interface, which facilitated the post-synthesis transfer of the assemblies to any other type of substrate.

In another work by Singh et al. [46] iron oxide nanocubes were used and the magnetic field applied was perpendicular to the liquid-air interface ($\theta = 90^{\circ}$ in Fig. 10.6a). At a lower magnetic field, hexagonally packed pillars formed. On further increasing the magnetic field (H > 1000G), the superstructures were separated at a considerable distance and assumed C-shaped geometry as seen in Fig. 10.6b. At higher fields, there is a tendency to minimize magnetic dipole-dipole repulsion. These assemblies bear similarities to previously reported microring structure [47].

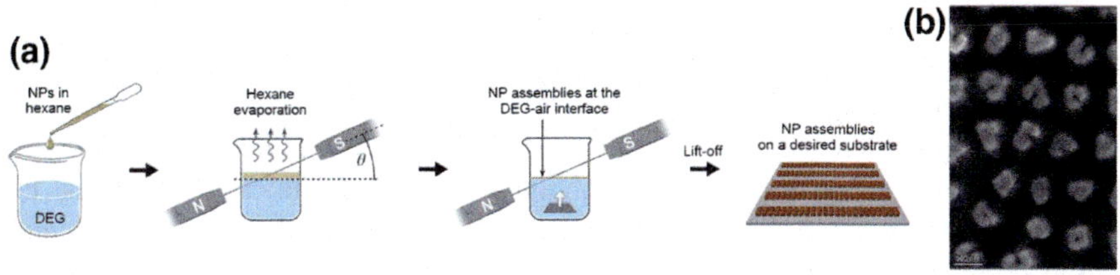

Fig. 10.6 **a** Schematic representation of the assembly of iron oxide cubes when θ = 90°, **b** top view of C-shaped nanocubes using SEM. Reprinted with permission from Ref. [46]. Copyright 2015, The Royal Society of Chemistry

10.2.5 Superficial Forces

The goal of using ligands during nanoparticle synthesis is three-fold: to passivate their surfaces, prevent agglomeration which occurs spontaneously in order to decrease their surface energy and render them dispersible in solutions with a polarity similar to that of the capping ligands thereby achieving a homogenous solution. The nature of the binding ligand will define the type of forces that play a role in the assembly process, and it is the intermolecular attractive forces among the ligands that give rise to the superstructures [48]. For example, nanorods form chain-like structure by using hydrogen bonding [49], linker molecules like DNA cause selective binding [50] and dipole-dipole interactions within particles induces their rapid assembly into superlattices. When certain functional groups like amines, carboxylic acids are used as surface modifiers, then hydrogen bonding plays an important role [51]. The magnitude of intermolecular forces is proportional to the density and strength of the bonds formed by the interaction of the functional group whereas the length scale is defined by the dimension of the modifier molecules. In a colloidal assembly consisting of oleic acid/oleylamine-stabilized Fe_3O_4 nanoparticles, superlattice formation is the result of the hybridization of assembled fatty acid molecules and nanoparticles. The average length of the modifier molecules is about 2 nm, and they both are known to arrange themselves parallel in the alternating directions [52]. The interaction between them is due to van der Waals forces, and they arrange themselves on the outer edges owing to the hydrophilic nature of the functional group. The functional groups have an affinity towards the nanoparticle through hydrogen bonding and hence create superlattice structures.

10.3 The Functionality of Nanoparticle Superstructures

For various fundamental and industrial applications, superstructures of single or binary nanoparticles are studied. But in the case of binary superstructures, several new functionalities are known to arise. The functionalities can be well explained with an example of semiconductor (SC), wherein electrons and holes inside a single semiconductor undergo recombination and generate luminescence only on photo-illumination [53]. In contrast to this, multiple different properties arise when semiconductor nanoparticles blend with other types of nanoparticles in the same superstructure. When the noble metals combine with semiconductors, photogenerated electrons transfer from semiconductor to metal thereby increasing the recombination time [54]. The superstructures built using single or binary nanoparticles find application in the field that ranges from flexible electronics, catalysis, semiconductors, magnetism, etc. and will be summarized below.

10.3.1 Mechanical Strength

The new class of nanomembranes consists of free-standing nanoparticle arrays that have a young modulus value of several GPa and differ from the traditional solid-supported film [55, 56]. The fabrication of nanomembranes has seen exciting progress, but the difficulty lies in fabricating defect-free membranes on a larger area. Chen et al. [57] are successful in fabricating nanomembranes using single crystalline, ultrathin Au nanowires via LB technique. Each nanowire is single crystalline with a width of 2.5 nm and is stabilized by oleic acid molecules. Figure 10.7a represents the as-synthesized wires, while Fig. 10.7b depicts the superlattice nanomembrane film formed using the LB technique.

The membrane possesses enough strength and flexibility to be transferred to any substrate, and the force-displacement curve exhibits the mechanical behavior. The applied force was increased in a stepwise manner until the membrane completely ruptured. Fitting the curve as shown in Fig. 10.7c, Young's modulus was obtained at 5.2 ± 0.4 GPa using an existing model [58]. The membrane is similar to monolayer graphene [59] and has the characteristic of a 2D material given its ultrathin nature and large aspect ratio. This nanomembrane shows a high optical

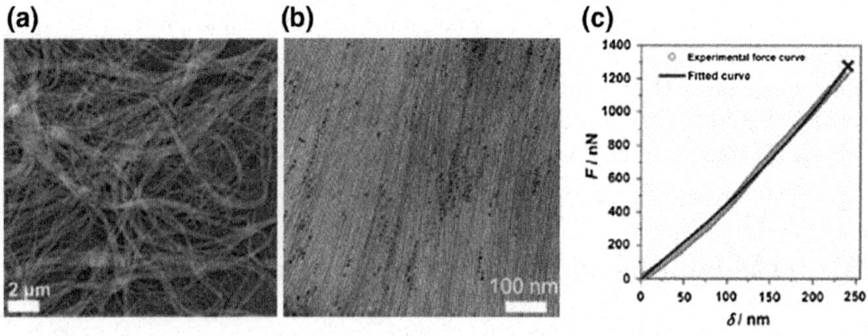

Fig. 10.7 **a** SEM image of Au nanowire (NW) bundles. **b** TEM image of monolayer nanomembranes synthesized using LB method. **c** Mechanical properties of monolayered superlattice nanomembranes obtained using Force displacement curve (AFM indentation) and theoretical fitting. Reprinted with permission from Ref. [57]. Copyright 2013, Wiley

transmittance of 90–97% which correlates to the dimensions of the film. In addition to this, the membranes are conductive in nature. This method is scalable and proposes a promising methodology for practical applications in lightweight and foldable electronics [60].

10.3.2 Photoluminescence

Carbone and coworkers [61] synthesized CdSe/CdS nanorods having a narrow distribution of rod length and diameters. They assembled the rods and measured their photoluminescence (PL) to understand how different superstructure organization and coupling effects can alter this physical property. Evaporation mediated process brings about the assembly of the nanorods while, the vertical alignment of rods uses an electric field in that direction. Vertically oriented rods are non-emissive and show a blue shift as compared to disordered structure while the ribbon-like assembly is characterized by a blue shift of 30 meV as compared to PL of disordered rods. The observed blue shift is related to the internal electric field that is produced by photogenerated charge distributions in neighboring rods which are more efficient in ordered assemblies than in disordered ones. In another example, Xing et al. [62] studied the photoluminescence behavior of the 3D superstructure of cubic β-In_2S_3. The stacks composed of thin micro-flakes which were luminescent as compared to the non-luminescent behavior seen in bulk In_2S_3. The emission at ~436 nm corresponds to the presence of deep traps in the structure while, luminescence at ~545 nm is due to the presence of interstitial sites.

Superstructures of rare earth oxides have long luminescent lifetimes (~1 ms) and are considered for application in the field of luminescent probes as seen in the case of Eu_2O_3, an oxide phosphor that has its central emission band at about 612 nm. Pol and coworkers [63] fabricated a 2D structure of Eu_2O_3 plates which shows luminescent characteristic. Upon excitation at 300 nm, red emission at 612 nm is measured. The red emission is because of the stabilization of europium ions in the +3 state, and this structure is known to have a thermal stability even at temperatures as high as 750 °C.

10.3.3 Catalysis

The requirement for a new generation of catalysts imposes a uniform composition of the nanoparticles along with a well-defined atomic arrangement at the surface. Metal oxides are well known as a catalyst support [64] and regulating their shape and size will let control the interfaces thereby increasing selectivity and activity for the desired reaction [65]. Yamada et al. [66] used the existing theory of integrating binary nanocrystals to form multiple interfaces [67]. However they were successful in obtaining cubic nanocrystals and thereby forming highly ordered superlattices of the binary components. The design of the tandem catalyst is as shown in Fig. 10.8a which consists of metal Pt, sandwiched between two oxides CeO_2 and SiO_2. The two separate and sequential reactions are known to occur at the catalyst interfaces. The assembly of the nanocubes is as shown in the TEM image in Fig. 10.8b. The two reactions that occur on the catalyst surface include decomposition of methanol as shown in Fig. 10.8c to yield CO and H_2 and ethylene hydroformylation with CO and H_2 yielding propanol. The better performance of this catalyst is due to two reasons, the occurrence of methanol decomposition at the Pt-CeO_2 interface and in situ generation of CO and H_2 at the Pt surface.

In another study by Tsao and coworkers [68], Au-Ag core-shell nanocrystals with cubic, octahedral and some intermediate geometries were synthesized as can be seen in Fig. 10.9. The plasmonic and the catalytic behavior of these are monitored to understand how shape and presence of binary component will affect the properties. The localized surface plasmon resonance (LSPR) band shows a red shift as compared to pristine Ag of same size and shape due to LSPR effect from the gold core. Depending on the shell thickness, both blue and red shift can be observed.

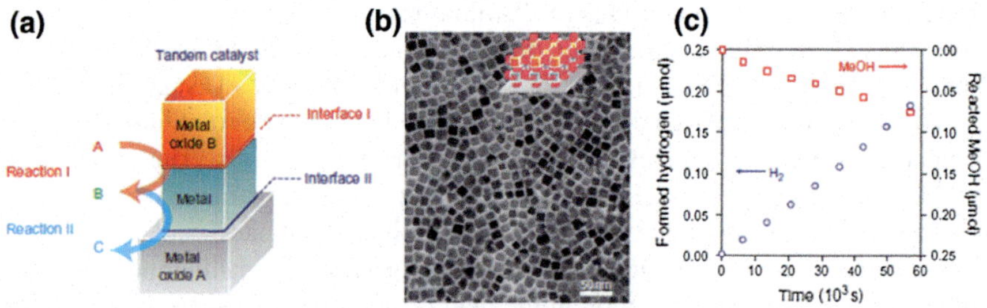

Fig. 10.8 **a** Schematic of sequential deposition of single layers of metal and metal oxide nanocubes. **b** TEM micrograph of a monolayer of CeO_2 colloidal nanocrystals assembled onto a Pt monolayer. **c** Time dependence of H_2 generated by the thermal decomposition of MeOH over a Pt/CeO_2 interface. Reprinted with permission from Ref. [66]. Copyright 2011, The Nature Publishing Group

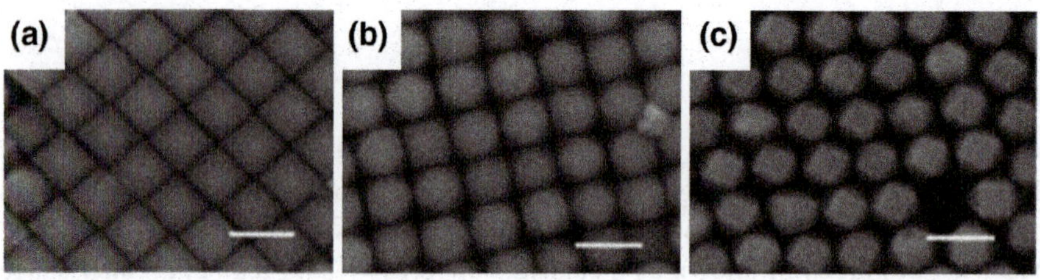

Fig. 10.9 SEM images of Au-Ag core shell **a** nanocubes, **b** truncated cubes, **c** cuboctahedra All scale bars are equal to 100 nm. Reprinted with permission from Ref. [68]. Copyright 2014, American Chemical Society

Both the Au-Ag cubes as well as octahedra showed catalytic activity towards $NaBH_4$ reduction of 2-amino-5-nitrophenol with the octahedra exhibiting highest catalytic activity.

Developing new catalysts for environmental application has its own set of challenges with the combustion of methane being one of the several [69]. PdO particles dispersed in oxide matrix are well-known methane combustion catalysts but, they get deactivated at high temperatures [70]. Colussi et al. were able to make Pd-O-Ce surface superstructures as it was previously reported that insertion of Pd into CeO_2 surfaces provides lower energy barrier for dissociative adsorption of methane [71]. This catalyst showed a high activity towards methane oxidation. This work opened up a new perspective towards stabilizing noble metal centers in an oxide moiety and thereby developing a range of efficient and durable catalysts [72]. Yin et al. studied the catalytic and photoluminescence behavior of superstructures formed of hollow ZnO nanorods. The number of hollow cages varied from single, double, multi to connected ones. The nanorods were synthesized using the hydrothermal method in conjugation with carboxymethyl cellulose (CMC) as the crystal growth modifier. Photoluminescence (PL) studies revealed that increase in cage number increased PL intensity around 500–700 nm which is also ascribed to increase in defects at the interface of the hollow structure. Although all the superstructures showed good catalytic activity towards decomposition of ammonium phosphate, the highest was recorded for single cage ZnO superstructure [73].

10.3.4 Plasmonics

The emergence of novel optical properties in hybrid systems consisting of semiconductor and noble metal nanocrystals is due to interparticle interactions particularly, the coupling of excitons in semiconductor along with plasmon resonances in the metal [74]. Shevchenko and coworkers used this concept to build a long range ordered binary nanoparticle superlattices using CdSe and Au nanocrystal [75]. The semiconductor particles are 8.7 nm and emit in the visible region while Au nanocrystals are 5.5 nm and have plasmon resonance in the visible range. Quenching of fluorescence in semiconductor crystal is well known. But in the hybrid structure, fluorescence intensity is only 5–15% as that obtained for CdSe sample and fluorescence lifetime is shortened. This behavior is due to transfer of energy to the surrounding Au nanocrystals which influences the radiative rate of the CdSe. Plasmonic response occurs due to the generation of free carriers in nanocrystals caused either due to defects or through doping. Indium doped cadmium oxide (ICO) films show higher conductivity than ITO and these conducting oxides are an ideal candidate for plasmonics. Gordon et al. [76] synthesized nanocrystals of ICO having a spherical and octahedral morphology that resulted in 3D superstructures having long range ordering. The nanoparticle films reveal negative permittivity in near IR (NIR) range indicating its relevance to be used as optical metamaterials. The design of 3D superstructures without the use of ligand exchange further paves a way to new metamaterials with a negative index. Metallic supercrystals present a new model that explains fabrication of a new class of plasmonic materials. In another work demonstrated by Tao et al. [77] 3D assembly of Ag nanoparticles was constructed using bottom-up approach and scaled up for the dimension of few millimeters. The limiting value in these self-assembly process is the volume of the starting material. These metallic crystals exhibit different electromagnetic field distribution, and they are ideal frequency selective materials for functioning in the visible and near-IR range. This work denotes the magnitude of bottom-up assembly process which offers control to achieve 3D structures which are otherwise achievable only via lithography.

10.3.5 Surface Enhanced Raman Spectroscopy (SERS)

SERS (Surface Enhanced Raman spectroscopy) of proteins is consistently examined to attain the right enhancement factor (EF) which will help in direct detection of low concentration of proteins [78]. It is crucial to develop ultrasensitive substrates that are capable of generating high EF as in the case of nanoantennas [79] and nanorod dimers [80]. Alvarez-Puebla et al. [81] were successful in fabricating Au nanorods supercrystals as SERS substrates for detection of complex biomolecules. The substrate was stable, reproducible, and showed a homogenous distribution of the electric field. Using such substrates they were able to detect prions in a complex biological medium such as serum or blood. These nanorod supercrystals are excellent SERS substrates in comparison to nanoparticle array as there is increased plasmon concentrations on top surfaces as layer of rods get added. This approach used for detection of prions is promising as it does not increase the time and reduces the uncertainty of final results. SERS substrates which are 2D planar systems have their active area limited to a single Cartesian plane. To overcome this limitation the developed 3D substrates have extended SERS hotspots to the third dimension, as large number of hotspots imply high enhancement factors. Zhang and coworkers [82] constructed 3D SERS-active substrate of Ag nanocubes using laser photolithography. This fabrication technique in conjugation with Langmuir-Blodgett enables a range of 3D microstructures such that it possesses high hotspot density in all three directions. The different microstructures compared to their SERS signal include pyramidal, right prism, truncated cone and square block. The pyramidal microstructure is the most efficient in comparison to all the other geometries thereby showing uniform signal due to contribution from all four facets. This prismatic microstructure presents promising application as micro-barcodes and will aid in the future design of substrates that find their use in the field of information tagging and storage.

In the field of detection of bioanalytes, the concentration limit ranging from nano to picomolar [83] can be achieved by using Au nanorods [84]. Hamon et al. used Au nanorods to understand how the stacked nanorod layers will affect the enhancement factor and will standardize the architectural layout [85]. The SERS signal obtained for Au nanorods were at two excitation wavelength namely 633 and 785 nm. On exciting with a wavelength of 633 nm, no variations in SERS signals were observed irrespective of the number of Au layers but at 785 nm some changes were seen. The theoretical data was used to explain the experimentally obtained results and confirmed that the analyte penetration was only into the topmost layer. To understand the enhancement produced by the stacked nanorods, it is essential for the diffusion of the analyte to the interior of the supercrytals.

10.4 Superlattice Formation

Colloidal assembly bears similarity with microscopic systems and it serves as an important model for understanding the atomic world. The broad classification of colloidal particles includes isotropic, anisotropic, and amphiphilic. An amphiphilic colloid that consists of Janus particles can impose specific interparticle interactions while keeping the geometry simple [86]. These colloidal particles form superlattices which are dependent on the quality of the nanocrystals as well as the environment used for the assembly. The superlattice formation is a process wherein distinct nanocrystals present in any solvent destabilize by forming long-range ordered structures [87]. This process is typically associated with thermodynamic equilibrium, with the structure being characterized by minimum free energy [88]. In the case of spherical particles, the superlattice structure results in simple cubic (*sc*) [89, 90] face-centered cubic (*fcc*) [89–91] and hexagonal close packing (*hcp*) [92]. They most commonly pile up to a close-packed pattern resulting in *fcc* or *hcp* arrangement to minimize potential energy and have highest packing efficiency of 74.04% [74, 93, 94]. For spherical particles, the crystal orientations in the blocks are random, and the whole pattern is isotropic. At the nanoscale, the interactions between particles become strongly anisotropic [95]. This anisotropic interaction is very crucial for the assembly and can result in complex morphologies [35]. Depending upon the various force fields around the nanoparticles that cause the assembly, it can be classified into 1D, 2D, and 3D structures [96, 97]. To achieve the nanoscale assemblies, two mostly used approaches include top-down and bottom-up methods. Top-down approaches involves use of lithography to generate patterned structures and it gave rise to Si-integrated chip technology. But this method has its set of disadvantages which gives more importance to a bottom-up approach which is currently under investigation.

There are different ways in which one can classify the superlattice assembly process. One way includes classification based on the constituents such as single [89, 98], binary [99–101] and even ternary [101, 102] components or by the shapes such as nanocubes, nano octahedron, nanoplates, nanorods and other morphologies. The assembly process is due to the direct interaction between the particles resulting in self-assembly, or it could be a directed self-assembly which includes using templates or external stimuli. In this section, we focus on various methods applied for assemblies of non-spherical nanoparticles used for formation of superlattices.

10.4.1 Nanocubes

Recent developments in nanoparticle synthesis have permitted the fabrication of cubes in various systems [103, 104] providing excellent building blocks for arrangements spanning from 2D to 3D array. A cube is known to have 6 faces, 12 edges and covered by six [100] facets which can thereby serve as a model to study the surface property of this particular side. Nanocubes with flat faces and sharp edges assemble in a simple cubic superlattices [105] while, with rounded corners and non-flat faces results in rhombohedral structure [106]. In this work, Yang et al. [107] obtained monodisperse Cu nanoparticles using two different shape controlling agent. Cu is known to be the second best conductor and exhibits LSPR in the nanoregime. The Cu nanocubes have truncated cubic shapes and organize themselves into 2D and 3D supercrystals with rhombohedral structure. The ligands play a crucial role in selectively stabilizing specific crystal planes thereby achieving monodisperse Cu nanocubes.

10.4.2 Nano-octahedra

Octahedra are composed of 12 edges, 8 equivalent triangular faces and their surface are terminated with [111] facets and side edges along [110]. [108, 109] Lu and co-workers [110] synthesized octahedral In_2O_3 in cubic phase and were able to create 2D and 3D assembly of these nanocrystals. In_2O_3 is a transparent conductor and exists in cubic or hexagonal phase. In two-dimensions, three types of projected assembly along [001], [110] and [111] were obtained. The packing efficiency achieved was as high as 66.67% without any tetrahedral holes. By controlling the solvent evaporation rate the nanocrystals have sufficient time to find equilibrium superlattice sites on the growing structure. Three different types of packing resulted in 3D superlattice structures by maintaining three different solvent evaporation rates.

10.4.3 Nanoplates and Nanostars

Huang and coworkers [111] reported large area 3D and 2D hcp array of PbS nanostars on Si substrate using drop casting and vertical deposition process. PbS is a narrow band gap semiconductor in the nanocrystalline regime and shows large quantum size effects. They used MCC template to bring about a non-close packed (*ncp*) array of [111] oriented PbS nanostars. The reflectance spectrum of MCC-PbS composite showed a red shift in the reflectance peak as compared to the original MCC template. In a review Bouet et al. [112] fabricated 2D assemblies of semiconducting nanoplatelets such as In [113], Sn [114], Cu [115] and Cd [116]. These 2D crystals form a key component of modern microelectronics and opto-electronics. Nanocrystals are known to form well-defined facets resulting in 3D assembly and on suppressing the growth in one direction, results in a 2D lattice. Lanthanide fluoride nanocrystals are chosen as model system by Ye et al. [117] due to their anisotropic crystal structures [117]. The shape and composition of the nanocrystals can be varied upon selecting the lanthanide element and/or adding LiF salts or lithium trifluoroacetate. The nanocrystals LnF_3 (Ln = La to Sm) adopt trigonal a tysonite structure and have the aspect of round nanoplates. For heavier Ln^{+3} ions (Ln = Er to Lu), nanocrystals with a tetragonal-bipyramidal shapes are formed. When Ln ranges from Eu to Dy, irregular nanoplates with orthorhombic β-YF_3 structures are produced. As for the morphology of the nanoplates, Saunders and coworkers [118] synthesized copper sulfide with two different compositions by using two different methods. The solventless synthesis yields nanodisks of Cu_2S having a hcp structure while, CuS nanodisks having a hexagonal structure is formed by arrested precipitation in a hot solvent. They also reported two stacking orientations of the nanoplates that include face down on the substrate in the case of CuS and Cu_2S nanoplates orient in an edge down fashion.

10.4.4 Nanorods

The self-assembly of nanorods is not an entropically favorable process as it requires both positional and orientational ordering. For a single layer of nanorods both nematic or smectic-A ordering has been observed. However, when more than one monolayer is present, it exhibits only nematic order. Talapin and coworkers [119] reported that CdSe nanorods dispersed in toluene results in only nematic order while the addition of a high boiling solvent to the above mixture followed by a drying under reduced pressure results in smectic-A ordered superstructures. In another work, Diroll et al. [120] built polycrystalline films of superlattices of CdSe/CdS nanorods on polar liquids that vary in viscosity, surface tension and dielectric constant. The various solvents that formed part of this study include ethylene glycol, diethylene glycol, triethylene glycol, tetraethylene glycol, dimethylformamide, dimethyl sulfoxide, dimetylacetimide, dimethyl sulfoxide, dimethylacetimide, formamide and acetonitrile. The oriental ordering is strongly related to the surface tension of the sub-phase. The optical properties of the nanorod assemblies consisted of strong absorption near 500 nm due to CdS shell and the emission originates from CdSe core. The optical anisotropy which is a projection of excitation dipole on the emission dipole is also measured. In case of disordered structures, random excitation dipoles result in small absorption anisotropy. When the nanorods are vertical they show no excitation polarization as now the excitation dipole is symmetrical along the long axis of the rods while, when the nanorods are parallel to the substrate strong excitation anisotropy is observed. These assemblies are in turn known to enhance the long-range electronic coupling.

10.5 Methods Used for the Directed Assembly of Nanoparticles

10.5.1 The Langmuir-Blodgett (LB) Method

The directed assembly of colloidal nanocrystals can be performed by several methodologies that includes LB (Langmuir-Blodgett) [121] technique, layer-by-layer (LBL) [122] assembly and self-assembled methods [123]. When amphiphilic molecules are dissolved in an immiscible solvent and spread onto the water surface, a Langmuir film is formed. A barrier used in this technique will define the nature of the monolayer. The fluid consisting of nanoparticles or nanowires are compressed to achieve nanoscale assembly by adjusting the surface pressure. This film is transferred to a substrate by dip-coating [124].

This technique though is limited for amphiphilic molecules, is currently being explored for inorganic materials that have an organic capping layer [125]. This method can vary particle density, spacing between particles which is followed by the arrangement. In the work by Liu and coworkers [126], LB technique was used to fabricate mesostructured assemblies of nanowires or nanotubes. The nanowires are dispersed in a mixture consisting of $CHCl_3$ and DMF. The π-A isotherm is as shown in Fig. 10.10a represents different surface pressures corresponding to 0, 8.5, 19.5 and 23 mN/m, respectively. The corresponding TEM images collected at different surface pressures are given in Fig. 10.10b. The same tellurium wires were further tailored into nanomesh like arrangement wherein the second layer of

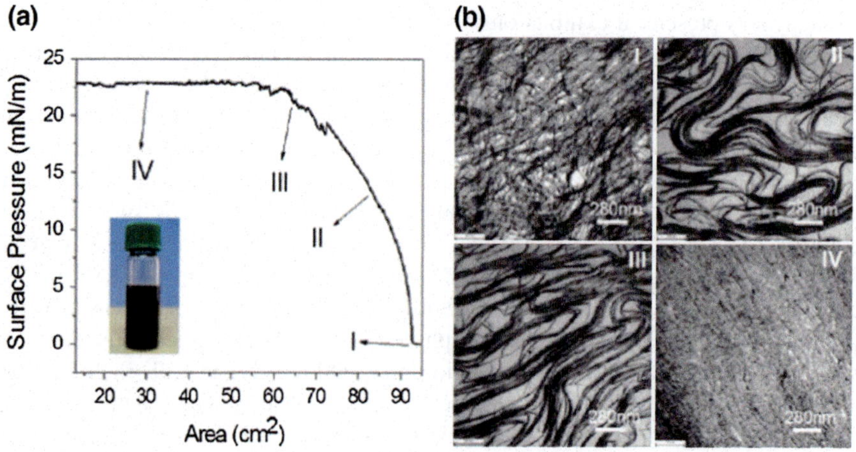

Fig. 10.10 **a** (*inset*) Solution of Te nanowires dispersed in $CHCl_3$ and DMF mixture. π versus area of the Te nanowire monolayer obtained at air water interface at room temperature. **b** TEM images of Te nanowires collected at different surface pressure corresponding to I, II, III and IV, respectively. Reprinted with permission from Ref. [126]. Copyright 2010, The American Chemical Society

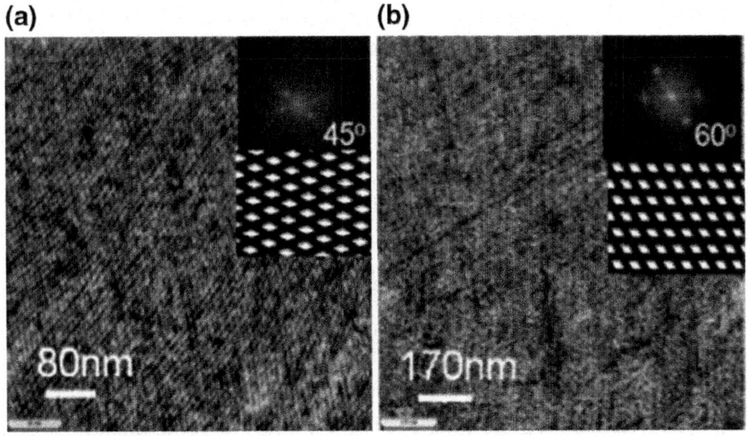

Fig. 10.11 **a–b** TEM images of nanomesh-like assemblies formed by packing two layers of Te nanowires. The *inset* shows the angle between the monolayers as well as the fast Fourier transform of the TEM images. Reprinted with permission from Ref. [126]. Copyright 2010, The American Chemical Society

nanowires were perpendicular to the first layer and different assemblies were obtained by varying the angle between two monolayers as seen in Fig. 10.11a, b. The above method led to the assemblies of Ag_2Te nanowires and Pt nanotubes thereby proving to be a flexible and promising technique for fabrication of integrated and periodic nanostructures. The tellurium films displayed reversibly switched photoelectric properties and the photocurrent was influenced by number of monolayers of nanowire.

In another work by Moon and coworkers [127], hydrophilic particles without any treatment were assembled into 2D monolayers using LB technique. The range of particles assembled includes SiO_2, polystyrene, Ag nanocubes, pNIPAM hydrogel particles, graphene oxide nanosheet, single-walled carbon nanotubes, tellurium nanowires and $Co(OH)_2$ nanoplates. Alcohols used in this assembly process are thermodynamically miscible with water, but their kinetics will depend on the viscosity of the alcohol. Alcohols on contact with water are known to spread creating a temporary alcohol-water bilayer. Ethanol tends to mix with water at an early stage indicating that particles submerge in water before aggregating in the alcohol phase. In the case of butanol, a thin alcohol layer formed is because it can spread over longer distances. Thus the particles now flow along the alcohol layer and make a floating monolayer. This research presents a new way to use traditional LB method in creating 2D colloidal assembly.

Mahmoud and others [128] used Au nanorods to create a 2D arrays using the Langmuir-Blodgett technique and to see its effect on the localized surface plasmon resonance (LSPR). The Au nanorods were stabilized with CTAB (cetyltrimethylammonium bromide) and were dispersed in thiolated polyethylene glycol (PEG). The structure of the assembly was varied by changing the surface pressure and the chain length of the polymer (PEG). When Au nanorods functionalized with 6k-thiolated PEG was assembled on a Si surface at a surface pressure of 6 mN m^{-1}, voids were observed through the array as can be seen in the SEM image in Fig. 10.12a. With the increase in surface pressure, the packing increased and both end-to-end and side-to-side assemblies were formed, the dominant one being the end-to-end assembly. With Au nanorods functionalized with 2k-thiolated PEG assembled on a Si surface at a surface pressure of 5 mN m^{-1}, voids were observed through the array as can be seen in Fig. 10.12b. With the increase in surface pressure, a tightly packed side-by-side assembly of nanorods was obtained. Au nanorods are known to have two localised surface plasmon resonance (LSPR) peaks. When non-polarized light is used for excitations both the longitudinal and transverse plasmons are excited. Now if the incident light is polarized along the long axis of the Au nanorod, the longitudinal mode will be excited. In order to excite the transversal mode the incident light should be perpendicular to the long axis of the nanorod. Figure 10.12c displays the LSPR spectrum for Au nanorods functionalized with 6k thiolated-PEG and assembled using LB at surface pressure of 6 mN m^{-1}. The polarization of light is at 0° and 90° and a variation in peak intensity and peak position was observed. When the incident light was polarized at 0°, the longitudinal peak was more intense while, on increasing the angle of polarization to 90° the intensity of longitudinal peak decreased and transverse peak increased. The transverse and longitudinal peaks are observed at 540 and 834 nm.

Au NCs (gold nanocubes) are assembled by Mahmoud [129] and coworkers at a lower surface pressure and the polymer micelles, polyethylene glycol (PEG) is used to drive the assembly. The distance between the cubes is tuned by varying the chain length of the PEG molecules, and these micelles void behave as a soft template. When Au nanocubes functionalized with 6k PEG polymer were used to design the assembly at a surface pressure of 11 mN m^{-1}, the resulting assembly is as shown in Fig. 10.13a. The circular voids are being surrounded by a highly packed array of

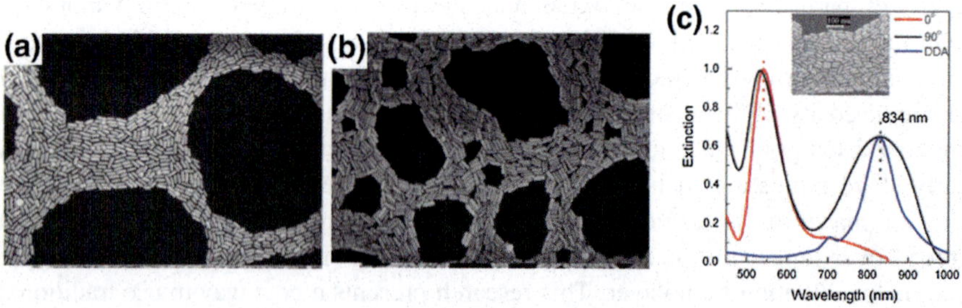

Fig. 10.12 SEM images of 2D array of gold nanorods functionalized with thiolated PEG on silicon substrate. **a** end-to-end assembly in case of molecular weight of thiolated-PEG is 6k and the surface pressure at 6 mN m^{-1}. **b** Side-by-side assembly obtained when molecular weight of thiolated-PEG is 2k and the surface pressure at 5 mN m^{-1}. The LSPR spectrum obtained for Au nanorod array at different angles of polarization of the incident excitation light. **c** Au nanorods functionalized with 6k thiolated-PEG and the surface pressure at 6 mN m^{-1}. Reprinted with permission from Ref. [128]. Copyright 2014, The Royal Society of Chemistry

Au NCs. When the same experiment is repeated for Au NCs functionalized with 2k PEG, a similar kind of assembly is formed and is shown in Fig. 10.13b. Atomic force microscopy (AFM) imaging (Fig. 10.13c) was used to decipher the constituents of the voids and to understand the topography of the 2D array. To further support and understand the composition of the voids; micro-Raman spectroscopy was used. The Raman spectra were collected from the surface of the voids and from the surface of the Au NCs. The above spectra were compared to that obtained from pure CTAB and 6k PEG films cast on Si as can be seen in Fig. 10.13d. The Raman band intensities obtained for pure CTAB were found to be higher than that measured for pure PEG film. The spectrum obtained from the surface of the NC's and the voids resemble the Raman bands obtained from both CTAB and PEG. This confirms that the nanodisks present in the void are composed of both PEG and

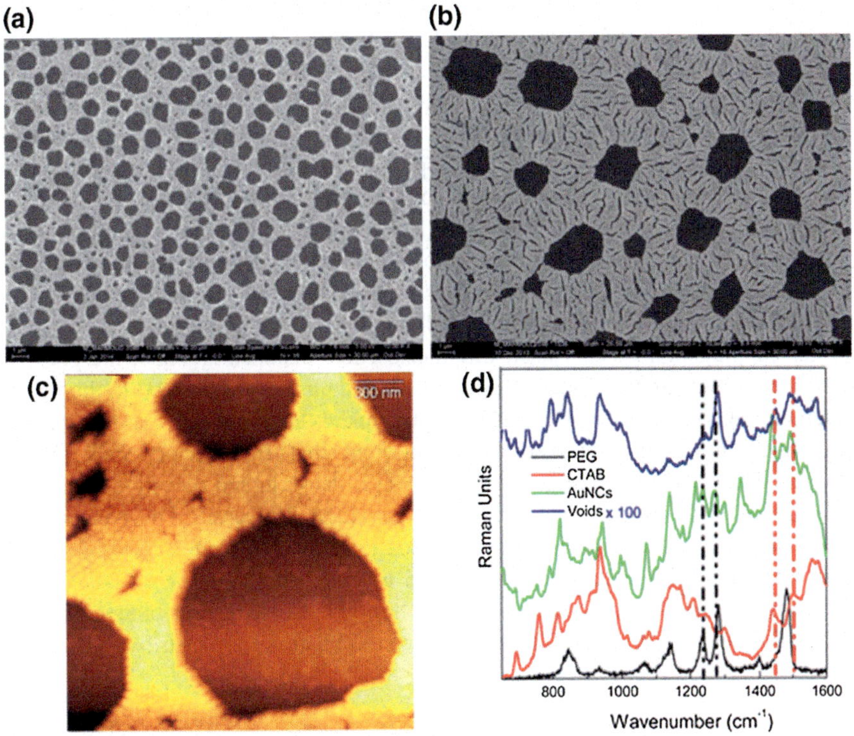

Fig. 10.13 SEM images of gold nanocubes functionalized with PEG on silicon substrate. **a** Assembly of cubes obtained when molecular weight of PEG is 6k and the surface pressure is at 11.5 mN m^{-1}. **b** Assembly when molecular weight of PEG is 2k and the surface pressure is at 6.5 mN m^{-1}. **c** AFM image of AuNCs functionalized with 6k PEG and the surface pressure at 11.5 mN m^{-1}. The image shows that the void area is filled with nanodiscs. **d** Micro-Raman spectra collected for 2D AuNC functionalized with 6k PEG from the void areas and surface of the nanocubes. Reprinted with permission from Ref. [129]. Copyright 2015, The American Chemical Society

CTAB. The balance between the attractive forces that bring the cubes together along with the repulsive forces is responsible for the 2D array of NCs.

10.5.2 Ligand Stabilization

CTAB stabilized nanoparticles are very common as they provide flexibility to synthesize any shapes and sizes as well as stability thereby preventing aggregation. The post-synthesis functionalization of nanoparticles is for its transfer to any solvent. When nanoparticles try to assemble itself into supercrystals, they minimize the total free energy. The two well-known superlattice polymorphs that control the assemblies include *fcc* and *bcc* structures. It is known that the assembly not only involves a complex interplay of both entropic and enthalpic components but, additional features such as temperature, pressure, evaporation rate, surface capping ligand play a pivotal role. Wang et al. [130] tried to explain the structural phase transformations as a function of packing density of nanoparticles and inter-nanoparticle distances. They synthesized 3.5 nm PbS nanoparticles capped with oleic acid (OA) to be used as a model system for the study. After the thermal treatment, they observed the superlattice phase transformation from amorphous to bcc and then to *fcc*. As shown in the plot in Fig. 10.14, the increase in inter-nanoparticle distance is accompanied with a decrease in packing density and OA modifies the surface from a disordered to ordered state. This shows the evolution from amorphous to bcc and finally to *fcc* structure. This transformation is driven by an increased ratio of structure ordering and enhanced interactions of surface-capping molecules. When the inter-particle distance is small, the packing density and the configurational entropy are maximum. In a similar work, the ligand coverage on the nanocrystal surface was linked to directing the superlattice assembly. Choi and coworkers [131] reported that PbS nanocrystals (NC's) assembled into fcc superlattice when there was a dense ligand coverage while, *bcc* superlattice was formed when the ligand coverage decreased. The nanocrystals were stabilized with oleic acid and suspended in hexane. A combination of surface techniques and DFT calculations was used to understand the role of the surface ligands in preserving a particular lattice structure. Pb-oleate ligands bind stronger to the [111] facets than to the [100] facets. Since the Pb atoms are distributed on [111]

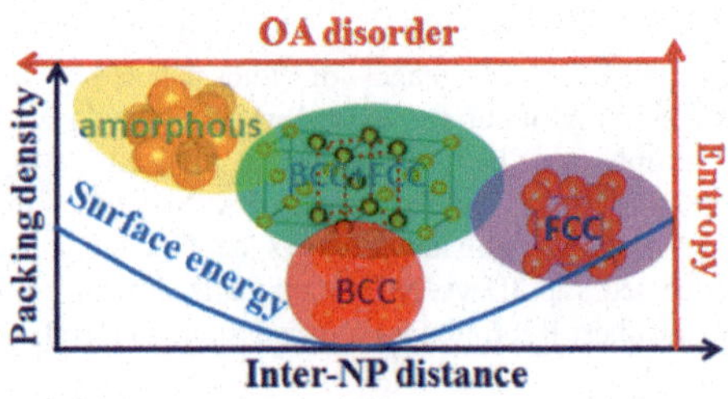

Fig. 10.14 Model used to interpret nucleation and growth of nanoparticles in assembled superlattices. Reprinted with permission from Ref. [130]. Copyright 2013, The American Chemical Society

facets they bury the S atom underneath thereby preventing them from oxidation. The facet specific ligand coverage can be understood by assuming an eight-fold coordination of nanocrystals in bcc superlattice and eight-fold degeneracy of [111]. When a surfactant is used, the formation of 3D gold nanorod (NR) superstructures has a limited control on both the dimensionality and directionality of the assembly [132].

CTAB (cetyltrimethylammonium bromide) is a well-known surfactant which functions as shape inducing agent along with preventing aggregation. In case of Au nanorods, post-synthesis functionalization and transfer into organic solvents are proposed for construction of superlattice structure. Martinez et al. [133] found a unconventional surfactant known as the gemini surfactant [134] to produce monodisperse nanorods that can be used to produce 2D and 3D superlattices. Gemini surfactants are made up of two hydrophobic tails and two hydrophilic tails linked by a spacer chain and are known for their amphiphilic properties. In their work, the spacer was made up of ethylene oxide group and the assembly spans from 1 to 10 μm across as can be seen in Fig. 10.15a.

The top view of the long range assembly of Au nanorods is seen in Fig. 10.15b, c. The Fourier transform confirms the formation of hexagonal geometry. The 2D assembly results due to formation of van der Waals and electrostatic bonds at the interfaces. For 3D assembly, smectic-B liquid is formed wherein the bilayer of the ethoxide binds to two adjacent nanorods but there exist strong van der Waals interactions of the alkyl chains and electrostatic interactions at the nanorod surfaces. An alternate synthesis route for the preparation of nanocrystalline metal oxides consists of avoiding the use of a surfactant. However, as expected, the control of the distribution of the size of the nanocrystals has proven difficult.

Pucci et al. [48] demonstrated in situ 3D assemblies of surfactant free ZrO_2 and rare-earth stabilized ZrO_2 particles. The nanocrystals are benzoate-capped which is in situ generated. ZrO_2 synthesized at 230 °C composed of particles which are about 4 nm in size. Temperature is known to play a crucial role in defining the

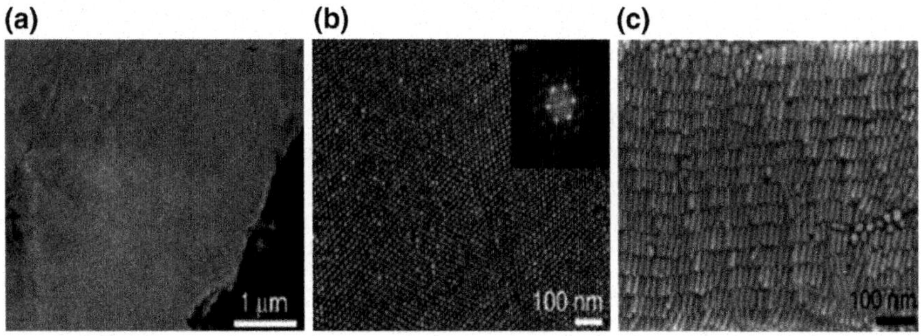

Fig. 10.15 SEM images of gold nanorods functionalized with gemini surfactant on silicon substrate. **a** *Top view* of the rods and its long range ordering. **b** *Top view* of AuNRs and the *inset* is the fourier transform of the image. **c** Vertically aligned nanorods. Reprinted with permission from Ref. [133]. Copyright 2009, Wiley

morphology. At 325 °C, Tb-doped ZrO_2 mainly show flower-like morphology whereas when the temperature is increased to 350 °C rhombic dodecahedron supercrystals are formed. When ytterbium was used as a doping ion, it resulted in bipyramid supercrystals. The important characteristic observed in this method is that irrespective of the temperature and nature of the doping ion used; the nanoparticles have sizes ranging from 3 to 4.4 nm. When the surface of the nanoparticles was studied it showed the presence of benzoates. They form bridging bidentate coordination to the oxide layers and π-π stacking is responsible for the formation of periodic organic-inorganic structure. The formation of uniform particle size is due to limiting the growth as accumulated benzoate groups poison the growth faces. When the benzoate amount is low, supercrystals formation is not seen. This one-pot approach is used for fabricating grams of superstructure made up of 3–4 nm building blocks.

10.5.3 The Solvent Evaporation Technique

The solvent evaporation of highly concentrated solution of Au colloidal nanocrystals seemingly results in an ordered assembly of Au nanostructures [135]. The Au particles are stabilized by CTAB and depending upon molar ratio of the binary mixtures, nematic or smectic superstructures are formed. A minimum of surfactant concentration is important as it balances the entropic and electrostatic potential which is crucial in superstructure formation. When surfactants are present in excess then phase separation occurs resulting in their own superstructure which is difficult to be imaged. Besides the surfactant concentration, the formation of ordered superstructure depends on temperature at which the solvent evaporates. A temperature of above 60 °C results in only disordered assembly due to fast evaporation rate. The droplet evaporation assembly is applied for Au particles with different shapes, including rods, cubes and polyhedra. Gold polyhedra assemble into hexagonally packed superstructure (Fig. 10.16a) while, nanocubes assemble into hexagonally packed superstructure (Fig. 10.16b). Au nanorods assemble into nematic structure (end-to-end) resulting in long chains nearly parallel to each other forming 3D lattices as seen in Fig. 10.16c. When nanorods assembled in side-by-side fashion, it results in smectic-A structure. Au nanorods of different sizes were mixed and let to assemble, nematic superstructures were formed and the diameters of the two samples played an important role in the assembly.

Solvent evaporation induced assembly was applied for fabrication of large area 3D and 2D hcp arrays of PbS nanostars on clean Si substrate by Huang and co-workers [111]. PbS is a semiconductor with a narrow band gap of 0.41 eV and shows great promises in nanometer-size regime. The star shaped nanocrystals are obtained in presence of both cationic as well as anionic surfactants and their 2D and 3D assembly is brought about by using drop casting and vertical deposition. When a drop of PbS solution was drop-casted onto Si substrate the particles attained

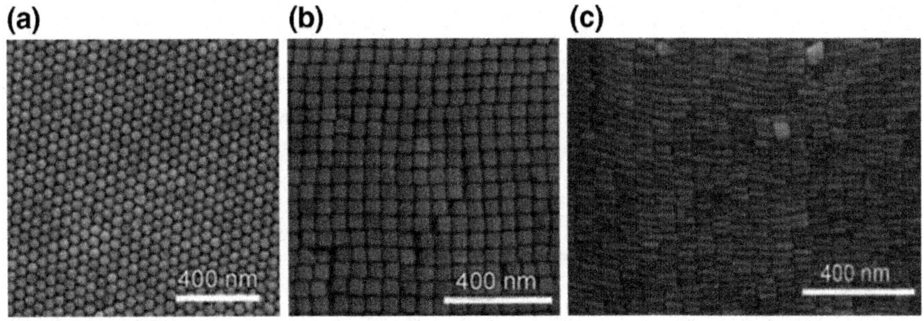

Fig. 10.16 SEM images showing the assembly obtained from differently shaped Au nanostructures **a** Au polyhedral in hexagonal packing, **b** packing of Au nanocubes, **c** nematic superstructure assembly of Au nanorods (the aspect ratio for Au nanorods is 4.4). Reprinted with permission from Ref. [135]. Copyright 2008, Wiley

ordered structure with slow evaporation resulting in ring like structure. The formation of 3D structure with hexagonal arrangement was confirmed using SEM and fast Fourier transform. Using vertical deposition, 2D hcp assembly was formed. The top layer hcp assembly was similar in both 2D and 3D structures with the difference seen in average spacing between the two nanostars. The vertical deposition is limited to colloidal particles as they sediment slower than the solvent evaporation. The PbS nanostars have high density and their deposition on vertical substrate was competitive causing orientational deviation.

To further elucidate the effect of the solvent on the structural transition from a *fcc* to a *hcp* crystal structure, Courty et al. [136] synthesized 5 nm silver nanocrystals capped with alkyl chain length varying from C10 to C12. The nanocrystals are coated on HOPG (highly oriented pyrolytic graphite) substrate and the temperature for assembly is varied from 15 to 50 °C. When the temperature was at 15 °C, disordered 3D assemblies were observed irrespective of the nature of the solvent used. When the temperature increased to 25 °C, *hcp* supracrystals were obtained for less volatile solvent. A further increase in temperature to 50 °C saw a transition to bcc structure in all the solvents used. To understand these transitions, a model was developed by their group which employs the van der Waals interaction between the Ag nanocrystals. The two important parameters that play a role in defining the kind of packing is the length of the ligand (L) and nanocrystal radius (R). From the model they predicted that the bcc structure is dependent on the deposition temperature and on the length of coating agent while the *hcp* to *fcc* transition is dependent on the nature of the solvent and is attributed to stacking processes which is in turn dependent on evaporation kinetics.

In a related work by Xiao et al. [137] vertical arrays of Au nanorods were obtained when the bulk solution was evaporated. The aqueous solution of CTAB stabilized rods were allowed to assemble itself on Si substrate. The concave meniscus forms when the solution evaporates (Fig. 10.17a) and contact line recedes from interior towards the edge. When all of the solution completely evaporates,

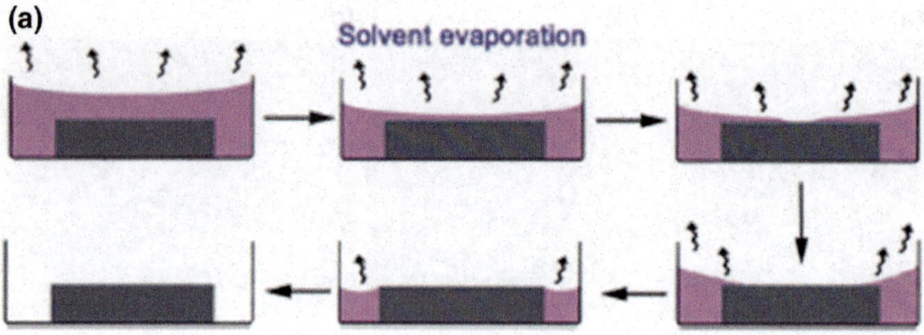

Fig. 10.17 a Schematic representation of the assembly process using bulk evaporation process on a Si substrate. The view of the concave meniscus formed along with solvent evaporation direction. Reprinted with permission from Ref. [137]. Copyright 2014, The Royal Society of Chemistry

microplates like structures were observed. The typical coffee ring formation was not seen in this case as the contact line was moving continuously causing a uniform distribution of the nanorods over the Si substrate.

A further SEM analysis suggested that the rectangular microplates are aligned perpendicular to the substrate. A closer view of the microplates signifies nanorods pack in a side-by-side fashion over a larger area. The packing is in a parallel fashion resulting in a smectic superstructure as seen in Fig. 10.18a, b which depicts the hexagonally close-packed nanorods. The surfactant plays a crucial role in the vertical assembly. When the concentration was lower than 1 mM, no assembly was observed. Thus a minimum of 1 mM concentration is required to balance the van der Waals attraction, electrostatic repulsion and capillary forces. The formation of the unique supercrystal arrays is based on movement of contact line which is accompanied by continuation of nucleation and growth process.

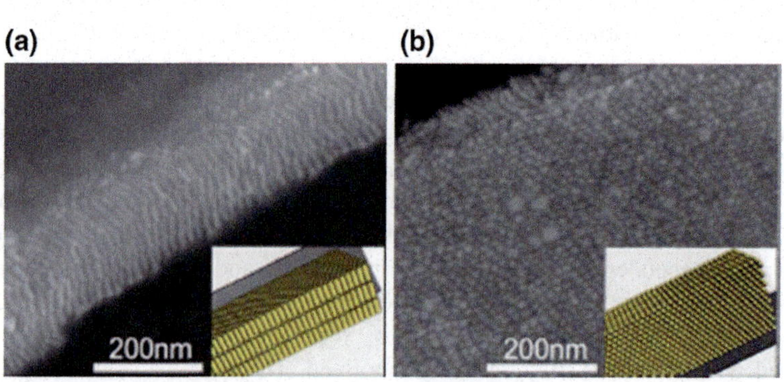

Fig. 10.18 SEM images of Au nanorods. a Smectic assembly of nanorods. b Hexagonal packing of standing nanorods. Insets in (a) and (b) are the 3D simulated models of the microplates viewed along different observation angles. Reprinted with permission from Ref. [137]. Copyright 2014, The Royal Society of Chemistry

10.5.4 The DNA-Template Method

The first work exploiting DNA for the assembly of nanoparticles was reported in 1996 [50, 138]. Single stranded DNA oligonucleotides of defined sequence and length were employed in these studies. The advantages over the use of linker molecules include use of DNA synthesizer to control the length and surface binding functionality in automated fashion. The interparticle distances and stability of the structure can be controlled by the choice of oligonucleotide sequence, solvent, temperature and electrolyte concentration. In a work by Park and coworkers [139], single or binary DNA sequence was attached to same inorganic nanoparticles resulting in assembly into different crystalline states. 15 nm Au nanoparticles were modified with synthetic oligonucleotides and a linker DNA. The unique part of this system is that without any chemical alteration of the Au-oligonucleotide conjugate both single as well as binary system can be simulated. Single component assembly implies using one DNA linker while, binary component uses two different DNA linkers. In a single component, each particle can bind to other particle forming a closed packed *fcc* structure. Alternately binary system results in a non-close packed *bcc* structure. The formation of different structure can be ascertained to competition between the enthalpic and entropic terms at different temperatures. If the entropic term dominates, it results in close-packed structure as smallest volume fraction implies highest entropy. If the assembly happens at temperature near the melting temperature T_m of DNA, a close-packed structure is obtained. At temperatures below T_m, the assembly is governed by an enthalpic term and, therefore will result into a non-closed packed structure.

To further enhance the versatility of DNA based assembly, nanoparticle crystallization was studied by Mirkin et al. [140]. They observed that when DNA functionalized nanoparticles (gold in their case) was heated above the melting temperature of the DNA links and slowly cooled to room temperature; it yielded faceted crystals thereby mimicking atomic crystallization. The microcrystals made from gold nanoparticles of varied sizes such as 5, 10 and 20 nm resulted in stabilizing rhombic dodecahedron shape with bcc packing as shown in Fig. 10.19a–c.

The formation of rhombic dodecahedron microcrystals is explained on the basis of surface energy of the exposed facets. The facets that enclose rhombic dodecahedron are (110) which in case of bcc represents the closest packed plane. When closest packed plane is exposed, the process is thermodynamically favored as less number of particle-particle interactions per unit cell are broken. Surface energy calculations along with molecular dynamics simulation predict the Wulff polyhedron for bcc metal to be rhombic dodecahedron enclosed by (110). This work demonstrates the utilization of DNA not only for nanoparticle assembly but also in formation of macroscopic structures.

DNA has been used in assembling elementary plasmonic nanoparticles. To create well-defined nanostructures, the bonding interactions of nanoparticles need to be understood. The complex forces involved are van der Waals, capillary, electrostatic, solvation to name a few [141]. The assembly obtained for most

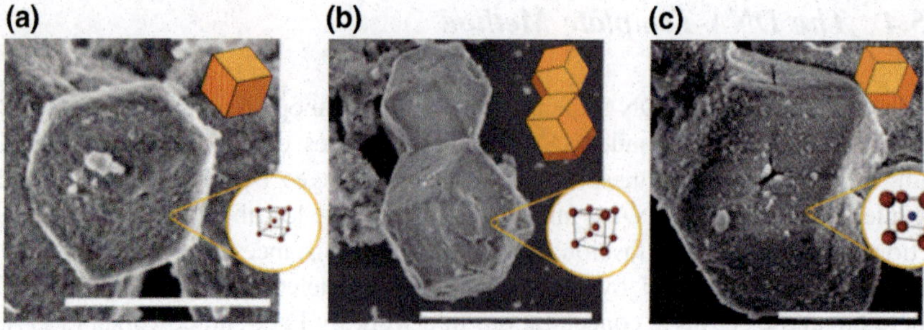

Fig. 10.19 SEM images of rhombic dodecahedron microcrystals synthesized using gold nanoparticles of various sizes **a** 5 nm (scale bar-2 μm), **b** 10 nm (scale bar-4 μm) and **c** 15 nm (scale bar-2 μm). Reprinted with permission from Ref. [140]. Copyright 2014, Nature Publications Group

Fig. 10.20 Schematic displays the combination of various DNA motifs and plasmonic atoms to form range of assemblies. Reprinted with permission from Ref. [142]. Copyright 2011, Nature Publications Group

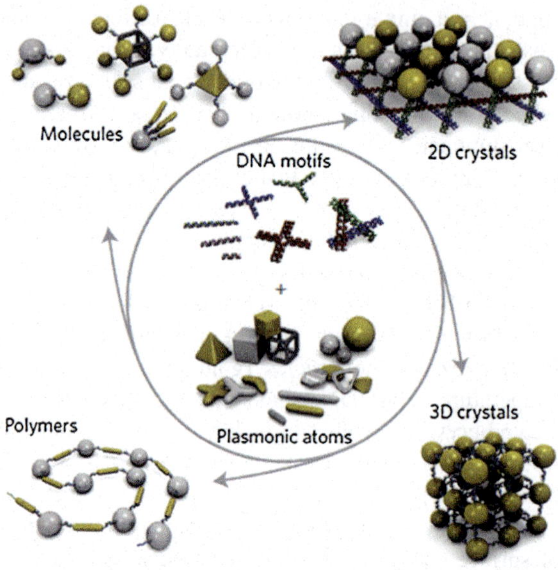

nanoparticles does not signify thermodynamically lowest energy state but kinetically trapped non-equilibrium structure. DNA has the ability for molecular recognition and renders structural versatility. As compared to the organic ligand, DNA offers manipulative ligand length thereby influencing the assembly. The vast variety of DNA motif in conjugation with various shapes of plasmonic atoms results in range of plasmonic nanostructures as can be seen in Fig. 10.20. When DNA is used, the complex interplay of forces gets balanced such as van der Waals forces are balanced by steric hindrance, electrostatic interaction is balanced by ionic strength and the attractive Watson-Crick base pairing forces are activated by controlling the sequence, ligand density and number of DNA bases. This biological moiety can not only be used as linkers, templates but also as spacers to promote positioning of metallic nanoparticles [142].

10.5.5 Template Assembly

Highly-ordered architectures can be obtained by self-assembling colloidal nanocrystals. To this end, a new concept involves combination of disassembly and self-reassembly synthesis. This method is based on constructing 3D ordered macroporous (3DOM) structure using colloidal crystals template followed by spontaneous disassembly of these structures by introducing amphiphilic surfactants. Colloidal crystals are 3D periodic arrays built from monodisperse colloids and the interstitial space is filled with another substance. By removal of the spheres after fillings are done forms 3D ordered macroporous structures. This is followed by controlled heating and the structure spontaneously disassembles into well-defined shapes. The size of the particles will depend on the colloidal spheres and the shapes ranges from cubes to spheres to multipods. Li and co-workers [143] synthesized 3d block transition metal (Cr, Mn, Fe, Co, Ni, Cu and Zn) oxides to highlight the generality of this assembly method. They were successful in synthesis of single as well as mixed metal oxides. In their work, PMMA spheres were used to define the shape, phosphates served as surfactants and metal acetates were the source of the metal ion. Uniform nanocubes were formed for various materials and assembled in 3D and 2D morphology.

10.5.6 The Sedimentation Method

Ag nanocrystals ranging in size from 100 to 300 nm in various shapes were synthesized using polyol method and stabilized with PVP (Polyvinylpyrrolidone) [144]. These polyhedra behave like quasi-hard particles and assemble into 3D supercrystals by sedimentation process. Calculations on these systems imply that gravity acts as a driving force for these assemblies due to their sizes whereas, an evaporation driven assembly is typical in the case of smaller nanoparticles. When solvent driven assembly was applied for these nanocrystals, long range order was missing. Thus sedimentation is known to provide more homogenous as well as a gentle force for the assembly process as compared to solvent evaporation process. The assembly was done using a PDMS reservoir that was filled with solutions of particle. When the substrate was tilted causing the particles to gradually sediment and assemble at the bottom of the reservoir. Figure 10.21a displays the range of shapes for Ag nanocrystal that can be stabilized using this method. The long-range ordering is obtained for nanocrystals of different shapes as can be seen in Fig. 10.21b, c. The thickness of the polymer layer is 20 nm which is sufficient to prevent close approach of the surfaces. In here the strong entropic repulsion from

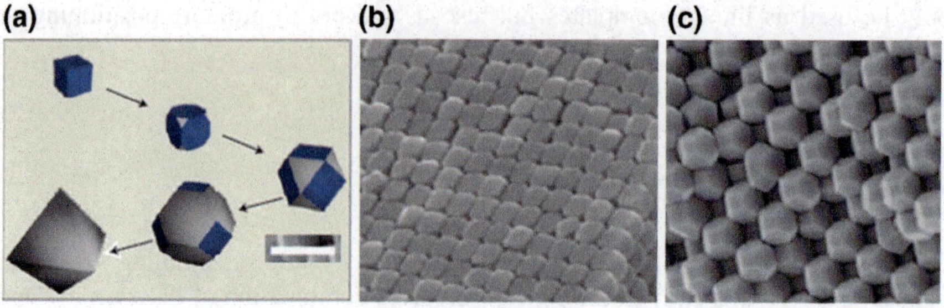

Fig. 10.21 Self-assembly of polyhedron lattices **a** Schematic representation of different shapes of Ag polyhedral. SEM of different colloidal lattices, **b** truncated cubes, **c** truncated octahedra. Scale bar in (**a**) is 500 nm and used for all the SEM micrographs. Reprinted with permission from Ref. [144]. Copyright 2012, Nature Publications Group

the polymer layer competes with the van der Waals attraction from the Ag core. The solvent used in this case is DMF which has lower van der Waals energies as compared to ethanol and water. Thus a disordered assembly is observed in water and ethanol. This method can be used to design large scalable 3D plasmonic materials that find applications in various fields.

10.5.7 Pressure Induced Growth

Oriented growth is an effective route to design nanostructures in both 2D as well as 3D [145]. But a major limitation with this method is that the oriented growth is only for nanoparticles (<10 nm). Wong et al. have used high pressure synthesis of 200 nm SnO_2 particles and constructed its 3D oriented assembly. When the general solvothermal method was employed, SnO_2 dodecahedra crystals were obtained that composed of growth along [110] and [111] facets. Just by tuning the concentration of TMAH (tetramethyl ammonium hydroxide) used in the synthesis, the microstructure ranged from dodecahedra to octahedral to egg like structure. This morphological evolution only implies that the growth is stabilized along certain facet due to its favorable adsorption energy. In their work, another parameter that contributes towards the assembly is the pressure term. As can be seen in Fig. 10.22, different structures can be stabilized as a function of varying pressure and temperature. Region 1 corresponds to low temperature and pressure and stabilizes SnO_2 in octahedral structure. In region 2, the transition from octahedron to dodecahedron is obtained along with the formation of a 2D superstructure while at a particular pressure and temperature 3D ordered superstructures were obtained. Oriented growth requires the combination of long range forces that bring crystallites into close proximity and short range forces that allow the rearrangement. The crystallites are able to approach each other due to the high pressure supplied.

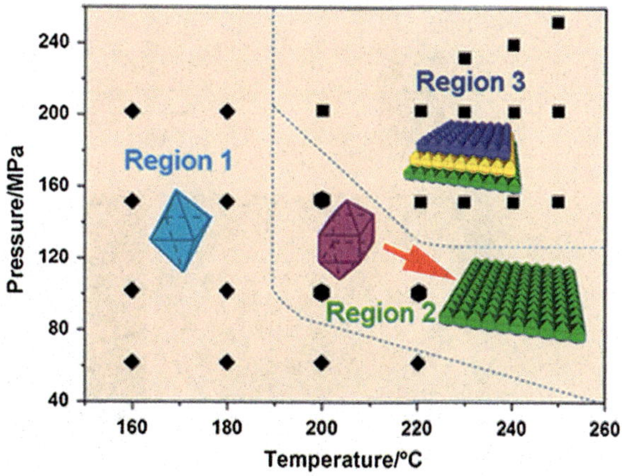

Fig. 10.22 P versus T diagram explaining the different structures of SnO_2 and their superstructure formation. Reprinted with permission from Ref. [145]. Copyright 2016, The American Chemical Society

10.5.8 Light-Induced Assembly

Light-induced self-assembly (LISA) is seen at colloidal and macromolecular scales but not very common at the nanoscale. Klajn and coworkers [146] have overcome this limitation wherein light induced 3D suprastructures are created by photoisomerization of dithiol molecules that are bound onto metal nanoparticles surface. In their work, Au nanoparticles are stabilized by a capping agent and to this; the photoactive ligand (azobenzene, ADT) is added. The isomerization of the ADT attached to the nanoparticle is significant for self-assembly when >75% of the nanoparticle surface was covered, no aggregation was observed even upon prolonged UV irradiation. Alternatively, when no ADT was adsorbed on the particle surface, the trans-cis isomerization occurred readily. To elucidate the role of the photoisomer and the cross-linker, model compounds were considered. The light-induced assembly is not only brought about by dipole-dipole interactions, but it is crucial to have cross-linking of the nanoparticles so as to overcome the thermal fluctuations and form crystalline phases. The number of azobenzene ligands per nanoparticle is vital as this will define the nature of the crystals formed. When the concentration of the ligands was low as seen in Fig. 10.23a, the crystals would be stable until being irradiated with UV light and free nanoparticles were beginning to be observed when exposed to visible light. This process was reversible and could be repeated a multiple number of times. Sparse ADT cross-links and light induced dipoles were weaker when the cis-trans isomerization occurred on being irradiated with UV light. Thus the thermal fluctuations were sufficient to cause the disintegration of the crystals. Now when the amount of ADT cross-links increased as seen in Fig. 10.23b, permanent crystals formed were stable both to light and heat. Thus light can be used as a guide to construct larger architecture using nanoscopic components that show controllable stability and tunable mechanical properties.

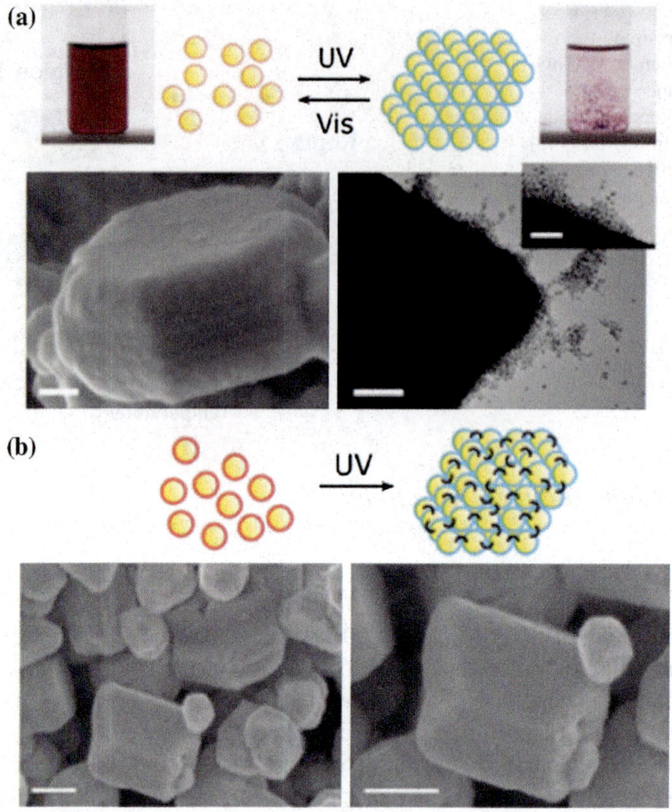

Fig. 10.23 **a** Nanoparticles that have a low concentration of ADT ligands (*red* solution) on being exposed to UV light result in the formation of faceted crystal as shown in the SEM image on *left*. The reversible process is seen on being exposed to visible light, results in clear solution with few dispersed particles. The corresponding TEM image is seen on the *right*. (scale bar: 100 nm in the main images and 50 nm in the *inset*. **b** Higher concentration of ADT ligands, irreversible permanent crystals are formed after being irradiated. SEM images show crystals of various morphologies (scale bars: 200 nm). Reprinted with permission from Ref. [146]. Copyright 2007, National Academy of Sciences, U.S.A.

10.6 Conclusions and Perspectives

The assembly of nanoparticles into highly organized hierarchical structures emerged in the past few years as a viable approach for the bottom-up fabrication of novel functional materials with controlled morphologies and tailored properties. This is not only because solution-based techniques are low cost and overcome the challenges associated to lithographic techniques, but also because metamaterials obtained via self-assembly techniques exhibit novel coupling properties from the collective interactions between the nanoparticles which can be exploited in optoelectonics, plasmonics, high areal density data storage, catalysis and the medical field. Such a progress in material fabrication by soft-solution routes can be hardly ignored since these structures have been recently proposed for the design of all-nanocrystal electronic devices for energy storage/conversion devices and flexible electronics. However, before such self-assembled materials can be fully incorporated into functional devices, several major obstacles should be overcome. These include the complete understanding of the types of forces that govern the interactions between the particles at nanoscale, the development of novel synthetic routes allowing the large scale synthesis of monodisperse nanocrystals with tunable internal structure, size, shape and the control of the surface composition of the nanocrystals to ensure a strong coupling between the constituting nanoparticles. If these critical aspects will be addressed, metamaterials obtained by the self-assembly of uniform colloidal nanocrystals will play a pivotal role in the development of cutting-edge technologies in the years to come.

Bibliography

1. Walter, P., E. Welcomme, P. Hallégot, N.J. Zaluzec, C. Deeb, J. Castaing, P. Veyssière, R. Bréniaux, J.-L. Lévêque, and G. Tsoucaris. 2006. Early Use of PbS Nanotechnology for an Ancient Hair Dyeing Formula. *Nano Letters* 6 (10): 2215–2219.
2. Reibold, M., N. Pätzke, A.A. Levin, W. Kochmann, I.P. Shakhverdova, P. Paufler, and D.C. Meyer. 2009. Structure of Several Historic Blades at Nanoscale. *Crystal Research and Technology* 44 (10): 1139–1146.
3. Brun, N., L. Mazerolles, and M. Pernot. 1991. Microstructure of Opaque Red Glass Containing Copper. *Journal of Materials Science Letters* 10 (23): 1418–1420.
4. Padeletti, G., and P. Fermo. 2003. How the Masters in Umbria, Italy, Generated and Used Nanoparticles in Art Fabrication During the Renaissance Period. *Applied Physics A* 76 (4): 515–525.
5. Davis, M.E., Z. Chen, and D.M. Shin. 2008. Nanoparticle Therapeutics: An Emerging Treatment Modality for Cancer. *Nature Reviews Drug Discovery* 7 (9): 771–782.
6. Daniela, C., C. Gabriel, and J.O.C. Charles. 2007. Magnetic Properties of Variable-Sized Fe_3O_4 Nanoparticles Synthesized from Non-Aqueous Homogeneous Solutions of Polyols. *Journal of Physics D: Applied Physics* 40 (19): 5801.
7. Adireddy, S., C. Lin, B. Cao, W. Zhou, and G. Caruntu. 2010. Solution-Based Growth of Monodisperse Cube-Like $BaTiO_3$ Colloidal Nanocrystals. *Chemistry of Materials* 22 (6): 1946–1948.
8. Park, J., J. Joo, S.G. Kwon, Y. Jang, and T. Hyeon. 2007. Synthesis of Monodisperse Spherical Nanocrystals. *Angewandte Chemie International Edition* 46 (25): 4630–4660.
9. Mourdikoudis, S., and L.M. Liz-Marzán. 2013. Oleylamine in Nanoparticle Synthesis. *Chemistry of Materials* 25 (9): 1465–1476.
10. Quan, Z., and J. Fang. 2010. Superlattices with Non-Spherical Building Blocks. *Nano Today* 5 (5): 390–411.
11. Pileni, M.P. 2011. Supra- and Nanocrystallinities: A New Scientific Adventure. *Journal of Physics: Condensed Matter* 23 (50): 503102.
12. Zhuang, J., H. Wu, Y. Yang, and Y.C. Cao. 2008. Controlling Colloidal Superparticle Growth Through Solvophobic Interactions. *Angewandte Chemie International Edition* 47 (12): 2208–2212.

13. Caruntu, D., T. Rostamzadeh, T. Costanzo, S. Salemizadeh Parizi, and G. Caruntu. Solvothermal Synthesis and Controlled Self-Assembly of Monodisperse Titanium-Based Perovskite Colloidal Nanocrystals. *Nanoscale* 7 (30): 12955–12969.
14. Ung, D., L.D. Tung, G. Caruntu, D. Delaportas, I. Alexandrou, I.A. Prior, and N.T.K. Thanh. 2009. Variant Shape Growth of Nanoparticles of Metallic Fe-Pt. Fe-Pd and Fe-Pt-Pd alloys. *CrystEngComm* 11 (7): 1309–1316.
15. Gupta, M.K., T. König, R. Near, D. Nepal, L.F. Drummy, S. Biswas, S. Naik, R.A. Vaia, M. A. El-Sayed, and V.V. Tsukruk. 2013. Surface Assembly and Plasmonic Properties in Strongly Coupled Segmented Gold Nanorods. *Small* 9 (17): 2979–2990.
16. Wiley, B.J., Y. Chen, J.M. McLellan, Y. Xiong, Z.-Y. Li, D. Ginger, and Y. Xia. 2007. Synthesis and Optical Properties of Silver Nanobars and Nanorice. *Nano Letters* 7 (4): 1032–1036.
17. Klajn, R., A.O. Pinchuk, G.C. Schatz, and B.A. Grzybowski. 2007. Synthesis of Heterodimeric Sphere-Prism Nanostructures Via Metastable Gold Supraspheres. *Angewandte Chemie International Edition* 46 (44): 8363–8367.
18. Manna, L., E.C. Scher, and A.P. Alivisatos. 2000. Synthesis of Soluble and Processable Rod-, Arrow-, Teardrop-, and Tetrapod-Shaped CdSe Nanocrystals. *Journal of the American Chemical Society* 122 (51): 12700–12706.
19. Daniel, M.-C., and D. Astruc. 2004. Gold Nanoparticles: Assembly, Supramolecular Chemistry, Quantum-Size-Related Properties, and Applications Toward Biology, Catalysis, and Nanotechnology. *Chemical Reviews* 104 (1): 293–346.
20. Chen, J., B. Wiley, Z.Y. Li, D. Campbell, F. Saeki, H. Cang, L. Au, J. Lee, X. Li, and Y. Xia. 2005. Gold Nanocages: Engineering Their Structure for Biomedical Applications. *Advanced Materials* 17 (18): 2255–2261.
21. Wei, Y., R. Klajn, A.O. Pinchuk, and B.A. Grzybowski. 2008. Synthesis, Shape Control, and Optical Properties of Hybrid Au/Fe_3O_4 "Nanoflowers". *Small* 4 (10): 1635–1639.
22. Gupta, M.K., D.D. Kulkarni, R. Geryak, S. Naik, and V.V. Tsukruk. 2013. A Robust and Facile Approach to Assembling Mobile and Highly-Open Unfrustrated Triangular Lattices from Ferromagnetic Nanorods. *Nano Letters* 13 (1): 36–42.
23. Trindade, T., P. O'Brien, and N.L. Pickett. 2001. Nanocrystalline Semiconductors: Synthesis, Properties, and Perspectives. *Chemistry of Materials* 13 (11): 3843–3858.
24. Lee, K.H., J. Kao, S.S. Parizi, G. Caruntu, and T. Xu. 2014. Dielectric Properties of Barium Titanate Supramolecular Nanocomposites. *Nanoscale* 6 (7): 3526–3531.
25. Martin, C.R. 1994. Nanomaterials: A Membrane-Based Synthetic Approach. *Science* 266 (5193): 1961–1966.
26. Luo, D., C. Yan, and T. Wang. 2015. Interparticle Forces Underlying Nanoparticle Self-Assemblies. *Small* 11 (45): 5984–6008.
27. Perepichka, D.F., and F. Rosei. 2007. Metal Nanoparticles: From "Artificial Atoms" to "Artificial Molecules". *Angewandte Chemie International Edition* 46 (32): 6006–6008.
28. Henderson, D., D.-M. Duh, X. Chu, and D. Wasan. 1997. An Expression for the Dispersion Force between Colloidal Particles. *Journal of Colloid and Interface Science* 185 (1): 265–268.
29. Swope, W.C., H.C. Andersen, P.H. Berens, and K.R. Wilson. 1982. A computer Simulation Method for the Calculation of Equilibrium Constants for the Formation of Physical Clusters of Molecules: Application to Small Water Clusters. *The Journal of Chemical Physics* 76 (1): 637–649.
30. Wang, C., C. Siu, J. Zhang, and J. Fang. 2015. Understanding the Forces Acting in Self-Assembly and the Implications for Constructing Three-Dimensional (3D) Supercrystals. *Nano Research* 8 (8): 2445–2466.
31. Bishop, K.J.M., C.E. Wilmer, S. Soh, and B.A. Grzybowski. 2009. Nanoscale Forces and Their Uses in Self-Assembly. *Small* 5 (14): 1600–1630.
32. Harfenist, S.A., Z.L. Wang, M.M. Alvarez, I. Vezmar, and R.L. Whetten. 1996. Highly Oriented Molecular Ag Nanocrystal Arrays. *The Journal of Physical Chemistry* 100 (33): 13904–13910.

33. Leidinger, P., R. Popescu, D. Gerthsen, and C. Feldmann. 2013. Nanoscale Ag2S Hollow Spheres and Ag2S Nanodiscs Assembled to Three-Dimensional Nanoparticle Superlattices. *Chemistry of Materials* 25 (21): 4173–4180.
34. Gu, Z., Y. Chen, and D.H. Gracias. 2004. Surface Tension Driven Self-Assembly of Bundles and Networks of 200 nm Diameter Rods Using a Polymerizable Adhesive. *Langmuir* 20 (26): 11308–11311.
35. Xu, L., W. Ma, L. Wang, C. Xu, H. Kuang, and N.A. Kotov. 2013. Nanoparticle Assemblies: Dimensional Transformation Of Nanomaterials and Scalability. *Chemical Society Reviews* 42 (7): 3114–3126.
36. Bodnarchuk, M.I., M.V. Kovalenko, W. Heiss, and D.V. Talapin. 2010. Energetic and Entropic Contributions to Self-Assembly of Binary Nanocrystal Superlattices: Temperature as the Structure-Directing Factor. *Journal of the American Chemical Society* 132 (34): 11967–11977.
37. Quan, Z., H. Xu, C. Wang, X. Wen, Y. Wang, J. Zhu, R. Li, C.J. Sheehan, Z. Wang, D.-M. Smilgies, Z. Luo, and J. Fang. 2014. Solvent-Mediated Self-Assembly of Nanocube Superlattices. *Journal of the American Chemical Society* 136 (4): 1352–1359.
38. Hong, L., A. Cacciuto, E. Luijten, and S. Granick. 2006. Clusters of Charged Janus Spheres. *Nano Letters* 6 (11): 2510–2514.
39. Caruntu, D., B.L. Cushing, G. Caruntu, and C.J. O'Connor. 2005. Attachment of Gold Nanograins onto Colloidal Magnetite Nanocrystals. *Chemistry of Materials* 17 (13): 3398–3402.
40. Zhang, S., Y. Shao, G. Yin, and Y. Lin. 2010. Electrostatic Self-Assembly of a Pt-around-Au Nanocomposite with High Activity towards Formic Acid Oxidation. *Angewandte Chemie International Edition* 49 (12): 2211–2214.
41. Kalsin, A.M., M. Fialkowski, M. Paszewski, S.K. Smoukov, K.J.M. Bishop, and B.A. Grzybowski. 2006. Electrostatic Self-Assembly of Binary Nanoparticle Crystals with a Diamond-Like Lattice. *Science* 312 (5772): 420–424.
42. Tohver, V., J.E. Smay, A. Braem, P.V. Braun, and J.A. Lewis. 2001. Nanoparticle Halos: A New Colloid Stabilization Mechanism. *Proceedings of the National Academy of Sciences* 98 (16): 8950–8954.
43. Singh, G., H. Chan, A. Baskin, E. Gelman, N. Repnin, P. Král, and R. Klajn. 2014. Self-Assembly of Magnetite Nanocubes into Helical Superstructures. *Science* 345 (6201): 1149–1153.
44. Dong, L., Y. Liu, Y. Lu, L. Zhang, N. Man, L. Cao, K. Ma, D. An, J. Lin, Y.-J. Xu, W.-P. Xu, W.-B. Wu, S.-H. Yu, and L.-P. Wen. 2013. Tuning Magnetic Property and Autophagic Response for Self-Assembled Ni–Co Alloy Nanocrystals. *Advanced Functional Materials* 23 (47): 5930–5940.
45. Mehdizadeh Taheri, S., M. Michaelis, T. Friedrich, B. Förster, M. Drechsler, F.M. Römer, P. Bösecke, T. Narayanan, B. Weber, I. Rehberg, S. Rosenfeldt, and S. Förster. 2015. Self-Assembly of Smallest Magnetic Particles. *Proceedings of the National Academy of Sciences* 112 (47): 14484–14489.
46. Singh, G., H. Chan, T. Udayabhaskararao, E. Gelman, D. Peddis, A. Baskin, G. Leitus, P. Kral, and R. Klajn. 2015. Magnetic field-Induced Self-Assembly of Iron Oxide Nanocubes. *Faraday Discussions* 181: 403–421.
47. Zhong, S.-L., J.-M. Song, S. Zhang, H. Yao, A.-W. Xu, W.-T. Yao, and S.-H. Yu. 2008. Template-Free Hydrothermal Synthesis and Formation Mechanism of Hematite Microrings. *The Journal of Physical Chemistry C* 112 (50): 19916–19921.
48. Pucci, A., M.-G. Willinger, F. Liu, X. Zeng, V. Rebuttini, G. Clavel, X. Bai, G. Ungar, and N. Pinna. 2012. One-Step Synthesis and Self-Assembly of Metal Oxide Nanoparticles into 3D Superlattices. *ACS Nano* 6 (5): 4382–4391.
49. Xiaoge, H., C. Wenlong, W. Tie, W. Erkang, and D. Shaojun. 2005. Well-Ordered End-to-End Linkage of Gold Nanorods. *Nanotechnology* 16 (10): 2164.

50. Mirkin, C.A., R.L. Letsinger, R.C. Mucic, and J.J. Storhoff. 1996. A DNA-Based Method for Rationally Assembling Nanoparticles into Macroscopic Materials. *Nature* 382 (6592): 607–609.
51. Steiner, T. 2002. The Hydrogen Bond in the Solid State. *Angewandte Chemie International Edition* 41 (1): 48–76.
52. Harada, T., and T.A. Hatton. 2009. Formation of Highly Ordered Rectangular Nanoparticle Superlattices by the Cooperative Self-Assembly of Nanoparticles and Fatty Molecules. *Langmuir* 25 (11): 6407–6412.
53. Smith, A.M., and S. Nie. 2010. Semiconductor Nanocrystals: Structure, Properties, and Band Gap Engineering. *Accounts of Chemical Research* 43 (2): 190–200.
54. Govorov, A.O., G.W. Bryant, W. Zhang, T. Skeini, J. Lee, N.A. Kotov, J.M. Slocik, and R. R. Naik. 2006. Exciton–Plasmon Interaction and Hybrid Excitons in Semiconductor–Metal Nanoparticle Assemblies. *Nano Letters* 6 (5): 984–994.
55. Cheng, W., N. Park, M.T. Walter, M.R. Hartman, and D. Luo. 2008. Nanopatterning Self-Assembled Nanoparticle Superlattices by Moulding Microdroplets. *Nature Nanotechnology* 3 (11): 682–690.
56. Courty, A., A. Mermet, P.A. Albouy, E. Duval, and M.P. Pileni. 2005. Vibrational Coherence of Self-Organized Silver Nanocrystals in f.c.c. Supra-Crystals. *Nature Materials* 4 (5): 395–398.
57. Chen, Y., Z. Ouyang, M. Gu, and W. Cheng. 2013. Mechanically Strong, Optically Transparent, Giant Metal Superlattice Nanomembranes From Ultrathin Gold Nanowires. *Advanced Materials* 25 (1): 80–85.
58. Cheng, W., M.J. Campolongo, J.J. Cha, S.J. Tan, C.C. Umbach, D.A. Muller, and D. Luo. 2009. Free-Standing Nanoparticle Superlattice Sheets Controlled by DNA. *Nature Materials* 8 (6): 519–525.
59. Lee, C., X. Wei, J.W. Kysar, and J. Hone. 2008. Measurement of the Elastic Properties and Intrinsic Strength of Monolayer Graphene. *Science* 321 (5887): 385–388.
60. Hu, L., H.S. Kim, J.-Y. Lee, P. Peumans, and Y. Cui. 2010. Scalable Coating and Properties of Transparent, Flexible, Silver Nanowire Electrodes. *ACS Nano* 4 (5): 2955–2963.
61. Carbone, L., C. Nobile, M. De Giorgi, F.D. Sala, G. Morello, P. Pompa, M. Hytch, E. Snoeck, A. Fiore, I.R. Franchini, M. Nadasan, A.F. Silvestre, L. Chiodo, S. Kudera, R. Cingolani, R. Krahne, and L. Manna. 2007. Synthesis and Micrometer-Scale Assembly of Colloidal CdSe/CdS Nanorods Prepared by a Seeded Growth Approach. *Nano Letters* 7 (10): 2942–2950.
62. Xing, Y., H. Zhang, S. Song, J. Feng, Y. Lei, L. Zhao, and M. Li. 2008. Hydrothermal synthesis and photoluminescent properties of stacked indium sulfide superstructures. *Chemical Communications* 12: 1476–1478.
63. Pol, V.G., and J.M. Calderon-Moreno. 2010. Fabrication of Luminescent Eu_2O_3 Superstructures. *The Journal of Physical Chemistry Letters* 1 (1): 319–322.
64. Stakheev, A.Y., and L.M. Kustov. 1999. Effects of the Support on the Morphology and Electronic Properties of Supported Metal Clusters: Modern Concepts and Progress in 1990s. *Applied Catalysis, A: General* 188 (1–2): 3–35.
65. Zhou, Z., S. Kooi, M. Flytzani-Stephanopoulos, and H. Saltsburg. 2008. The Role of the Interface in CO Oxidation on Au/CeO_2 Multilayer Nanotowers. *Advanced Functional Materials* 18 (18): 2801–2807.
66. Yamada, Y., C.-K. Tsung, W. Huang, Z. Huo, S.E. Habas, T. Soejima, C.E. Aliaga, G.A. Somorjai, and P. Yang. 2011. Nanocrystal Bilayer for Tandem Catalysis. *Nature Chemistry* 3 (5): 372–376.
67. Shevchenko, E.V., D.V. Talapin, N.A. Kotov, S. O'Brien, and C.B. Murray. 2006. Structural Diversity in Binary Nanoparticle Superlattices. *Nature* 439 (7072): 55–59.
68. Tsao, Y.-C., S. Rej, C.-Y. Chiu, and M.H. Huang. 2014. Aqueous Phase Synthesis of Au–Ag Core-Shell Nanocrystals with Tunable Shapes and Their Optical and Catalytic Properties. *Journal of the American Chemical Society* 136 (1): 396–404.

69. Armaroli, N., and V. Balzani. 2007. The Future of Energy Supply: Challenges and Opportunities. *Angewandte Chemie International Edition* 46 (1–2): 52–66.
70. Gélin, P., and M. Primet. 2002. Complete Oxidation of Methane at Low Temperature Over Noble Metal Based Catalysts: A Review. *Applied Catalysis, B: Environmental* 39 (1): 1–37.
71. Mayernick, A.D., and M.J. Janik. 2008. Methane Activation and Oxygen Vacancy Formation over CeO_2 and Zr, Pd Substituted CeO_2 Surfaces. *The Journal of Physical Chemistry C* 112 (38): 14955–14964.
72. Colussi, S., A. Gayen, M. Farnesi-Camellone, M. Boaro, J. Llorca, S. Fabris, and A. Trovarelli. Nanofaceted Pd-O Sites in Pd-Ce Surface Superstructures: Enhanced Activity in Catalytic Combustion of Methane. *Angewandte Chemie International Edition* 48 (45): 8481–8484.
73. Yin, J., Q. Lu, Z. Yu, J. Wang, H. Pang, and F. Gao. 2010. Hierarchical ZnO Nanorod-Assembled Hollow Superstructures for Catalytic and Photoluminescence Applications. *Crystal Growth & Design* 10 (1): 40–43.
74. Urban, J.J., D.V. Talapin, E.V. Shevchenko, C.R. Kagan, and C.B. Murray. 2007. Synergism in Binary Nanocrystal Superlattices Leads to Enhanced p-Type Conductivity in Self-Assembled PbTe/Ag2Te Thin Films. *Nature Materials* 6 (2): 115–121.
75. Shevchenko, E.V., M. Ringler, A. Schwemer, D.V. Talapin, T.A. Klar, A.L. Rogach, J. Feldmann, and A.P. Alivisatos. 2008. Self-Assembled Binary Superlattices of CdSe and Au Nanocrystals and Their Fluorescence Properties. *Journal of the American Chemical Society* 130 (11): 3274–3275.
76. Gordon, T.R., T. Paik, D.R. Klein, G.V. Naik, H. Caglayan, A. Boltasseva, and C.B. Murray. 2013. Shape-Dependent Plasmonic Response and Directed Self-Assembly in a New Semiconductor Building Block, Indium-Doped Cadmium Oxide (ICO). *Nano Letters* 13 (6): 2857–2863.
77. Tao, A.R., D.P. Ceperley, P. Sinsermsuksakul, A.R. Neureuther, and P. Yang. 2008. Self-Organized Silver Nanoparticles for Three-Dimensional Plasmonic Crystals. *Nano Letters* 8 (11): 4033–4038.
78. Keating, C.D., K.M. Kovaleski, and M.J. Natan. 1998. Protein: Colloid Conjugates for Surface Enhanced Raman Scattering: Stability and Control of Protein Orientation. *The Journal of Physical Chemistry B* 102 (47): 9404–9413.
79. Rodríguez-Lorenzo, L., R.A. Álvarez-Puebla, I. Pastoriza-Santos, S. Mazzucco, O. Stéphan, M. Kociak, L.M. Liz-Marzán, and F.J. García de Abajo. 2009. Zeptomol Detection Through Controlled Ultrasensitive Surface-Enhanced Raman Scattering. *Journal of the American Chemical Society* 131 (13): 4616–4618.
80. Aizpurua, J., G.W. Bryant, L.J. Richter, F.J. García de Abajo, B.K. Kelley, and T. Mallouk. 2005. Optical Properties Of Coupled Metallic Nanorods for Field-Enhanced Spectroscopy. *Physical Review B* 71 (23): 235420.
81. Alvarez-Puebla, R.A., A. Agarwal, P. Manna, B.P. Khanal, P. Aldeanueva-Potel, E. Carbó-Argibay, N. Pazos-Pérez, L. Vigderman, E.R. Zubarev, N.A. Kotov, and L.M. Liz-Marzán. 2011. Gold Nanorods 3D-Supercrystals as Surface Enhanced Raman Scattering Spectroscopy Substrates for the Rapid Detection of Scrambled Prions. *Proceedings of the National Academy of Sciences* 108 (20): 8157–8161.
82. Zhang, Q., Y.H. Lee, I.Y. Phang, C.K. Lee, and X.Y. Ling. 2014. Hierarchical 3D SERS Substrates Fabricated by Integrating Photolithographic Microstructures and Self-Assembly of Silver Nanoparticles. *Small* 10 (13): 2703–2711.
83. Peng, B., G. Li, D. Li, S. Dodson, Q. Zhang, J. Zhang, Y.H. Lee, H.V. Demir, X. Yi Ling, and Q. Xiong. 2013. Vertically Aligned Gold Nanorod Monolayer on Arbitrary Substrates: Self-Assembly and Femtomolar Detection of Food Contaminants. *ACS Nano* 7 (7): 5993–6000.
84. Burda, C., X. Chen, R. Narayanan, and M.A. El-Sayed. 2005. Chemistry and Properties of Nanocrystals of Different Shapes. *Chemical Reviews* 105 (4): 1025–1102.

85. Hamon, C., S.M. Novikov, L. Scarabelli, D.M. Solís, T. Altantzis, S. Bals, J.M. Taboada, F. Obelleiro, and L.M. Liz-Marzán. 2015. Collective Plasmonic Properties in Few-Layer Gold Nanorod Supercrystals. *ACS Photonics* 2 (10): 1482–1488.
86. Li, F., D.P. Josephson, and A. Stein. 2011. Colloidal Assembly: The Road from Particles to Colloidal Molecules and Crystals. *Angewandte Chemie International Edition* 50 (2): 360–388.
87. Murray, C.B., C.R. Kagan, and M.G. Bawendi. 2000. Synthesis and Characterization of Monodisperse Nanocrystals and Close-Packed Nanocrystal Assemblies. *Annual Review of Materials Science* 30 (1): 545–610.
88. Whitesides, G.M., and B. Grzybowski. 2002. Self-Assembly at All Scales. *Science* 295 (5564): 2418–2421.
89. Zhang, J., C. Rowland, Y. Liu, H. Xiong, S. Kwon, E. Shevchenko, R.D. Schaller, V.B. Prakapenka, S. Tkachev, and T. Rajh. 2015. Evolution of Self-Assembled ZnTe Magic-Sized Nanoclusters. *Journal of the American Chemical Society* 137 (2): 742–749.
90. Bendova, M., M. Puchberger, and U. Schubert. 2010. Characterization of "Cd10S4(SPh)12", the Thermal Decomposition Product of $(NMe_4)_4[Cd_{10}S_4(SPh)_{16}]$: Synthesis of a Neutral Cd_{54} Sulfide Cluster and of a Polymeric Chain of Thiolate-Bridged Cd_{17} Sulfide Clusters. *European Journal of Inorganic Chemistry* 2010 (21): 3299–3306.
91. Lisiecki, I., P.A. Albouy, and M.P. Pileni. 2003. Face-Centered-Cubic "Supracrystals" of Cobalt Nanocrystals. *Advanced Materials* 15 (9): 712–716.
92. Stoeva, S.I., B.L.V. Prasad, S. Uma, P.K. Stoimenov, V. Zaikovski, C.M. Sorensen, and K. J. Klabunde. 2003. Face-Centered Cubic and Hexagonal Closed-Packed Nanocrystal Superlattices of Gold Nanoparticles Prepared by Different Methods. *The Journal of Physical Chemistry B* 107 (30): 7441–7448.
93. Shevchenko, E.V., D.V. Talapin, S. O'Brien, and C.B. Murray. 2005. Polymorphism in AB13 Nanoparticle Superlattices: An Example of Semiconductor—Metal Metamaterials. *Journal of the American Chemical Society* 127 (24): 8741–8747.
94. Shevchenko, E.V., D.V. Talapin, C.B. Murray, and S. O'Brien. 2006. Structural Characterization of Self-Assembled Multifunctional Binary Nanoparticle Superlattices. *Journal of the American Chemical Society* 128 (11): 3620–3637.
95. Srivastava, S., and N.A. Kotov. 2009. Nanoparticle Assembly for 1D and 2D Ordered Structures. *Soft Matter* 5 (6): 1146–1156.
96. Yi, L., A. Tang, M. Niu, W. Han, Y. Hou, and M. Gao. 2010. Synthesis and Self-Assembly of Cu1.94S-ZnS Heterostructured Nanorods. *CrystEngComm* 12 (12): 4124–4130.
97. Lin, S., M. Li, E. Dujardin, C. Girard, and S. Mann. 2005. One-Dimensional Plasmon Coupling by Facile Self-Assembly of Gold Nanoparticles into Branched Chain Networks. *Advanced Materials* 17 (21): 2553–2559.
98. Lee, N., D. Yoo, D. Ling, M.H. Cho, T. Hyeon, and J. Cheon. 2015. Iron Oxide Based Nanoparticles for Multimodal Imaging and Magnetoresponsive Therapy. *Chemical Reviews* 115 (19): 10637–10689.
99. Chen, Z., J. Moore, G. Radtke, H. Sirringhaus, and S. O'Brien. 2007. Binary Nanoparticle Superlattices in the Semiconductor–Semiconductor System: CdTe and CdSe. *Journal of the American Chemical Society* 129 (50): 15702–15709.
100. Chen, J., X. Ye, and C.B. Murray. 2010. Systematic Electron Crystallographic Studies of Self-Assembled Binary Nanocrystal Superlattices. *ACS Nano* 4 (4): 2374–2381.
101. Evers, W.H., H. Friedrich, L. Filion, M. Dijkstra, and D. Vanmaekelbergh. 2009. Observation of a Ternary Nanocrystal Superlattice and Its Structural Characterization by Electron Tomography. *Angewandte Chemie International Edition* 48 (51): 9655–9657.
102. Dong, A., X. Ye, J. Chen, and C.B. Murray. 2011. Two-Dimensional Binary and Ternary Nanocrystal Superlattices: The Case of Monolayers and Bilayers. *Nano Letters* 11 (4): 1804–1809.
103. López-Ortega, A., E. Lottini, C.D.J. Fernández, and C. Sangregorio. 2015. Exploring the Magnetic Properties of Cobalt-Ferrite Nanoparticles for the Development of a Rare-Earth-Free Permanent Magnet. *Chemistry of Materials* 27 (11): 4048–4056.

104. Ren, J., and R.D. Tilley. 2007. Preparation, Self-Assembly, and Mechanistic Study of Highly Monodispersed Nanocubes. *Journal of the American Chemical Society* 129 (11): 3287–3291.
105. Ni, R., A.P. Gantapara, J. de Graaf, R. van Roij, and M. Dijkstra. 2012. Phase Diagram of Colloidal Hard Superballs: From Cubes Via Spheres to Octahedra. *Soft Matter* 8 (34): 8826–8834.
106. Zhang, Y., F. Lu, D. van der Lelie, and O. Gang. 2011. Continuous Phase Transformation in Nanocube Assemblies. *Physical Review Letters* 107 (13): 135701.
107. Yang, H.-J., S.-Y. He, H.-L. Chen, and H.-Y. Tuan. 2014. Monodisperse Copper Nanocubes: Synthesis, Self-Assembly, and Large-Area Dense-Packed Films. *Chemistry of Materials* 26 (5): 1785–1793.
108. Niu, W., L. Zhang, and G. Xu. 2010. Shape-Controlled Synthesis of Single-Crystalline Palladium Nanocrystals. *ACS Nano* 4 (4): 1987–1996.
109. Zhang, J., H. Yang, J. Fang, and S. Zou. 2010. Synthesis and Oxygen Reduction Activity of Shape-Controlled Pt_3Ni Nanopolyhedra. *Nano Letters* 10 (2): 638–644.
110. Lu, W., Q. Liu, Z. Sun, J. He, C. Ezeolu, and J. Fang. 2008. Super Crystal Structures of Octahedral c-In_2O_3 Nanocrystals. *Journal of the American Chemical Society* 130 (22): 6983–6991.
111. Huang, T., Q. Zhao, J. Xiao, and L. Qi. 2010. Controllable Self-Assembly of PbS Nanostars into Ordered Structures: Close-Packed Arrays and Patterned Arrays. *ACS Nano* 4 (8): 4707–4716.
112. Bouet, C., M.D. Tessier, S. Ithurria, B. Mahler, B. Nadal, and B. Dubertret. 2013. Flat Colloidal Semiconductor Nanoplatelets. *Chemistry of Materials* 25 (8): 1262–1271.
113. Park, K.H., K. Jang, and S.U. Son. 2006. Synthesis, Optical Properties, and Self-Assembly of Ultrathin Hexagonal In2S3 Nanoplates. *Angewandte Chemie International Edition* 45 (28): 4608–4612.
114. Vaughn, D.D., S.-I. In, and R.E. Schaak. 2011. A Precursor-Limited Nanoparticle Coalescence Pathway for Tuning the Thickness of Laterally-Uniform Colloidal Nanosheets: The Case of SnSe. *ACS Nano* 5 (11): 8852–8860.
115. Li, X., H. Shen, J. Niu, S. Li, Y. Zhang, H. Wang, and L.S. Li. 2010. Columnar Self-assembly of Cu2S Hexagonal Nanoplates Induced by Tin(IV)–X Complex as Inorganic Surface Ligand. *Journal of the American Chemical Society* 132 (37): 12778–12779.
116. Ithurria, S., and B. Dubertret. 2008. Quasi 2D Colloidal CdSe Platelets with Thicknesses Controlled at the Atomic Level. *Journal of the American Chemical Society* 130 (49): 16504–16505.
117. Ye, X., J. Chen, M. Engel, J.A. Millan, W. Li, L. Qi, G. Xing, J.E. Collins, C.R. Kagan, J. Li, S.C. Glotzer, and C.B. Murray. 2013. Competition of Shape and Interaction Patchiness for Self-assembling Nanoplates. *Nature Chemistry* 5 (6): 466–473.
118. Saunders, A.E., A. Ghezelbash, D.-M. Smilgies, M.B. Sigman, and B.A. Korgel. 2006. Columnar Self-Assembly of Colloidal Nanodisks. *Nano Letters* 6 (12): 2959–2963.
119. Talapin, D.V., E.V. Shevchenko, C.B. Murray, A. Kornowski, S. Förster, and H. Weller. 2004. CdSe and CdSe/CdS Nanorod Solids. *Journal of the American Chemical Society* 126 (40): 12984–12988.
120. Diroll, B.T., N.J. Greybush, C.R. Kagan, and C.B. Murray. 2015. Smectic Nanorod Superlattices Assembled on Liquid Subphases: Structure, Orientation, Defects, and Optical Polarization. *Chemistry of Materials* 27 (8): 2998–3008.
121. Ariga, K., J.P. Hill, and Q. Ji. 2007. Layer-by-Layer Assembly as a Versatile Bottom-Up Nanofabrication Technique for Exploratory Research and Realistic Application. *Physical Chemistry Chemical Physics* 9 (19): 2319–2340.
122. Borges, J., and J.F. Mano. 2014. Molecular Interactions Driving the Layer-by-Layer Assembly of Multilayers. *Chemical Reviews* 114 (18): 8883–8942.
123. Love, J.C., L.A. Estroff, J.K. Kriebel, R.G. Nuzzo, and G.M. Whitesides. 2005. Self-Assembled Monolayers of Thiolates on Metals as a Form of Nanotechnology. *Chemical Reviews* 105 (4): 1103–1170.

124. Tao, A.R., J. Huang, and P. Yang. 2008. Langmuir—Blodgettry of Nanocrystals and Nanowires. *Accounts of Chemical Research* 41 (12): 1662–1673.
125. Acharya, S., J.P. Hill, and K. Ariga. 2009. Soft Langmuir-Blodgett Technique for Hard Nanomaterials. *Advanced Materials* 21 (29): 2959–2981.
126. Liu, J.-W., J.-H. Zhu, C.-L. Zhang, H.-W. Liang, and S.-H. Yu. 2010. Mesostructured Assemblies of Ultrathin Superlong Tellurium Nanowires and Their Photoconductivity. *Journal of the American Chemical Society* 132 (26): 8945–8952.
127. Moon, G.D., T.I. Lee, B. Kim, G. Chae, J. Kim, S. Kim, J.-M. Myoung, and U. Jeong. 2011. Assembled Monolayers of Hydrophilic Particles on Water Surfaces. *ACS Nano* 5 (11): 8600–8612.
128. Mahmoud, M.A. 2014. Controlling the Orientations of Gold Nanorods Inside Highly Packed 2D Arrays. *Physical Chemistry Chemical Physics* 16 (47): 26153–26162.
129. Mahmoud, M.A. 2015. Dynamic Template for Assembling Nanoparticles into Highly Ordered Two-Dimensional Arrays of Different Structures. *The Journal of Physical Chemistry C* 119 (1): 305–314.
130. Wang, Z., C. Schliehe, K. Bian, D. Dale, W.A. Bassett, T. Hanrath, C. Klinke, and H. Weller. 2013. Correlating Superlattice Polymorphs to Internanoparticle Distance, Packing Density, and Surface Lattice in Assemblies of PbS Nanoparticles. *Nano Letters* 13 (3): 1303–1311.
131. Choi, J.J., C.R. Bealing, K. Bian, K.J. Hughes, W. Zhang, D.-M. Smilgies, R.G. Hennig, J. R. Engstrom, and T. Hanrath. 2011. Controlling Nanocrystal Superlattice Symmetry and Shape-Anisotropic Interactions through Variable Ligand Surface Coverage. *Journal of the American Chemical Society* 133 (9): 3131–3138.
132. Gupta, S., Q. Zhang, T. Emrick, and T.P. Russell. 2006. "Self-Corralling" Nanorods Under an Applied Electric Field. *Nano Letters* 6 (9): 2066–2069.
133. Guerrero-Martínez, A., J. Pérez-Juste, E. Carbó-Argibay, G. Tardajos, and L.M. Liz-Marzán. 2009. Gemini-Surfactant-Directed Self-Assembly of Monodisperse Gold Nanorods into Standing Superlattices. *Angewandte Chemie International Edition* 48 (50): 9484–9488.
134. Zana, R., and Y. Talmon. 1993. Dependence of Aggregate Morphology on Structure of Dimeric Surfactants. *Nature* 362 (6417): 228–230.
135. Ming, T., X. Kou, H. Chen, T. Wang, H.-L. Tam, K.-W. Cheah, J.-Y. Chen, and J. Wang. 2008. Ordered Gold Nanostructure Assemblies Formed By Droplet Evaporation. *Angewandte Chemie International Edition* 47 (50): 9685–9690.
136. Courty, A., J. Richardi, P.-A. Albouy, and M.-P. Pileni. 2011. How To Control the Crystalline Structure of Supracrystals of 5-nm Silver Nanocrystals. *Chemistry of Materials* 23 (18): 4186–4192.
137. Xiao, J., Z. Li, X. Ye, Y. Ma, and L. Qi. 2014. Self-Assembly of Gold Nanorods into Vertically Aligned, Rectangular Microplates with a Supercrystalline Structure. *Nanoscale* 6 (2): 996–1004.
138. Alivisatos, A.P., K.P. Johnsson, X. Peng, T.E. Wilson, C.J. Loweth, M.P. Bruchez, and P.G. Schultz. 1996. Organization of 'Nanocrystal Molecules' Using DNA. *Nature* 382 (6592): 609–611.
139. Park, S.Y., A.K.R. Lytton-Jean, B. Lee, S. Weigand, G.C. Schatz, and C.A. Mirkin. 2008. DNA-Programmable Nanoparticle Crystallization. *Nature* 451 (7178): 553–556.
140. Auyeung, E., T.I.N.G. Li, A.J. Senesi, A.L. Schmucker, B.C. Pals, M.O. de la Cruz, and C. A. Mirkin. 2014. DNA-Mediated Nanoparticle Crystallization into Wulff Polyhedra. *Nature* 505 (7481): 73–77.
141. Min, Y., M. Akbulut, K. Kristiansen, Y. Golan, and J. Israelachvili. 2008. The Role of Interparticle and External Forces in Nanoparticle Assembly. *Nature Materials* 7 (7): 527–538.
142. Tan, S.J., M.J. Campolongo, D. Luo, and W. Cheng. 2011. Building Plasmonic Nanostructures with DNA. *Nature Nanotechnology* 6 (5): 268–276.
143. Li, F., Y. Qian, and A. Stein. 2010. Template-Directed Synthesis and Organization of Shaped Oxide/Phosphate Nanoparticles. *Chemistry of Materials* 22 (10): 3226–3235.

144. Henzie, J., M. Grünwald, A. Widmer-Cooper, P.L. Geissler, and P. Yang. 2012. Self-assembly of Uniform Polyhedral Silver Nanocrystals into Densest Packings and Exotic Superlattices. *Nature Materials* 11 (2): 131–137.
145. Wang, J., G. Lian, H. Si, Q. Wang, D. Cui, and C.-P. Wong. 2016. Pressure-Induced Oriented Attachment Growth of Large-Size Crystals for Constructing 3D Ordered Superstructures. *ACS Nano* 10 (1): 405–412.
146. Klajn, R., K.J.M. Bishop, and B.A. Grzybowski. 2007. Light-Controlled Self-Assembly of Reversible and Irreversible Nanoparticle Suprastructures. *Proceedings of the National Academy of Sciences* 104 (25): 10305–10309.

Chapter 11
Nanostructured Catalysts for the Electrochemical Reduction of CO_2

Ming Ma and Wilson A. Smith

Abstract The electrochemical conversion of CO_2 into carbon-based fuels has attracted considerable attention as a promising strategy for closing the anthropogenic carbon cycle. A key challenge for achieving this goal is to develop selective, stable, and efficient electrocatalysts for the electrocatalytic reduction of CO_2. Nanostructured catalysts can provide many advantages compared to bulk materials, including the increase of active sites, the change of the local pH of the electrolyte, and improved stability. This chapter reviews the recent development of nanostructured metal catalysts for the electrocatalytic reduction of CO_2, mainly focusing on the fabrication, characterization, catalytic performance, and the reaction mechanism of these materials. In addition, the recent utilization of nanostructured bimetallic catalysts are introduced and a fundamental understanding of the reaction mechanism for their ability to reduce CO_2 is discussed. Finally, nanostructured carbon is shortly reviewed due to its low cost and improved catalytic activity and stability for the electroreduction of CO_2.

Keywords Electrochemical reduction of CO_2 · Conversion of CO_2 into fuels · Electrocatalysts · Metal nanocatalysts · Bimetallic nanocatalysts · Nano-carbon catalysts

M. Ma · W.A. Smith (✉)
Materials for Energy Conversion and Storage (MECS),
Department of Chemical Engineering, Faculty of Applied Sciences,
Delft University of Technology, 2629 HZ Delft, The Netherlands
e-mail: w.smith@tudelft.nl

S.E. Hunyadi Murph et al. (eds.), *Anisotropic and Shape-Selective Nanomaterials*,
Nanostructure Science and Technology, DOI 10.1007/978-3-319-59662-4_11

11.1 Introduction

11.1.1 Background

The global atmospheric concentration of carbon dioxide (CO_2) has increased dramatically due to human activity, particularly after the industrial revolution. The accumulation of CO_2 (greenhouse gas) in the atmosphere is a serious environmental threat and a leading cause of global climate change [1, 2]. The increased global mean temperature may trigger the rise of sea level and the frequency and intensity of extreme weather events. In addition, the increase of the global CO_2 concentration has resulted in an increased ocean acidification, leading to the demise of many natural habitats and aquatic species [3, 4]. The oceans are a principal sink for dissolving the anthropogenic CO_2 where it is estimated to have caused a rise of 30% in the concentration of H^+ in the ocean surface waters since the early 1900s, which may adversely affect many marine ecosystems in the near future [4].

For mitigating the atmospheric CO_2 concentration, carbon capture and sequestration (CCS) at large emission sources such as industrial power plants has been proposed. However, there are some unknown ecological and environmental impacts and risks associated with the sequestration methods including geological and deep sea storage. Thus, carbon capture and utilization (CCU) could be a feasible strategy. For the utilization of the captured CO_2, the electrochemical reduction of CO_2 to fuels and value-added chemicals at mild conditions has attracted considerable attention as a promising solution [5–10]. In this process, the captured CO_2 could be utilized as a chemical feedstock and converted into carbon monoxide (CO), methane (CH_4), ethylene (C_2H_4), and even liquid products such as formic acid (HCOOH), methanol (CH_3OH) and ethanol (C_2H_5OH) [9–11]. The high energy density hydrocarbons can be directly and conveniently utilized as fuels within the current energy infrastructure. In addition, the production of CO is very interesting since it can be used as a reagent in the Fischer–Tropsch process, a well-developed technology that has been widely used in industry to convert syngas (CO and hydrogen (H_2)) into valuable chemicals such as methanol and synthetic fuels (such as diesel fuel) [5, 6].

In order to avoid extra CO_2 emission, the electrochemical reduction of CO_2 should be powered by electricity from renewable energy sources such as solar energy, hydropower and wind energy [6, 9]. In this way, the anthropogenic carbon cycle can be closed by the conversion of CO_2 into fuels and useful products (Fig. 11.1). A key technological challenge for achieving this goal is to develop cheap and earth-abundant catalysts that are capable of electrochemically reducing CO_2 in cost-effective process with high efficiency, controllable selectivity and long-term stability [5, 9, 12–14].

Hori and other researchers have investigated the electrocatalytic reduction of CO_2 using various metal foil electrodes in CO_2-saturated aqueous solutions [8, 9, 15–20]. Recent studies indicate that a nanostructured metal catalyst has a significant effect on the catalytic selectivity, efficiency and stability in the electrochemical reduction of CO_2 [14, 21–24].

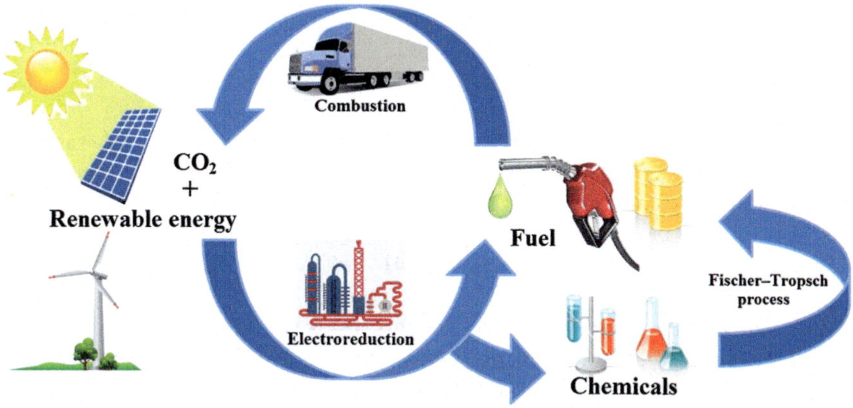

Fig. 11.1 The carbon-neutral fuel production from the electrochemical reduction of CO_2 coupled to renewable energy sources such as solar energy, wind energy and hydropower, offers a promising route to close the anthropogenic carbon cycle

A review of nanostructured catalysts for the electrocatalytic reduction of CO_2 is presented in this chapter, mainly focusing on the fabrication, characterization and catalytic performance of the nanostructured metal catalysts. In addition, a fundamental understanding of the reaction mechanism for the electrochemical CO_2 reduction on nanomaterial-based catalysts is also introduced.

11.1.2 Bulk Metal Catalysts for CO_2 Reduction

In the electrocatalytic CO_2 reduction process, CO_2 can be converted into many different products when combined with water on metal surfaces, and each product is formed by a different number of electrons at different equilibrium potentials. The electrochemical reduction of CO_2 to CO, CH_4, C_2H_4, HCOOH, CH_3OH and C_2H_5OH with their corresponding equilibrium potentials (with respect to the reversible hydrogen electrode (RHE)) are given in the following reaction equations:

$$CO_2 + H_2O + 2e^- \rightarrow CO + 2OH^- \quad (-0.11\ \text{V vs. RHE}) \tag{11.1}$$

$$CO_2 + H_2O + 2e^- \rightarrow HCOO^- + OH^- \quad (-0.03\ \text{V vs. RHE}) \tag{11.2}$$

$$2CO_2 + 8H_2O + 12e^- \rightarrow C_2H_4 + 12OH^- \quad (0.08\ \text{V vs. RHE}) \tag{11.3}$$

$$CO_2 + 6H_2O + 8e^- \rightarrow CH_4 + 8OH^- \quad (0.17\ \text{V vs. RHE}) \tag{11.4}$$

$$CO_2 + 5H_2O + 6e^- \rightarrow CH_3OH + 6OH^- \quad (0.03\ \text{V vs. RHE}) \tag{11.5}$$

$$2CO_2 + 9H_2O + 12e^- \rightarrow C_2H_5OH + 12OH^- \quad (0.09\ \text{V vs. RHE}) \tag{11.6}$$

H_2 evolution is a competing reaction with CO_2 reduction in CO_2-saturated electrolytes. Thus, water could be reduced to H_2 during the CO_2 reduction according to the below reaction:

$$2H_2O + 2e^- \rightarrow H_2 + 2OH^- \quad (0\ \text{V vs. RHE}) \tag{11.7}$$

It is important to note two key insights from the aforementioned reaction equations. Firstly, the thermodynamic potential for each product formation are very close to that of hydrogen (0 V vs. RHE), which indicates that it theoretically should not add a substantial amount of extra potential to drive these reactions compared to simple water reduction. However, it is also important to understand that many of these reactions require many electron (and proton coupled electron) transfer steps, as shown by the number of electrons needed to drive each reaction (Eqs. 11.1–11.7). Secondly, the number of electrons required to form certain product generally scales with the increased amount of reaction intermediates, and thus the activation energy for each intermediate needs to be overcome to form the final product. Therefore, while the overall thermodynamic potential for each product formation seems to be small and reasonable, the formation of many intermediates and electron transfer could significantly increase overpotentials, and the overpotential of ~ 1 V is generally needed to drive each reaction effectively and selectively in many practical cases. The actual required overpotentials for driving the reduction of CO_2 are strongly materials dependent.

Hori and coworkers carried out the electrochemical CO_2 reduction over various transition metal electrodes (Table 11.1), discovering several metallic electro \catalyst materials with the capability of reducing CO_2 [8]. It has been demonstrated that polycrystalline Ag and Zn show relatively high faradaic efficiency (FE) for the electrocatalytic reduction of CO_2 to CO, although these occur at high overpotentials. In addition, Au is the most efficient and selective metal surface for the electrocatalytic conversion of CO_2 into CO at a moderate overpotential, as presented in Table 11.1. Further reduction of CO to more complex products is rare on Au, Ag and Zn catalysts due to the weak binding strength of CO with these metallic surface [11, 25]. However, the strong binding of CO on Ni, Pt and Fe metallic catalysts limits the desorption of CO, [11, 25] resulting in suppressed CO_2 reduction with dominant H_2 evolution, as shown in Table 11.1. Among the identified transition metal materials, copper electrodes can uniquely reduce CO_2 to CH_4, C_2H_4, and alcohols at significant amounts in CO_2-saturated aqueous electrolytes at ambient pressure and temperature [15, 26].

The synthesis of products from CO_2 on metal catalysts is a complex multistep reaction with multiple adsorbed intermediates, most notably adsorbed CO [10, 27, 28]. The reaction mechanisms for the various products are attributed to the adsorption strength of intermediates (such as CO, COH, CHO and CH_3) on the

Table 11.1 Faradaic efficiencies of products in CO_2 reduction at various metal electrodes. Electrolyte: CO_2-saturated 0.1 M $KHCO_3$, T = 18.5 ± 0.5 °C. Reprinted with permission from Ref. [8]

Electrode	V versus SHE	Faradaic efficiency (%)							
		CH_4	C_2H_4	Ethanol	n-propanol	CO	HCOOH	H_2	Total
Au	−1.14	0	0	0	0	87.1	0.7	10.2	98.0
Ag	−1.37	0	0	0	0	81.5	0.8	12.4	94.6
Zn	−1.54	0	0	0	0	79.4	6.1	9.9	95.4
Pd	−1.20	2.9	0	0	0	28.3	2.8	26.2	60.2
Ga	−1.24	0	0	0	0	23.2	0	79.0	102.0
Pb	−1.63	0	0	0	0	0	97.4	5.0	102.4
Hg	−1.51	0	0	0	0	0	99.5	0	99.5
Tl	−1.60	0	0	0	0	0	95.1	6.2	101.3
In	−1.55	0	0	0	0	2.1	94.9	3.3	100.3
Sn	−1.48	0	0	0	0	7.1	88.4	4.6	100.1
Cd	−1.63	1.3	0	0	0	13.9	78.4	9.4	103.0
Bi	−1.56	–	–	–	–	–	77	0	–
Cu	−1.44	33.3	25.5	5.7	3.0	1.3	9.4	20.5	103.5
Ni	−1.48	1.8	0.1	0	0	0	1.4	88.9	92.4
Fe	−0.91	0	0	0	0	0	0	94.8	94.8
Pt	−1.07	0	0	0	0	0	0.1	95.7	95.8
Ti	−1.60	0	0	0	0	0	0	99.7	99.7

catalyst surface during the electrochemical reduction of CO_2 [24, 29, 30]. The catalyst provides the reactive sites, and the product selectivity in the electrocatalytic reduction of CO_2 is affected by whether the reactants and other related species are adsorbed at the active sites.

Because H_2 evolution is always a competing reaction with CO_2 reduction in CO_2-saturated electrolytes, H_2 was also identified and quantified during CO_2 reduction electrolysis. Polycrystalline Pt, Ni and Fe, shown in the Table 11.1, exhibit an extremely high selectivity for H_2 (water reduction is always a competing reaction) along with very little catalytic activity for the electrochemical reduction of CO_2. On Pt, Ni and Fe, it is easy for CO_2 activation and conversion of CO_2 into adsorbed CO, but the rate determining step for CO generation is the desorption of CO due to the strong binding of CO on the catalyst surface [11, 25, 31]. The coverage of CO adsorbed on polycrystalline Pt, Ni and Fe blocks the reactive sites for CO_2 reduction, influencing the catalytic activity for the electrocatalytic reduction of CO_2. Recently, the catalytic activity and selectivity for the electroreduction of CO_2 on seven transition metal surfaces (Au, Ag, Zn, Cu, Ni, Pt and Fe) were further investigated by Jaramillo [9, 32]. Au, Ag and Zn exhibit a high faradaic efficiency for the reduction of CO_2 to CO due to the surface of Au, Ag and Zn bind CO weakly, while Ni, Pt, and Fe with binding CO strongly have a very low catalytic activity for CO_2 reduction along with dominant H_2 evolution. Cu, with a moderate CO binding energy, uniquely shows a total of 16 different CO_2 reduction products [9].

11.1.3 Nanostructured Metal Catalysts for CO_2 Reduction

Catalytic behavior is not only strongly materials-dependent, but also extremely morphology-dependent [33]. The development of nanoscience research provides the opportunity to synthesize nanocatalysts with controllable size and shape, which has led to the discovery of the altered or enhanced catalytic activity and selectivity with the size and shape of nanostructured catalyst. Researchers have demonstrated that a nanostructured catalyst can exhibit very different behavior compared to its bulk counterpart. Much work in the field has focused on the elucidation of the effects of nanostructured catalysts.

Recently nanostructured metallic catalysts have been the object of an increasing interest for the electrocatalytic reduction of CO_2. The main advantage of metallic nanocatalysts is that nanostructured catalysts are capable of providing more active sites on the high surface area compared to bulk metal catalysts. The increased active surface sites on nanocatalysts may cause an enhanced catalytic performance due to the fact that the catalytic activity is proportional to the number of the active surface sites [34]. In addition, the morphology of nanostructured surface contains more low-coordinated sites (such as edge sites and corner sites) that are more active for CO_2 reduction in comparison with smooth or planar surface [35]. Furthermore, the catalytic stability of the electrochemical reduction of CO_2 has been improved on nanostructured catalysts, which can be attributed to the enhanced tolerance to heavy metal impurities in an electrolyte. Hori has demonstrated that unavoidable ppm concentration of heavy metal impurities (such as Fe or Pb) in the electrolytes can be reduced and deposited on the cathodic electrode surface, which could poison a flat catalyst, resulting in a significant influence in the electrocatalytic activity for CO_2 reduction [8]. The large surface area of nanostructured electrocatalysts could adapt the contamination or impurities, having a better catalytic stability [34].

This chapter mainly focuses on the recent development of heterogeneous nanocatalysts for the electrochemical reduction of CO_2, including the fabrication, characterization, catalytic performance and reaction mechanism of nanostructured metallic catalysts.

11.2 Nanostructured Metal Catalysts for CO_2 Reduction to CO

The selective conversion of CO_2 into CO is a promising route for clean energy generation. CO can be used as feedstock in the Fischer–Tropsch process, a well-developed technology that has been used in industry to produce chemicals (such as methanol) and synthetic fuels (such as diesel fuel) from syngas (CO + H_2) for many decades [5, 6]. While several electrocatalyst materials (such as Au, Ag and Zn) are capable of reducing CO_2 to CO in CO_2-saturated aqueous solutions at ambient temperature and pressure, all suffer from one or more of the following problems: high overpotential requirement, low current density and rapid deactivation of CO_2 reduction activity in favor of H_2 evolution [8, 13, 17]. The main challenge for the electrochemical syngas production is the development of electrocatalysts that are capable of reducing CO_2 to CO efficiently and selectively at low overpotential over a long period of time.

11.2.1 Nanostructured Au

Au is currently the most efficient electrocatalytic surface for the reduction of CO_2 to CO (Table 11.1). It has been demonstrated that polycrystalline Au is capable of reducing CO_2 to CO with a high faradaic efficiency of $\sim$87% at -0.74 versus RHE [32]. While the low abundance and high cost of Au may prevent its large-scale applications, it remains an interesting catalytic material for fundamental investigation due to a high selectivity for CO formation at a relatively low overpotential in the electroreduction of CO_2 (Table 11.1).

11.2.1.1 Au Nanoparticles

Min et al. reported the electrocatalytic CO_2 conversion into CO by nanostructured gold catalysts deposited on carbon papers by an e-beam evaporator using an Au pellet (99.99%) source [36]. The morphology of the deposited Au was observed to vary from the small nanoparticles (NPs) to aggregated clusters to layered film (Fig. 11.2).

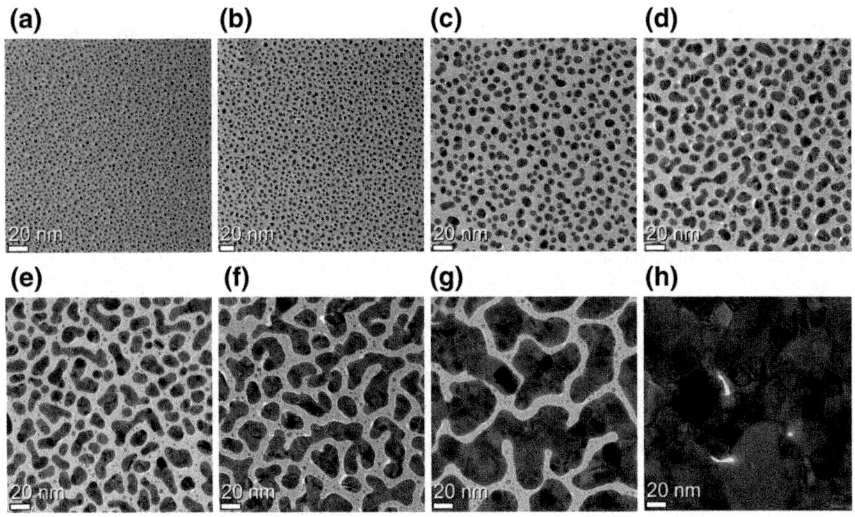

Fig. 11.2 TEM images of Au samples, showing the morphology changes with the dependence of Au coverages which were presented by Au film thickness (increased Au coverages from **a** to **h**). Reprinted with permission from Ref. [36]

The CO faradaic efficiency and production rate were found to increase with increasing the Au amount. The FE for CO formation reached the saturation point at samples thicker than Au-4 (Au-4 represents that Au amount corresponds to 4 nm thick Au film), which is comparable with a commercial Au foil. In addition, Nam et al. synthesized a concave rhombic dodecahedron (RD) gold nanoparticle by using 4-aminothiophenol as an additive and demonstrated that this concave RD had a superior electrocatalytic performance for reduction CO_2 to CO in CO_2-saturated aqueous solutions [37]. For instance, concave RD achieved a FE of approximately 52% for CO formation at a potential of −0.26 V versus RHE, which corresponds to an overpotential of 0.15 V relative to the CO_2/CO equilibrium potential (−0.11 V vs. RHE).

Au clusters have also been explored by the combination of experimental and computational studies, providing insights into the electronic interactions between Au_{25} clusters and weakly bound adsorbates [38]. It was discovered that that the Au_{25} clusters could perform as superior catalysts for the electrochemical conversion of CO_2 into CO and promote the reduction of CO_2 to CO within 90 mV of the formal potential (thermodynamic limit), which represents an approximate 200–300 mV reduction in potential compared to the larger Au NPs and bulk Au tested.

Recently, Sun et al. reported that the electrocatalytic reduction of CO_2 on Au NPs with a series of sizes (4, 6, 8 and 10 nm) in CO_2-saturated 0.5 M $KHCO_3$ [39]. The crystallite diameters of 4, 6, 8 and 10 nm NPs were estimated to be 2.0, 2.3, 4.0 and 5.9 nm, respectively. Among 4, 6, 8 and 10 nm Au NPs, the 8 nm NPs exhibited the highest selectivity for the reduction of CO_2 to CO (FE = 90% at −0.67 V vs. RHE). Density functional theory (DFT) calculations on different crystalline faces suggested that edge sites on Au NPs are active for CO formation while corner sites favor the competitive H_2 evolution reaction. To understand the size-dependent electroreduction of CO_2 on Au NPs, the correlation of the density of catalytically active surface sites with the cluster diameter was investigated, as shown in Fig. 11.3. It was found that more edge sites than corner sites on the Au NP surface could facilitate the stabilization of $COOH^{\cdot}$ intermediate, resulting in the improved reduction of CO_2 to CO with suppressed H_2 evolution.

11.2.1.2 Au Nanowires

The previous study on monodispersed Au NPs indicated that the edge sites of Au NPs are favorable for the reduction CO_2 to CO while corner sites are active for the H_2 evolution [39]. Thus, a one-dimensional morphology may offer an abundance of edge sites which is preferred for the reduction of CO_2. Zhu et al. demonstrated ultrathin Au nanowires (NWs) for the selective electrochemical reduction of CO_2 to CO [40]. A facile seed-mediated growth method was developed to fabricate the ultrathin Au NWs by reducing $HAuCl_4$ in the presence of 2 nm nanoparticles.

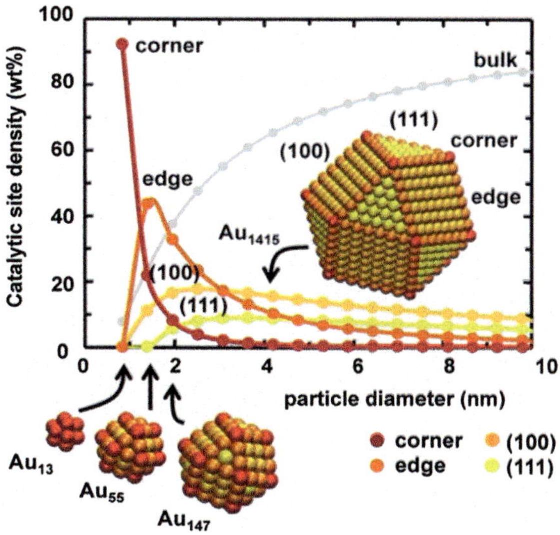

Fig. 11.3 Density of adsorption sites (*yellow*, *light orange*, *dark orange*, or *red symbols* for (111), (001), *edge*, or *corner* on-top sites, respectively) on closed-shell cuboctahedral Au clusters versus the cluster diameter. The weight fraction of Au bulk atoms is marked with *gray dots*. Reprinted with permission from Ref. [39]

These Au NWs are capable of catalyzing CO_2 to form CO with FE of 94% at −0.35 V versus RHE in CO_2-saturated 0.5 M $KHCO_3$. The modeling work (DFT) revealed that the excellent catalytic performance for the reduction of CO_2 to CO on Au NWs is ascribed to a high edge-to-corner ratio in ultrathin Au NWs and the weak CO binding on these reactive edge sites.

11.2.2 Nanostructured Ag

Metallic Ag has attracted considerable attention due to its relatively low cost compared to Au, and its high selectivity for the conversion of CO_2 into CO [8, 11]. As presented in Table 11.1, Ag foils are capable of electrochemically reducing CO_2 to CO with a high FE at a relatively smaller overpotential than other metallic counterparts (except for Au). Recently, Jaramillo investigated the catalytic activity and selectivity of the electrochemical reduction of CO_2 as a function of potential on metallic silver surfaces under ambient conditions [20]. The authors found that the applied potential can significantly influence the selectivity for the two major products (CO and H_2). Specifically, the optimal selectivity for CO formation on the polycrystalline silver catalyst was observed at the range from −1.0 to −1.2 V versus RHE and hydrogen evolution dominated at very high and low potentials (Fig. 11.4). By using high sensitivity of the experimental methods for identifying and quantifying products of CO_2 reduction, formate, methane, methanol, and ethanol were observed as minor products (methanol and ethanol formation had never been reported on Ag before this study).

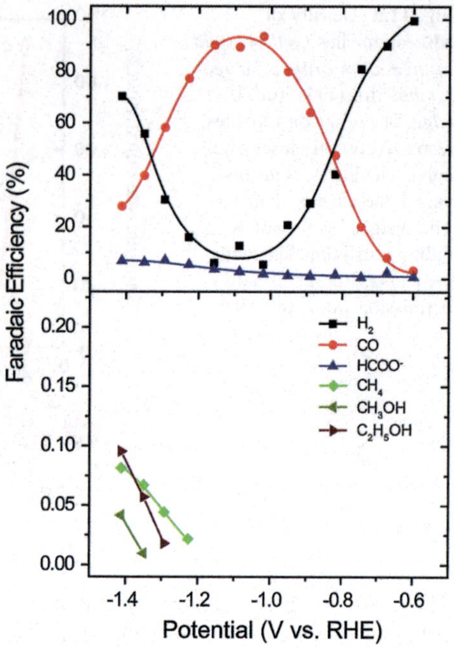

Fig. 11.4 Faradaic efficiency for each product as a function of potential on polycrystalline Ag foils. Reprinted with permission from Ref. [20]

On Ag catalysts, the overpotential (>0.9 V) required for driving the electrocatalytic reduction of CO_2 efficiently and selectively with suppressed H_2O reduction remains relatively high [32]. The large overpotential required for the reduction of CO_2 is attributed to the hindrance for the initial electron transfer to a CO_2 molecule to form a $CO_2^{\cdot-}$ or $COOH^{\cdot}$ intermediate [5, 12, 25, 35]. Nanostructured Ag catalysts have been considered to be a promising candidate for the electroreduction of CO_2 to CO. It has been demonstrated that the surface of nanostructured Ag catalysts offers the low-coordinated surface Ag sites, which is favorable for CO_2 fixation (to stabilize the formed $COOH^{\cdot}$ intermediate) through reducing the activation energy barrier of the initial electron transfer [35].

In addition, Hori et al. has investigated the electroreduction of CO_2 on single crystalline silver electrodes, showing that the CO_2 conversion into CO is favorable on Ag(110) compared to Ag(111) or Ag(100) [17]. Recently, a DFT simulation study (Fig. 11.5) exhibited that Ag(110) or Ag(211) surface has a lower free energy change for the first proton-coupled electron transfer for $COOH^{\cdot}$ stabilization than Ag(111) or Ag(100), resulting in better catalytic activity for the reduction of CO_2 to CO on Ag(110) or Ag(211) [35]. It was believed that stepped Ag(110) or Ag(211) surfaces are much prevalent on the nanostructured Ag catalysts than that on bulk Ag electrodes [35].

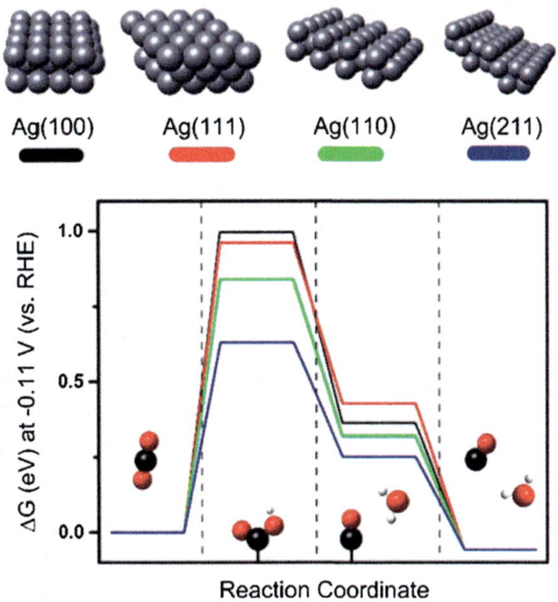

Fig. 11.5 Free energy diagrams for the electrochemical reduction of CO_2 to CO on flat (Ag(100) and Ag(111)) and edge (Ag (221) and Ag(110)) surfaces. The first two steps include a simultaneous proton/electron transfer, with the final molecular surface configuration at each step depicted on the *bottom* of the graph. Values of ΔG are reported with an applied potential of −0.11 V versus RHE. Sphere colors: *white*, H; *black*, C; *red*, O; *silver*, Ag. Reprinted with permission from Ref. [35]

11.2.2.1 Ag Nanoparticles

Recently Kim et al. have examined silver nanoparticles with different sizes for the electrocatalytic reduction of CO_2 in CO_2-saturated 0.5 M $KHCO_3$ [41]. Silver nanoparticles with three different sizes (3, 5 and 10 nm) were fabricated on carbon substrates, respectively, by a facile one-pot method using cysteamine as an anchoring agent. The authors discovered that immobilized Ag nanoparticles supported on carbon had an increased FE and a lower overpotential for selective reduction of CO_2 to CO. Notably, the particle size of 5 nm exhibited the highest catalytic activity for the selective reduction of CO_2 to CO. As shown in Fig. 11.6b, a FE of 84.4% for CO formation and a decrease of the overpotential by 300 mV were observed on NPs with diameter of 5 nm at −0.75 V versus RHE compared to polycrystalline Ag foils. The DFT studies suggested that the specific interaction (Ag-S interaction) between Ag nanoparticles and the anchoring agents modified the catalyst induces a selectively higher affinity to the $COOH^{\cdot}$ intermediate, which effectively lowers the overpotential and improves the catalytic activity for the reduction of CO_2 to CO. In addition, for identifying the carbon source for the CO_2 reduction, a $^{13}CO_2$ isotope experiment was performed, which confirmed that the carbon source for CO formation was completely derived from the dissolved CO_2 in electrolytes, and carbon from carbon support or cysteamine were not involved in the reduction reaction.

A previous report also evaluated the Ag NP size effect, suggesting that that the catalytic activity increases with decreasing particle size until a certain particle size

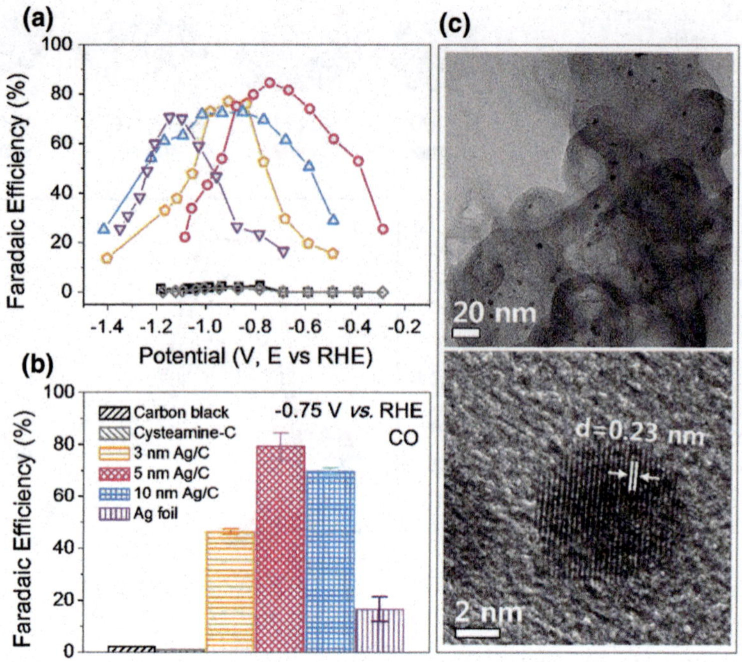

Fig. 11.6 CO faradaic efficiency **a** depending on applied potential and **b** fixed potential at −0.75 V (vs. RHE). CO_2 electrochemical reduction was performed in CO_2-saturated 0.5 M $KHCO_3$. **c** TEM image and HR-TEM image of 5 nm Ag/C after CO_2 reduction. Reprinted with permission from Ref. [41]

of 5 nm, and that the activity decreases with going to even smaller nanoparticle size (1 nm) [42]. This observation is consistent with the results from Kim (5 nm NPs offer the maximum catalytic activity for CO_2 reduction) [41].

11.2.2.2 Nanoporous Ag

Lu et al. reported that a nanoporous Ag catalyst prepared by the de-alloying of an Ag-Al precursor is capable of reducing CO_2 electrochemically to CO with a high FE of >90% at a moderate potential of −0.6 V versus RHE in CO_2-saturated 0.5 M $KHCO_3$ [5]. To explore the electrokinetics of CO_2 reduction, Tafel analysis was performed on the nanoporous Ag catalysts and the polycrystalline Ag electrode, respectively.

It is known that a two electron-transfer process is involved in the electroreduction of CO_2 to CO on Ag catalysts. Firstly, one electron is transferred to a CO_2 molecule to produce a $CO_2^{\cdot-}$ intermediate, and then one proton is coupled to form a $COOH^{\cdot}$ intermediate. Subsequently, the $COOH^{\cdot}$ intermediate takes another electron and one proton, resulting in the CO formation on catalyst surface. The initial

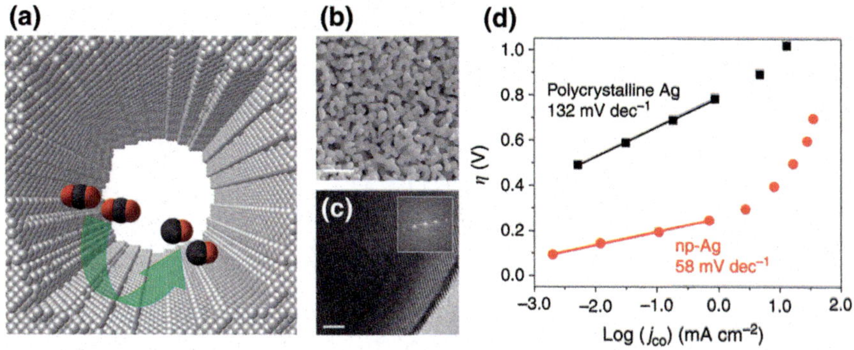

Fig. 11.7 **a** A schematic diagram of a nanopore of the silver electrocatalyst with highly curved internal surface. **b** Scanning electron micrograph of np-Ag dealloyed in 5 wt% HCl for 15 min and further in 1 wt% HCl for 30 min (scale bar, 500 nm). **c** Corresponding high-resolution transmission electron micrograph with visible lattice fringes. Inset: the Fourier transform shows that the np-Ag is composed of an extended crystalline network (scale bar, 2 nm). **d** Tafel plots of the partial current density for CO production. Reprinted with permission from Ref. [5]

electron transfer to a CO_2 molecule is the rate determining step for the overall process because the first electron transfer requires a much more negative potential compared to the following steps [5, 12, 25, 35]. A Tafel slope of 132 mV/dec for polycrystalline Ag shown in Fig. 11.7d indicates the rate determining step is the initial electron transfer to CO_2 for the formation of the $CO_2^{\cdot -}$ adsorbed on polycrystalline Ag. In contrast, the nanoporous Ag with a Tafel slope of 58 mV/dec reveals a fast initial electron-transfer step, suggesting that a better stabilization for $CO_2^{\cdot -}$ intermediate on nanoporous-Ag in comparison to the polycrystalline Ag. This study directly shows a clear advantage of CO_2 reduction catalysis on nanostructured electrodes.

11.2.3 Nanostructured Zn

Zn is an earth-abundant material, which is roughly 3 orders of magnitude more abundant in the Earth's crust than Ag. Hori et al. reported that polycrystalline Zn electrodes are capable of catalytically reducing CO_2 to CO with a FE of 79.4% [8]. However, the Zn electrodes require a relatively higher overpotential for driving CO_2 reduction selectively compared to Au and Ag (Table 11.1).

Nanostructured Zn dendrite electrodes were synthesized on Zn foil electrodes using an electrochemical Zn deposition [43]. As shown in Fig. 11.8a, the scanning electron microscope (SEM) images reveal that the dendrite layer was about 50–100 μm thick and uniformly deposited on Zn foils. The X-ray diffraction

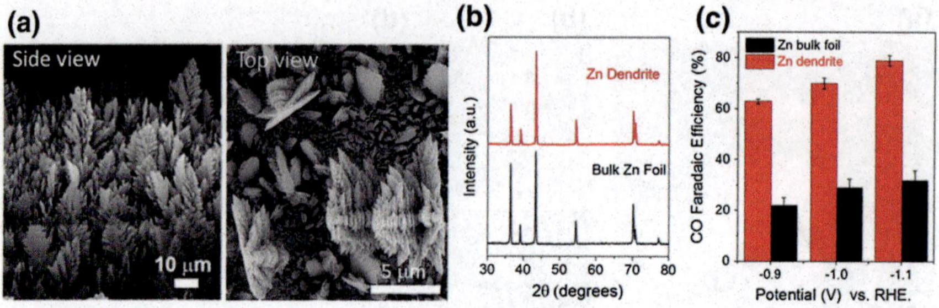

Fig. 11.8 **a** Typical SEM image for a Zn dendrite catalyst. **b** PXRD patterns of bulk and dendritic Zn electrodes. **c** CO faradaic efficiency for bulk and dendritic Zn electrocatalysts. Reprinted with permission from Ref. [43]

(XRD) patterns in Fig. 11.8b show that all the diffraction peaks of the nanostructured Zn dendrite and the Zn bulk foil are aligned precisely, indicating a pure metallic Zn. This nanostructured Zn dendrite catalyst was able to electrochemically reduce CO_2 to CO in an aqueous bicarbonate electrolyte, with a CO faradaic efficiency of around 3-fold higher than that of bulk Zn counterparts, as displayed in Fig. 11.8c.

11.2.4 Nanostructured Pd

Bao et al. investigated a particle size dependence for the electrocatalytic reduction of CO_2 on Pd NPs with the size range from 2.4 to 10.3 nm [44]. The authors discovered that an increase in the selectivity for CO production with reducing the size of Pd NPs, which varies from the FE of 5.8% on 10.3 nm Pd NPs to that of 91.2% on 3.7 nm NPs at −0.89 V versus RHE. DFT calculations suggested that corner and edge sites on small Pd NPs facilitate the adsorption of CO_2 and the stabilization of $COOH^{\cdot}$ intermediates during CO_2 reduction compared with terrace one on large Pd NPs.

11.2.5 Metal Organic Frameworks

Metal Organic Frameworks (MOFs) have backbones constructed from metals and organic ligands, which are structured and porous materials with nano-scale pores for heterogeneous and homogeneous catalysis [45, 46]. Fe-based MOF catalysts have been recently used for the electrochemical reduction of CO_2 to CO (faradaic efficiencies are 41 ± 8 and $60 \pm 4\%$ for CO and H_2) [47]. The well-defined nanoscale porosity of the MOF facilitates access of the solvent, reactant, and electrolyte to the surface of catalytically active sites.

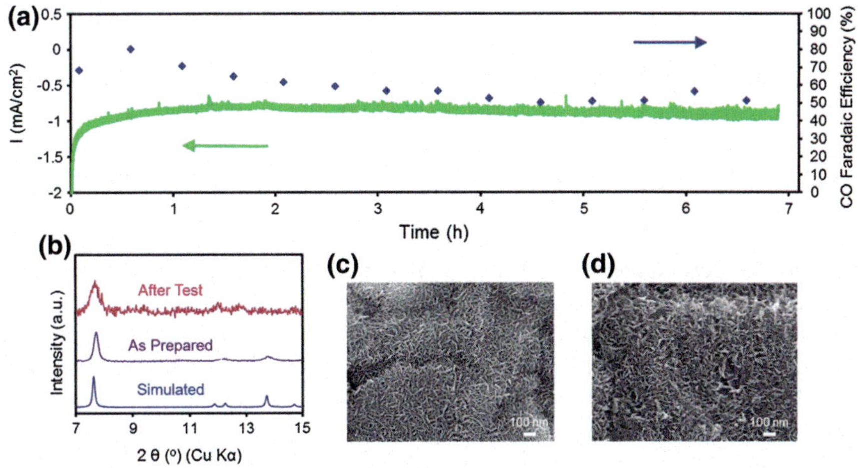

Fig. 11.9 Stability of the MOF catalyst is evaluated through chronoamperometric measurements in combination with faradaic efficiency measurements (**a**). XRD analysis indicates that the MOF retains its crystalline structure after chronoamperometric measurement (**b**). SEM images of the MOF catalyst film before (**c**) and after electrolysis (**d**) reveal the retention of the platelike morphology. Reprinted with permission from Ref. [46]

Yang et al. reported an efficient and selective nano-scale MOFs catalyst for the electrochemical generation of CO from CO_2 in CO_2-saturated aqueous bicarbonate solutions [46]. A cobalt–porphyrin MOF, $Al_2(OH)_2$TCPP-Co (TCPP-H2 = 4,4′,4″,4‴-(porphyrin-5,10,15,20-tetrayl)tetrabenzoate) exhibited a FE of exceeding 76% for CO formation at −0.7 V versus RHE and a stable current of MOF catalysts was observed after initial electrolysis, as presented in Fig. 11.9a. The crystallinity of the MOF catalyst was still maintained after CO_2 reduction electrolysis (Fig. 11.9b) and the SEM images show that the retention of the platelike morphology after CO_2 reduction (Fig. 11.9c, d).

11.3 Nanostructured Metal Catalysts for CO_2 Reduction to Hydrocarbons

The conversion of CO_2 and H_2O into hydrocarbons on electrocatalysts has attracted considerable attention [9, 13, 14, 21]. The main advantages of directly forming hydrocarbons are their high energy density and the ease of utilization as fuels in our existing energy infrastructure. However, a major challenge of this goal is to develop electrocatalysts that are capable of reducing CO_2 efficiently and selectively at low potentials. Hori and his coworkers carried out CO_2 reduction over various metal electrodes (Table 11.1), and it was demonstrated that many electrocatalyst materials such as polycrystalline Cu, Au, Ag and Zn show high faradaic efficiency for the electrochemical reduction of CO_2, although these occur at very high overpotentials [8]. It was found that copper electrodes can uniquely reduce CO_2 to significant amounts of CH_4, C_2H_4, and alcohols in aqueous electrolytes at ambient pressure and temperature [8, 15, 26]. One main reason for this unique catalytic activity of copper is attributed to the appropriate binding strength of intermediates (such as CO, COH, CHO and CH_3,) on the copper surface during the electrochemical reduction of CO_2 [24, 29, 30].

Recently, Jaramillo et al. [9] reported new insights into the electrocatalytic reduction of CO_2 on metallic Cu surface, finding as many as 16 different products from the electroreduction of CO_2 (five of them have never been reported previously). At low negative potential of >−0.75 V versus RHE, only H_2, CO and HCOOH were detected on Cu. At −0.75 V versus RHE, C_2H_4 and CH_4 formation started to be observed. In addition, other C_2 and C_3 products were initially detectable at −0.95 V versus RHE.

Hori et al. [48] investigated the electrochemical reduction of CO_2 on Cu single crystal electrodes (100), (110), and (111) in aqueous electrolytes at ambient pressure, showing that C_2H_4 is favorably produced on Cu(100) and CH_4 is preferred on Cu(111). In addition, the similar product selectivity was also observed for the electroreduction of CO on Cu single crystal electrodes [48, 49], due to the fact that CO is the intermediate in the reduction of CO_2 to hydrocarbons [29, 30, 50, 51]. Furthermore, a more detailed study on the effect of Cu crystal faces were carried out [18, 52].

Until now, copper is the only known material which is capable of catalyzing the formation of significant amounts of hydrocarbons at high reaction rates in aqueous solutions at ambient conditions of temperature and pressure [9, 22, 50]. However, controlling the product selectivity of the CO_2 reduction effectively at low overpotential with high current density for the formation of hydrocarbons is a major scientific challenge for the practical use of this technology [12–14, 53]. It has been demonstrated that the surface morphology and roughness of copper electrodes have a dramatic influence on the catalytic activity and product selectivity for the electrochemical reduction of CO_2 in an aqueous solution [14, 21–24].

11.3.1 Cu Nanoparticles

Tang et al. reported that Cu nanoparticle covered electrodes exhibited a better selectivity towards C_2H_4 and CO formation in comparison with an electropolished Cu electrode and an argon gas sputtered Cu electrode [24]. In that study, the increased catalytic selectivity for C_2H_4 formation in CO_2 reduction was explained by the roughened Cu surface which was able to provide a greater abundance of undercoordinated sites. In addition, Alivisatos et al. has demonstrated that Cu NPs supported on glassy carbon (n-Cu/C) exhibited a 4-fold higher methanation current densities compared to the high-purity Cu foil counterpart [54]. A faradaic efficiency of 80% for CH_4 was achieved on the n-Cu/C electrocatalysts. The rate-limiting step

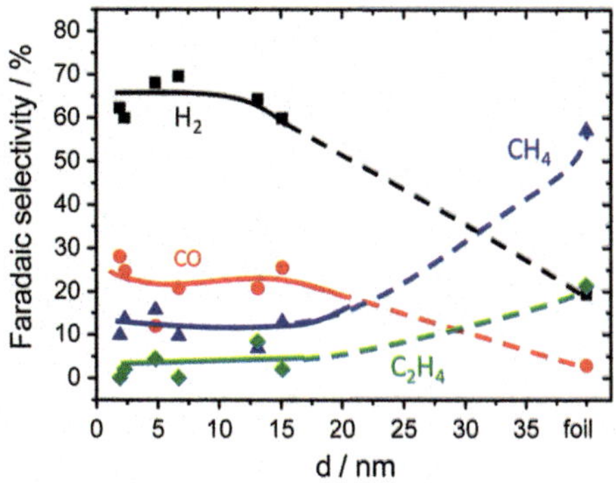

Fig. 11.10 Catalytic selectivity of reaction products during the CO_2 electroreduction on Cu NPs in CO_2-saturated 0.1 M $KHCO_3$ at −1.1 V versus RHE under 25 °C. Reprinted with permission from Ref. [21]

for CH_4 on Cu foils is a single electron transfer to CO_2, which corresponds to a Tafel slope of 120 mV/dec. The Tafel slope for CH_4 is 60 ± 4.2 mV/dec on the n-Cu/C, which is consistent with the theoretical value of 59 mV/dec, indicating a one-electron pre-equilibrium step prior to a rate-determining non-electrochemical step.

Recently, the particle size dependence for the electrochemical reduction of CO_2 has been investigated on Cu NPs with the size range from 2 to 15 nm [21]. Cu NPs exhibited that an increase in the catalytic activity and selectivity for CO and H_2 with decreasing Cu particle size, as presented in Fig. 11.10. In addition, catalytic selectivity for hydrocarbons (CH_4 and C_2H_4) was suppressed on the nanoscale Cu surface compared to the bulk Cu foil. To gain insight into the Cu particle size effect, the atomic coordination of model spherical Cu particles from 1 to 18 nm was utilized, suggesting low-coordinated sites and the strong chemisorption are linked to the acceleration for H_2 evolution and the CO_2 reduction to CO.

11.3.2 Cu Nanowires

Smith et al. reported a new synthesis method to prepare a Cu nanowire catalyst for the electroreduction of CO_2 at room temperature and atmospheric pressure [55]. Cu nanowire array electrodes were synthesized through a two-step synthesis of Cu $(OH)_2$ and CuO nanowire arrays on Cu foil substrates and a subsequent electrochemical reduction of the CuO nanowire arrays. $Cu(OH)_2$ nanowire arrays (Fig. 11.11a) were first synthesized on Cu foils using a simple wet chemical method [56, 57]. After this step, CuO nanowire arrays were formed by annealing the Cu $(OH)_2$ nanowires at 150 °C for 2 h in an air atmosphere [55]. The resulting CuO nanowire arrays were directly utilized in the electrochemical reduction of CO_2 in CO_2-saturated 0.1 M $KHCO_3$ electrolytes, and were reduced to Cu nanowire during electrolysis. The CuO-derived Cu nanowire arrays are able to electrochemically reduce CO_2 to CO with a FE of ~50% at a moderate overpotential of 490 mV, which is significantly higher than that of polycrystalline Cu foil catalysts at identical

Fig. 11.11 SEM images of $Cu(OH)_2$ nanowires before (**a**) and after (**b**) annealing in air at 150 °C for 2 h, respectively. **c** Tafel plots of the CO partial current density for polycrystalline Cu and Cu nanowires. Reprinted with permission from Ref. [55]

conditions. The Tafel slope of ~110 mV/dec for Cu nanowire arrays shown in Fig. 11.11c indicates a fast initial electron tranfer to CO_2 for the formation of adsorbed $CO_2^{\cdot-}$ intermediate. Thus, the improved catalytic selectivity for the reduction of CO_2 to CO is ascribed to the enhanced stabilization for the $CO_2^{\cdot-}$ intermediate on the high surface area Cu nanowire arrays.

It was also recently demonstrated that highly dense Cu nanowires for the electrochemical reduction of CO_2 [58]. CuO nanowires were prepared by the oxidation of Cu mesh in air and then reduced by annealing in the presence of hydrogen or applying a cathodic electrochemical potential to produce Cu nanowires. These nanowires showed the distinct catalytic selectivity for CO_2 reduction, which is linked to the different nanoscale crystalline and surface structures. The authors found that the Cu nanowires fabricated by the electrochemical reduction of CuO exhibited a FE toward CO as high as ~60% at overpotentials of less than 0.4 V [58].

At more negative potentials, hydrocarbon gas phase products on Cu nanowire arrays were observed [55]. Smith et al. tuned the selectivity of hydrocarbon products on Cu nanowire arrays by systematically varying the length and density of the Cu nanowires which can provide a high local pH within the nanowire arrays (Fig. 11.12) [59]. With increasing the Cu nanowire length ($\geq 2.4 \pm 0.56$ μm), the formation of n-propanol was detected along with CO, HCOOH and C_2H_4. On longer Cu nanowires ($\geq 7.3 \pm 1.3$ μm), C_2H_6 formation appeared, accompanying with the ethanol formation. The formation of C_2H_6 has never been reported in the electrocatalytic reduction of CO_2 on smooth Cu, [8, 9, 11] but was detected on nanostructured Cu catalysts as a minor product [14, 23, 24]. The authors proposed a reaction pathway towards C_2H_6 from the intermediate (CH_3CH_2O) in path (ii), as shown in Scheme 11.1 (a route to C_2H_6 had never been reported before this study). Furthermore, an increased selectivity for C_2H_4 was observed on Cu nanowire arrays with increasing the Cu nanowire length and density (Fig. 11.12a), which is due to the improved formation of C_2H_4 through a CO coupling mechanism caused by a high local pH within the Cu nanowire arrays.

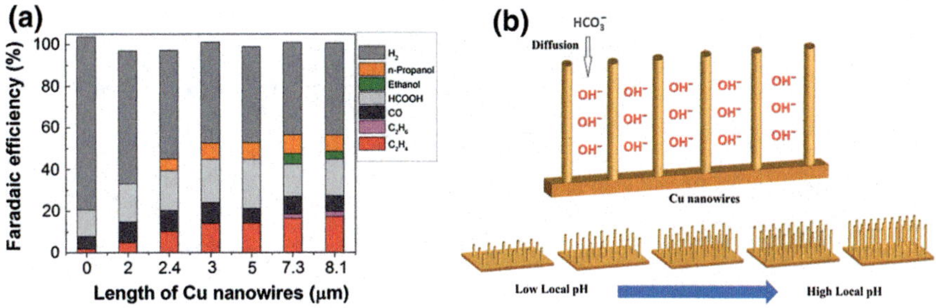

Fig. 11.12 **a** Faradaic efficiency for C_2H_4, C_2H_6, CO, HCOOH, ethanol, n-propanol and H_2 on Cu nanowire arrays with different lengths at −1.1 V versus RHE in CO_2-saturated 0.1 M $KHCO_3$ electrolytes (0 μm nanowire represents Cu foil). **b** Schematic illustration of the diffusion of electrolytes into Cu nanowire arrays. Reprinted with permission from Ref. [59]

Scheme 11.1 Proposed reaction paths for electrocatalytic reduction of CO_2 on Cu nanowire arrays, with path (i), *left*, showing COH formation, and path (ii), *right*, showing CO dimerization. In path (i), 2 CH_2 and 2 CH_3 intermediates are required for C_2H_4 and C_2H_4 formation, respectively. Reprinted with permission from Ref. [59]

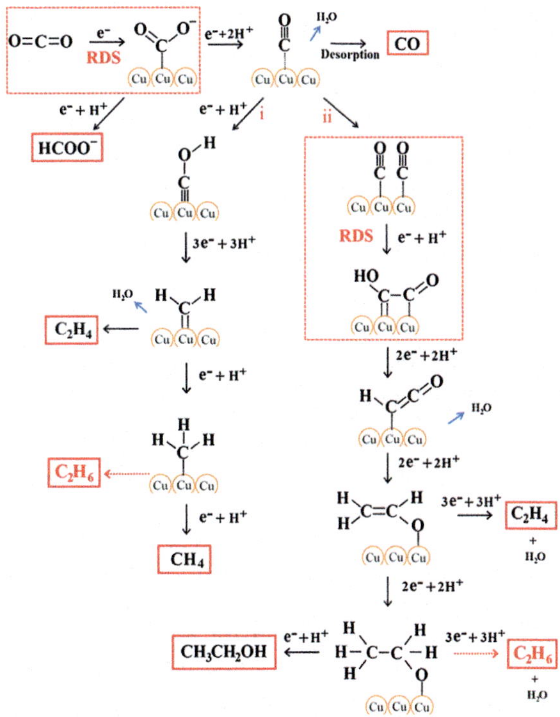

11.3.3 Cu Nanofoam

Sen et al. investigated the electrochemical reduction of CO_2 over copper foams with hierarchical porosity [23]. Three-dimensional foams of copper were electrodeposited onto mechanically polished copper substrates. SEM images of copper foams

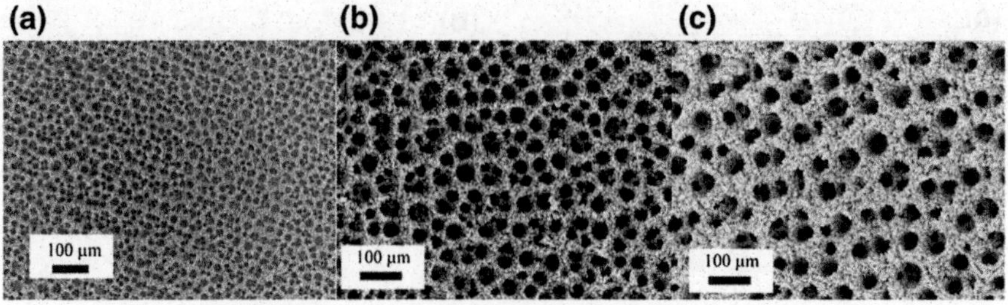

Fig. 11.13 SEM images of electrodeposited copper foams on a copper substrate for **a** 10 s; **b** 15 s; **c** 30 s. Reprinted with permission from Ref. [23]

prepared at different amounts of electrodeposition time are shown in Fig. 11.13. The electrocatalytic studies exhibited that the copper nanofoams are capable of reducing CO_2 to HCOOH, H_2, and CO as major products along with small amounts of C_2H_4, C_2H_6, CH_4, and C_3H_6, which indicates a distinct product distribution (ethane and propylene were not observed on smooth copper) in comparison with that on smooth copper. The authors believed that these differences for the CO_2 reduction performance are attributed to high surface roughness, hierarchical porosity, and confinement of reactive species. In addition, it was observed that a gradual increase in the thickness the copper nanofoams with increasing electrodeposition time enhanced the FE of HCOOH formation by suppressing the electrochemical reduction of adsorbed $H^{\cdot}$ to H_2.

11.4 Oxide-Derived Metallic Nanocatalysts for CO_2 Reduction

While the reported metallic catalysts (Table 11.1) are competent for the electrocatalytic reduction of CO_2, all have one or more following drawbacks: (1) high overpotential requirement for driving the CO_2 reduction. (2) poor catalytic selectivity, and (3) fast deactivation of electroreduction CO_2 process with favorable competing H_2 evolution. In order to improve the CO_2 reduction performance, an interesting way of preparing catalysts through the electrochemical reduction of metal oxides was recently proposed. In the process, the bulk metal is firstly oxidized, and then the metal oxide layer is directly utilized in the electrochemical CO_2 reduction in CO_2-saturated electrolytes, reducing to metallic nanocatalysts during electrolysis. The recent development and progress of these oxide-derived (OD) catalysts are discussed in the following section.

11.4.1 Oxide-Derived Cu

Kanan et al. demonstrated the electrochemical reduction of CO_2 on a metallic Cu catalyst resulting from the reduction of thick Cu_2O films prepared by annealing Cu foil in air [14]. This sample formed by the electroreduction of a Cu_2O layer during the CO_2 reduction electrolysis was analyzed by XRD and XPS, indicating the complete reduction of the copper oxide layer. By systematically varying the original annealing temperature and time, it was found that annealing the Cu foil in air at 500 °C for 12 h results in a thick Cu_2O layer, which then forms a nanostructured Cu (OD-Cu) during electrocatalysis that are able to produce CO with ~40% FE and HCCOH with ~33% FE at −0.5 V versus RHE (Fig. 11.14a, b). Notably, the selectivity for CO (~40% FE) was maintained during the whole electrolysis of 7 h, revealing a dramatically improved catalytic stability for CO_2 reduction on OD-Cu. As shown in Fig. 11.14b, the OD-Cu exhibited a high FE for CO_2 reduction at much lower overpotential in comparison with the bulk Cu electrodes. The Tafel slope of ~116 mV/dec for OD-Cu in the low overpotential range (Fig. 11.14c) indicates the Cu formed by the electrochemical reduction of Cu_2O during electrolysis is in favor of the formation of $CO_2^{\cdot-}$ intermediate, resulting in the improved catalytic activity for CO_2 reduction.

Furthermore, Kas et al. investigated the electrocatalytic CO_2 reduction on copper nanoparticles derived from different orientation of electrodeposited Cu_2O ([110][111][100]) [22]. The authors reported that the initial crystal orientation of Cu_2O had a minor effect on the catalytic selectivity for CO_2 reduction. However, it was found that the thickness of the initial Cu_2O layer strongly influenced the product selectivity, which is consistent with the previous study on OD-Cu [14].

The electrochemical reduction of CO_2 into fuels such as hydrocarbons on Cu catalysts is a complex multistep reaction with adsorbed intermediates, most notably adsorbed CO. Thus, CO reduction activity was investigated on OD-Cu, producing liquid products (ethanol, acetate and n-propanol) with nearly 50% FE at moderate potentials [53]. In order to correlate a particular structural feature of oxide-derived Cu with the catalytic activity, the authors proposed that the Cu catalysts

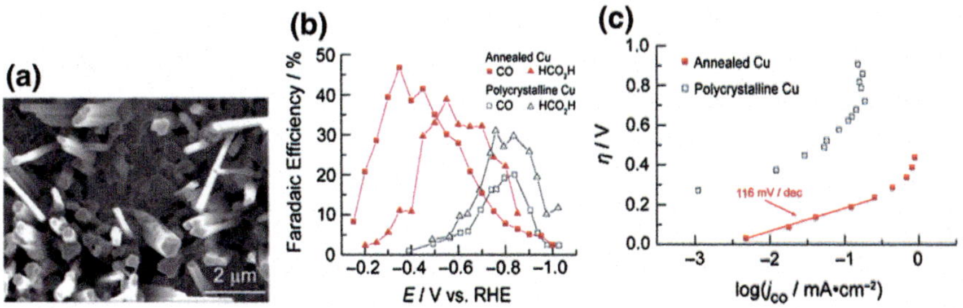

Fig. 11.14 **a** SEM image of OD-Cu. **b** Faradaic efficiencies for CO and HCOOH versus potential. And **c** CO partial current density Tafel plots for polycrystalline Cu and Cu annealed at 500 °C for 12 h. Reprinted with permission from Ref. [14]

electrochemically reduced from copper oxide may obtain a high density of grain boundary surfaces, likely resulting in a highly active sites for CO_2 reduction [53]. A temperature-programmed desorption (TPD) of CO was utilized to probe the surface chemistry of OD-Cu, revealing a high CO reduction activity on OD-Cu is linked to the active sites that bond CO more strongly in comparison with low-index and stepped Cu facets [60]. It was proposed that the strong binding sites with CO are supported by grain boundaries on the OD-Cu, which are only accessible in this nanostructured platform.

11.4.2 Oxide-Derived Au

It was reported that Au was the most efficient metal catalyst for CO_2 reduction, exhibiting the best catalytic selectivity and activity for the electrocatalytic reduction of CO_2 to CO among the identified bulk metals, as shown in Table 11.1. The Cu NP films prepared by the electrochemical reduction of thick Cu_2O layers exhibited a dramatically improved catalytic activity for the electroreduction of CO_2 to CO at lower overpotential with high resistance to deactivation, compared to Cu foils. Similar to this OD-Cu, OD-Au nanocatalysts were also studied for CO_2 reduction [13]. Initially, the Au oxide layers were prepared on Au foil electrodes by applying periodic square-wave pulsed potentials in 0.5 M H_2SO_4. After the pulsed anodization on Au, the Au oxide layers were synthesized on Au foils.

The Au oxide layers on Au foils were directly utilized for CO_2 reduction in CO_2-saturated electrolytes and were electrochemically reduced to metallic Au at the beginning of electrolysis (Fig. 11.15a shows the SEM images of the oxide-derived Au). The significant distinct CO_2 reduction activity for OD-Au and polycrystalline Au were observed at low overpotentials, as shown in Fig. 11.15b. At a low overpotential of 0.24 V, the FE for CO formation on the OD-Au was maintained at ~96% over the course of an 8 h electrolysis experiment. Furthermore, the Tafel slope of 114 mV/dec (Fig. 11.15b) on polycrystalline Au indicates the rate determining step is the initial electron transfer to form an adsorbed $CO_2^{\cdot-}$ intermediate

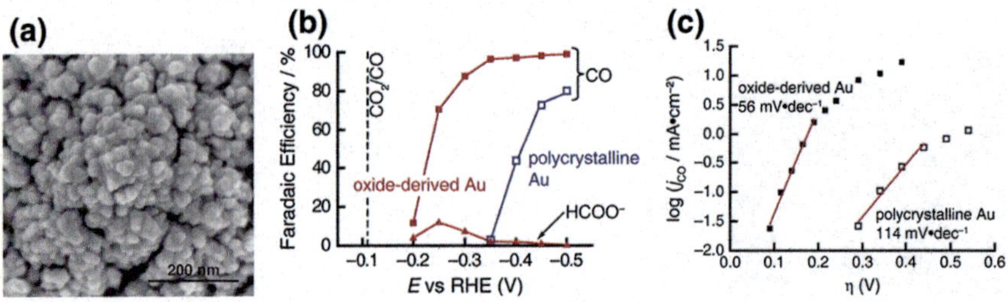

Fig. 11.15 **a** SEM image of oxide-derived Au NPs. **b** Faradaic efficiencies for CO and HCOOH, and **c** Tafel plots of CO partial current density of oxide-derived Au NPs and polycrystalline Au. Reprinted with permission from Ref. [13]

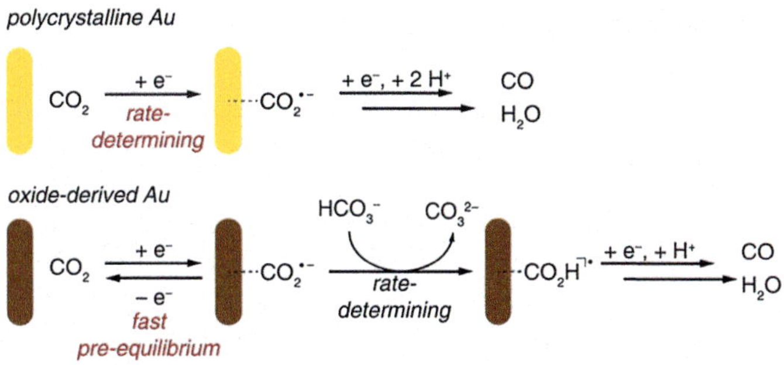

Scheme 11.2 Proposed mechanisms for CO_2 reduction to CO on polycrystalline Au and Oxide-Derived Au. Reprinted with permission from Ref. [13]

(Scheme 11.2). The poor stabilization of $CO_2^{\cdot-}$ intermediate is correlated with the high overpotential required on polycrystalline Au. In contrast, the active sites on OD-Au nanocatalysts offer the fast initial electron transfer (low Tafel slope in Fig. 11.15b), which results in the dramatically enhanced stabilization of $CO_2^{\cdot-}$ intermediate. The HCO_3^- concentration dependence studies on OD-Au showed an increase in the CO partial current density with gradually increasing the $NaHCO_3$ concentration, implying that the first proton donation from HCO_3^- is rate determining step on OD-Au (Scheme 11.2). Furthermore, a study on grain boundary effect showed that the catalytic activity for the electroreduction of CO_2 is linearly correlated with the grain boundary density on Au NPs and the high density of grain boundaries on OD-Au was thought to be linked to its enhanced catalytic performance [61].

11.4.3 *Oxide-Derived Pb*

Nanocrystalline Pb prepared by the electroreduction of PbO_2 exhibited a 700-fold lower H_2 selectivity compared to polycrystalline Pb, with maintaining the catalytic activity for CO_2 reduction [62]. OD-Pb showed a nearly 100% FE for HCOOH formation at the potential range from −1.0 to −0.75 V versus RHE in CO_2-saturated electrolytes, which is much better than the polycrystalline Pb (FF for HCOOH was 30% at −0.8 V vs. RHE). Notably, the high catalytic activity and selectivity for the electroreduction of CO_2 to HCOOH was more or less maintained during a prolonged electrolysis of 75 h. The dramatically improved CO_2 reduction performance on OD-Pb is attributed to its morphology and microstructure [62].

11.4.4 Oxide-Derived Ag

Recently, a porous-like nanostructured Ag catalyst (Fig. 11.16a, b) was prepared by electrochemically reducing Ag_2O which was synthesized by anodization of a Ag foil in alkaline solutions [63]. The results of the XRD patterns (Fig. 11.16c) and the Ag 3d XPS spectrum analysis (Fig. 11.16d) reveal that the electroreduction of Ag_2O to metallic Ag was complete within the detection limit of these characterization techniques. This OD-Ag exhibited an enhanced catalytic activity for the reduction of CO_2, and the high selectivity for CO was shifted by >400 mV towards a lower overpotential than that of untreated polycrystalline Ag. Notably, the OD-Ag is able to reduce CO_2 to CO with ~80% EF at a moderate overpotential of 0.49 V, which is dramatically higher compared to that (~4%) of untreated polycrystalline Ag at identical conditions. The dramatically improved catalytic activity and

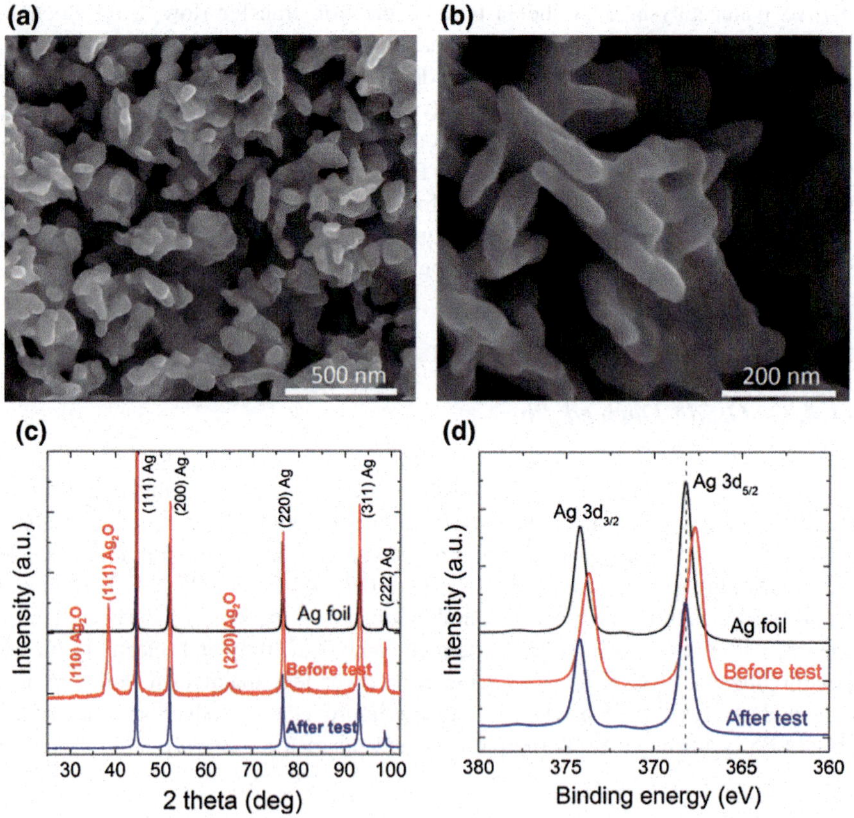

Fig. 11.16 **a** and **b** SEM images of oxide-derived Ag. **c** XRD patterns and **d** XPS spectrum of the polycrystalline Ag electrode (*black line*) and Ag oxide electrode before (*red line*) and after (*blue line*) CO_2 reduction electrolysis, respectively. Reprinted with permission from Ref. [63]

selectivity for CO_2 reduction to CO is likely correlated with of the nanostructured surfaces, resulting in highly active sites for stabilizing COOH˙ intermediate. In addition, it was demonstrated that the porous-like nanostructured Ag catalyst is likely capable of generating a high local pH near the catalyst surface, which may also play a role in the improved CO_2 reduction along with suppressed H_2 evolution.

11.5 Bimetallic Nanocatalysts

Electrode surfaces provide the active sites for CO_2 reactions, and the reactants and other related species adsorbed on the electrode play an important role in the product selectivity of the CO_2 reduction. CO_2 can be electrochemically reduced to CO, HCOOH, and hydrocarbons on metal electrodes in aqueous solution. The intermediates formed on the electrode surface directly affect the product formed in the final steps of the catalytic reaction. Therefore, it is important to know the binding energy at the surface of different metals for different reaction intermediates.

Table 11.2 shows a comparison of the binding strength of CO_2 reduction intermediate species on FCC (111) single metal electrodes (adapted from Ref. [31]). The product distribution and selectivity of CO_2 reduction could be influenced by whether or not the related intermediates are adsorbed. Alloying an element that that has high oxygen affinity into a catalyst may allow CHO to bind to the surface through the carbon and oxygen atoms, thus increasing the stability of CHO, which may affect the final product distribution and selectivity [30]. Thus, the interaction of the two different metallic atoms in the bimetallic alloy may significantly influence the catalytic activity and selectivity of alloy surface.

The electrocatalytic CO_2 reduction was performed on a nanostructured Cu-Au alloy prepared through electrochemical deposition with a nanoporous Cu film (NCF) as template, showing an improved catalytic selectivity for alcohols (methanol and ethanol) [64]. The authors found that the FE for alcohols was dependent on the nanostructured morphology and composition of Cu-Au alloys, as shown in Table 11.3. In addition, the FE of 15.9% for methanol was detected on $Cu_{63.9}Au_{36.1}$/NCF, which is ~19 times higher than that of pure Cu.

Yang et al. assembled monodisperse Au-Cu bimetallic NPs (Fig. 11.17a) with different compositions by implementing the solvent evaporation-mediated self-assembly approach and then the nanoparticle monolayer was transferred onto

Table 11.2 Comparison of the binding strength on FCC (111) transition metal facets. Reprinted with permission from Ref. [31]

Species on FCC(111) transition metal facets binding strength	
*CO	Rh > Pd > Ni > Pt > Cu > Au > Ag
*COH	Rh > Pt > Pd > Ni > Cu > Au > Ag
*CHO	Rh > Pt > Pd > Ni > Au > Cu > Ag
*OCH_3	Ni > Rh > Cu > Ag > Pd > Pt > Au
*CH_3	Pt > Rh > Ni > Pd > Cu > Au > Ag

Table 11.3 Faradic efficiencies of main liquid products with different electrodes. Reprinted with permission from Ref. [64]

Electrode	Faradaic efficiency (%)	
	HCOOH	Alcohols (methanol and ethanol)
Bulk Cu	11.5	4.8
Nanostrucutred Cu	27.4	9.7
$Cu_{63.9}Au_{36.1}$/NCF	12.6	28
$Cu_{70.6}Au_{29.4}$/NCF	18.1	17.2
$Cu_{81.3}Au_{19.7}$/NCF	21.2	11.5

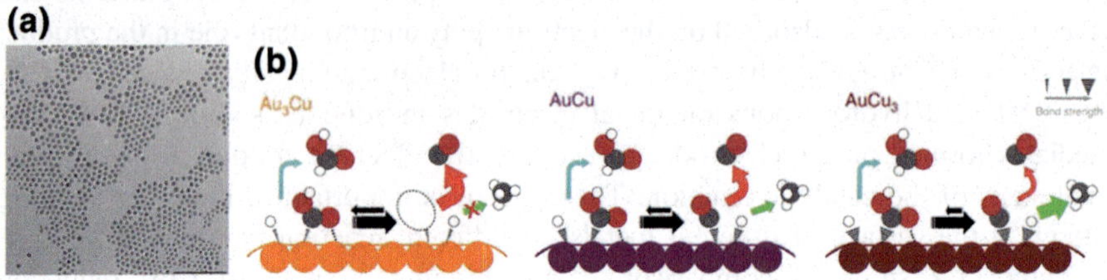

Fig. 11.17 **a** TEM image of $AuCu_3$ nanoparticles (scale bar, 100 nm). Average size 11.20 ± 1.65 nm. **b** Schematic showing the proposed mechanism for CO_2 reduction on the catalyst surface of Au-Cu bimetallic nanoparticles. *Filled circles* with *grey* is C, *red* is O and *white* is H. The relative intermediate binding strength is indicated by the stroke weight (on the *top right corner*). Additional binding between the COOH and the catalyst surface is presented as a *dotted line*. *Arrows* between the adsorbed COOH and adsorbed CO is to show the difference in probability of having COOH adsorbed on different types of surfaces. *Coloured arrows* indicate the pathway to each product: *red* for CO, *blue* for formate and *green* for hydrocarbons. *Larger arrows* indicate higher turnover. Reprinted with permission from Ref. [65]

a glassy carbon substrate for the electrochemical reduction of CO_2 [65]. With increasing the Cu content, different products were observed and the largest number in product distribution were found on pure Cu nanoparticles. The authors believed the degree of stabilization of the intermediates on these nanoparticle surfaces can be tuned by varying the composition of Au-Cu bimetallic NPs, which results in the different final products (Fig. 11.17b). Furthermore, the catalytic performance of these bimetallic catalysts was attributed to two factors: (1) the electronic effect on the binding strength of intermediates, which is linked to the change of electronic structure that is tuned with surface composition of the Au-Cu, and (2) the geometric effect that is correlated with the atomic arrangement at the active site that have a significant influence on the binding strength of intermediates [66].

The electrochemical CO_2 reduction was performed on a nanostructured Cu-In bimetallic alloy prepared by the in-situ electrochemical reduction of Cu_2O in $InSO_4$ electrolytes, and it was found that the In incorporation exhibited an enhanced selectivity for the electroreduction of CO_2 to CO with high catalytic stability [67]. At potential from −0.3 to −0.7 V versus RHE, CO was generated on Cu-In alloy catalysts as almost the only products in CO_2 reduction, and notably a FE of ~90%

for CO formation was achieved on the alloy catalysts at −0.5 V versus RHE. A DFT study suggested that the incorporation of In may cause both local electronic effect and local geometric effect. The presence of In atoms has a significant effect on the binding energy of the associated intermediates adsorbed on Cu. Furthermore, the dramatically enhanced FE for the conversion of CO_2 into CO is also achieved on nanostructured Cu-Sn bimetallic electrocatalysts at decreased overpotentials [68].

Recently, a mesoporous nanostructured Pd-Cu bimetallic electrocatalyst was fabricated by an electrodeposition method for CO_2 reduction and Pd_7-Cu_3 exhibited a FE of more than 80% for selective CO formation at −0.8 V versus RHE [69]. The mesoporous nanostructure has a roughened surface, which could offer more active sites for the selective reduction of CO_2 to CO. In addition, first principle calculations suggested that Pd atoms on the surface of the bimetallic catalysts act as reactive centers with an increased adsorption ability of CO_2 and $COOH^{\cdot}$, and the presence of Cu could modify the CO desorption ability. The authors believed that the catalytic activity is ascribed to the change of electronic structure of their neighboring element and adjustment of atomic arrangement in the active sites.

Koper et al. have used bimetallic Pd-Pt nanoparticles for the electrocatalytic CO_2 reduction [70]. They found that Pd-Pt nanoparticles had a low onset potential for the reduction of CO_2 to HCOOH, starting at 0 V versus RHE for HCOOH formation, which is close to the theoretical equilibrium potential of producing HCOOH (Eq. 11.2). In addition, the FE for HCOOH formation was dependent on the composition of the NPs and a high FE of 88% towards HCOOH formation was reached on Pd_{70}-Pt_{30} at an applied potential of −0.4 V versus RHE.

While the progress on bimetallic catalysts for the electrochemical CO_2 reduction performance has been made, the fundamental understanding of the relationship between the bimetallic catalyst and catalytic activity of CO_2 reduction is still not very clear. The further understanding of the correlation of bimetallic catalyst with catalytic activity of CO_2 reduction is critical for reasonable designing the high-performance bimetallic catalysts.

11.6 Nano Carbon Catalysts

Carbon is a remarkable element, which can exist in a variety of stable forms such as diamond (or diamond-like amorphous film), [71] graphite, fullerenes (C_{60}) and carbon nanotubes (CNTs), [72] as shown in Fig. 11.18. Nanostructured carbon materials (such as graphene and CNTs) have attracted a great deal of interest in the past because of their unique structure and interesting electronic, physical and chemical properties [73]. Recently, nanostructured carbon materials including graphene, nano-diamond, carbon nanofibers (CNFs) and CNTs have been applied as the electrodes in the electrocatalytic CO_2 reduction due to the high catalytic selectivity, low cost and high stability.

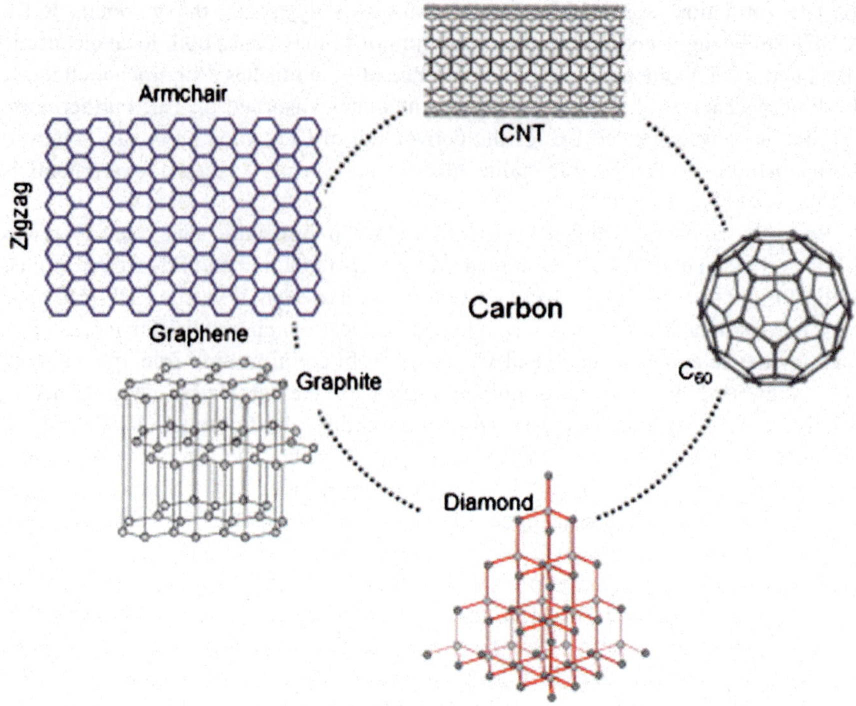

Fig. 11.18 Polymorphs of carbon. Reprinted with permission from Ref. [73]

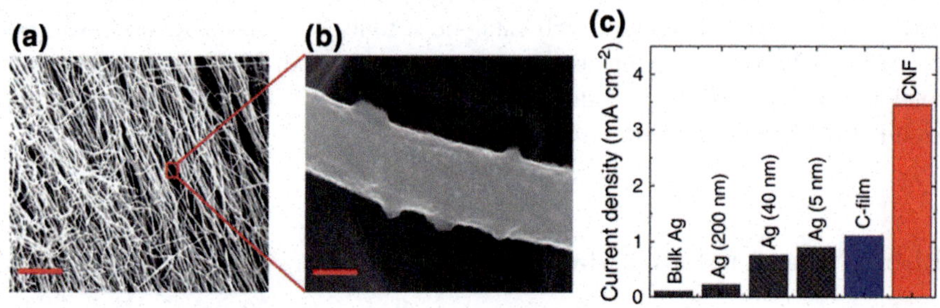

Fig. 11.19 **a** SEM image of the CNF mat displaying entangled fibers (scale bar, 5 mm), **b** high-resolution SEM image on individual fibers (scale bar, 200 nm). **c** Current density for CO_2 reduction at different (bulk Ag, Ag nanoparticles and CNF) electrodes in pure EMIM-BF4 electrolyte. Reprinted with permission from Ref. [74]

Kumar et al. reported the electrocatalytic CO_2 reduction on metal-free carbon nanofibers prepared by pyrolysis of electrospun nanofibre mats of heteroatomic polyacrylonitrile (PAN) polymer (Fig. 11.19) [74]. It was confirmed that the presence of nitrogen atoms (defects) in CNFs due to the existence of nitrogen atoms in the backbone of the precursor (PAN). The catalytic activity of CO_2 reduction was performed in CO_2-saturated 1-ethyl-3-methylimidazolium tetrafluoroborate (EMIM-BF4) electrolyte due to its high CO_2 solubility. The CNFs catalyst showed a very small overpotential of 0.17 V for driving CO_2 reduction to CO (the CO_2/CO equilibrium potential is −0.11 V vs. RHE) and more than an order of magnitude higher current density in comparison with the bulk silver catalyst under similar experimental conditions (Fig. 11.19). Furthermore, a FE of 98% for CO formation was detected at the potential of −0.57 V versus SHE. The high CO_2 reduction performance of CNFs is attributed to its nanofibrillar morphology and rough surface (due to the fractal-like corrugations on the surface) which offers high number of active catalytic sites and high binding energies of key intermediates on the CNF surface.

Nitrogen-doped carbon nanotubes (NCNTs) followed with the adsorption of Polyethylenimine (PEI) functions as a co-catalyst were synthesized by ammonia plasma treatment of CNTs on glassy carbon substrates for CO_2 reduction [75]. The combination of N-doping of CNTs and PEI exhibited a significant reduction for overpotential and improved catalytic performance for the selective reduction of CO_2 to HCOOH with a high FE (87%) and current density (9.5 mA/cm^2) in CO_2-saturated 0.1 $KHCO_3$ aqueous solutions. The rate determining step for CO_2 reduction is the initial electron transfer to CO_2 for the $CO_2^{\cdot -}$ intermediate and CO_2 is adsorbed to the basic nitrogen binding sites in NCNT for reducing to $CO_2^{\cdot -}$ in the proposed mechanism. The PEI works as a co-catalyst to improve the stabilization of $CO_2^{\cdot -}$ intermediate by a H-bond interaction, $\mathrm{NCNT{-}N{-}C(O)O^{\cdot -} \cdots H{-}N{-}PEI}$. Thus, the PEI overlayer and N-doping could provide a synergistic effect, creating a local environment for stabilizing $CO_2^{\cdot -}$ intermediate from CO_2 with significantly reduced overpotential.

N incorporated CNT arrays with a total N content of 5 atom% were fabricated by liquid chemical vapor deposition (CVD) method for CO_2 reduction, acting as a highly selective, efficient and stable electrocatalyst for the catalytic reduction of CO_2 to CO (Fig. 11.20) [76]. The high selectivity toward the production of CO was achieved on NCNTs (The FE of ~80% for CO formation), as shown in Fig. 11.20b. In contrast, pristine CNTs show much high overpotential for CO_2 reduction with low catalytic activity and selectivity for the electroreduction of CO_2 to CO (Fig. 11.20). The improved catalytic activity for CO_2 reduction on NCNTs is attributed high electrical conductivity, pyridinic N defects (preferable catalytic sites), low free energy for CO_2 activation and high barrier for hydrogen evolution. In addition, the reaction mechanism of electroreduction of CO_2 to CO on the pristine CNT and NCNT surfaces through the adsorbed intermediates was studied by DFT calculations, suggesting that the suitable binding energy of the key intermediates enables strong adsorption of COOH and feasible CO desorption that contributes to the high selectivity toward CO formation (Fig. 11.21).

N-doped three-dimensional (3D) graphene foam (NG) fabricated by CVD has been used as a electrocatalyst for the electrocatalytic CO_2 reduction [77]. The various N-defect structures and content incorporated graphene foams were synthesized by a post growth doping process performed at various temperatures for CO_2 reduction (Fig. 11.22). N-doped NG processed at 800 °C (denoted as NG-800) exhibited that the selective reduction of CO_2 to CO started to be observed at −0.30 V versus RHE with a FE of 25%, which corresponds to an overpotential of 0.19 V for CO production. In addition, the maximum FE for CO production on NG-800 was reached to be ~85% with high stability during the electrolysis of 5 h at −0.58 V versus RHE, corresponding to an overpotential of 0.47 V. The correlation between catalytic activity and the N-defects along with its corresponding potential versus concentrations of different N species (Fig. 11.22c) shows that the maximum FE for CO formation increases and the corresponding potential shifts anodically with increasing the pyridinic-N content. These results suggested pyridinic-N is the most active site for CO_2 reduction. The high catalytic selectivity for CO and low overpotential on the NG-800 could be linked to the high pyridinic-N concentration which acts as the active site for CO_2 reduction.

N-doped nanodiamond coated on Si rod array has been demonstrated as an efficient and stable nonmetallic catalyst for the electrocatalytic CO_2 reduction in CO_2-saturated 0.5 $NaHCO_3$ electrolytes [78]. The onset potential for CO_2 reduction was found to be −0.36 V versus RHE. The FE for acetate (CH_3COOH) formation

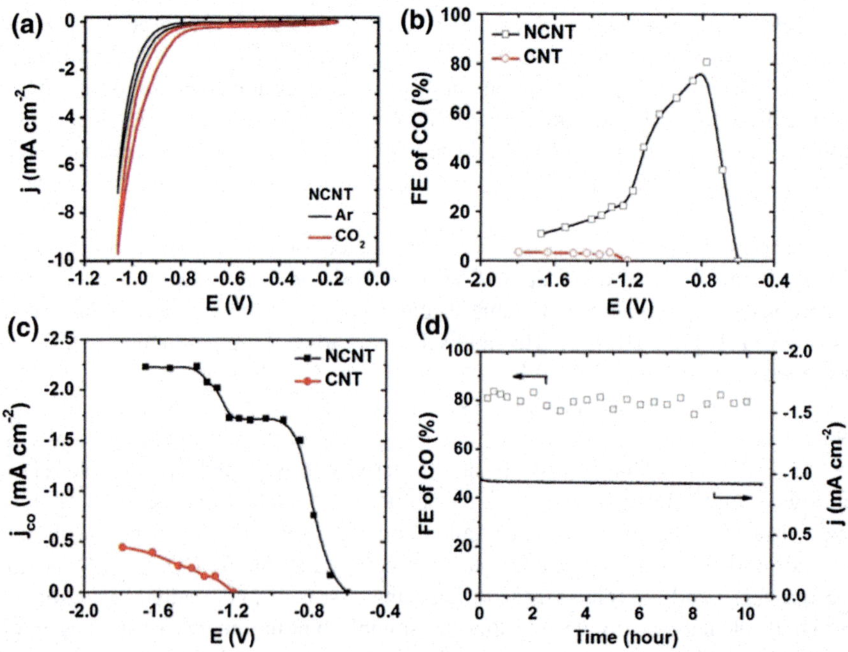

Fig. 11.20 Performance of NCNTs for electrochemical reduction of CO_2. **a** CVs for NCNTs in Ar- and CO_2-saturated 0.1 M $KHCO_3$ electrolyte, 50 mVs^{-1}. **b** Dependence of FE of CO on applied cell potential during electrocatalysis of CO_2 reduction for both NCNTs and CNTs catalysts. **c** Partial current density of CO versus applied cell potential for NCNTs and CNTs catalysts. **d** Stability of performance of NCNTs for CO_2 reduction operated at potentiostatic mode of −0.8 V for 10 h. Both current density and FE of CO remain steady over the duration of the test. Reprinted with permission from Ref. [76]

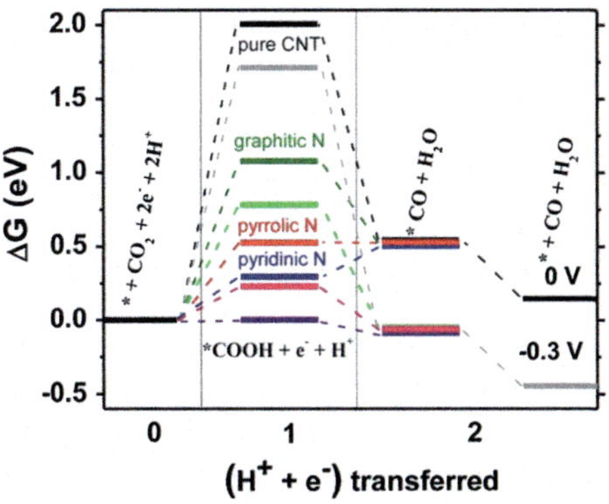

Fig. 11.21 Calculated free energy diagram for CO_2 electroreduction to CO on pristine CNTs and NCNTs. Reprinted with permission from Ref. [76]

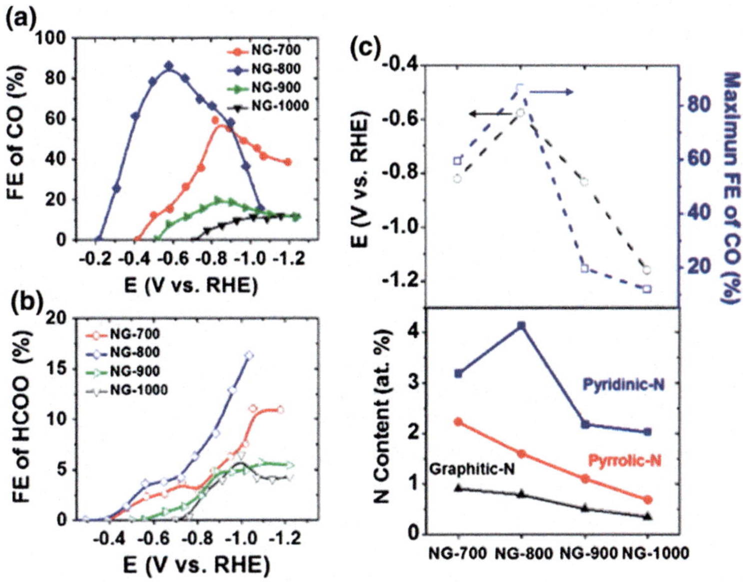

Fig. 11.22 Comparison of the electrocatalytic activities of nitrogen-doped graphene with doping temperature ranging from 700 to 1000 °C. **a** Faradaic efficiency of CO versus potential. **b** Faradaic efficiency of $HCOO^-$ versus potential. **c** Maximum faradaic efficiency of CO and its corresponding potential versus N functionality content. Reprinted with permission from Ref. [77]

(77.3–77.6%) was detected along with HCOOH (13.6–14.6%) at the applied potentials from −0.55 to −1.3 V versus RHE. It was found that the total FE of CH_3COOH and HCOOH reached 91.8% at −1.0 V versus RHE. The improved CO_2 reduction activity was ascribed to N-doped sp^3 carbon which was highly active for the electrocatalytic CO_2 reduction.

11.7 Summary and Outlook

This chapter provides a review of the development and recent trends on nanostructured catalysts for the electrocatalytic reduction of CO_2. Nanostructured Au, Ag, Zn, Pd, Cu, bimetallic and oxide-derived metal catalysts, and the recent discoveries in the application of metal-free nanostructured carbon for the electroreduction of CO_2 were reviewed.

According to the results discussed in this chapter, nanostructured catalysts are capable of electrochemically reducing CO_2 with high selectivity and efficiency. It is generally accepted that nanostructured metal catalysts could offer an abundance of edge sites which are active for certain intermediate formation (e.g. $COOH^{\cdot}$), resulting in the enhanced catalytic CO_2 reduction performance. In other words, the nanostructured catalysts could offer low-coordinated surface sites, which may facilitate the key intermediates formation by reducing the activation barrier of CO_2 reduction. Furthermore, the large surface area (roughened surface) provided by nanostructured catalysts is capable of adapting contamination and enhancing tolerance to heavy metal impurities, resulting in a better catalytic stability performance in CO_2 reduction on nanostructured catalysts in comparison with bulk catalysts.

The interaction of the two different metallic atoms in the bimetallic alloys may significantly influence the catalytic activity for the electrocatalytic CO_2 reduction. The changed catalytic activity on these bimetallic catalysts could be attributed to: (1) the electronic effect on the binding strength of intermediates, which is due to that electronic structure may be tuned by surface composition of two different metallic elements, (2) the geometric effect that is linked to the atomic arrangement that plays a significant role in the binding strength of intermediates. Thereby, it is critical to rationally design a bimetallic catalysts that could facilitate certain intermediate formation, leading to selective CO_2 reduction.

Nanostructured carbon could be a promising candidate as a metal-free electrocatalyst due to its low cost and high stability and catalytic activity for CO_2 reduction. N defects (or N doping) were generally presented in nanostructured carbon catalysts with the significantly enhanced catalytic activity and selectivity because N species were suggested to be the active site for CO_2 reduction. However, pristine nanostructured carbon catalysts have poor catalytic activity for CO_2 reduction. Different element doping could be designed to improve the catalytic activity on nanostructured carbon catalysts. We believe that the nanostructured carbon catalysts are also highly important for CO_2 reduction due to their high stability, low cost, earth abundance and comparable catalytic performance with expensive noble metal catalysts.

References

1. Davis, S.J., K. Caldeira, and H.D. Matthews. 2010. Future CO_2 emissions and climate change from existing energy infrastructure. *Science* 329: 1330–1333. doi:10.1126/science.1188566.
2. Hoegh-Guldberg, O., P.J. Mumby, A.J. Hooten, R.S. Steneck, P. Greenfield, E. Gomez, C.D. Harvell, P.F. Sale, A.J. Edwards, K. Caldeira, N. Knowlton, C.M. Eakin, R. Iglesias-Prieto, N. Muthiga, R.H. Bradbury, A. Dubi, and M.E. Hatziolos. 2007. Coral reefs under rapid climate change and ocean acidification. *Science* 318: 1737–1742. doi:10.1126/science.1152509.
3. Orr, J.C., V.J. Fabry, O. Aumont, L. Bopp, S.C. Doney, R.A. Feely, A. Gnanadesikan, N. Gruber, A. Ishida, F. Joos, R.M. Key, K. Lindsay, E. Maier-Reimer, R. Matear, P. Monfray, A. Mouchet, R.G. Najjar, G.-K. Plattner, K.B. Rodgers, C.L. Sabine, J.L. Sarmiento, R. Schlitzer, R.D. Slater, I.J. Totterdell, M.-F. Weirig, Y. Yamanaka, and A. Yool. 2005. Anthropogenic ocean acidification over the twenty-first century and its impact on calcifying organisms. *Nature* 437: 681–686. doi:10.1038/nature04095.
4. Hall-Spencer, J.M., R. Rodolfo-Metalpa, S. Martin, E. Ransome, M. Fine, S.M. Turner, S. J. Rowley, D. Tedesco, and M.-C. Buia. 2008. Volcanic carbon dioxide vents show ecosystem effects of ocean acidification. *Nature* 454: 96–99. doi:10.1038/nature07051.
5. Lu, Q., J. Rosen, Y. Zhou, G.S. Hutchings, Y.C. Kimmel, J.G. Chen, and F. Jiao. 2014. A selective and efficient electrocatalyst for carbon dioxide reduction. *Nature Communications* 5: 3242. doi:10.1038/ncomms4242.
6. Whipple, D.T., and P.J.A. Kenis. 2010. Prospects of CO_2 utilization via direct heterogeneous electrochemical reduction. *The Journal of Physical Chemistry Letters* 1: 3451–3458. doi:10.1021/jz1012627.
7. DiMeglio, J.L., and J. Rosenthal. 2013. Selective conversion of CO_2 to CO with high efficiency using an inexpensive bismuth-based electrocatalyst. *Journal of the American Chemical Society* 135: 8798–8801. doi:10.1021/ja4033549.
8. Hori, Y. 2008. Modern aspects of electrochemistry. In *Electrochemical CO_2 Reduction on Metal Electrodes*, vol. 42, ed. Vayenas, E.C.G., R.E. White, and M.E. Gamboa-Aldeco, 89. New York: Springer. doi:10.1007/978-0-387-49489-0_3.
9. Kuhl, K.P., E.R. Cave, D.N. Abram, and T.F. Jaramillo. 2012. New insights into the electrochemical reduction of carbon dioxide on metallic copper surfaces. *Energy & Environmental Science* 5: 7050–7059. doi:10.1039/c2ee21234j.
10. Qiao, J., Y. Liu, F. Hong, and J. Zhang. 2014. A review of catalysts for the electroreduction of carbon dioxide to produce low-carbon fuels. *Chemical Society Reviews* 43: 631–675. doi:10.1039/c3cs60323g.
11. Gattrell, M., N. Gupta, and A. Co. 2006. A review of the aqueous electrochemical reduction of CO_2 to hydrocarbons at copper. *Journal of Electroanalytical Chemistry* 594: 1–19. doi:10.1016/j.jelechem.2006.05.013.
12. Chen, Y., and M.W. Kanan. 2012. Tin oxide dependence of the CO_2 reduction efficiency on tin electrodes and enhanced activity for tin/tin oxide thin-film catalysts. *Journal of the American Chemical Society* 134: 1986–1989. doi:10.1021/ja2108799.
13. Chen, Y., C.W. Li, and M.W. Kanan. 2012. Aqueous CO_2 reduction at very low overpotential on oxide-derived Au nanoparticles. *Journal of the American Chemical Society* 134: 19969–19972. doi:10.1021/ja309317u.
14. Li, C.W., and M.W. Kanan. 2012. CO_2 reduction at low overpotential on Cu electrodes resulting from the reduction of thick Cu_2O films. *Journal of the American Chemical Society* 134: 7231–7234. doi:10.1021/ja3010978.
15. Hori, Y., K. Kikuchi, and S. Suzuki. 1985. Production of CO and CH_4 in electrochemical reduction of CO_2 at metal electrodes in aqueous hydrogencarbonate solution. *Chemistry Letters* 1695–1698. doi:10.1246/cl.1985.1695.

16. Hori, Y., H. Wakebe, T. Tsukamoto, and O. Koga. 1994. Electrocatalytic process of CO selectivity in electrochemical reduction of CO_2 at metal electrodes in aqueous media. *Electrochimica Acta* 39: 1833–1839. doi:10.1016/0013-4686(94)85172-7.
17. Hoshi, N., M. Kato, and Y. Hori. 1997. Electrochemical reduction of CO_2 on single crystal electrodes of silver Ag(111), Ag(100) and Ag(110). *Journal of Electroanalytical Chemistry* 440: 283–286. doi:10.1016/S0022-0728(97)00447-6.
18. Hori, Y., I. Takahashi, O. Koga, and N. Hoshi. 2002. Selective formation of C2 compounds from electrochemical reduction of CO_2 at a series of copper single crystal electrodes. *The Journal of Physical Chemistry B* 106: 15–17. doi:10.1021/jp013478d.
19. Xiao, J., A. Kuc, T. Frauenheim, and T. Heine. 2014. CO_2 reduction at low overpotential on Cu electrodes in the presence of impurities at the subsurface. *Journal of Materials Chemistry A* 2: 4885. doi:10.1039/c3ta14755j.
20. Hatsukade, T., K.P. Kuhl, E.R. Cave, D.N. Abram, and T.F. Jaramillo. 2014. Insights into the electrocatalytic reduction of CO_2 on metallic silver surfaces. *Physical Chemistry Chemical Physics: PCCP* 16: 13814–13819. doi:10.1039/C4CP00692E.
21. Reske, R., H. Mistry, F. Behafarid, B. Roldan Cuenya, and P. Strasser. 2014. Particle size effects in the catalytic electroreduction of CO_2 on Cu nanoparticles. *Journal of the American Chemical Society* 136: 6978–6986. doi:10.1021/ja500328k.
22. Kas, R., R. Kortlever, A. Milbrat, M.T.M. Koper, G. Mul, and J. Baltrusaitis. 2014. Electrochemical CO_2 reduction on Cu2O-derived copper nanoparticles: Controlling the catalytic selectivity of hydrocarbons. *Physical Chemistry Chemical Physics: PCCP* 16: 12194. doi:10.1039/c4cp01520g.
23. Sen, S., D. Liu, and G.T.R. Palmore. 2014. Electrochemical reduction of CO_2 at copper nanofoams. *ACS Catalysis* 4: 3091–3095. doi:10.1021/cs500522g.
24. Tang, W., A.A. Peterson, A.S. Varela, Z.P. Jovanov, L. Bech, W.J. Durand, S. Dahl, J.K. Nørskov, and I. Chorkendorff. 2012. The importance of surface morphology in controlling the selectivity of polycrystalline copper for CO_2 electroreduction. *Physical Chemistry Chemical Physics* 14: 76–81. doi:10.1039/c1cp22700a.
25. Hansen, H.A., J.B. Varley, A.A. Peterson, and J.K. Nørskov. 2013. Understanding trends in the electrocatalytic activity of metals and enzymes for CO_2 reduction to CO. *The Journal of Physical Chemistry Letters* 4: 388–392. doi:10.1021/jz3021155.
26. Y. Hori, K. Kikuchi, A. Murata, and S. Suzuki. 1986. Production of methane and ethylene in electrochemical reduction of carbon dioxide at copper electrode in aqueous hydrogencarbonate solution. *Chemistry Letters* 897–898. doi:10.1246/cl.1986.897.
27. Li, Y., S.H. Chan, and Q. Sun. 2015. Heterogeneous catalytic conversion of CO_2: A comprehensive theoretical review. *Nanoscale* 7: 8663–8683. doi:10.1039/c5nr00092k.
28. Kortlever, R., J. Shen, K.J.P. Schouten, F. Calle-Vallejo, and M.T.M. Koper. 2015. Catalysts and reaction pathways for the electrochemical reduction of carbon dioxide. *The Journal of Physical Chemistry Letters* 6: 4073–4082. doi:10.1021/acs.jpclett.5b01559.
29. Peterson, A.A., F. Abild-Pedersen, F. Studt, J. Rossmeisl, and J.K. Nørskov. 2010. How copper catalyzes the electroreduction of carbon dioxide into hydrocarbon fuels. *Energy & Environmental Science* 3: 1311. doi:10.1039/c0ee00071j.
30. Peterson, A.A., and J.K. Nørskov. 2012. Activity descriptors for CO_2 electroreduction to methane on transition-metal catalysts. *The Journal of Physical Chemistry Letters* 3: 251–258. doi:10.1021/jz201461p.
31. Shi, C., H.A. Hansen, A.C. Lausche, and J.K. Nørskov. 2014. Trends in electrochemical CO_2 reduction activity for open and close-packed metal surfaces. *Physical Chemistry Chemical Physics* 16: 4720. doi:10.1039/c3cp54822h.
32. Kuhl, K.P., T. Hatsukade, E.R. Cave, D.N. Abram, J. Kibsgaard, and T.F. Jaramillo. 2014. Electrocatalytic conversion of carbon dioxide to methane and methanol on transition metal surfaces. *Journal of the American Chemical Society* 136: 14107–14113. doi:10.1021/ja505791r.
33. Mohamed, R.M., D.L. McKinney, and W.M. Sigmund. 2012. Enhanced nanocatalysts. *Materials Science and Engineering: R: Reports* 73: 1–13. doi:10.1016/j.mser.2011.09.001.

34. Lu, Q., J. Rosen, and F. Jiao. 2015. Nanostructured metallic electrocatalysts for carbon dioxide reduction. *ChemCatChem* 7: 38–47. doi:10.1002/cctc.201402669.
35. Rosen, J., G.S. Hutchings, Q. Lu, S. Rivera, Y. Zhou, D.G. Vlachos, and F. Jiao. 2015. Mechanistic insights into the electrochemical reduction of CO_2 to CO on nanostructured Ag surfaces. *ACS Catalysis* 5: 4293–4299. doi:10.1021/acscatal.5b00840.
36. Nursanto, E.B., H.S. Jeon, C. Kim, M.S. Jee, J.H. Koh, Y.J. Hwang, and B.K. Min. 2016. Gold catalyst reactivity for CO_2 electro-reduction: From nano particle to layer. *Catalysis Today* 260: 107–111. doi:10.1016/j.cattod.2015.05.017.
37. Lee, H.-E., K.D. Yang, S.M. Yoon, H.-Y. Ahn, Y.Y. Lee, H. Chang, D.H. Jeong, Y.-S. Lee, M.Y. Kim, and K.T. Nam. 2015. Concave rhombic dodecahedral Au nanocatalyst with multiple high-index facets for CO_2 reduction. *ACS Nano* 9: 8384–8393. doi:10.1021/acsnano.5b03065.
38. Kauffman, D.R., D. Alfonso, C. Matranga, H. Qian, and R. Jin. 2012. Experimental and computational investigation of Au_{25} clusters and CO_2: A unique interaction and enhanced electrocatalytic activity. *Journal of the American Chemical Society* 134: 10237–10243. doi:10.1021/ja303259q.
39. Zhu, W., R. Michalsky, Ö. Metin, H. Lv, S. Guo, C.J. Wright, X. Sun, A.A. Peterson, and S. Sun. 2013. Monodisperse Au nanoparticles for selective electrocatalytic reduction of CO_2 to CO. *Journal of the American Chemical Society* 135: 16833–16836. doi:10.1021/ja409445p.
40. Zhu, W., Y.-J. Zhang, H. Zhang, H. Lv, Q. Li, R. Michalsky, A.A. Peterson, and S. Sun. 2014. Active and selective conversion of CO_2 to CO on ultrathin Au nanowires. *Journal of the American Chemical Society* 136: 16132–16135. doi:10.1021/ja5095099.
41. Kim, C., H.S. Jeon, T. Eom, M.S. Jee, H. Kim, C.M. Friend, B.K. Min, and Y.J. Hwang. 2015. Achieving selective and efficient electrocatalytic activity for CO_2 reduction using immobilized silver nanoparticles. *Journal of the American Chemical Society* 137: 13844–13850. doi:10.1021/jacs.5b06568.
42. Salehi-Khojin, A., H.M. Jhong, B.A. Rosen, W. Zhu, S. Ma, P.J.A. Kenis, and R.I. Masel. 2013. Nanoparticle silver catalysts that show enhanced activity for carbon dioxide electrolysis. *Journal of Physical Chemistry C* 117: 1627–1632. doi:10.1021/jp310509z.
43. Rosen, J., G.S. Hutchings, Q. Lu, R.V. Forest, A. Moore, and F. Jiao. 2015. Electrodeposited Zn dendrites with enhanced CO selectivity for electrocatalytic CO_2 reduction. *ACS Catalysis* 5: 4586–4591. doi:10.1021/acscatal.5b00922.
44. Gao, D., H. Zhou, J. Wang, S. Miao, F. Yang, G. Wang, J. Wang, and X. Bao. 2015. Size-dependent electrocatalytic reduction of CO_2 over Pd nanoparticles. *Journal of the American Chemical Society* 137: 4288–4291. doi:10.1021/jacs.5b00046.
45. Hinogami, R., S. Yotsuhashi, M. Deguchi, Y. Zenitani, H. Hashiba, and Y. Yamada. 2012. Electrochemical reduction of carbon dioxide using a copper rubeanate metal organic framework. *ECS Electrochemistry Letters* 1: H17–H19. doi:10.1149/2.001204eel.
46. Kornienko, N., Y. Zhao, C.S. Kley, C. Zhu, D. Kim, S. Lin, C.J. Chang, O.M. Yaghi, and P. Yang. 2015. Metal-organic frameworks for electrocatalytic reduction of carbon dioxide. *Journal of the American Chemical Society* 137: 14129–14135. doi:10.1021/jacs.5b08212.
47. Hod, I., M.D. Sampson, P. Deria, C.P. Kubiak, O.K. Farha, and J.T. Hupp. 2015. Fe-porphyrin-based metal-organic framework films as high-surface concentration, heterogeneous catalysts for electrochemical reduction of CO_2. *ACS Catalysis* 5: 6302–6309. doi:10.1021/acscatal.5b01767.
48. Hori, Y., H. Wakebe, T. Tsukamoto, and O. Koga. 1995. Adsorption of CO accompanied with simultaneous charge transfer on copper single crystal electrodes related with electrochemical reduction of CO_2 to hydrocarbons. *Surface Science* 335: 258–263. doi:10.1016/0039-6028(95)00441-6.
49. Schouten, K.J.P., Z. Qin, E.P. Gallent, and M.T.M. Koper. 2012. Two pathways for the formation of ethylene in CO reduction on single-crystal copper electrodes. *Journal of the American Chemical Society* 134: 9864–9867. doi:10.1021/ja302668n.

50. Hori, Y., R. Takahashi, Y. Yoshinami, and A. Murata. 1997. Electrochemical reduction of CO at a copper electrode. *The Journal of Physical Chemistry B* 101: 7075–7081. doi:10.1021/jp970284i.
51. Schouten, K.J.P., Y. Kwon, C.J.M. van der Ham, Z. Qin, and M.T.M. Koper. 2011. A new mechanism for the selectivity to C1 and C2 species in the electrochemical reduction of carbon dioxide on copper electrodes. *Chemical Science* 2: 1902. doi:10.1039/c1sc00277e.
52. Hori, Y., I. Takahashi, O. Koga, and N. Hoshi. 2003. Electrochemical reduction of carbon dioxide at various series of copper single crystal electrodes. *Journal of Molecular Catalysis A: Chemical* 199: 39–47. doi:10.1016/S1381-1169(03)00016-5.
53. Li, C.W., J. Ciston, and M.W. Kanan. 2014. Electroreduction of carbon monoxide to liquid fuel on oxide-derived nanocrystalline copper. *Nature* 508: 504–507. doi:10.1038/nature13249 .
54. Manthiram, K., B.J. Beberwyck, and A.P. Alivisatos. 2014. Enhanced electrochemical methanation of carbon dioxide with a dispersible nanoscale copper catalyst. *Journal of the American Chemical Society* 136: 13319–13325. doi:10.1021/ja5065284.
55. Ma, M., K. Djanashvili, and W.A. Smith. 2015. Selective electrochemical reduction of CO_2 to CO on CuO-derived Cu nanowires. *Physical Chemistry Chemical Physics: PCCP* 17: 20861–20867. doi:10.1039/C5CP03559G.
56. Zhang, W., X. Wen, S. Yang, Y. Berta, and Z.L. Wang. 2003. Single-crystalline scroll-type nanotube arrays of copper hydroxide synthesized at room temperature. *Advanced Materials* 15: 822–825. doi:10.1002/adma.200304840.
57. Lin, C., Y. Lai, D. Mersch, and E. Reisner. 2012. $Cu_2O|NiOx$ nanocomposite as an inexpensive photocathode in photoelectrochemical water splitting. *Chemical Science* 3: 3482–3487. doi:10.1039/c2sc20874a.
58. Raciti, D., K.J. Livi, and C. Wang. 2015. Highly dense Cu nanowires for low-overpotential CO_2 reduction. *Nano Letters* 15: 6829–6835. doi:10.1021/acs.nanolett.5b03298.
59. Ma, M., K. Djanashvili, and W.A. Smith. 2016. Controllable hydrocarbon formation from the electrochemical reduction of CO_2 over Cu nanowire arrays. *Angewandte Chemie International Edition* 55: 6680–6684. doi:10.1002/anie.201601282.
60. Verdaguer-Casadevall, A., C.W. Li, T.P. Johansson, S.B. Scott, J.T. McKeown, M. Kumar, I. E.L. Stephens, M.W. Kanan, and I. Chorkendorff. 2015. Probing the active surface sites for CO reduction on oxide-derived copper electrocatalysts. *Journal of the American Chemical Society* 137: 9808–9811. doi:10.1021/jacs.5b06227.
61. Feng, X., K. Jiang, S. Fan, and M.W. Kanan. 2015. Grain-boundary-dependent CO_2 electroreduction activity. *Journal of the American Chemical Society* 137: 4606–4609. doi:10.1021/ja5130513.
62. Lee, C.H., and M.W. Kanan. 2015. Controlling H+ vs CO_2 reduction selectivity on Pb electrodes. *ACS Catalysis* 5: 465–469. doi:10.1021/cs5017672.
63. Ma, M., B.J. Trześniewski, J. Xie, and W.A. Smith. 2016. Selective and efficient reduction of carbon dioxide to carbon monoxide on oxide-derived nanostructured silver electrocatalysts. *Angewandte Chemie International Edition* 55: 9748–9752. doi:10.1002/anie.201604654.
64. Jia, F., X. Yu, and L. Zhang. 2014. Enhanced selectivity for the electrochemical reduction of CO_2 to alcohols in aqueous solution with nanostructured Cu-Au alloy as catalyst. *Journal of Power Sources* 252: 85–89. doi:10.1016/j.jpowsour.2013.12.002.
65. Kim, D., J. Resasco, Y. Yu, A.M. Asiri, and P. Yang. 2014. Synergistic geometric and electronic effects for electrochemical reduction of carbon dioxide using gold–copper bimetallic nanoparticles. *Nature Communications* 5: 4948. doi:10.1038/ncomms5948.
66. Siahrostami, S., A. Verdaguer-Casadevall, M. Karamad, D. Deiana, P. Malacrida, B. Wickman, M. Escudero-Escribano, E.A. Paoli, R. Frydendal, T.W. Hansen, I. Chorkendorff, I.E.L. Stephens, and J. Rossmeisl. 2013. Enabling direct H_2O_2 production through rational electrocatalyst design. *Nature Materials* 12: 1137–1143. doi:10.1038/nmat3795.
67. Rasul, S., D.H. Anjum, A. Jedidi, Y. Minenkov, L. Cavallo, and K. Takanabe. 2015. A highly selective copper-indium bimetallic electrocatalyst for the electrochemical reduction of

aqueous CO_2 to CO. *Angewandte Chemie International Edition* 54: 2146–2150. doi:10.1002/anie.201410233.
68. Sarfraz, S., A.T. Garcia-Esparza, A. Jedidi, L. Cavallo, and K. Takanabe. 2016. Cu–Sn bimetallic catalyst for selective aqueous electroreduction of CO_2 to CO. *ACS Catalysis* 6: 2842–2851. doi:10.1021/acscatal.6b00269.
69. Li, M., J. Wang, P. Li, K. Chang, C. Li, T. Wang, B. Jiang, H. Zhang, H. Liu, Y. Yamauchi, N. Umezawa, and J. Ye. 2016. Mesoporous palladium–copper bimetallic electrodes for selective electrocatalytic reduction of aqueous CO_2 to CO. *Journal of Materials Chemistry A* 4: 4776–4782. doi:10.1039/C6TA00487C.
70. Kortlever, R., I. Peters, S. Koper, and M.T.M. Koper. 2015. Electrochemical CO_2 reduction to formic acid at low overpotential and with high faradaic efficiency on carbon-supported bimetallic Pd–Pt nanoparticles. *ACS Catalysis* 5: 3916–3923. doi:10.1021/acscatal.5b00602.
71. Robertson, J. 2002. Diamond-like amorphous carbon. *Materials Science and Engineering: R: Reports* 37: 129–281. doi:10.1016/S0927-796X(02)00005-0.
72. Iijima, S. 1991. Helical microtubules of graphitic carbon. *Nature* 354: 56–58. doi:10.1038/354056a0.
73. Zhu, H., J. Wei, K. Wang, and D. Wu. 2009. Applications of carbon materials in photovoltaic solar cells. *Solar Energy Materials and Solar Cells* 93: 1461–1470. doi:10.1016/j.solmat.2009.04.006.
74. Kumar, B., M. Asadi, D. Pisasale, S. Sinha-Ray, B.A. Rosen, R. Haasch, J. Abiade, A.L. Yarin, and A. Salehi-Khojin. 2013. Renewable and metal-free carbon nanofibre catalysts for carbon dioxide reduction. *Nature Communications* 4: 2819. doi:10.1038/ncomms3819.
75. Zhang, S., P. Kang, S. Ubnoske, M.K. Brennaman, N. Song, R.L. House, J.T. Glass, and T. J. Meyer. 2014. Polyethylenimine-enhanced electrocatalytic reduction of CO_2 to formate at nitrogen-doped carbon nanomaterials. *Journal of the American Chemical Society* 136: 7845–7848. doi:10.1021/ja5031529.
76. Wu, J., R.M. Yadav, M. Liu, P.P. Sharma, C.S. Tiwary, L. Ma, X. Zou, X.-D. Zhou, B.I. Yakobson, J. Lou, and P.M. Ajayan. Achieving highly efficient, selective, and stable CO_2 reduction on nitrogen-doped carbon nanotubes. *ACS Nano* 150423123413002. doi:10.1021/acsnano.5b01079.
77. Wu, J., M. Liu, P.P. Sharma, R.M. Yadav, L. Ma, Y. Yang, X. Zou, X.-D. Zhou, R. Vajtai, B. I. Yakobson, J. Lou, and P.M. Ajayan. 2016. Incorporation of nitrogen defects for efficient reduction of CO_2 via two-electron pathway on three-dimensional graphene foam. *Nano Letters* 16: 466–470. doi:10.1021/acs.nanolett.5b04123.
78. Liu, Y., S. Chen, X. Quan, and H. Yu. 2015. Efficient electrochemical reduction of carbon dioxide to acetate on nitrogen-doped nanodiamond. *Journal of the American Chemical Society* 137: 11631–11636. doi:10.1021/jacs.5b02975.

Chapter 12
Strategies for the Synthesis of Anisotropic Catalytic Nanoparticles

Hazim A. Al-Zubaidi, Chartanay D.J. Bonner, Minghong Liu and Sherine O. Obare

Abstract The design and synthesis of well-defined nanoscale anisotropic particles is opening new avenues toward developing a fundamental understanding of their chemical and physical properties. Obtaining nanoparticles that are homogenous in size and shape can be a challenging process, particularly for particles that consist of reactive metals. Anisotropy further adds to the complexity in controlling the kinetics and thermodynamics of the nucleation and growth processes. Access to well-defined nanoparticles provides the ability to develop succinct pathways toward elucidating reaction mechanisms they mediate as well as understanding their stability in various environments. Thus, significant effort in the field of nanoscale science and technology has focused on developing procedures that are reproducible in yielding well-defined nanoparticles. This chapter reviews various methods for the synthesis of catalytic metal nanoparticles and the impact of their shape on the reactivity. Methodology for the characterization of the nanostructures is also described.

Keywords Catalysis · Seed-mediated nanoparticle synthesis · Nucleation and growth · Nanowires

12.1 Introduction

Designing nanoparticles with well-defined size, shape, and morphology is essential toward understanding their size- and shape-dependent physical and chemical properties, and consequently being able to rationaly apply them to advancing specific technological applications physiochemical. There has been growing interest in developing strategies toward the synthesis of nanoscale materials, particularly anisotropic nanoparticles that display unique electronic, optical, magnetic and catalytic properties relative to their isotropic and bulk counterparts [1, 2].

H.A. Al-Zubaidi · C.D.J. Bonner · M. Liu · S.O. Obare (✉)
Department of Chemistry, Western Michigan University, Kalamazoo MI 49008, USA
e-mail: sherine.obare@wmich.edu

S.E. Hunyadi Murph et al. (eds.), *Anisotropic and Shape-Selective Nanomaterials*,
Nanostructure Science and Technology, DOI 10.1007/978-3-319-59662-4_12

Studies have focused on tuning the morphology of different nanostructures include one-dimensional (1D) nanoparticles (nanowires [3], nanorods [4], nanotubes [5], nanobelts [6], and nanoribbons [7]), two-dimensional (2D) nanoparticles (branched structures [8], nanoprisms [9], nanoplates [10], nanosheets [11], nanowalls [12], and nanodisks [13]), and three-dimensional (3D) nanoparticles (nanodendrites [14], nanocoils [15], nanopillars [16], and nanoflowers [17]), using a number of techniques. Anisotropic metal nanostructures have proven to be highly effective catalysts due to the reactivity of the surface atoms at the different facets. This chapter focuses on synthetic routes of catalytic anisotropic nanoparticles.

Advances in the design and synthesis of anisotropic nanoparticles have been accomplished by controlling the growth of facets, steps, kinks, and edges on the nanocrystal surfaces [18, 19]. Noble metals including silver (Ag), gold (Au), platinum (Pt), and palladium (Pd) that have high symmetry with face centered cubic (fcc) packing, spontaneously grow through the (111), (100), and (110) facets [20, 21]. Manipulation at the surface level allows the expansion of nanostructure properties. For example, the unique optical properties that accompany anisotropic nanoparticles have extensively been studied [22–26]. Plasmonic Ag, Au, and copper (Cu) nanoparticles have shown shape-specific surface plasmon resonance (SPR) that have resulted in the development of novel optical and Surface-Enhanced Raman Scattering (SERS) sensors [27]. The unique plasmonic properties of anisotropic 2D nanoparticles arise from their ability to absorb energy along more than one axis, unlike the isotropic counterparts that absorb along one axis [28, 29].

Similarly, the catalytic properties of anisotropic metal nanoparticles are unique relative to isotropic nanoparticles [30–33]. Reaction parameters can be tuned to provide catalytic nanoparticles with desired size, shape, and surface energy. In particular, noble metals have played a significant role in catalytic applications that have shaped industrial processes. Semagina et al. studied the catalytic efficiency of isotropic and anisotropic palladium nanoparticles for the hydrogenation of olefinic alcohols [31]. Tetrahedron-shaped Pd nanoparticles displayed twice the reactivity in terms of rates, when compared to their isotropic counterparts, which could be attributed to the tetrahedron-shape having edge atoms less restricted surface [20, 34]. Research is currently focused understanding the growth mechanisms so that a consensus can be reached on the parameters that influence nanostructure growth [35, 36].

Several synthetic methods have been developed for noble metals, metal oxides, and semiconductor anisotropic nanostructures. Studies have highlighted the influence of capping agents and additives that are used to dictate the growth process of nanoparticles in a specific direction. Moreover, standard reaction kinetics including concentration, temperature, reaction time, type of precursor, and the synthetic process used can influence the products formed. Unfortunately, several challenges still exist in establishing facile protocols for the formation of monodisperse anisotropic nanoparticles with high yield, relative to the established procedures for uniform isotropic nanoparticles. Many review articles have exclusively focused on the optical response of anisotropic structures, yet neglect the influence of anisotropic nanoparticles have expanded the field of catalysis. Below we highlight synthetic strategies for the design of anisotropic catalytic nanoparticles.

12.2 Synthesis of Catalytic Nanoparticles

Anisotropic nanoparticle growth can be controlled based on the reaction kinetics and thermodynamics. By tuning the size, shape, and morphology of the nanoparticles, the catalyst selectivity and efficiency can be controlled. For example, We and others, have studied how Pd, one of the most effective metal catalysts for C–C coupling reactions, dehydrogenation, hydrogen storage, and environmental remediation, can be manipulated technological applications [37–41]. Studies have shown that selective adsorption of capping agents, surfactants, ligand species, or polymers can direct particle growth along desired facets [42]. The presence of additives alter metal surface energy at selective facets and promote changes in morphology. When synthesizing Au and Ag nanorods, the capping agent cetyltrimethylammonium bromide (CTAB) is used to control particle growth and uniformity. Sokolov et al. discovered that when synthesizing Pd nanorods, trisodium citrate concentration contributed more to nanostructure than CTAB capping agent [43]. It is proposed that changing the concentration of citrate present with CTAB can induce co-adsorption responsible for direction nanostructure growth versus separate entities. When synthesizing Ag and Pd nanostructures, Lin et al. used oligosaccharide β-cyclodextrin as capping agents to influence the morphology of Ag by limiting the reaction kinetics and controlling the thermodynamics [39]. The reaction resulted in Pd NPs nanodendrites and multipods [39]. Moreover, the reaction parameters also have influence on the nucleation and growth of anisotropic nanostructures. Various synthetic methods have provided an understanding of the growth mechanisms and their influence on the resulting nanostructures. The most common synthetic routes for anisotropic nanoparticles include the formation of seeds and the use of templates for shape control. In the case of metal nanoparticles, seed-mediated growth can occur via wet-chemical synthesis method, template-mediated growth, thermal decomposition, galvanic replacement and photochemistry [44]. This section will highlight the most predominant synthesis methods, primary advancements, challenges, and promising studies.

12.2.1 Seed Mediated Growth

Seed mediated growth is one of the most commonly used synthesis process due to the ability to control facet direction growth. In a two-step procedure, the metal precursor is stabilized with surfactants in the presence of reducing agent to develop seed for growth. In the second step, nucleation and growth is achieved by growth control [45]. To develop anisotropic shapes, researchers have focused on varying standard parameters such as the surfactant, reducing agent, and the metal precursor [45]. One of the first established anisotropic synthetic reactions was based on the reduction of Ag and Au ions using citrate to form the corresponding nanoparticles [46]. The synthetic method for Ag and Au nanorods has expanded to

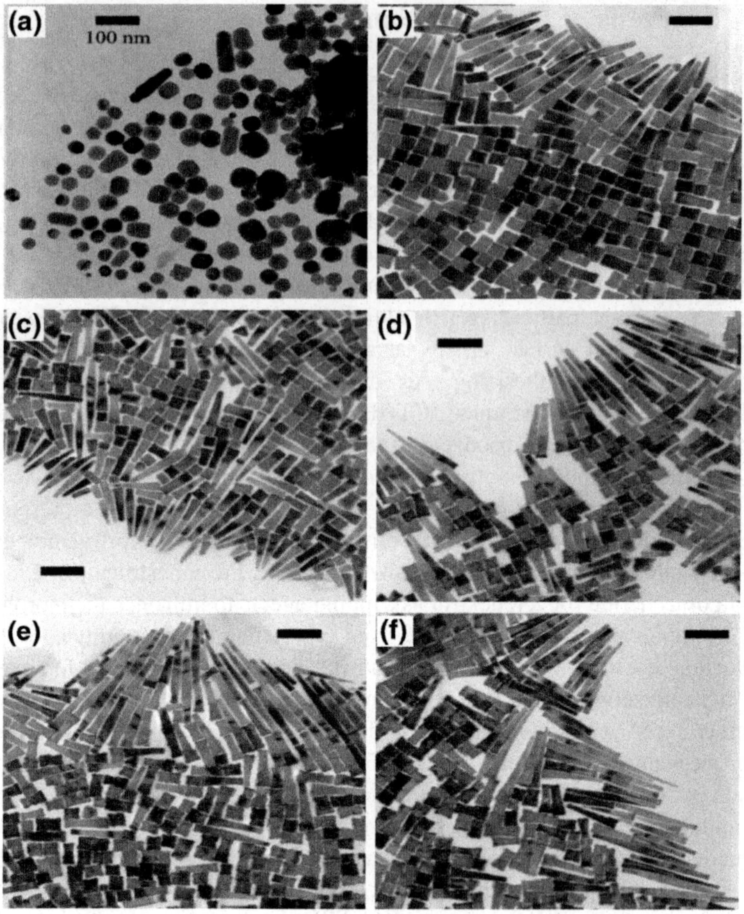

Fig. 12.1 TEM images of Pd NPs at **a** 0 mM, **b** 0.2 mM, **c** 0.4 mM, **d** 0.6 mM, **e** 0.08 mM, and **f** 1.0 mM of trisodium citrate. Reproduced with permission from Ref. [43]

the development of other catalytic anisotropic metal nanostructures. Sun et al. followed a one-pot synthesis method to develop palladium (Pd) nanostructures with capping agent CTAB and using ascorbic acid as a reducing agent at room temperature in the presence of trisodium citrate [43]. The particle morphology was tuned by varying the concentration of trisodium citrate added (Fig. 12.1).

In the absence of sodium citrate, the Pd NPs that formed appeared to have a spherical shape. Increasing the citrate concentration resulted in selective facet growth at (100) and the formation of nanocubes and nanorods. Xie et al. determined growth of platinum (Pt) nanoparticles could be tuned by varying the citrate ion concentration and adjusting the reaction temperature. The use of carbon monoxide (CO) was also found to be effective in tuning the nanoparticle shape [47]. At elevated temperatures, isotropic spherical particles were formed, yet at lower temperatures slower kinetic reduction occurred leading to anisotropic nanostructures. Various reaction conditions led to the formation of Pt nanospheres, nanorods, multibranched and hyperbranched nanostructures. The presence of a capping agent such as citrate plays a significant role in tuning the directional growth of the resulting particle. Controlling the overall homogeneity of the nanoparticles was challenging and highly dependent on the concentration of the capping agent.

By following a seed mediated growth method, varying metal species can be used for the seed to direct the growth of variant metal species. Fan et al. used a Au octahedral core with Au @ Pd nanocubes over grown on the core [48]. H_2PdCl_4 was reduced by ascorbic acid (AA) with CTAB capping agent. Altering the dimensional growth of nanostructures improved the efficiency and selectivity of catalyst. However, depending on the metal adjustments must be made to synthesis routes. Bakshi et al. reviewed the influence that the hydrophobic surfactant, 1,2-diheptanoyl-sn-glycero-3-phosphocholine (DHPC) had on the formation of Pd nanoparticles [49]. Hydrophobic surfactants were found to promote colloidal stability and led to the formation of 1D flower-like and nanowire structures. Increasing the reaction temperature promoted interparticle fusion due to Ostwald ripening.

Due to its success in developing nanorod structures, CTAB is a commonly used capping agent. Consequently, CTAB ligand strong interfacial adsorption and attachment at selective facets limit the yield and uniformity of nanostructures [49]. Therefore, various capping agents have been utilized to influence directional growth of different nanostructures. For example, Lu et al. used an imidazolium salt (Im salts) as a capping agent to synthesize Pd nanocubes that were effective catalysts for a Suzuki coupling reaction [39]. Im salts are heterocyclic structures with high polarity and biological significance. In that study, Im salts behaved as an easily altered capping agent with similar binding as CTAB for the formation of nanorods. The interaction energy between the Im salts and metal surface was smaller in comparison to ammonium salts, thus allowing the formation of cube-like structure. By changing the salt capping agent, formation of palladium nanoparticles were protected and led to high yield during reaction as seen in Fig. 12.2.

1,3-di(dodecyl)-imidazolium chloride led to the formation of cube-like Pd NPs with an average length of 7 nm. The Pd nanocubes were characterized by electron diffraction.

To further tune the morphology, ionic species have been used to promote particle shape and size. Some of the most commonly used ionic species consist of halides and silver (Ag^+) ions [36]. Iodide (I^-), bromide, (Br^-) and chloride (Cl^-) anions have also been used to influence the particle growth in seed mediated synthesis of nanoparticles. Due to smaller size for surface binding, halides have reduced steric hindrance at selective facets. At room temperature in the presence of CTAB and ascorbic acid, Fan et al. was able to tune morphology of palladium nanostructures with the addition of halide ions including Cl^- and Br^- [14]. The CTAB stabilizer binds to specific facets and the presence of the halide ions can contribute to surface adsorption. Growth was based on competitive adsorption

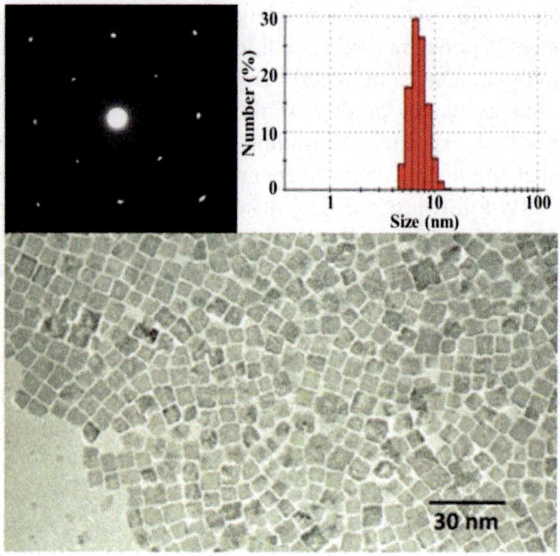

Fig. 12.2 TEM image of Pd NCs with imidazolium salt capping agent. Reproduced with permission from Ref. [39]

between ions on the nanoparticles surface. In the case of Pd NPs, Br^- and I^- are known to strongly adsorb on Pd surface compared to Cl^-. In Fan et al. study, nanocubes and nanodendrites were formatted by varying the NaCl concentration.

Bimetallic nanostructures have become a promising catalyst for several chemical operations due to the synergetic effect of the metals complex. Noble metals such as Au and Pd are considered one of the most common nanostructures used and developed. Au has been dominant in optical studies as Pd is highly utilized in catalysis. By combining their unique characteristics, anisotropic synthesis can exploit properties for enhancement in efficiency. Zheng et al. used a seed mediated method to prepare Au NRs with Pd deposited either at the tips of the rods or coated on the surface [37]. Using CTAB as a capping agent and ascorbic acid as a mild reducing agent, Au NRs were synthesized in high yield. Following the synthesis of the Au NRs, Pd nanocrystals were selectively deposited on the rod surface. Consequently, the Pd nanocrystals at the surface resulted in promoting electron/hole separation in the dehydrogenation reaction of formic acid.

Huang et al. also developed core–shell Au–Pd nanostructure. The electrocatalysis morphology was tuned by changing the Au core (rod, sphere and cube) to influence nanostructure growth [50]. Changing the structure of the seed led to varying the active surface area for Pd growth and electrocatalytic activity. Wang et al. synthesized PtSn bimetallic nanowires by using capping agent dodecylamine to promote directional growth [51]. Consequently, only a low yield of the product was obtained. By reviewing the influence of anions and pH, the growth of bimetallic Au and Pd was controlled in a seed mediated co-reduction procedure [45]. Bower et al. reviewed the influence of pH, precursor concentration, and presence of chloride and iodide directed formation of nanostructures [45].

Seed mediated growth has expanded the development of anisotropic nanostructures of various metal species. By directing aggregation on facets, multi-dimensional nanostructures are formed. Seed mediated growth is limited based on low yield, shape uniformity, cost effectiveness and toxicity of harmful reducing agents. Seed mediated growth by standard synthesis is often restricted to formation of nanorods. With other metal nanoparticles low yields and uniformity are reviewed. Therefore, studies have highlighted the incorporation of templates to improve synthesis methods.

12.2.2 Template Mediated Growth

To improve the yield and uniformity of anisotropic nanostructures researchers have focused on using template species. By varying template species, new metal structures can be determined. Using a template to establish growth process can promote stability and tune morphology. Templated structures are usually based on a hard or soft-template for synthesis [47]. Some of the most common used hard templates consist of mesoporous silica as soft templates are more diverse with the use of surfactants, polymers, and biomacromolecules. Soft-templates are more favorable because they are easier to remove and can be more environmentally friendly. Development of catalytic stability has been improved by using templates for assembly and stability of nanostructures. 2D Pd nanobars and nanorods have been prepared by using templates [52]. Song et al. used a soft template formed by CTAB in a two-phase water chloroform system for the chemical reduction of platinum complexes leading to the formation of nanowires [53]. Different templates for growth can also present easy control over nanostructure synthesis. Tan et al. studied the synthesis of monometallic Pt hollow nanoboxed with Ag–Pt core shell nanocube for template [54]. In this study, Bis(*p*-sulfonatonphenyl)phenyl phosphine (BSPP) behaved as an oxidant. BSPP was able to preserve the template geometry and control shell composition at room temperature. A self-assembly method with template has led to the development of In_2O_3 nanocubes [55]. By using a magnetic template such as Fe_2O_3 and noble metal can led to advancement in tuning magnetic and catalytic response. Kim et al. developed a hybrid nanostructure with Pd or Au on Fe_3O_4 [56]. Nucleation and growth was controlled by using mixture of capping agent of oleylamine and oleic acid. This led to the formation of unique structures such as dumbbell like structures with Pd at the surface.

Polyol synthesis is based on a self-assembly process that is easy and efficient for nanoparticle production that can promote the formation of 1D, 2D, and 3D nanostructures. Metal containing compounds are developed in the presence of polymer and ethylene glycol. Ethylene glycol can behave as both as solvent and reducing agent presenting less harsh metals. Polymers are predominately used as capping agent which leads to the formation of nanostructures at elevated

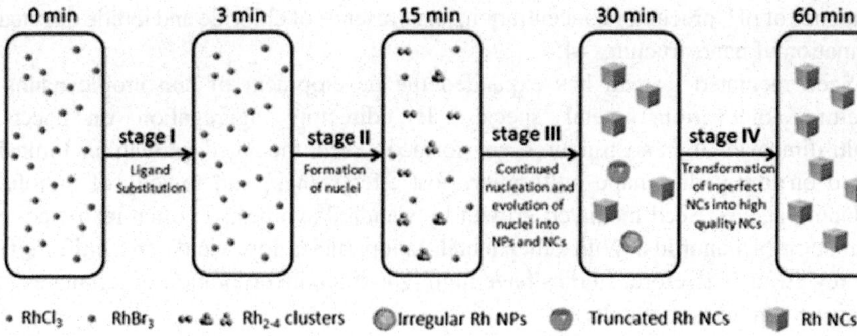

Fig. 12.3 Nucleation and growth process of Rh NCs. Reproduced with permission from Ref. [60]

temperatures with varying size and shapes. Single crystalline platinum nanowires have exclusively been developed by using polyol process method [57]. Herrick et al. synthesized anisotropic platinum nanoparticles following a polyol synthesis method [58]. Platinum precursor was reduced by ethylene glycol at 160 °C. The addition of various amounts of sodium nitrate and Pt precursor allowed the tuning of nanoparticles. Xiao et al. tuned the morphology of Pd nanorods by adjusting the ratio of surfactants CTAB and PVP [59]. The Br^- counter ion (to the CTAB) behaved as the etchant at the metal surface. The PVP stabilizer led to increasing in the Pd nanorod. In a similar manner, Yao et al. was able to synthesize rhodium (Rh) nanocubes using a polyol method and tetradecyltrimethylammonium bromide (TTAB) as a stabilizer as depicted in Fig. 12.3 [60].

Yao and colleagues reduced $RhCl_3{\cdot}3H_2O$ with PVP and TTAB in the presence of ethylene glycol [60]. Tuning morphology was developed in a four-step process at 130 °C over a 1 h timeframe as shown in Fig. 12.3. Controlling rate of the reaction allowed nuclei formation and the evolution of cubical growth. Wang et al. varied the Pd precursor to develop seed structure using the polyol method [61]. Cuboctahedral seeds were produced by reducing Na_2PdCl_4 and $Pd(acac)_2$. Different growth patterns were observed based on the type of precursor used. Furthermore, by adjusting reaction parameters, the growth process could be controlled, resulting in changes in surface energy, and consequently particle shape.

Biological synthesis provides an environmentally friendly approach for developing anisotropic nanoparticles. The most promising aspects of biological synthesis is due to rapid production, controlled toxicity, cost effectiveness, and environmental friendly approach. It substitutes the use of aggressive capping agents, solvents, reducing agents, and surfactants for less toxic components that are not harmful when present in the environment and also present a cost effective synthesis route. A standard biological synthesis follows a seed source with template to expand nucleation growth from set facets. Common sources of use are bacteria, fungi, yeast, and plant extracts. Kochkar et al. developed silver and Pd nanoparticles on TiO_2 by using β-cyclodextrin to promote hydrogentation [42]. For more environmental friendly synthesis and high efficiency Pd nanoparticles, Kennan et al. incorporated biomolecule 5-hydroxytryptophan (HTP) as a stabilizer and reducing agent [40]. This allowed the reaction to be achieved at room temperature with the elimination of harsh reaction conditions. By changing concentration ratio between palladium precursor and HTP biomolecule, changes in morphology were accomplished leading to the formation of palladium nanodendrites used for electrocatalytic oxidation of formic acid. Fu et al. used amino acid arginine to synthesize palladium tetrapods for catalytic reduction [62]. The reaction was able to take place inwater solvent with PVP surfactant. The concentration of amino acid controlled the morphology of nanostructures. Chen et al. utilized microorganism's template to develop Au/Pd/Ag nanowires. CTAB, palladium, and gold was adsorbed by pichia pastoris cells template [63]. Growth was dependent on the mass of yeast and concentration of CTAB capping agent.

Hard templates have led to the development of promising anisotropic structures. Mono- and bicomponents metallic nanowires were synthesized by our research group [64]. Monometallic Au and bimetallic Au–Ni nanowires were prepared using anodic aluminum oxide membranes (AAO). The membranes' surface was coated with 80 nm of Au to form a conductive layer. The deposition of Au to form the nanowires took place for 1 h. The membrane-loaded Au nanowires were immersed in a 6 M NaOH for 20 min to dissolve the membrane.

The prepared sample was sonicated and the nanowires were centrifuged, washed with ethanol, and dried under vacuum. Au nanowires were characterized using scanning electron microscope (SEM) as shown in Fig. 12.4. In order to prepare Au–Ni bimetallic nanowires, a sequential electrodeposition method was used. Ni nanowires were formed in AAO membranes via a reduction procedure. A constant current of 3 mA was applied for 30 min. Au nanowires were deposited on Ni nanowires by applying a current of 26 mA for 30 min. Figure 12.5 shows a SEM image of Ni–Au nanowires.

Our group has synthesized several metallic nanowires for catalytic and biological applications. Examples include, the synthesis of Ag–Au metallic alloy

Fig. 12.4 SEM image for Au nanowires, synthesized using AAO membranes

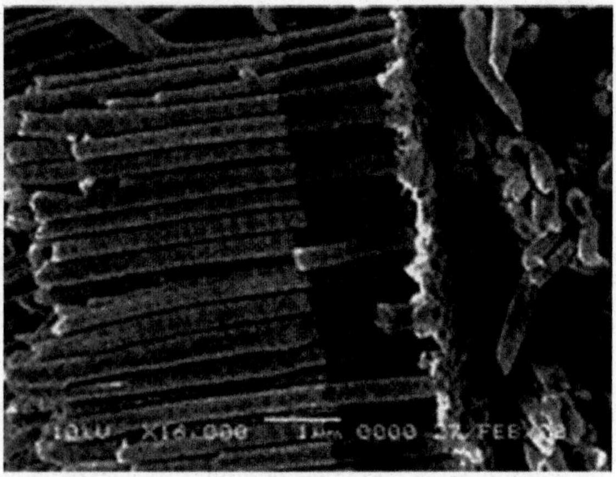

Fig. 12.5 SEM image for Ni–Au nanowires

nanowires that were produced using electrolyte solution of 20% Ag and 80% Au solution. An applied constant current of 2 mA was used for 1 h. Three solutions consisting of 0.2 M $NiCl_2$, 0.5 M H_3BO_3 and 0.02 M $CuSO_4$, were used to synthesize Ni–Cu alloy nanowires. An applied voltage of −1.5 V was used for 30 min. Au plating solution was used with 0.04 M $CuSO_4$ to obtain Au–Cu alloy with 2 mA current and reaction time of 1 h. In addition, Ti–Ni alloy were also produced using 0.2 $NiCl_2$, 0.05 M $TiCl_3$ and HCl. The reduction process took 1 h using a current of −1.5 V. Figure 12.6 shows SEM images of the nanowires produced [64].

12.2.3 Thermal Decomposition

Thermal decomposition has led to the synthesis of metallic and semiconductor nanostructures with varying morphologies. Elevated temperatures allow directional growth of the nanoparticles [65, 66]. Thermal decomposition is usually selected as the method of choice as the elevated reaction temperatures allow the reduction of metals that are hard to reduce. Zhou et al. used thermal decomposition to produce PbS semiconductors in the presence of thioacetamide (TAA) and CTAB stabilizers [17]. Particle growth was controlled by changing the initial reaction temperature over time. Copper NCs were prepared in a similar manner and were found to have an important role as cathodes in electrocatalytic reactions [33]. Wu et al. was able to prepare the Cu NCs by using copper (II) nitride as a precursor and dissolving it in a mixed solvent containing 1-octadecylamine and 1-octadecene at a temperature of 150–250 °C. Primary amines have served as effective capping agents that also allow shape control of the particles. Xie et al. synthesized CdSe nanocrystals by seed mediated method on zinc chalcogenide [67]. The morphology was altered by

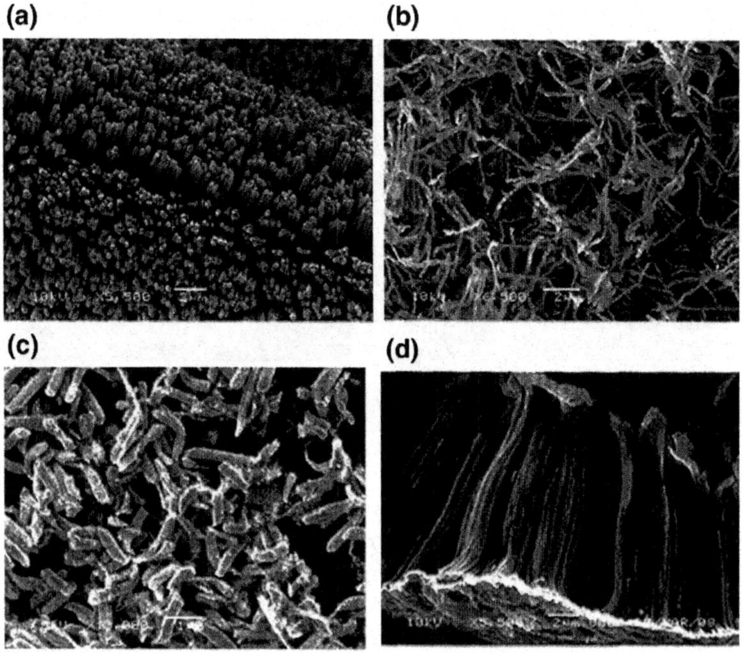

Fig. 12.6 SEM images for two components alloy nanowires of **a** Ag–Au, **b** Ni–Cu, **c** Au–Cu and **d** Ti–Ni

changing the temperature and the injection rate of the precursor during the growth process.

Hydrothermal and solvothermal synthetic routes have been promising in the production of metal oxides and carbon nanostructures. Teng et al. synthesized 3D platinum multipods by solvothermal decomposition [68]. Silver acetylacetonate was used to promote the formation of branched platinum. Hexadecylamine, diphenyl ether, and adamantanecarboxylic acid was used to promote stabilization and reduction of precursor. The application of heat over time promoted alteration in nanostructures. Zhou et al. used a hydrothermal method to synthesize Cu NCs and bimetallic Cu and Pd core–shell nanostructures [69].

As seen in Fig. 12.7a, Cu NCs were monodispersed with high yield. The bimetallic complex of Cu/Pd were formed by the reduction of H_2PdCl_4 with Cu NCs as seeds. Mankin et al. developed platinum nanocubes and nanopods following a solvothermal method that promoted change in morphology by adjusting reaction time [70]. Low boiling solvents hexane and acetone were used to control the first growth phase. Platinum was reduced by morpholine borane in the presence of *N*-methyl-2pyrrolidone. Oleylamine was able to serve as solvent and surfactant. LaGrow et al. developed nickel nanocubes following a hydrothermal procedure [71].

Fig. 12.7 **a** TEM of Cu NCs and **b** SEM of Cu/Pd core–shell structures. Reproduced with permission from Ref. [69]

Hexadecylamine and trioctylphosphine favor cubic morphology. Variation in metal precursor to trioctlyphosphine ratio led to alteration in size and shape.

Vapor phase is based on a pyrolysis method that is predominately used in the synthesis of semiconductors and metal oxides. A vapor precursor established from liquid or gas is forced through a hole or opening of a high pressure apparatus to burn. The resulting product is used as particles. The process is simple, cost effective with continuous operation that promotes high yield synthesis of nanoparticles. Park et al. synthesized hematite rhombohedra with NP9 surfactant and NaCl at molar ratios [72]. By elevated reaction temperatures 240–820 °C led to the various morphological nanostructures of iron. A carbothermal reaction allows the formation of nanoparticles at elevated temperatures. Platinum atoms conventionally form facet spheres and there are limited studies that grow within a 1D and 2D nanostructures. Using vapor phase method allows the formation of platinum nanostructures by controlling reaction temperature. Ma et al. used a carbothermal sapphire surface to develop platinum nanoplatelets and nanobelts [57]. The reaction took place within high temperature furnace at 1250–1500 °C to tune morphology. Cu NRDs upon a carbon substrate was synthesized by Prunier et al. at elevated temperatures to control morphology [73]. Chemical vapor deposition was used to form the controllable synthesis of In_2O_3 at 1000 °C under N_2 to control morphology [74].

12.2.4 Electrochemical and Galvanic Replacement

Electrochemical synthesis or galvanic replacement is one of the first developed methods for the formation of nanoparticles. The method provides an easy synthesis procedure that is cost effective. Some past studies have used varying metal starters to form semiconductors, metal oxides, and noble metal nanoparticles. A number of methods including lithography, chemical vapor deposition, physical vapor deposition, sol-gel deposition, chemical polymerization, and vapor liquid solid growth are becoming increasingly attractive for the synthesis of new materials and nanostructures. Traditionally, electrochemistry has been used to grow thin films on conductive substrates. Electrochemical deposition could be a convenient way of generating on dimensional nanostructures by simultaneously controlling the dimensions, properties and morphology. The use of mild reaction conditions (low temperature and low cost) makes electrochemical deposition a competitive technique.

Most electrochemical deposition methods for nanowire synthesis are based on the use of templates. A notable method was pioneered by Charles R. Martin for the fabrication of metallic and semiconductor nanowires and nanotubes [75–77]. In this case the templates used were nanoporous membranes having cylindrical shaped pores of controllable pore diameter and with a narrow size distribution. The metal atoms were electrochemically reduced and deposited inside the nanopores and then formed 1D nanostructures.

The particle morphology was tuned by adjusting the reaction temperature and the concentration of the Pd precursor [78]. Wang et al. prepared Ag/Pd alloy nanodendrites using an electrochemical synthesis method. Fang et al. synthesized palladium nanospikes and nanourchins determined by controlled potential and time [30]. Chen et al. synthesized bimetallic nanostructures with copper metal and noble metal nanoparticles for nanorods [44]. Metal precursor was dissolved in selective amine with the removal of O_2 and heated up to160 °C. Venkatasubramanian et al. discovered that using KBr as an additive caused the formation of plate-like Cu nanostructures [79]. Nam et al. used a template-free method to synthesize tin (Sn) nanofibers [80]. Electrochemical deposition of the Sn salt was achieved by its strong adsorption to Triton X-100. Pigozzi et al. analyzed thermal effects on bimetallic Ni_3Si/silica core shell semiconductor [81]. Changing the precursor and heat treatment can lead to change in size and change in morphology.

Our research group has utilized electrochemical methods for the synthesis of monodisperse Pd and Ru nanowires [64]. Bimetallic Pd–Ru nanowires were also obtained by an electrochemical deposition method. Briefly, one surface of an AAO membrane was sputter coated with a thin film of gold, and was attached to an ITO coated glass slide to serve as the working electrode. Platinum gauze was used as both reference electrode and counter electrode. Ru plating solution was 0.04 M Ru salt aqueous solution. Pd plating solution was 0.03 M for $PdCl_3$ aqueous solution. Pd–Ru alloy plating solution consisted of an aqueous solution of 0.003 M Pd salt and 0.01 M Ru salt. For all three kinds of nanowires, 3 mA constant current was applied for the reduction of metal salts, which deposited on the bottom of the membrane channel. After deposition, the alumina was dissolved in 6 M NaOH solution, releasing the nanowires from the membranes. In the water suspension, the nanowires could be collected by centrifugation, washed with milli-Q water (Millipore) five times, and twice in 70% ethanol before drying in the vacuum. Electrochemical deposition was used to fabricate Pd, Ru and Pd–Ru nanowires as shown in Fig. 12.8. The nanowires produced by this method were monodisperse and were either vertically orderly aligned on a substrate or were free standing in

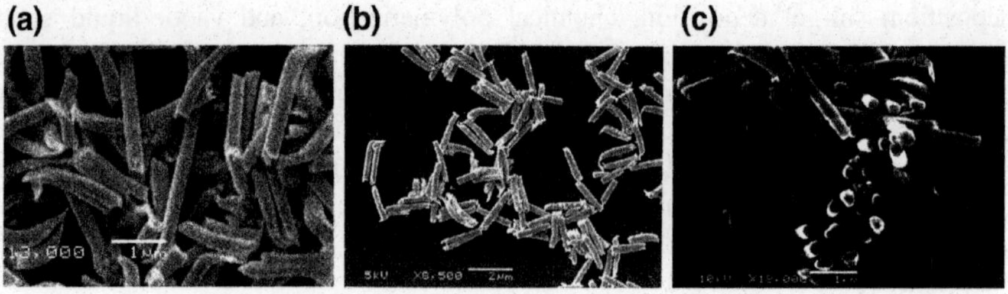

Fig. 12.8 SEM images of **a** Pd nanowires, **b** Ru nanowires, **c** Pd–Ru alloy nanowires

solution. Control of the nanowire length was accomplished by monitoring the pore-filling process (time vs. current). Diameters are dependent on the selective membrane pore size. The advantage of using electrodeposition to obtain nanowires is that the nanowires are always continuous and have good crystalline structures.

By altering the synthetic method, various anisotropic catalytic nanoparticles can be developed. Kinetic and thermodynamic conditions directly influence the yield, monodispersity, and stability of nanostructures. By changing reactions conditions, selectivity and efficiency of catalytic response can be tuned. Developing nanostructures is primarily influenced by reduction methods and stability influenced by surfactant and capping agents.

12.3 Anisotropic Metal Nanoparticles Catalytic Applications

12.3.1 Catalytic Applications

Nanoscale metal-based catalysts have been used for different industrial applications since last century [82]. Homogeneous catalysis of nanoscale metals were reported for different processes, for example, the decomposition of hydrogen peroxide and cross coupling reactions [83]. For the heterogeneous catalysis, both size and shape are crucial for the catalytic reactivity and selectivity. Surface to volume ratio for the solid metal catalyst control the catalyst active sites, then the catalytic activity. Controlling the surface shapes, lead to high catalytic selectivity and reactivity for many reactions including hydrogenations reactions [84, 85], oxidation reaction [86, 87], cross coupling reactions [88, 89], electron transfer [90].

As mentioned earlier, anisotropic metal nanocrystals exhibit special catalytic behavior due to the sharp edges and corners on their surface. For instance, Xiong et al., reported that branched Pt nanoparticles (with 3, 4, 6 and 8 branches) showed high reactivity for the oxidation of formic acid comparing with spherical Pt as depicted in Fig. 12.9. Octapod nanostructures showed higher catalytic activity

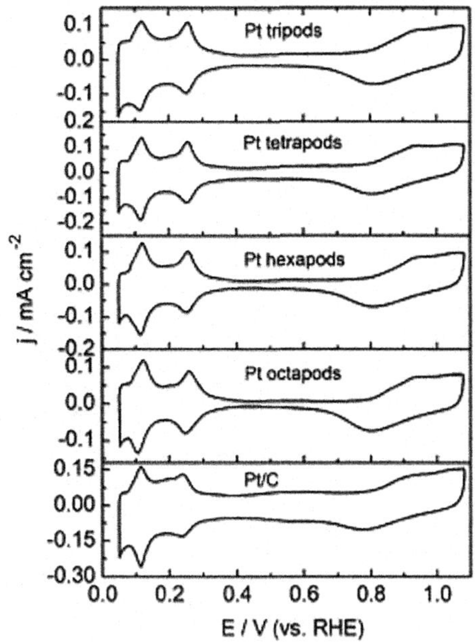

Fig. 12.9 Cyclic voltammograms of Pt tripods, tetrapods, hexapods, and octopods, respectively. Reproduced with permission from Ref. [91]

because of the high branching which leads to higher surface area, and more Pt atoms on the catalyst surface [91].

In this section we will discuss the catalytic activity of anisotropic mono- and bimetallic nanoparticles. Anisotropic nanostructures have significantly enhanced the catalytic reactivity and response, thus looking at some of the leading discoveries and applications will give insight to further development.

Platinum: Platinum nanoparticles based catalyst is widely used for many chemical reactions. Different shapes of Pt nanoparticles were synthesized, and have shown various catalytic activities [92]. Fukuoka et al. found that Pt nanowires showed more catalytic reactivity relative to Pt nanospheres (both supported on FSM-16) for the butane hydrogenolysis reaction [93]. El-Sayed and coworkers, investigated the change in Pt nanoparticles catalytic activity based on their anisotropic shape. The catalytic activity could be tuned by changing the ratio of Pt atoms numbers on the edges and the corners on the facets of different shapes. They found that tetrahedral and cubic shapes of Pt nanoparticles became spherical in shape during the catalytic reaction because it is favorable thermodynamically [94]. Platinum tetrahedral shaped nanoparticles showed less stability relative to cubic structures during the electron transfer reactions between hexacyanoferrate (III) and thiosulfate ions [95]. Moreover, El-Sayed et al. synthesized Pt nanostars in high yield. The multiarms nanostar Pt showed high catalytic activity in the reduction of ferricyanide by thiosulfate [96]. Wang et al. compared the electrocatalytic activity of Pt tetrahexahedral and Pt nanospheres. The anisotropic structure of Pt (tetrahexahedral structure) showed higher electrocatalytic activity (double) than Pt nanospheres for the oxidation of formic acid and ethanol [97]. In another study, the catalytic activity of Pt multipods, discs and hexagons, were examined for the formic acid and ethanol oxidation. Hexagonal structures showed higher electrocatalytic activity for the oxidation of formic acid [98]. Furthermore, formic acid and ethanol were used as a model compounds to examine the catalytic activity of Pt Y-junction, nanowires and commercial Pt/C. Higher catalytic activity was obtained with Pt Y-junction structure than the other structures [99]. Sun et al., showed that cubic structure of Pt showed twice catalytic reactivity comparing with the spherical Pt nanoparticles for the oxygen reduction [100].

Palladium: Palladium is one of the most valuable catalysts for many organic reactions [101]. Palladium with nanorods shape was synthesized and utilized for the hydrogenation of styrene. The designed anisotropic catalyst showed an excellent hydrogenation conversion for styrene, with moderate reaction conditions as 40 °C as temperature and pressure of 0.2 MPa (Scheme 12.1) [102].

The catalytic activity of chain-like Pd nanocubes were examined and compared to Pd nanospheres. Chain-like Pd nanocubes showed catalytic reactivity 11.5 times higher than the isotropic Pd nanoparticles for the oxidation of formic acid and methanol [103]. Palladium nanowires was obtained by self-assembly approach. High catalytic reactivity was demonstrated the oxidation of formic acid as well [104]. Ramaprabhu et al., reported the synthesis of Pd nanoparticles with triangular shape and support it on nitrogen doped graphene. In a moderate reaction conditions (25 °C and 2 MPa), the anisotropic catalyst showed high activity for the oxygen reduction [105]. Gas phase hydrogenation of butadiene, carried out using anisotropic Pd nanoparticles. Palladium nanocubes and nanoprisms exhibited high catalytic reactivity for the butadiene hydrogenation [93]. Palladium nanorods and branched palladium nanocrystals were designed by Huang et al., by the seed-mediated method. Both Pd nanostructures showed high catalytic reactivity and selectivity for Suzuki coupling reaction between phenylboronic acid and iodobenzene [106]. The catalytic reactivity of highly branched Pd nanoparticles was assisted for the hydrogenation of nitrobenzene to aniline with an excellent conversion percentage [107]. Suzuki cross coupling reactions were carried out with high efficiency, using hollow Pd nanospheres and furthermore, the anisotropic Pd nanoparticles synthesized using uniform silica spheres as a template were highly effective catalyts [108].

Gold: Gold is well known as unreactive noble metal, however, this metal becomes very reactive with unique optical and electrical properties with the nanoscale size. Au nanoparticles showed high catalytic reactivity for different

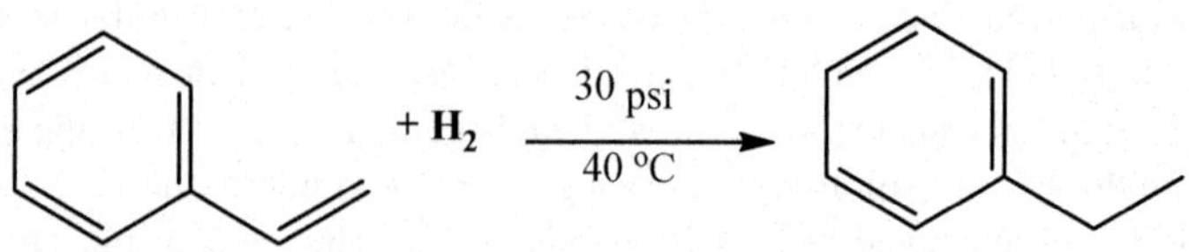

Scheme 12.1 Hydrogenation of styrene to form ethylbenzene catalyzed by Pd nanorods

organic reactions such as the oxidation of alcohol [109], hydrocarbon [110], hydrogenation of aldehydes [111, 112], alkyne [113], and 4-nitrophenol reduction [114]. Many studies have shown that anisotropic Au nanoparticles have more catalytic reactivity than the spherical shape. Mandal et al., reported the designing of polygonal Au nanoparticles. Polygonal Au nanoparticles was showed high catalytic efficiency for the reduction of nitrophenols isomers [115]. Another study showed that anisotropic Au nanoparticles exhibited high catalytic reactivity for both organic and inorganic reactions. The organic transformations included the hydrogenation of *p*-nitrophenol to *p*-aminophenol, while the inorganic reaction illustrated the transformation of ferricyanide to ferrocyanide by thiosulphate [116]. Misra et al., synthesized shape controlled Au nanoparticles with two morphologies as rice and dendrimer shapes. The catalyst showed high potential for azo bond cleavage in Sudan-1 compound [117]. Catalytic CO oxidation was carried out successfully at room temperature with Au nanocubes based catalyst [118]. Dendrites gold nanoparticles showed high catalytic performance for the *p*-nitraniline reduction [119]. Gold nanotubes were used for the catalytic conversion of 4-nitrophenol to 4-aminophenol with an excellent conversion percentage (Scheme 12.2) [120].

Cobalt: Cobalt nanostructure with controlled morphology, has been used for important application such as: data storage [121], magnetic separation [122], and catalysis [123]. Cobalt nanorods exhibited higher catalytic reactivity and selectivity for the hydrogenolysis of glycerol to 1,3-propanediol. Anisotropic cobalt nanoparticles also showed higher catalytic reactivity than spherical cobalt [124]. Cobalt with hollow sphere shape was used as a catalyst for the coupling of alkenes with aryl iodide or aryl bromide via Heck reaction. The catalytic reaction yield was moderate to good and the catalyst was recyclable [125]. Cobalt nanowires demonstrated good potential for both citral hydrogenation and hydrogenolysis of glycerol [126]. Glycerol hydrogenolysis was obtained as well with good yield with hollow Co nanostructures [127] and Co nanowires [128].

12.3.2 Bimetallic Anisotropic Nanoparticles

It has been reported that bimetallic nanostructures exhibit high catalytic reactivity, selectivity, and life time [129]. In this section we review the catalytic reactivity and selectivity for different anisotropic bimetallic systems. Core–shell Pd–Pt nanocubes were synthesized by Han and coworkers. The catalytic activity for the hollow Pd–Pt

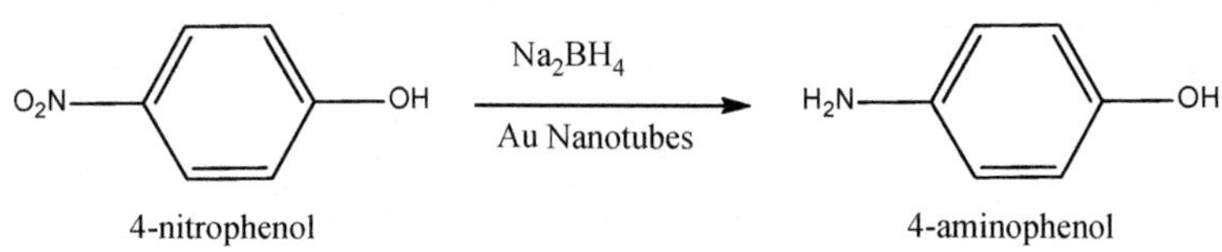

Scheme 12.2 Reduction of the 4-nitrophenol to 4-aminophenol catalyzed by Au nanotubes

nanocubes was higher than the isotropic core–shell and the spherical Pd/C for the oxygen reduction reaction [130]. Two anisotropic Pd–Pt were prepared as nanocubes and nanotetrahedrons shapes. The catalytic reactivity and durability for these two shapes were examined and compared with commercial Pd/C. The nanocubes structure demonstrate high reactivity, while the nanotetrahedrons shape showed high durability [131]. The catalytic elecroreduction of oxygen was carried out using Pt–Pd/Cu core–shell bimetallic system. The shell contains monolayer of Pt and the core is porous/hollow Pd/Cu alloy. Pt–Pd/Cu showed higher catalytic reactivity than Pt-bulk Pd by 3.5 times [132]. Xia et al., designed Pt–Pd nanodendritic bimetallic catalysts. The catalytic reactivity was 2.5 times higher than Pt/C for the oxygen reduction reaction [133]. Nanowire structure of Pd–Pt core shell, showed high electrocatalytic efficiency for oxygen reduction compared to commercial isotropic Pt based catalyst [134]. Anisotropic Pd–Ni nanowires, was prepared and used for the catalytic hydrogenation of *p*-nitrophenol. The catalytic performance of the anisotropic Pd–Ni nanowires was higher than both monometallic Pd nanowires and Pd–Ni nanoparticles [135]. Cavity-conformed Pd–Ni nanocatalysts were used for the hydrogenation dialkyl ketones, with high catalytic efficiency due to the synergistic effect between the two metals [136]. Tripods Pd–Cu bimetallic nanoparticles, reported as an active catalyst for the oxidation of formic acid [137]. Bimetallic Pd–Au with anisotropic shapes, illustrate high catalytic reactivity for CO oxidation [138], hydrogen peroxide oxidation [139], and 1,2-propanediol oxidation [140].

Recently, Pt–Au anisotropic bimetallic based catalyst has attracted the attention for different catalytic transformations. High catalytic reactivity and selectivity was observed for oxygen reduction [141], CO and methanol oxidation [142], ethanol oxidation [143], and formic acid oxidation [144]. Anisotropic Pt–Ru bimetallic system demonstrate an excellent electrocatalytic oxidation for CO [145–147]. Moreover, Pt–Ru with nanofiber [148, 149] and core–shell shapes [150] have shown high catalytic reactivity for methanol oxidation. Many reports have shown that anisotropic Au–Ni nanoparticles have shown higher catalytic reactivity, selectivity and stability comparing with the analogous Ni and Au monometallic nanoparticles [151–153].

12.4 Conclusion

In the last decade, anisotropic nanoparticles have attracted the attention in both academia and industry. Modifying the nanoparticles from the isotropic to anisotropic structure, enhance the optical, electrical, magnetic and catalytic properties. Different shapes and dimensions of anisotropic nanoparticles were reported as: one dimension (nanowires, nanorods, nanotubes, nanobelts and nanoribbons), two dimensions (branched structure, nanoprisms, nanoplates, nanosheets, nanowalls, nanodisks) and three dimensions (nanodendrites, nanocoils, nanopillars and nanoflowers). Several anisotropic structures of metals, oxides and semiconductors were designed using different synthetic strategies. Anisotropic noble metals like Au, Ag, Pt and Pd were synthesized and their applications were studied extensively.

The catalytic efficiency can be tuned by controlling the shape and size of the catalyst. Nanoscale based catalysts showed enhanced catalytic performance relative to their bulk counterparts. The enhanced catalytic activity can be attributed to the high surface area and the large fraction of the small particle atoms on the surface [82]. Furthermore, nanoscale materials have unique physical, chemical, optical and electrical properties. These unique properties are related to the electrons confinement in the nanostructure crystals [85, 154]. Metal nanoparticles are widely used for many catalytic applications because they are electrons rich elements [86, 155]. As previously mentioned, catalytic reactivity of metal nanoparticles can be affected by different parameters such as tuning size, morphology, composition and structure. The catalytic reactivity can be directly influenced by size, while, selectivity can be promoted by altering the shape of the metal nanocrystals [87, 88, 156, 157]. Furthermore, anisotropic nanoparticles have illustrated high catalytic reactivity for a wide range of chemical processes and transformations including: oxidation, reduction and hydrogenation reactions. The outstanding optical and catalytic reactivity is due to the presence of sharp edges in the anisotropic nanoparticles, which interact with the electromagnetic radiation. Moreover, anisotropic bimetallic nanoparticles have exhibited an excellent catalytic and optical activities comparing with the monometallic nanoparticles.

References

1. Demortiere, A., P. Launois, N. Goubet, P.-A. Albouy, and C. Petit. 2008. *Journal of Physical Chemistry B* 112: 14583–14592.
2. Viau, G., R. Brayner, L. Poul, N. Chakroune, E. Lacaze, F. Fievet-Vincent, and F. Fievet. 2003. *Chemistry of Materials* 15: 486–494.
3. Bromly, K.M., A.J. Patil, A.W. Perriman, G. Stubbs, and S. Mann. 2008. *Journal of Materials Chemistry* 18: 4796–4801.
4. Huang, X., X. Qi, Y. Huang, S. Li, C. Xue, C.L. Gan, F. Boey, and H. Zhang. 2010. *ACS Nano* 4: 6196.
5. Wang, Y., M. Lingenfelder, T. Classen, G. Costantini, and K.J. Kern. 2007. *Journal of the American Chemical Society* 129 (51): 15742–15743.
6. Liang, J., Q. Peng, X. Wang, X. Zheng, R. Wang, X. Qiu, C. Nan, and Y. Li. 2005. *Inorganic Chemistry* 44 (25): 9405–9415.
7. Zao, Y., Z.-G. Chen, Y. Huang, L. Yang, J. Drennan, and J. Zou. 2014. *Journal of Physical Chemistry C* 118: 20620–20626.
8. Ortiz, N., and S.E. Skrabalak. 2011. *Crystal Growth & Design*. 11: 3545–3550.
9. DuChene, J.S., W. Niu, J.M. Abendroth, Q. Sun, W. Zhao, F. Huo, and W.D. Wei. 2013. *Chemistry of Materials* 25: 1392–1399.
10. Wang, L., X. Chen, Z. Sun, and Y. Chai. 2007. *Canadian Journal of Chemical Engineering* 85: 598–601.
11. Junggeburth, S.C., L. Diehl, S. Werner, V. Duppel, W. Sigle, and B.V. Lotsch. 2013. *Journal of the American Chemical Society* 135: 6157–6164.

12. Banerjee, D., S. Mukherjee, and K.K. Chattopadhyay. 2011. *Applied Surface Science* 257: 3717–3722.
13. Bryks, W., M. Wette, N. Velez, S.-W. Hsu and A.R. Tao. 2014. *Journal of the American Physical Society* 136: 6175–6178.
14. Fan, F.-R., A. Attia, U.K. Sur, J.-B. Chen, Z.-X. Xie, J.-F. Li, B. Ren and Z.-Q. Tian. 2009. *Crystal, Growth, and Design* 9: 2335–2340.
15. Liu, Q., Z.-H. Cui, Z. Ma, S.-W. Bian, and W.-G. Song. 2007. *Journal of Physical Chemistry C* 111: 12420–12424.
16. Yeom, B., H. Zhang, H. Zhang, J. Park, K. Kim, A.O. Govorov, and N.A. Kotov. 2013. *Nano Letters* 13: 5277–5283.
17. Zhou, G., M. Lii, Z. Xiu, S. Wang, H. Zhang, Y. Zhou, and S. Wang. 2006. *Journal of Physical Chemistry B.* 110: 6543–6548.
18. Nakagawa, Y., H. Kageyama, Y. Oaki, and H. Imai. 2014. *Journal of the American Chemical Society* 136: 3716–3719.
19. Zhang, X., C. Li, C. Zhang, J. Yang, Z. Quan, P. Yang, and J. Lin. 2008. *Crystal Growth & Design* 12: 4564–4570.
20. Kim, I.Y., Y.K. Jo, J.M. Lee, L. Wang, and S.-J. Hwang. 2014. *Journal of Physical Chemistry Letters* 5: 4149–4161.
21. Burda, C., X. Chen, R. Narayanan, and M.A. El-Sayed. 2005. *Chemical Reviews* 105 (4): 1025–1102.
22. Guo, H., Y. Chen, M.B. Cortie, X. Liu, X. Wang, and D.-L. Peng. 2014. *Journal of Physical Chemistry C* 118: 9801–9808.
23. Wang, M., T. Odoom-Wubah, H. Chen, X. Jing, T. Kong, D. Sun, J. Huang, and Q. Li. 2013. *Nanoscale* 5: 6599–6606.
24. Cheng, G., J. Di, and Y. Wang. 2015. *Journal of Physical Chemistry C* 119: 22122–22130.
25. Ma, Y., W. Li, E.C. Cho, Z. Li, T. Yu, J. Zeng, Z. Xia, and Y. Xia. 2010. *ACS Nano* 4: 6725–6734.
26. Pedireddy, S., A. Li, M. Bosman, I.Y. Phang, S. Li, and X.Y. Ling. 2013. *Journal of Physical Chemistry C* 117: 16640–16649.
27. Gerard, V.A., and Y.K. Gunko. 2013. *Nanoscience* 2: 1–22.
28. Murphy, C.J., A.M. Gole, S.E. Hunyadi, and C.J. Orendorff. 2006. *Inorganic Chemistry* 45: 7544–7554.
29. El-Sayed, M.A. 2001. *Accounts of Chemical Research* 34: 257–264.
30. Fang, Y., S. Guo, C. Zhu, S. Dong, and E. Wang. 2010. *Langmuir* 26: 17816–17820.
31. Ma, R., and N. Semagina. 2010. *Journal of Physical Chemistry C* 114: 15417–15423.
32. Shanmugasundaram, A., B. Ramireddy, P. Basak, S.V. Manorama, and S. Srinath. 2014. *Journal of Physical Chemistry* 118: 6909–6921.
33. Wu, H., and W. Chen. 2011. *Journal of the American Chemical Society* 133: 15236–15239.
34. Burrows, N.D., A.M. Vartanian, N.S. Abadeer, E.M. Grzincic, L.M. Jacob, W. Lin, J. Li, J. M. Dennison, J.G. Hinman, and C.J. Murphy. 2016. *Journal of Physical Chemistry Letters* 7: 632–641.
35. Adair, J.H., and E. Suvaci. 2000. *Current Option in Colloid and Interface Science* 5: 160–167.
36. Lohse, S.E., N.D. Burrows, L. Scarbelli, L.M. Liz-Marzan, and C.J. Murphy. 2014. *Chemistry of Materials* 26: 34–43.
37. Zheng, Z., T. Tachikawa, and T. Majima. 2015. *Journal of the American Chemical Society* 137: 948–957.
38. Endo, O., M. Nakamura, K. Amemiya, and H. Ozaki. 2016. *Journal of Physical Chemistry C* 120: 5495–5502.
39. Lu, J.-T., J.C.Y. Lin, D. Khupse, and J.B. Lin. 2014. *Langmuir* 30: 10440.
40. Kannan, P., J. Dolinska, T. Maiyalagan, and M. Opallo. 2014. *Nanoscale* 6: 11169.
41. Freemantle, R.G., M. Liu, W. Guo and S.O. Obare. 2009. *Approaches to synthesis and characterization of spherical & anisotropic palladium nanomaterials, in metallic nanomaterials for life sciences* Kumar, C. S. S. R., ed. Wiley-VCH: Weinheim, vol. 1, 305.

42. Kochkar, H., M. Aouine, A. Ghorbel, and G. Berhault. 2011. *Journal of Physical Chemistry C* 115: 11364.
43. Sun, Y., L. Zhang, H. Zhou, Y. Zhu, E. Sutter, Y. Ji, M.H. Rafailovich, and J.C. Sokolov. 2007. *Chemistry of Materials* 19: 2065.
44. Chen, S., S.V. Jenkins, J. Tao, Y. Zhu, and J. Chen. 2013. *Journal of Physical Chemistry C* 117: 8924–8932.
45. Bower, M.M., C.J. DeSantis, and S.E. Skrabalak. 2014. *Journal of Physical Chemistry C* 118: 18762–18770.
46. Turkevich, J., P.C. Stevenson, and J. Hiller. 1951. *Discuussion of the Faraday Society* 11: 55–75.
47. Xie, J., Q. Zhang, W. Zhou, J.Y. Lee, and D.I.C. Wang. 2009. *Langmuir* 25: 6454–6459.
48. Fan, F.-R., D.-Y. Liu, S. Duan, Z.-X. Xie, Z.-Y. Jiang, and Z.-Q. Tian. 2008. *Journal of the American Chemical Society* 130: 6949–6951.
49. Bakshi, M.S. 2009. *Journal of Physical Chemistry C* 113: 10921–10928.
50. Huang, Y., A.R. Ferhan, A. Dandapat, C.S. Yoon, J.E. Song, E.C. Cho, and D.-H. Kim. 2015. *Journal of Physical Chemistry C* 119: 26164–26170.
51. Wang, D., J. Stover, V. Zielasek, L. Altmann, K. Thiel, K. Al-Shamery, M. Baumer, H. Borchert, J. Parisi, and J. Kolny-Olesiak. 2011. *Langmuir* 27: 11052–11061.
52. Xiong, Y., H. Cai, B.J. Wiley, J. Wang, M.J. Kim, and Y. Xia. 2007. *Journal of the American Chemical Society* 129: 3665.
53. Song, Y., R.M. Garcia, R.M. Dorin, H. Wang, Y. Qui, E.N. Coker, W.A. Steen, J.E. Miller, and J.A. Shelnutt. 2007. *Nano Letters* 7: 3650.
54. Tan, Y.-N., J. Yang, J.Y. Lee, and D.I.C. Wang. 2007. *Journal of Physical Chemistry C* 111: 14084–14090.
55. Srinath, et al. 2014. *Journal of Physical Chemistry* 118: 6909–6921.
56. Kim, M., and H. Song. 2014. *Journal of Materials Chemistry* 2: 4997.
57. Ma, D.-L., and H.-L. Chen. 2009. *Crystal Growth & Design* 9: 2030.
58. Herricks, T., J. Chen, and Y. Xia. 2004. *Nano Letters* 4: 2367.
59. Xiao, C., H. Ding, C. Shen, T. Yang, C. Hui, and H.-J. Gao. 2009. *Journal of Physical Chemistry C* 113: 13466.
60. Yao, S., Y. Yuan, C. Xiao, W. Li, Y. Kou, P.J. Dyson, N. Yan, H. Asakura, K. Teramura, and T. Tanaka. 2012. *Journal of Physical Chemistry C* 116: 15076–15086.
61. Wang, Y., S. Xie, J. Liu, C.Z. Huang, and Y. Xia. 2013. *Nano Letters* 13: 2276–2281.
62. Fu, G.-T., X. Jiang, S.-H. Wei, D.-M. Sun, Y.-W. Tang, T.-H. Lu, and Y. Chen. 2014. *ACS Applied Materials & Interfaces* 6: 22790.
63. Chen, X., G. Wang, T. Odoom-Wubah, X. Jing, J. Huang, Q. Li, and Y. Zheng. 2016. *Materials Letters* 165: 29–32.
64. Liu, M. 2010. Isotropic and anisotropic metallic and bimetallic nanoparticles and their potential applications in biology. Doctoral dissertation. Western Michigan University.
65. Gao, G., X. Liu, R. Sin, K. Zhou, Y. Shi and R. Ma. 2010. Takayama-Muromachi, E.; Qiu, G. *Crystal Growth & Design* 10: 2888–2894.
66. Zhen, G., B.W. Muir, B.A. Moffat, P. Harbour, K.S. Murray, B. Moubaraki, K. Suzuki, I. Madsen, N. Agron-Olshina, L. Waddington, P. Mulvaney, and P.G. Hartley. 2011. *Journal of Physical Chemistry C* 115: 327.
67. Xie, R., and M. Zhou. 2015. *Chemistry of Materials* 27: 3055–3064.
68. Teng, X., and H. Yang. 2005. *Nano Letters* 5: 885.
69. Zhou, G., M. Lu, and Z. Yang. 2006. *Langmuir* 22: 5900.
70. Mankin, M.N., V. Mazumder, and S. Sun. 2011. *Chemistry of Materials* 23: 132–136.
71. LaGrow, A.P., B. Ingham, S. Cheong, and G.V.M. Williams. 2012. *Journal of American Chemical Society* 134: 855–858.
72. Park, T.-J. and S.S. Wong. 2006. *Chemistry of Materials* 18: 5289–5295.
73. Prunier, H., C. Ricolleau, J. Nelayah, G. Wang and D. Alloyeau. 2014. *Crystal Growth & Design* 14: 6350–6356.
74. Yin, W., M. Cao, C. Luo and B. Wei. 2009. *Crystal Growth & Design* 9: 2173–2178.

75. Martin, C.R. 1994. *Nanomaterials* 266: 1961–1966.
76. Hulteen, J.C., and C.R. Martin. 1997. *Journal of Materials Chemistry* 1075: 7.
77. Collinson, M.M. 2007. *Accounts of Chemical Research* 40 (9): 777–783
78. Wang, D., Y. Li, Y. Liu, J. Huang, and T. You. 2009. *Crystal Growth & Design* 9: 4351–4355.
79. Venkatasubramanian, R., J. He, M.W. Johnson, I. Stern, D.H. Kim, and N.S. Pesika. 2013. *Langmuir* 29: 13135–13139.
80. Nam, D.-H., T.-H. Kim, K.-S. Hong, and H.-S. Kwon. 2014. *ACS Nano* 8: 11824–11835.
81. Pigozzi, G., D. Mukherji, Y. Elerman, P. Strunz, R. Gilles, M. Hoelzel, B. Barbier and P. Schmutz. 2014. *Journal of Alloys and Compounds* 584: 119–127.
82. Bell, A.T. 2003. *Science* 299: 1688.
83. Reetz, M.T., and W. Helbig. 1994. *Journal of the American Chemical Society* 116: 7401–7402.
84. Molnar, A., A. Sarkany, and M. Varga. 2001. *Journal of Molecular Catalysis A: Chemical* 173: 185–221.
85. Borodzinski, A. 2001. *Catalysis Letters* 71: 169.
86. Arenz, M., K.J.J. Mayrhofer, V. Stamenkovic, B.B. Bizanac, T. Tomoyuki, P.N. Ross, and N.M. Markovic. 2005. *Journal of the American Chemical Society* 127: 6819–6829.
87. Herzing, A.A., C.J. Kiely, A.F. Carley, P. Landon, and G.J. Hutchings. 2008. *Science* 321: 1331.
88. Li, Y., E. Boone, and A. El-Sayed. 2002. *Langmuir* 18: 4921–4925.
89. Bars, J.L., U. Specht, J.S. Bradley, and D.G. Blackmond. 1999. *Langmuir* 15: 7621–7625.
90. Fukuoka, A., N. Higashimoto, Y. Sakamoto, S. Inagaki, Y. Fukushima, and M. Ichikawa. 2001. *Microporous and Mesoporous Materials* 48: 171–179.
91. Ma, L., C. Wang, M. Gong, L. Liao, R. Long, J. Wang, D. Wu, W. Zhong, M.J. Kim, Y. Chen, Y. Xie, and Y. Xiong. 2012. *ACS Nano* 6: 9797–9806.
92. Ahmadi, T.S., Z.L. Wang, T.C. Green, A. Henglein, and M.A. El-Sayed. 1996. *Science* 272: 1924–1926.
93. Piccolo, L., A. Valcarcel, M. Bausach, C. Thomazeau, D. Uzio, and G. Berhault. 2008. *Physical Chemistry Chemical Physics: PCCP* 10: 5504–5506.
94. Narayanna, R., and M. El-Sayed. 2004. *Journal of the American Chemical Society* 109: 7194–7196.
95. Narayanna, R., and M.A. El-Sayed. 2004. *Journal of Physical Chemistry B* 108: 5726–5733.
96. Mahmoud, M.A., C.E. Tabor, M.A. El-Sayed, Y. Ding, and Z.L. Wang. 2008. *Journal of the American Chemical Society* 130: 4590–4591.
97. Tian, N., Z.-Y. Zhou, S.-G. Sun, Y. Ding, and Z.L. Wang. 2007. *Science* 316: 732.
98. Subhramannia, M., K. Ramaiyan, and V.K. Pillai. 2008. *Langmuir* 24: 3576–3583.
99. Mahima, S., R. Kannan, I. Komah, M. Aslam, and V.K. Pillai. 2008. *Chemistry of Materials* 20: 601–603.
100. Wang, C., H. Daimon, Y. Lee, J. Kim, and S. Sun. 2007. *Journal of the American Chemical Society* 129: 6974–6975.
101. Chen, A. and C. Ostrom. 2015. *Chemical Reviews* A.
102. Sebastian, V., S. Basak, and K.V. Jensen. 2016. *AlChE Journal* 62: 373.
103. Kannan, P., T. Maiyalagan, and M. Opallo. 2013. *Nano Energy* 2: 677–687.
104. Wang, S., X. Wang, and S.P. Jiang. 2008. *Nanotechnology* 19: 455602.
105. Vinayan, B.P., K. Sethupathi, and S. Ramaprabhu. 2013. *International Journal of Hydrogen Energy* 38: 2240–2250.
106. Chen, Y.-H., H.-H. Hung, and M.H. Huang. 2009. *Journal of the American Chemical Society* 131: 9114–9121.
107. Watt, J., S. Cheong, M.F. Toney, B. Ingham, J. Cookson, P.T. Bishop, and R.D. Tilley. 2009. *ACS Nano* 4: 396.
108. Kim, S.-W., M. Kim, W.Y. Lee, and T. Hyeon. 2002. *Journal of the American Chemical Society* 124: 7642–7643.
109. Wang, X., H. Kawanami, N.M. Islam, M. Chattergee, T. Yokoyama and Y. Ikushima. 2008. *Chemical Communications* 4442–4444.

110. Hughes, M.D., Y.-J. Xu, P. Jenkins, P. McMom, P. Landon, D.L. Enache, A.F. Carley, G.A. Attard, G.J. Hutchings, F. King, E.H. Stitt, P. Johnston, K. Griffen, and C.J. Kiely. 2005. *Nature* 437: 1132.
111. Bailie, J.E. and G.J. Hutchings. 1999. *Chemical Communications* 2151–2152.
112. Bailie, J.E., H.A. Abdullah, J.A. Anderson, C.H. Rochester, N.V. Richardson, N. Hodge, J.-G. Zhang, A. Burrows, C.J. Kiely, and G.J. Hutchings. 2001. *Physical Chemistry Chemical Physics: PCCP* 3: 4113–4121.
113. Jia, J., K. Haraki, J.N. Kondo, K. Domen, and K. Tamaru. 2000. *Journal of Physical Chemistry B* 104: 11153–11156.
114. Praharaj, S., S. Nath, S. Ghosh, S. Kundu, and T. Pal. 2004. *Langmuir* 20: 9889–9892.
115. Rashid, M.H., and T.K. Mandal. 2008. *Advanced Functional Materials* 18: 2261–2271.
116. Jana, D., A. Dandapat, and G. De. 2010. *Langmuir* 26: 12177–12184.
117. Sahoo, G.P., D.K. Bhui, and A. Misra. 2014. *Journal of Molecular Liquids* 198: 215–222.
118. Sanchez-Castillo, M.A., C. Couto, W.B. Kim, and J.A. Dumesic. 2004. *Angewandte Chemie International Edition* 43: 1140.
119. Huang, D., X. Bai, and L. Zheng. 2011. *Journal of Physical Chemistry C* 115: 14641.
120. Yu, Y., K. Kant, J.G. Shapter, J. Addai-Mensah, and D. Losic. 2012. *Microporous and Mesoporous Materials* 153: 131–136.
121. Comessana-Hermo, M., D. Ciuculescu, Z.-A. Li, S. Stienen, M. Spasova, M. Farle, and C. Amiens. 2012. *Journal of Materials Chemistry* 22: 8043.
122. Zeng, H., and S. Sun. 2008. *Advanced Functional Materials* 18: 391–400.
123. Atmane, K.A., C. Michel, J.-Y. Piquemal, P. Sautet, P. Beaunier, M. Giraud, M. Sicard, S. Nowak, R. Losno, and G. Viau. 2014. *Nanoscale* 6: 2682.
124. Liu, Q., X. Cao, T. Wang, C. Wang, Q. Zhang, and L. Ma. 2015. *RSC Advances* 5: 4861.
125. Zhou, P., Y. Li, P. Sun, J. Zhou and J. Bao. 2007. *Chemical Communication* 1418–1420.
126. Liu, Q., X. Guo, J. Chen, J. Li, W. Song, and W. Shen. 2008. *Nanotechnology* 19: 365608.
127. Liu, Q., X. Guo, Y. Li, and W. Shen. 2009. *Langmuir* 25: 6425–6430.
128. Liu, Q., X. Guo, Y. Li, and W. Shen. 2009. *Materials Letters* 63: 1407–1409.
129. Toshima, N. and T. Yonezawa. 1998. *New Journal of Chemistry* 1179–1201.
130. Hong, J.W., S.W. Kang, B.-S. Choi, D. Kim, and S.W. Han. 2012. *ACS Nano* 6: 2410–2419.
131. Yin, A.-X., X.-Q. Min, Y.-W. Zhang, and C.-H. Yan. 2011. *Journal of the American Chemical Society* 133: 3816–3819.
132. Shao, M., K. Shoemaker, A. Peles, K. Kaneko and L. Protsallo. 2010. *Journal of the American Chemical, Society* 132: 9253–9255.
133. Lim, B., M. Jiang, P.H.C. Camargo, E.C. Cho, J. Tao, X. Lu, Y. Zhu, and Y. Xia. 2009. *Science* 324: 1302.
134. Koenigsmann, C., A.C. Santulli, K. Gong, M.B. Vukmirovic, W-P. Zhou, E. Sutter, S.S. Wong and R.R. Adzic. 2011. *Journal of the American Chemical Society* 133: 9783–9795.
135. Imura, Y., K. Tsujimoto, C. Morita, and T. Kawaii. 2014. *Langmuir* 30: 5026–5030.
136. Hermannsdorfer, J., M. Friedrich, N. Miyajima, R.Q. Albuquerque, S. Kummel and R. Kempe. 2012. *Angewandte Chemie International* Edition 51: 11473–11477.
137. Zhang, L., S.-I. Choi, H.-C. Pemg, H.-C. Peng, S. Xie, Y. Zhu, Z. Xie, and Y. Xia. 2014. *Advanced Functional Materials* 24: 7520–7529.
138. Cheng, D., H. Xu, and A. Fortunelli. 2014. *Journal of Catalysis* 314: 47–55.
139. Son, J., S. Cho, C. Lee, Y. Lee, and J.H. Shim. 2014. *Langmuir* 30: 3579–3855.
140. Feng, Y., H. Yin, D. Gao, A. Wang, L. Shen, and M. Meng. 2014. *Journal of Catalysis* 316: 67–77.
141. Kaito, T., H. Mitsunoto, S. Sugawara, K. Shinohara, H. Uehara, H. Ariga, S. Takakusagi, Y. Hatakeyama, K. Nishikawa, and K. Asakura. 2014. *Journal of Physical Chemistry C* 118: 8481–8490.
142. Suntivich, S., Z. Xu, C.E. Carlton, J. Kim, B. Han, S.W. Lee, N. Bonnet, N. Marzari, L.F. Allard, H.A. Gasteiger, K. Hamad-Schifferli, and Y. Shao-Horn. 2013. *Journal of the American Chemical Society* 135: 7985–7991.

143. Song, H.M., D.H. Anjum, R. Sougrat, M. Hedhili, and N.M. Khashab. 2012. *Journal of Materials Chemistry* 22: 25003.
144. Liu, C.-W., Y.-C. Wei, C.-C. Liu, and K.-W. Wang. 2012. *Journal of Materials Chemistry* 22: 4641.
145. Johansson, A.C., J.V. Larsen, M.A. Verheijen, K.B. Haughshoj, H.F. Clausen, W.M.M. Kessels, L.H. Christensen, and E.V. Thomsen. 2014. *Journal of Catalysis* 311: 481–486.
146. Tong, Y., H.S. Kim, P.K. Babu, B.P. Waszczuk, A. Wieckowski, and E. Oldfield. 2001. *Journal of the American Chemical Society* 124: 468.
147. Alayoglu, S., A.U. Nilekar, M. Mavrkikakis, and B. Eichhorn. 2008. *Nature Materials* 7: 333.
148. Steigerwalt, E.S., G.A. Deluga, and C.M. Lukehart. 2002. *Journal of Physical Chemistry B* 106: 760–766.
149. Steigerwalt, E.S., G.A. Deluga, and C.M. Lukehart. 2001. *Journal of Physical Chemistry B* 105: 8097–8101.
150. Muthuswamy, N., J.L. Gomez de la Fuenta, D.T. Tran, J. Walmsley, M. Tsypkin, S. Raaen, S. Sunde, M. Ronning and D. Chen. 2013. *International Journal of Hydrogen Energy* 38: 16631–16641.
151. Jiang, H.-L., T. Umegaki, T. Akita, X.-B. Zhang, M. Haruta, and Q. Xu. 2010. *Chemistry-A European Journal* 16: 3132–3137.
152. Knudsen, J., L.R. Merte, G. Peng, R.T. Vang, A. Resta, E. Laegsgaard, J.N. Andersen, M. Mavrikakis and F. Besenbacher. 2010. *ACS Nano* 4: 4380.
153. Yuan, G., C. Louis, L. Delannoy, and M.A. Keane. 2007. *Journal of Catalysis* 247: 256–268.
154. Xia, Y., Y. Xiong, B. Lim, and S.E. Skrabalak. 2009. *Angewandte Chemie International Edition* 48: 60–103.
155. Daniel, M.-C., and D. Astruc. 2004. *Chemical Reviews* 104: 293–346.
156. Haruta, M. 1997. *Catalysis Today* 36: 153–166.
157. Narayan, R., and M. El-Sayed. 2005. *Journal of Physical Chemistry B* 109: 12663–12676.

Chapter 13
Biomedical Applications of Anisotropic Gold Nanoparticles

John W. Stone, Alaaldin M. Alkilany, Majd A. Hamaly and Stephanie Canonico-May

Abstract Synthesized colloidal suspensions of gold nanoparticles (GNP) have been around since the early 1950s. Since that time there has been an extremely large body of work dedicated to investigating both the chemical and physical properties of these fascinating materials. This work has by no means been limited to nanospheres, but over the last ~20 years included a wide variety of morphologies including nanorods, nanocubes, nanoshells, and nanocages to name a few. There has been much effort spent on functionalizing these nanomaterials for the purpose of novel nanotechnology-driven approaches to medicinal and biological challenges. Gold nanoparticles continue to be of interest due to their interesting optical properties, sizes, photothermal properties and long-term stability. More recently, there has been a more concerted effort to better understand the growth mechanism of the gold nanoparticles and to discover more efficient and/or greener alternatives for the synthesis of these materials. We cannot begin to provide a truly comprehensive explanation of all of the published work within the gold nanoparticle field in this chapter and so have chosen to outline a few key areas of study.

Keywords Gold nanoparticles · Optical properties · Surface functionalization · Therapeutics

J.W. Stone (✉) · S. Canonico-May
Department of Chemistry and Biochemistry,
Georgia Southern University, Statesboro, GA, USA
e-mail: jstone@georgiasouthern.edu

A.M. Alkilany
Department of Pharmaceutics and Pharmaceutical Technology,
University of Jordan, Amman, Jordan

M.A. Hamaly
King Hussein Cancer Center, Amman, Jordan

S.E. Hunyadi Murph et al. (eds.), *Anisotropic and Shape-Selective Nanomaterials*,
Nanostructure Science and Technology, DOI 10.1007/978-3-319-59662-4_13

13.1 Introduction

For hundreds of years noble metal nanoparticles have been used in art due to their vibrant colors. Objects from as far back as the Roman Empire have been found to contain gold and silver nanoparticles [1, 2]. These findings prompted researchers to investigate the source of the optical properties displayed by the nanoparticles. It is now well known that these metals behave very differently when on the nanoscale versus bulk metal. The size of the particles directly affect the optical properties, and it is those particles having one or more dimension ≤ 100 nm that exhibit these properties [3]. In 1908 Gustav Mie theorized solutions to Maxwell's equations which explain this phenomenon quantitatively [4]. On a basic level, Mie concluded that the colors were produced by light absorption and scattering done by the nanoparticle [3]. Another explanation is simply qualitative based on the known electron mean free path in metals which is ($\sim$10–100 nm) [5]. Any metallic nanoparticles in the size range of 5–200 nm can compare to this pathlength therefore radiation of the particles using certain frequencies of light cause electrons to oscillate adjacent to the surface of the metal. The electron oscillations have been labeled surface 'Plasmons' [6] or more specifically localized surface plasmon resonances (LSPR). A standard ultraviolet-visible spectrometer (UV-vis) can be used to detect plasmon bands by measuring the output when a sample is irradiated with white light.

In the last two decades, researchers have concentrated on how to synthesize different particle shapes and sizes as well as how the particle surfaces can be modified for maximum functionalization. Gold and silver nanoparticles are among the most highly used of the noble metal nanoparticles. Researchers discovered that by changing morphologies and sizes of the nanoparticles their optical properties can be tuned in both the visible and near-infrared spectrum (Fig. 13.1) [5]. While nanospheres have historically been the most common shape more recent research has determined the many attributes of anisotropic nanoparticles such as rods, cubes, stars, prisms, shells, and cages.

There are multiple methods for the synthesis of GNPs. For gold nanospheres the most commonly used is the Turkevich method (1951) which results in citrate capped 10–20 nm spheres [7]. The Brust method (1990s) is also a well-known synthesis method and used to grow $\sim$5–6 nm gold nanospheres in a surfactant solution [8]. Gold nanorods are typically prepared following a seed-mediated approach, a photochemical method, or one of two electrochemical methods [5, 9–11]. For more information on these methods please refer to the synthesis section of this report.

In more recent years research has been redirected toward finding ways to functionalize GNPs for biomedical uses. Gold has historically been utilized because of its use in the treatment of rheumatoid arthritis exhibiting the low cytotoxicity of the bulk gold. We have seen that the properties of nanoparticles can differ greatly from bulk material. For this reason, researchers have worked to determine the cytotoxicity of the numerous differently sized and shaped nanoparticles. This work

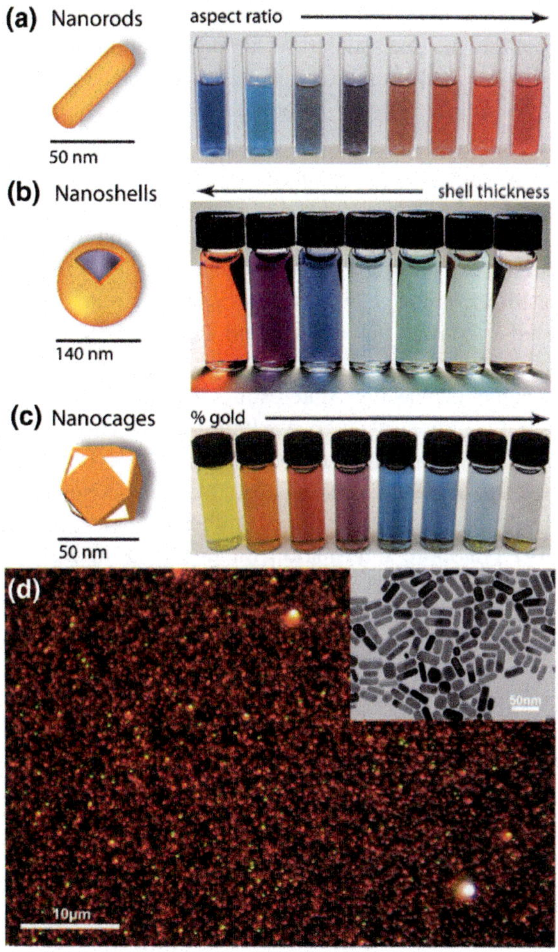

Fig. 13.1 Gold nanoparticles commonly applied in biomedical applications. **a** Gold nanorods, **b** silica–gold core–shell nanoparticles, and **c** gold nanocages. The intense color of these nanoparticles arises from the collective excitation of their conduction electrons, or surface plasmon resonance modes, which results in photon absorption at wavelengths which varies with **a** aspect ratio, **b** shell thickness, and/or **c** galvanic displacement by gold. **d** Optical dark-field scattering micrograph of gold nanorods (electron micrograph in the *inset*) showing resonant scattering from their transverse (short-axis) plasmon mode (*green*) and their lower energy, longitudinal (long-axis) plasmon mode (*red*) [5]

also expanded to include varying types of surface-capping agents, such as cetyltrimethylammonium bromide (CTAB) and sodium citrate [12]. This work determined that gold nanoparticles inherently have very low cytotoxicity rather it is surface functionality that drives their properties and toxicity. As such many uses for these GNPs have been discovered.

Gold nanorods (GNRs) are widely studied in biological applications due to their unique properties regarding LSPR. These anisotropic GNPs exhibit two plasmon bands. The oscillations of electrons along the short axis of the rod resulting from irradiation with light in resonance with these oscillations is referred to as the transverse band and the other being along the long axis is the longitudinal band [5]. Each peak corresponds to a wavelength and any changes in rod aspect ratio or morphology will be reflected in the peak position, peak shape and peak intensity.

Furthermore, since the process of surface modification is relatively facile, GNR coatings may be tailored to fit specific applications. Figure 13.2 shows the surface of the GNR being coated with polyacrylic acid (PAA) and polyallylamine hydrochloride (PAH) before subsequent bioconjugation to an antibody.

During this process the rod solution is continually characterized by using UV-vis to monitor changes in the observed LSPR. Clever surface functionalization allows the GNRs to be conjugated to various materials making them applicable to a variety of areas including sensing, imaging, drug delivery, and photothermal therapy. Biological sensing is accomplished by monitoring changes in plasmon bands as a result of minor changes to the local environment [13]. In imaging GNRs may be used as contrast agents based on their light scattering properties. Coated GNRs may be conjugated to linkers and thus to different types of drugs. As such GNRs have become increasingly interesting vehicles in the field of drug delivery. Another significant area of interest within biomedical community is the use of these materials as photothermal therapeutic agents. GNRs are conjugated to a "homing agent" on the rod surface. This agent finds the associated cells and when the rods are irradiated with a laser the photothermal heating results in irreparable cell damage. [14] For more information on this refer to the functionalization and plasmonic photothermal therapy sections of this report.

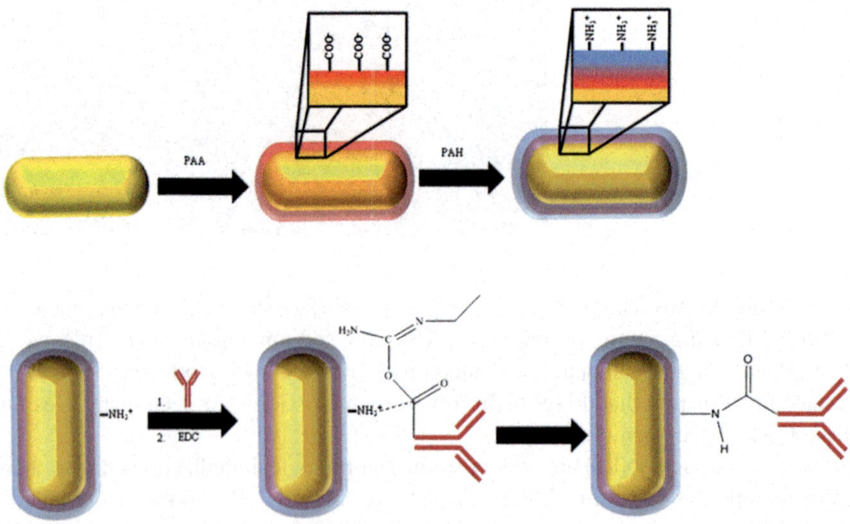

Fig. 13.2 Schematic outlining the bioconjugation of gold nanorods to antibodies via amide bond formation between free primary amines on the nanorod surface and carboxylic acids present on the antibodies [6]

13.2 Synthesis of Gold Nanorods

13.2.1 Synopsis

The first aqueous batches of colloidal gold nanoparticles were prepared in 1857 by Michael Faraday and still remain on display and optically active to this day in the Faraday Museum. Since that time, countless works have been published on a wide variety of gold nanoparticles having various sizes and shapes. Later work (some of which are described in this chapter) included the utilization of these optically fascinating materials for advances in the medicinal and biological communities. Gold remains a popular metal of study not only due its optical properties but also due to its low cytotoxicity and long-term stability. While the scope of this chapter primarily focuses on surface functionalization and biological applications of anisotropic gold nanoparticles, we have chosen to include a section on gold nanoparticle synthesis. Specifically, gold nanorods have been synthesized since the early 2000s with high monodispersity and reproducibility [5, 15–18]. The seed-mediated approach is commonly used to prepare these materials but suffers from an inherent problem with respect to yield. In fact only ~15% of the initial gold added is converted into nanorods. This problem prevailed for years, however, several recent investigations have reported alternative approaches to gold nanorod syntheses resulting in much higher gold conversion percentages [19–22]. Given this recent surge in gold nanorod synthetic improvements, we thought it deserved a brief discussion. Herein will be an overview of four such recent reports of improved gold conversion. For more detailed explanations, readers should refer to the provided references for each report.

13.2.2 Historical Synthetic Approaches

The most simple and commonly employed protocol for preparing gold nanospheres having a size range of 10–20 nm is the Turkevich method (1951) whereby sodium citrate is added to an aqueous boiling solution of chloroauric acid [7]. In this case, the citrate acts as both the reducing and capping agent resulting in particles with a net negative surface charge. These solutions are ruby red in color and extremely stable. Another well-known synthesis for preparing spherical gold nanoparticles is the Brust method (1990s). In this approach, chloroauric acid is mixed with tetraoctylammonium bromide, toluene and sodium borohydride [8]. Here, the sodium borohyride acts as both a reducing agent and anti-coagulant. These particles are smaller at ~5–6 nm and are stable in organic solvents that are not miscible with water.

There are several approaches to gold nanorod synthesis including: (1) the seed-mediated approach whereby gold seed particles initialize the reduction of gold ions in the presence of ascorbic acid, silver nitrate and the growth-directing agent cetyltrimethylammonium bromide (CTAB). Ascorbic acid is a weak reducing agent and results in the reduction of Au^{+3} to Au^{+1}. The seed (~1.5 nm gold spheres) serves to alter the reduction potential of the ascorbic acid allowing for the slow reduction of Au^{+1} to Au^{0}. CTAB directs growth by binding preferentially to the side faces of the gold as growth proceeds. The silver nitrate controls nanorod aspect ratio is the result of Ag deposition on the rod side and end faces. The mechanisms driving the role of silver are many and are beyond the scope of this chapter, (2) the electrochemical method in which nanorods are grown in an electrolytic solution under a 3 mA current, (3) the photochemical method where nanorods are grown under UV irradiation in the presence of gold salt, CTAB and tetradodecylammonium bromide, and (4) the electrochemical method where gold nanorods are synthesized by the electrochemical deposition of gold metal inside the pores of a membrane [5, 10, 11]. For the purpose of this report we will focus only on the seed-mediated protocol.

13.2.3 New Approaches to Nanorod Syntheses Via a Seed-Mediated Approach

While the seed-mediated synthesis results in highly monodisperse nanorod formation is it highly inefficient with respect to the conversion of gold from $HAuCl_4$ to Au^0. In fact only roughly 15% of the gold is utilized in the formation of gold nanorods. In recent years researchers have begun to offer synthetic alternatives to more efficiently convert the gold salt into gold nanorods. These approaches will be discussed briefly in the following 4 subsections.

13.2.3.1 Secondary Growth

In 2013, Kozek et al. report a 100% conversion of gold salt into gold nanorods based on optical density measurements [21]. Initially, gold nanorods were prepared following a very similar protocol as reported by Murphy et al. to a solution of these rods, ascorbic acid was added in a controlled manner resulting in continual conversion of gold ions into nanorods [5]. This process was continued until an increase in optical density was no longer observed (Fig. 13.3). They also report that the addition of benzyldimethylhexadecylammonium chloride (BDAC) to the "initial" rods followed by ascorbic acid addition resulted in rods having a larger aspect ratio compared to those prepared in the absence of BDAC. Furthermore, they demonstrate that the concentration and rate of ascorbic acid addition affects both the aspect ratio and monodispersity.

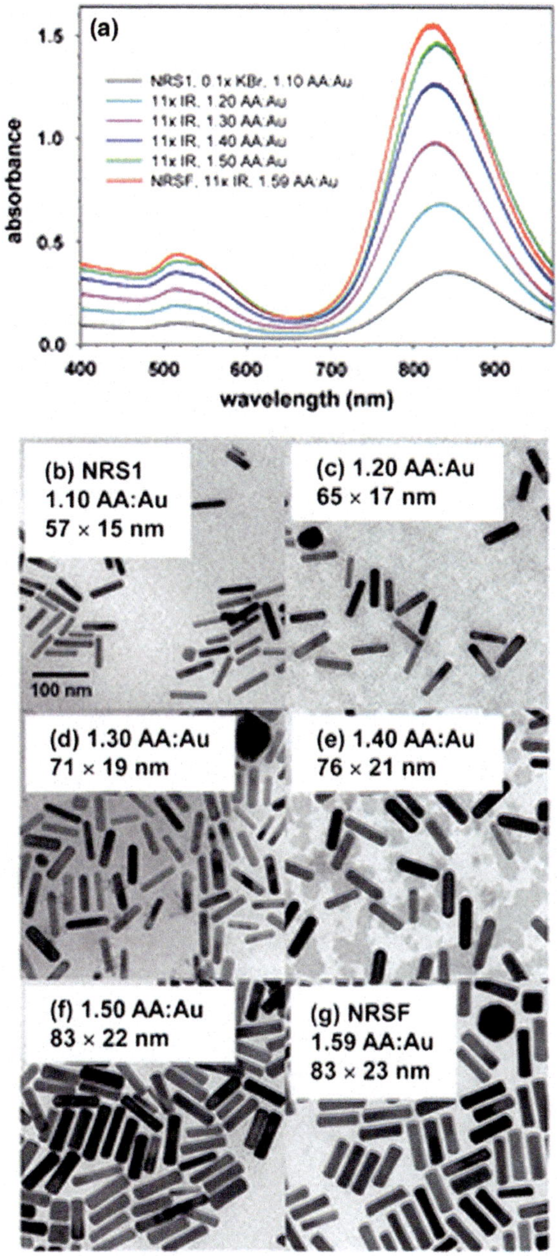

Fig. 13.3 Timed aliquots during secondary growth of the initial nanorods with a longitudinal band ~820 nm at an injection rate (IR) of 11X, which drives an increase in the AA:Au molar ratio: **a** optical absorbance spectra and **b–g** TEM images after adding different amounts of AA. Spectra and images for samples of **b** initial rods and **c** 31, **d** 62, **e** 94, **f** 125, and **g** 156 min after starting the secondary growth phase. **g** is the final rod solution

13.2.3.2 Pre-reduction with Salicylic Acid

Scarabelli et al. reported (2013) the use of 5-bromosalicylic acid (5-BrSA) as both a cofactor and reducing agent [22]. They looked at the ability of the acid to reduce Au^{+3} to Au^{+1} in a "prereduction" step followed by the addition of ascorbic acid to complete the intermediate reduction to Au^{+1}. Based on early literature reports, they monitored complete conversion of gold into nanorods by obtaining an absorption value of 1.2 at 400 nm [23]. Figure 13.4 shows the change in absorption with time, change in UV-vis spectra, and change in TEM images of gold nanorods as a function of various prereduction times. It is clear that a decrease in prereduction time accompanied by an increase in longitudinal band absorption as well as a decrease in aspect ratio. It should be noted that at the maximum time "f" (Fig. 13.5) the absorption at 400 nm has reached the desired value of 1.2 indicating complete conversion of gold ions.

The authors also demonstrated the ability to add gold onto presyntheisized gold nanorods by adding them to an "overgrowth" solution containing Au^{+1} that has been prereduced with 5BrSA followed by the addition of ascorbic acid. In this way

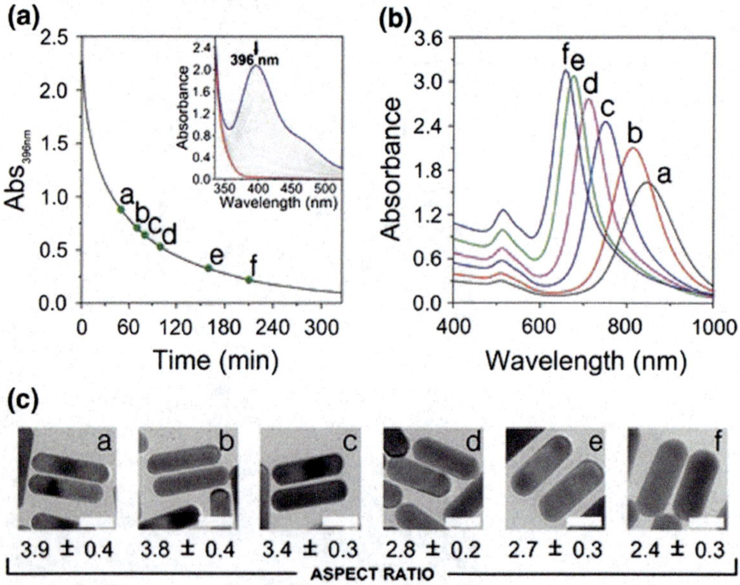

Fig. 13.4 The extent of prereduction by 5-BrSA affects both the aspect ratio and the amount of reduced gold. **a** Kinetic study of the prereduction step (*inset* full UV-vis spectra). **b** UV-vis NIR spectra of the various nanorod colloids obtained by changing the prereduction time. Abs396 nm: 0.88 (*a*); 0.71 (*b*); 0.64 (*c*); 0.53 (*d*); 0.33 (*e*); 0.22 (*f*) (Uv-vis spectra were multiplied by the corresponding dilution factor). **c** Representative TEM images of nanorods obtained with different prereduction times (the labels correspond to those on the spectra in **b**). Scale bars: 20 nm

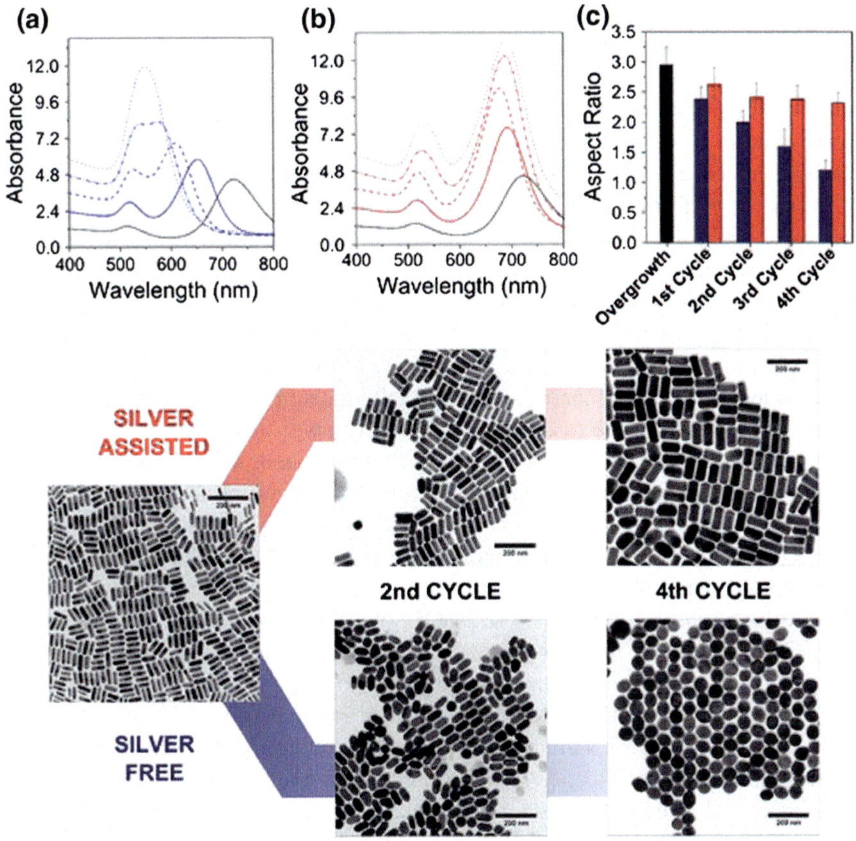

Fig. 13.5 Silver-free system, **b** silver *Upper panel* Visible-assisted system. The *black spectra* correspond—NIR spectra after each overgrowth step. **a** to the common starting point for the overgrowth reactions (UV-vis spectra were multiplied by the corresponding dilution factor). **c** Evolution of the aspect ratio for each overgrowth cycle in both silver-free (*blue*) and silver assisted (*red*) systems. *Lower panel* TEM images showing the shape evolution from the initial sample (centralcycles for silver-free (*blue line*) and the silver-left image) after two and four-assisted (*red line*) conditions. Scale bars: 200 nm

core-shell nanorods are formed. This process was repeated in a multicyclic fashion on preformed gold nanorods both in the presence and absence of Ag. Figure 13.5 is an overview of results obtained from the "overgrowth" process. It is clear that there is a marked blue shift as a well an increase in absorption with each cycle. It is also of particular interest, that when the overgrowth steps are performed in the absence of Ag the morphology becomes largely spherical by the 4th cycle.

13.2.3.3 Overgrowth of Gold Nanorods Via a Binary Surfactant Mixture

In 2014 Khlebtsov et al. published a report where presynthesized gold nanorods were added to a growth solution containing both CTAB and sodium oleate (NaOL) resulting in controlled overgrowth on the rod surface. This work was an extension of research performed by Murray and co-workers in 2013 [20, 24]. While improving gold conversion was not the focus of this work, the investigators do report a virtually 100% conversion of gold salt into nanorods again based on absorption data. In this approach gold nanorods were prepared by the reduction of $HAuCl_4$ in the presence of both CTAB and NaOL as well as ascorbic acid. These nanorods were then added to a secondary growth solution also containing the same concentrations of CTAB and NaOL and left undisturbed for 48 h resulting in overgrowth of Au^0 on the rod surface. This overgrowth was along the longitudinal axis and aspect ratios and thus longitudinal Plasmon bands were tuned over a wide range of wavelengths. Experiments were carried out in which the gold molar ratio (growth solution/nanorods) was varied between 1 and 20. It was found that as this gold molar ratio was increased there was a continued blue shift in the UV-vis spectra along with an increase in intensity of the transverse band (Fig. 13.6). In order to highlight the importance of NaOL, overgrowth experiments were carried out in the absence of the anionic surfactant. TEM images clearly show the formation of “dog-bones” rather than the “cigar” like shape desired. It was concluded that this occurs due to the faster kinetic growth rate when only CTAB is used as compared to the binary surfactant mixture. The authors point our that NaOL’s ability to drive slower kinetics results from its ability to improve stability on the nanorod crystal faces as well as act as a reductant for Au^{+3}. In this case the quantity

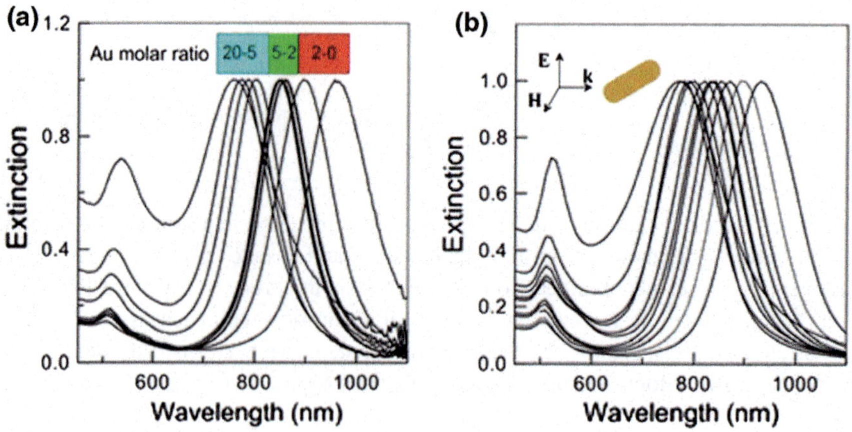

Fig. 13.6 Experimental (**a**) and simulated (**b**) extinction spectra of the nanorods obtained for overgrowth at gold molar rations of 0–20. For calculations, the geometrical parameters of the nanorods were taken from statistical analysis

of ascorbic acid used was reduced by 8 times. Overall the ability to tune the longitudinal Plasmon band using the binary surfactant mixture compared to CTAB alone was found to be ~75 nm verses ~20 nm respectively.

13.2.3.4 Improved Conversion of $HAuCl_4$ into Gold Nanorods Via Re-seeding Approach

In 2016 a report was published by Canonico-May et al. whereby gold salt conversion into gold nanorods was increased from ~15 to ~75% via a reseeding approach. Gold nanorods were grown following a typical seed-mediated protocol and purified via centrifugation [19]. Instead of discarding the supernantant, it was saved followed by the addition of more seed (~1.5 nm gold spheres) and ascorbic acid. This resulted in new nanorod growth and this process was repeated through 5 iterations resulting in monodisperse gold nanorods each time. Figure 13.7 is a UV-vis overlay showing the spectra of the original as well as the 5 re-seeded nanorod samples (left) and corresponding TEM images (right). The data shows that while as compared to the original sample, the remaining re-seeded solutions show a steady red-shift in the longitudinal Plasmon resonance. It should also be noted that an increase in aspect ratio was also observed in parallel with each reseeded solution. It is unclear as to why this particular pattern with respect to the absorption spectra and aspect ratio is observed other than to speculate that it may be function of varying kinetics associated with the rod growth. In order to further investigate the diversity of this synthetic approach, the authors added additional $AgNO_3$ to supernatants generated from the purification of the original rods in an effort to investigate the ability to tune the longitudinal Plasmon band independently within the re-seeded samples. The investigators found that with increasing concentrations of silver nitrate a red-shift was observed followed by a series of blueshifts (Fig. 13.8). This was an important discovery in that it demonstrated the ability to tune rod properties individually.

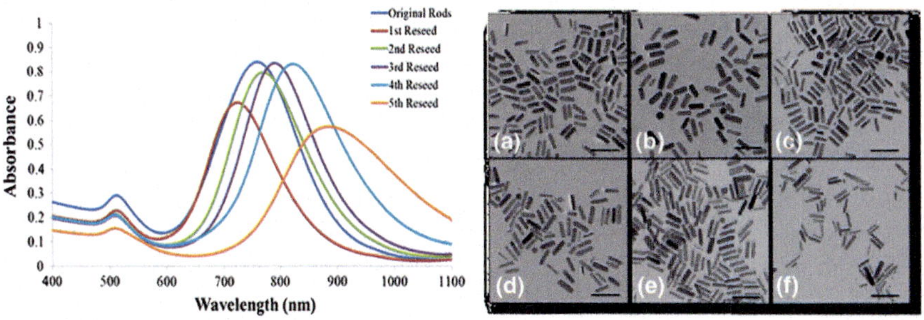

Fig. 13.7 Uv-vis spectral overlay of initial nanorod solutions (aspect ratio ~3) and the 5 re-seeded supernatant gold nanorod solutions (*left*) and TEM images of the initial and 5 re-seeded solutions (**a–f**) (*right*)

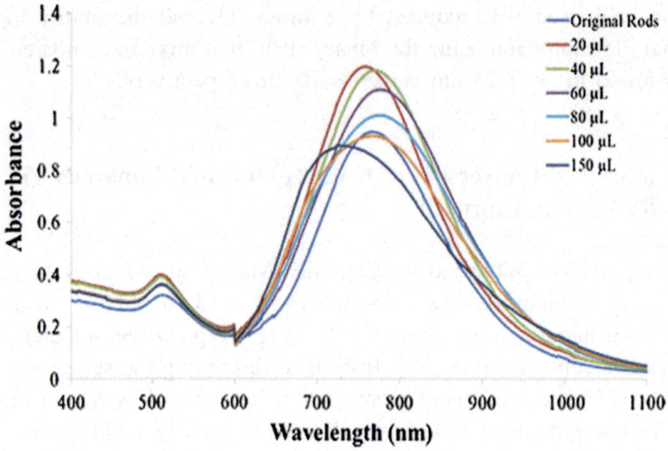

Fig. 13.8 Uv-vis spectral overlays for the varying additions of 0.01 M silver nitrate Scale bar: 100 nm

13.3 Functionalization of Gold Nanoparticles

13.3.1 Synopsis

In spite of the distinctive physical and chemical properties associated with nanoparticles, namely gold nanoparticles, surface fabrication is a rich area of research for the development of highly stable, selective and sensitive probes for biological applications [25]. Features include; optical properties, the availability of synthetic approaches for the generation of different sizes, shapes and morphologies and the relative biocompatibility.

GNPs exhibits unique particle-light interaction owing to the surface plasmon resonance (SPR) effect. Depending on the size, morphology and aspect ratio of GNPs and the type of laser light used for irradiation a number of possible interactions can be demonstrated including enhanced light (Rayleigh) scattering, fluorescence, Surface Enhanced Raman Scattering (SERS) and the plasmonic photothermal effect [26, 27].

Moreover, a substantial number of reproducible synthetic approaches toward the generation of GNPs with different attributes are highly addressed in literature, examples include: the synthesis of nanospheres [7], nanorods [28], nanocages [29], nanoshells [30] and others. Biocompatibility with the ligands in addition to the high surface area to volume allows for multi-functionalization of GNPs and accordingly GNPs can be fabricated to serve multiple roles at a time [30].

Bearing in mind the unique physico-chemical properties, structural diversity and surface attributes, GNPs appear as an appealing potential platform for fabrication for maximal utilization of these properties [31].

Herein, we aim to discuss the most widely used capping ligands, functionalization of GNPs using biomolecules (bioconjugation) addressing common conjugation techniques and applications described in literature.

13.3.2 Functionalization Using Capping Ligand

Today, GNPs are continuing to demonstrate their valuable contribution as key components for imaging and therapeutic applications [26]. To enable their utilization for such applications two main issues shall be appropriately tackled: the colloidal stability of the preparation and the successful fabrication with the necessary targeting moieties.

Colloidal stability of nanosized preparations is a major challenge, especially that their distinguished properties are directly correlated to their sizes and morphologies. Accordingly, aggregation resulting in structural deformations or size variations might be detrimental to the anticipated activity [32]. On the other hand and in terms of targeting, for a successful delivery of nanomaterials into the desired sites, homing moieties shall be introduced by surface functionalized of nanomaterials [3].

Herein, capping ligands come into the play; surface modification of GNPs were extensively addressed to promote higher colloidal stability and prevent aggregation. Upon attachment; capping agents demonstrate a steric, electrostatic stabilization effects or a combination of both. As for targeting, several capping ligands are addressed to facilitate crosslinking to biomolecules [33, 34].

Attachment of the capping ligands themselves is facilitated by several forces including: covalent, electrostatic, hydrophobic and van der walls interactions. Thiol containing molecules exhibit distinctive affinity to GNPs, alkyl and aryl thiol can form a highly stable self-assembled monolayer (SAM) on gold surfaces medicated by the sulfhydryl group [35]. Among the first approaches toward the formation of thiol SAM is the Brust method that utilizes the tightly packed SAM for facile phase transfer into toluene [8]. Produced GNPs are described as highly stable upon drying and thus can be stored in the powdered form. Moreover, thiol SAMs arising from different alkyl thiols can actually provide versatile platforms for further fabrication for biomedical applications [36].

Another common capping agent is poly ethylene glycol (PEG), thiol terminated PEG molecules display superior affinity to GNPs as a result of the gold thiol bond. PEG is extensively used in GNPs grafting for multiple reasons: (1) the superior stability of PEG grafted GNP over as prepared GNP, such stability was enhanced by increasing the concentration of PEG [37], not only this but PEG grafted GNP have demonstrated better colloidal stability over modification using thiol compounds including glutathione, mercaptopropionic acid and cystamine [34]. Such stabilizing effect is mainly attributed to the steric properties of the molecule with a still hydrated surface, rendering the composite less sticky, which will be particularly useful in high salt concentration media including biological systems [33]. (2) The remarkably longer blood circulation time and thereby plasma half-life for PEG-coated GNP compared to as prepared or plain GNP [38]. (3) Bi-functional PEG molecules can serve as a cross linker facilitating further grafting to biomolecules or other chemicals while maintaining the stability of the system under different conditions of grafting reactions [33].

Apart from the thiol-gold bond based stabilization, GNPs can be electrostatically stabilized. The simplest form of electrostatically stabilized GNP are citrate stabilized gold nanospheres (GNS). Citrate stabilized GNS are prepared according to the popular Turkevich method that involves the reduction of $HAuCl_4$ in aqueous media using sodium citrate. Citrate ions play a dual role in this chemistry where they act as a reducing and stabilizing agents for the formed nanospheres [7]. In spite of the relatively easy synthetic approach, citrate stabilized GNS are very susceptible to irreversible aggregation upon mechanical stressing or salt introduction [33, 39]. Accordingly, citrate stabilized GNS cannot be described as ideal candidates for functionalization with biomolecules, ligand exchange using PEG for example might be necessary to allow for further grafting [40].

Another interesting electrostatic stabilization technique is the polyelectrolyte coating applying the layer by layer deposition. Where layers of oppositely charge materials are adsorbed one over another resulting in surface charge flip and a hybrid form of nanocomposites with various functional groups that can be suitable for further functionalization such as carboxylic acid and amine groups [41]. For instance, carboxylic acid functionality was gained by the adsorption of the negatively charged polyacrylic acid over the positively charged surfactant CTAB used in the seed mediated synthesis of god nanorods, this group served as an anchor for the crosslinking of antibodies to the surface of GNPs [42, 43]. Furthermore, layer by layer coating can successfully transfer gold nanorods to a number of polar organic solvent and hydrophobic films while preserving their stability [44].

13.3.3 Functionalization Using Biomolecules

For a successful delivery and enhanced accumulation of nanoparticles in their designated sites, recognition moieties can be attached to the surface of GNPs. Examples include Oligonucleotides, antibodies and peptides [39].

13.3.3.1 Oligonucleotides

Viruses based delivery approach is a conventional approach for nucleic acid delivery. However, such approach can be unpredictably toxic with unexpected immune response. On the other hand, GNPs can be functionalized by thiol terminated oligonucleotides. In fact, nucleic acids are relatively easily thiolated for covalent conjugation to GNP [45, 46].

Such approach enable the formation of dense layer of oligonucleotides with distinctive affinity to complementary DNA sequence with association constants that is more than two orders of magnitude higher compared to the unconjugated oligonucleotides. Such distinctive behavior can be attributed to the dense packing of oligonucleotides on the surface of GNPs [39, 47]. Such dense packing is allowed by the small size and high surface area. It is worth mentioning that the nature of this layer can be fine-tuned in terms of hydrophobicity and surface charge to optimize the cellular uptake [46].

Heavily functionalized oligonucleotides-GNPs in combination with magnetic nanoparticles were used as an ultrasensitive method for the detection of the prostate-specific antigen; a tumor marker for the diagnosis prostate and breast cancer at concentration at the attomolar levels [48]. Another application is based on the relation that exist between the optical properties, namely the colors, the UV-Vis spectra and the inter-particular distance; a colorimetric approach was investigated for the sensitive detection of cancerous cells relying on aptamer functionalized GNPs. functionalized GNP were able to assemble at the surface of positive cancer cells and demonstrate spectral behavior compared to control cells. Absorbance was positively correlated with cellular concentration confirming the effect of assembled GNPs [49].

Furthermore, sharp melting transitions for the DNA-GNPs aggregates combined with the observed colorimetric changes can be used to detect the mercury ion Hg^{2+}. This assay mainly relies on thymidine-mercury-thymidine coordinate chemistry. Functionalized GNPs-Hg^{2+} aggregates were observed to melt at a higher temperature compared to the ones in the absence of Hg^{2+}, this is mainly attributed to the stabilizing effect of the coordination [50]. In a similar direction, the stronger interaction between Hg^{2+} enables the sequestering of mercury from the crosslinked thymidine functionalized GNP. Accordingly the decrease in melting point and the color change from purple confirmed the sequestering of mercury by cysteine coordination [51].

Moreover, the colloidal stability of GNP can be used as probes to detect nucleic acid in solution. Mixing DNA functionalized GNP with complementary DNA targets can result in the formation of GNP-DNA network and subsequent color and spectral changes [52].

Not only this, but oligonucleotide conjugated GNPs were also exploited for their interesting ability to assemble into different crystalline structures by modifying the DNA sequence [53].

13.3.3.2 Antibodies

Antibodies are characterized by their distinguished selectivity and binding affinity to matching antigens making them excellent candidates for tissue targeting and delivery for nano-scaffolds for both imaging and therapeutic purpose [27, 54, 55]. For instance, GNP were functionalized with specific tumor markers including epidermal growth factor receptor EGFR, Her2 and tumor necrosis factor [55–60].

GNP-antibody conjugates can be prepared by multiple approaches including passive adsorption at the isoelectric point of the protein relying on the non-covalent binding properties or a combination of electrostatic and hydrophobic interactions conjugation to anti-EGFR is an example, conjugation was evidenced by spectral changes and stability in phosphate buffer saline PBS [61–63]. However, such approach can have several drawbacks including the need for higher concentration of the antibody, the higher possibility of random orientation due to the variability in the electrostatic nature of different regions of the antibody and finally the possibility of the antibody of the protein getting replaced with higher affinity particles upon introduction to biological samples [64].

Alternatively, covalent conjugation via a directional linker can be employed to achieve consistent orientation of the active Fab portion of the antibody toward the outer surface of GNP. The covalent attachment was achieved via the attachment of a heterofunctional linker a thiol terminated amide-bonded adipic hydrazide and an alkane and to the non-targeting or Fc portion of the antibody and allowing the rest of the conjugation chemistry to occur here leaving the Fab region unhindered for maximal activity [64].

Other crosslinkers can be also employed for covalent conjugation, however, such conjugation and unlike the aforementioned techniques is rather nonselective in terms of orientation. Examples include ortho-pyridyldisulfide-polyethylene glycol N-succinimidyl propionate (OPSS-PEG-NHS); a carboxylic acid terminated thiol functionalized crosslinker. Carboxylic acid groups provide a suitable point for crosslinking applying ethyl dimethylaminopropyl carbodiimide (EDC) and N-hydroxysaccinmide coupling that result in the formation of amide bond [65, 66].

Alternatively, surface fabrication of GNP themselves can serve as an excellent platform for antibody functionalization [67] for instance, several thiol terminated ligands can be employed to functionalize GNP with amine groups providing suitable candidates for EDC-NHS coupling [35]. Furthermore, layer by layer coating can also promote GNP to successful EDC chemistry by either carboxylic acid or amine functionalization. For example, PAA coating was addressed for the facile formation of carboxylated GNP for anchorage of antibodies [42, 68]. Antibody functionalization is extensively employed for imaging applications where GNP can serve as suitable contrast agents and with the proper functionalization the are targeted on the molecular lever toward the tissue in question, for instance, the selective uptake of anti-HER2 functionalized GNP was revealed using confocal microscopy [62, 69]. Other imaging techniques such as dark field microscopy and Raman scattering were used hand in hand for successful visualization and probing of molecular targets. Antibody functionalized GNPs demonstrated considerable attenuation of X-rays and were considered as successful selective probes as computer topography contrast agents, compared to conventional agents such as iodine where such targeting wasn't possible [68, 70]. Such particles can be used as Raman tags for in vivo tumor imaging using surface enhancer Raman scattering [65, 71].

Moreover, vectorized Anti-EGFR conjugated GNPs demonstrated selective differentiation between cancerous cell that over express a matching tumor marker and benign ones and accordingly were used as optical agents for the photothermal therapy [72, 73]. Similarly, gold functionalized gold nanocages [74, 75] and gold nanoshells [76, 77] functionalized with an antiHER2 monoclonal antibody were addressed as well.

13.3.3.3 Peptides

Inspired by the colloidal stability of proteins in biological fluids, that is essentially attributed to the unique folding that orients the hydrophobic groups inside and the hydrophilic groups outside, peptide functionalized GNP were proposed [78]. Combinatorial synthetic approach was proposed for the determination of the optimal amino acids sequence that would passively adsorb to the surface of the particles and form a densely packed layer with hydrophilic terminus. The colloidal stability of the conjugates was affected by the amino acid sequence, chain length and the peptide density at the surface [79]. On a different direction, tyrosine containing bi-functional peptides was found to be able to strongly reduce chloroauric acid resulting in the formation of water soluble, mono-dispersed, with a relatively narrow size distribution and peptide functionalized GNPs [80].

Peptide functionalized GNP were exploited for their nuclear targeting properties. Nuclear targeting is an extremely challenging process, probes shall be able to enter the cell, escape the lysosomal pathway and they should be small enough to cross the nuclear membrane. This was achieved employing various viral peptide that are capable of fulfilling each of the aforementioned tasks and attaching them into a single platform. It was demonstrated that GNPs functionalization with peptides that are capable of activating receptor mediated endocytosis and possess a nuclear localization sequence achieved considerable nuclear accumulation [81]. Furthermore, GNPs functionalized with therapeutic peptide performed both a therapeutic and targeting roles, where the targeting role was mainly demonstrated by promoting receptor mediated endocytosis. Loading on GNPs was adopted to enhance the efficacy of the therapeutic peptide neuropilin-1 toward breast cancer cells that overexpress matching receptors [64].

Finally, amino acids are known to interact with various metal ions though non-covalent interactions employing their aromatic, hydroxyl and charged moieties. The complex formed between peptide functionalized GNPs is utilized through its colorimetric features for the detection of metal ions including Co^{2+}, Hg^{2+}, Pb^{2+}, Pd^{4+}, and Pt^{2+} where each metal ion displayed a distinct Plasmon peak upon interaction with peptide conjugated GNPs [82].

13.4 Plasmonic Photothermal Therapy

13.4.1 Synopsis

Hyperthermia is a therapeutic approach that relies on rising the temperature of tumor loaded tissues up to 41–43 °C to induce cellular damage and subsequent cell death [83]. Hyperthermia is frequently used in combination with other treatment modalities; for instance, a synergistic effect is demonstrated when hyperthermia is combined with radiation.

Furthermore, the activity of chemotherapeutic drugs is enhanced when combined with hyperthermia which can be attributed to the increased membrane permeability, altered cell metabolism and altered transport mechanisms. Examples of drugs combined with thermal enhancement are doxorubicin, cisplatin and bleomycin [84].

Hyperthermia induces apoptotic or necrotic cellular death mainly by acting on proteins as cellular targets. Intra-cellular protein denaturation will destabilize the cellular membrane, cytoskeleton, enzymes, alter the signal transduction and interfere with the DNA repair mechanism [85].

Hyperthermia can be achieved via several modalities including; whole body hyperthermia, regional, superficial and intra-cavity hyperthermia. Whole body hyperthermia is accomplished by raising the whole body temperature using heated fluids, blankets or air, moreover, such mode of hyperthermia can also be introduced via invasive approaches by circulating heated blood. Regional hyperthermia is when the temperature of deeper tissues is elevated, usually using an ultrasound equipment that can generate waves at a wavelength of 1–5 MHz. For deeper and larger tumors, longer wavelengths are usually applied namely at ranges of 10–120 MHz. Ultrasound based techniques demonstrated penetrating abilities down to 4–6 cm while microwaves are used to manage superficial cases at a depth around 2 cm [86].

Intra-cavity hyperthermia is capable of achieving more localized heating with reduced damage to normal tissues with the aid of an implant or usually described as antenna. This approach are especially applicable for solid tumors such as prostate and head and neck cancers and it enables the controlled application of heat in terms of degree of heating, sometimes cooling and the duration of applications [83].

Nevertheless, such modes of hyperthermia have several drawbacks; whole body hyperthermia may cause elevation of serum transaminases and bilirubin and can be associated with neurological manifestations. Other approaches, although more selective than whole body hyperthermia will still confer collateral damage to surrounding tissues due to their limited selectivity [87].

For a more selective targeting of the tumor environment the photothermal therapy was introduced. Using proper agents known as photosensitizes with the ability to dissipate the energy gained rapidly through vibrational mode resulting in local heating. Dyes such as cyanines, azodyes and porphyrins coordinated with transition metal ions are suitable agents for such applications [88]. One major drawback is their low absorption cross section as natural chromophors and the photo-bleaching of dye molecules upon lase irradiation [89].

In the past years, nanotechnology experienced a markedly rapid development and a wide variety of nanostructures with distinguished properties was produced and utilized for biomedical applications [3, 90, 91], herein, we wish to highlight the significance of GNPs as excellent candidates for photothermal treatment.

13.4.2 Optical Properties

13.4.2.1 Surface Plasmon Resonance SPR

Compared to nonmetallic nanomaterials, gold nanoparticles exhibit distinctive optical interaction with light as evidenced by their intense colors. Gold and silver free electrons are mainly responsible of their behavior, with a mean free path of 50 nm only electron present at the surface of nanoparticles with sizes smaller than the mean free path are capable of traveling and hence the term "surface". When gold nanomaterial are exposed to wavelengths larger than their dimensions, surface electrons will align and oscillate in a coherent manner in response to that, furthermore, this oscillation will exist in a resonance condition [92].

Upon interacting with metallic nanoparticles, electromagnetic waves experience two types of energy loss; absorption and scattering, collectively known as extinction. Absorption occurs when the energy is emitted through inelastic process, light photons are converted to heat. On the other hand, scattering occurs when electron oscillation results in photons are emitted at the same frequency and can be known as Rayleigh scattering or at a different frequency as Raman scattering [93].

What is really special about gold nanomaterials is their surface enhancement of plasmon properties with an optical cross section that is 5 order of magnitude larger than conventional dyes providing substantial improvement in sensitivity. Moreover, GNPs and unlike dyes are immune to photo-bleaching and can provide the desired optical behavior under high excitation and for longer periods of time [93].

13.4.2.2 Tunability of Optical Properties

Each different form of GNPs interact distinctively with light as characterized by their absorption band shape and maxima [94]. Figure 13.9 illustrates a number of GNPs morphologies and the corresponding plasmon peaks, for instance, for spherical GNP sized 30 nm a strong absorption peak is observed at around 520 nm and as the size increase a red shift is observed and can reach about 570 with slight peak broadening (D1), obviously, color change from red to pink will be observed as well [95].

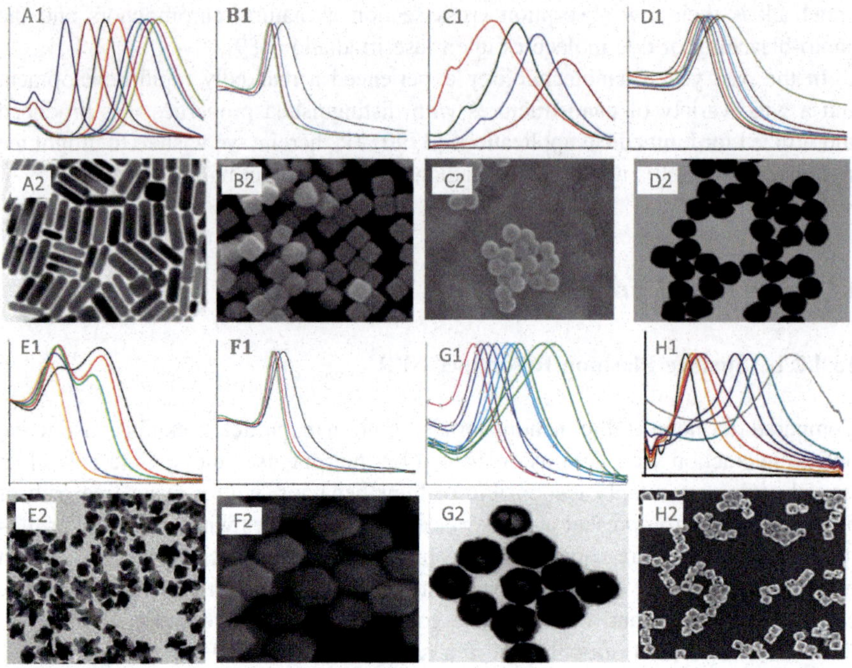

Fig. 13.9 Tunability of the optical and electrical properties of gold nanoparticles with changes in particles morphologies as evident by the TEM images, annotated as (1) and size as evident by the corresponding UV-VIS spectra for each annotated as (2). Such tunability is evident by spectral shifts of the surface plasmon resonance peak, spectral shift are due to changes in size, aspect ratio or shell thickness. 1: UV-Vis spectra and 2: TEM images for **A** nanorods **B** nanocubes **C** nanoshells **D** nanospheres **E** nanostars **F** rhombic dodecahedra **G** hollow gold nanospheres and **H** nanocages. Copyright: **A1** [26], **A2** [104], **B1**, **B2**, **F1** and **F2** [105], **C1** [27], **C2** [30], **D1** and **D2** [95], **E1** and **E2** [106], **G1** and **G2** [107], **H1** and **H2** [97]

The tunability of optical properties is not only size dependent, optical properties are also strongly affected by particles morphologies as well. Gold nanorod exhibit two absorption peaks transverse and longitudinal corresponding to the excitation of the surface electrons on the short and the long axes respectively. The short axis band resembles that of the nanospheres and displays a relatively weaker absorption than the longitudinal band. The size-sensitive longitudinal band is correlated to the aspect ratio (length/width) of the nanorods where a shift toward a longer wavelength is observed with the increase in aspect ratio as in (A1) [93, 94, 96]. Furthermore, and for gold nanoshells, the optical properties of the preparation is strongly correlated to both the relative size of the silica core and the thickness of the shell. For example Fig. 13.9 (C1) resembles the absorption band for particles with silica core of 60 nm and varying shell thickness ranging from 20–5 nm, it was observed that the decrease in the shell thickness was accompanied by a red shift toward the near infrared (NIR) region [30].

For another form of nanoparticles, Gold nanocages, optical properties can be tuned as well throughout the visible and into the near-infrared region. Gold nanocages are synthesized employing a galvanic replacement reaction that would sacrifice silver nanocubes with the lower reduction potential to reduce the chloroauric acid and allow it to get disposed gradually and ultimately resulting in cage formation.

Gold nanocages optical properties can be tuned by varying the amount of $HAuCl_4$ solution presented to the reaction media, figure (H1) demonstrate the shifting in the SPR band with an increasing volume of a 0.1 mM $HAuCl_4$ solution from 0 to 5.5 ml [29, 97].

The tenability of optical properties of GNP is of a paramount importance for biomedical application. For effective photothermal effect, laser light penetration and availability for cancerous tissue, where GNP are localized through various targeting approaches. Light at the NIR region was found to exhibit minimal absorption by tissues and is also known as "the transparent window" or the "therapeutic window", light at such wavelength exhibit minimal interaction with tissue chromophors including Hemoglobin and deoxyhemoglobin. Microwatt NIR laser light classified as FDA class 1 is reported to achieve 4–10 cm penetration depth depending on the tissue in question. High power laser, FDA class 3, can penetrated as deep as 7 cm into muscle and neonatal skull and brain tissues.

The above discussed nanostructures; the nanorods, nanoshells and nanocages are considered as promising candidate for biomedical applications owing to the tunability of their optical properties into this desirable region [31, 54, 98].

13.4.3 Targeting

Cancer cells are rabidly proliferating cells with high nutritional demands, oxygen supplies and waste excretion rates. Tumors at a size of 2 mm^2 and more become diffusion compromised and accordingly solid tumors respond by generating more blood vessels to increase the blood supply in a process known as angiogenesis. In contrast to normal blood vessels the vasculature around the tumor is described to be leaky and disordered with substantial gaps ranging from 100 nm to 2 μm depending on the tumor type.

Such phenomena is described as Enhanced Retention and Permeability (ERP) and is particularly important for the preferential accumulation of nanomaterials that are larger than the renal clearance threshold of 6 nm and smaller than 2 μm [27, 54, 58].

Nonetheless, several parameters shall be taken into consideration for tumor delivery; most importantly the hydrodynamic radius and the nature of the coat. For instance, smaller GNP with a thicker layer of PEG coat were found to achieved higher accumulation compared to larger particles with a thinner coat [59, 99]. One interesting work demonstrated the remarkable behavior of GNP with sizes between

20 and 100 nm with various grades of PEG in terms of the bold circulating half-life, tumor accumulation and vascular permeation [100].

Surface charge has an important contribution as well, positively charged particle achieved higher internalization for most of the cells in the tumors and subsequent payload dissociation. On the other hand, negatively charged particles demonstrated better diffusion to deeper tissues but with lower cellular uptake.

When the homing of GNPs into the tumor relies on the advantageous enhanced permeation and retention effect it will be described as passive targeting, nevertheless, the passive targeting is essentially hindered by its intrinsic lack of selectivity [46].

On the other hand, active targeting is achieved by attaching recognition moieties (antibodies, proteins, polypeptides, aptamars, etc.) to the surface of nanoparticles to enhance the target-ability and result in localized tissue accumulation. A number of ligand-functionalized nanoparticles are currently in different stages of clinical evaluation for various medical applications [27, 101, 102].

Examples include the successful targeting of the human epidermal receptors HER; a group of receptors that play a major role in cellular proliferation and thereby highly unregulated receptors in cancerous cells for both imaging and photothermal treatment [72, 73]. In spite of the increasing need of tumorous tissue to various nutrients such as iron and folic acid, transferrin and folate conjugated GNP were also exploited for active targeting [56, 57].

13.4.4 Examples

13.4.4.1 Gold Nanocages in the Photothermal Ablation of Breast Cancer

The enhanced light absorption for GNPs due to the LSPR effect is reflected upon their ability to emit that energy as heat. If proper laser light was used, such heating can result in coagulation and protein denaturation, thus cancer cell death with minimal invasiveness. In brief, gold nanocages, are hollow cage-shaped nanoparticles with thin and porous wall synthesized by galvanic replacement reaction through sacrificing silver nanocubes and chloroauric acid. Dimensions of gold nanocages can be tuned to absorb light at 810 nm, at the NIR range, achieving maximal breast tissue penetration. To investigate the photothermal effect of gold nanocages, breast cancer cell line SK-BR-3 overexpressing the HER2 receptor. Gold nanocages were functionalized using anti-HER2 antibodies for selective receptor targeting. Cells were irradiated with the proper laser light at the NIR. Cell viability was determined by incubating with calcein AM and Ethidium homodimer 1 (EthD-1) and then examined by fluorescence microscopy. Green fluorescence will be observed in viable cells as colourless calcein AM will be enzymatically converted to green fluorescence calcein, on the other hand in case of compromised cell

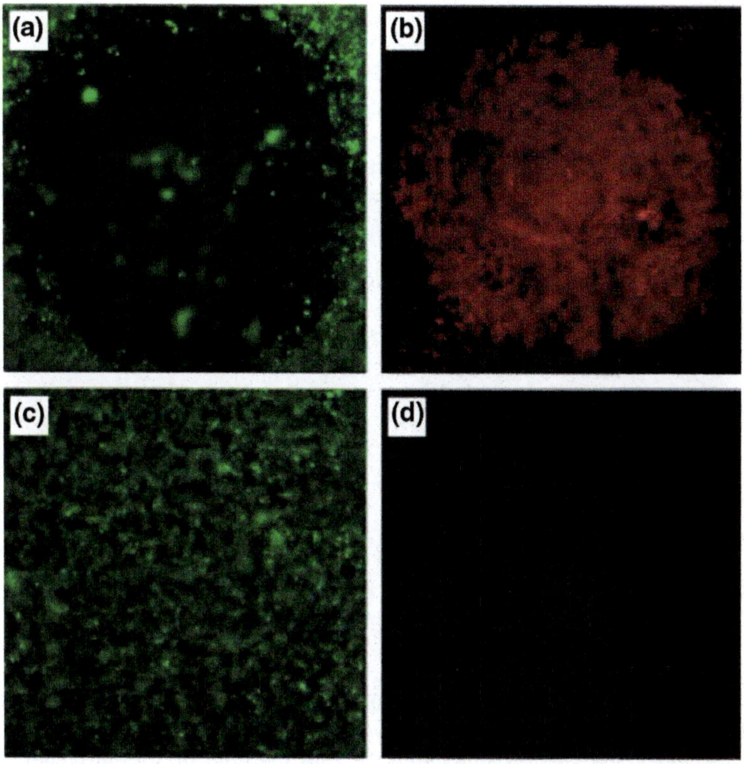

Fig. 13.10 Breast cancer cell line **a** the *green zone* represent the viable cells after adding the functionalized gold nanocages and irradiation with NIR laser light whereas **b** indicates Ethidium homodimer-1 (EthD-1) assay (dead cells) **c** and **d** represent control samples where cells were irradiated under the same conditions but without treatment with gold nanocages, as evident, cells maintain viability, adapted from [57]

membrane red fluorescence will result, which is in this case thermal damage. Using conjugated gold nanocages with NIR laser light irradiation resulted in a significant cell damage as indicated by (Fig. 13.10) [74].

13.4.4.2 Gold Nanorods in the Photothermal Ablation of Squamous Cell Carcinoma

Gold nanorods were prepared applying the seed-mediated growth technique. Seeds were prepared by fast reduction of $HAuCl_4$ using ice cold $NaBH_4$. The growth solution was consisted of CTAB, $HAuCl_4$, AgNO3 and ascorbic acid were gently mixed to form the transparent growth solution. Seeds were subsequently added to the growth solution and allowed to react for 2 h. Synthesized rods were of 12 nm in width and 50 nm in length and an aspect ratio of around 4, longitudinal plasmon

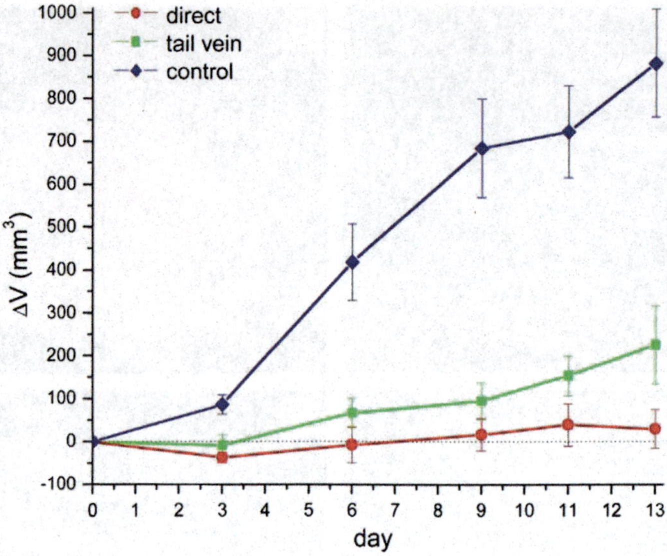

Fig. 13.11 Change in tumor size over a course of 13 days for control, direct administration into the tumor and administration into the tail vein, adapted from [103]

absorption band was observed at 800 nm. Excess CTAB was removed by double centrifugation and the cleaned-up rods were coated with thiolated poly (ethylene) glycol namely via covalent conjugation through the gold thiol bond. Two administration approaches were adopted, direct injection to the tumor and intravenous administration through the tail vain. For direct injection, irradiation with NIR light was done immediately done after injection to minimize the diffusion of pegylated gold nanorods beyond the tumor boundaries. For tail injection, tumor irradiation was 24 h post administration to allow for intra-tumor accumulation. Control mice were directly injected with phosphate buffer saline directly into the tumor with no NIR exposure. Changes in tumor size were recorded for 13 days, for direct injection more than 96% in tumor size was reported compared to more than 74% decrease in case of intravenous injection and all differences were statistically significant. The significant decrease in the tumor volume was mainly attributed to the localized hyperthermia effect [103] (Fig. 13.11).

13.5 Conclusion

Over the last roughly 20 years research focus with respect to gold nanoparticles has exploded. We now have a much better understanding of their optical properties, control over their synthesis, and the ability to prepare particles having a wide variety of morphologies, coatings and application-specific properties. Furthermore, there has been an increase in our understanding of the surface of these materials and the mechanism of particle growth. More recently there have been several reports on more efficient conversion of gold salt into gold nanorods resulting in much higher nanorods yields. We feel there is still fundamental science that should be pursued in an effort to better understand these processes. Additionally, there continues to be ongoing efforts to investigate the ability of these materials to act as therapeutic agents within the biological and medicinal communities. The future outlook for gold nanoparticles remains uncertain but with a continued increase in our knowledge and understanding, we will continue to gain insight into the potential of these fascinating nanomaterials.

References

1. Fedlheim, D.L., and C.A. Foss. 2001. *Metal nanoparticles: Synthesis, characterization, and applications*. CRC press.
2. Williamson, S.J., and H.Z. Cummins. 1983. *Color Research & Application* 10 (2): 2.
3. Dreaden, E.C., A.M. Alkilany, X. Huang, C.J. Murphy, and M.A. El-Sayed. 2012. *Chemical Society Reviews* 41 (7): 2740–2779.
4. Mie, G. 1908. *Annalen der Physik* 330 (3): 377–445.
5. Murphy, C.J., T.K. Sau, A.M. Gole, C.J. Orendorff, J. Gao, L. Gou, S.E. Hunyadi, and T. Li. 2005. *The Journal of Physical Chemistry B* 109 (29): 13857–13870.
6. Stone, J., S. Jackson, and D. Wright. 2011. *Wiley Interdisciplinary Reviews: Nanomedicine and Nanobiotechnology* 3 (1): 100–109.
7. Turkevich, J., P.C. Stevenson, and J. Hillier. 1951. *Discussions of the Faraday Society* 11: 55–75.
8. Brust, M., M. Walker, D. Bethell, D.J. Schiffrin, and R. Whyman. 1994. *Journal of the Chemical Society, Chemical Communications* (7): 801–802.
9. Martin, B.R., D.J. Dermody, B.D. Reiss, M. Fang, L.A. Lyon, M.J. Natan, and T.E. Mallouk. 1999. *Advanced Materials* 11 (12): 1021–1025.
10. Yu, Y.-Y., S.-S. Chang, C.-L. Lee, and C.C. Wang. 1997. *The Journal of Physical Chemistry B* 101 (34): 6661–6664.
11. Kim, F., J.H. Song, and P. Yang. 2002. *Journal of the American Chemical Society* 124 (48): 14316–14317.
12. Connor, E.E., J. Mwamuka, A. Gole, C.J. Murphy, and M.D. Wyatt. 2005. *Small* 1 (3): 325–327.
13. Yu, C., and J. Irudayaraj. 2007. *Analytical Chemistry* 79 (2): 572–579.
14. Norman, R.S., J.W. Stone, A. Gole, C.J. Murphy, and T.L. Sabo-Attwood. 2008. *Nano Letters* 8 (1): 302–306.
15. Sau, T.K., and C.J. Murphy. 2004. *Journal of the American Chemical Society* 126 (28): 8648–8649.
16. Gou, L., and C.J. Murphy. 2005. *Chemistry of Materials* 17 (14): 3668–3672.
17. Liu, M., and P. Guyot-Sionnest. 2005. *The Journal of Physical Chemistry B* 109 (47): 22192–22200.
18. Lohse, S.E., and C.J. Murphy. 2013. *Chemistry of Materials* 25 (8): 1250–1261.
19. Canonico-May, S.A., K.R. Beavers, M.J. Melvin, A.M. Alkilany, C.L. Duvall, and J.W. Stone. 2016. *Journal of Colloid and Interface Science* 463: 229–232.
20. Khlebtsov, B.N., V.A. Khanadeev, J. Ye, G.B. Sukhorukov, and N.G. Khlebtsov. 2014. *Langmuir* 30 (6): 1696–1703.

21. Kozek, K.A., K.M. Kozek, W.-C. Wu, S.R. Mishra, and J.B. Tracy. 2013. *Chemistry of Materials* 25 (22): 4537–4544.
22. Scarabelli, L., M. Grzelczak, and L.M. Liz-Marzán. 2013. *Chemistry of Materials* 25 (21): 4232–4238.
23. Rodríguez-Fernández, J., J. Perez-Juste, P. Mulvaney, and L.M. Liz-Marzan. 2005. *The Journal of Physical Chemistry B* 109 (30): 14257–14261.
24. Ye, X., C. Zheng, J. Chen, Y. Gao, and C.B. Murray. 2013. *Nano Letters* 13 (2): 765–771.
25. Rana, S., Y.-C. Yeh, and V.M. Rotello. 2010. *Current Opinion in Chemical Biology* 14 (6): 828–834.
26. Huang, X., P.K. Jain, I. El-Sayed, and M.A. El-Sayed. 2007. *Nanomedicine(Lond)* 2 (5): 681–693.
27. Pissuwan, D., S.M. Valenzuela, and M.B. Cortie. 2006. *Trends in Biotechnology* 24 (2): 62–67.
28. Pérez-Juste, J., I. Pastoriza-Santos, L.M. Liz-Marzán, and P. Mulvaney. 2005. *Coordination Chemistry Reviews* 249 (17): 1870–1901.
29. Skrabalak, S.E., J. Chen, Y. Sun, X. Lu, L. Au, C.M. Cobley, and Y. Xia. 2008. *Accounts of Chemical Research* 41 (12): 1587–1595.
30. Loo, C., A. Lin, L. Hirsch, M.-H. Lee, J. Barton, N. Halas, J. West, and R. Drezek. 2004. *Technology in Cancer Research & Treatment* 3 (1): 33–40.
31. Hu, M., J. Chen, Z.-Y. Li, L. Au, G.V. Hartland, X. Li, M. Marquez, and Y. Xia. 2006. *Chemical Society Reviews* 35 (11): 1084–1094.
32. Krumpfer, J.W., T. Schuster, M. Klapper, and K. Müllen. 2013. *Nano Today* 8 (4): 417–438.
33. Sperling, R.A., Parak, W. 2010. *Philosophical Transactions of the Royal Society, A.* 368 (1915): 1333–1383.
34. Gao, J., X. Huang, H. Liu, F. Zan, and J. Ren. 2012. *Langmuir* 28 (9): 4464–4471.
35. Häkkinen, H. 2012. *Nature Chemistry* 4 (6): 443–455.
36. Jin, R. 2010. *Nanoscale* 2 (3): 343–362.
37. Manson, J., D. Kumar, B.J. Meenan, and D. Dixon. 2011. *Gold Bulletin* 44 (2): 99–105.
38. Lipka, J., M. Semmler-Behnke, R.A. Sperling, A. Wenk, S. Takenaka, C. Schleh, T. Kissel, W.J. Parak, and W.G. Kreyling. 2010. *Biomaterials* 31 (25): 6574–6581.
39. Giljohann, D.A., D.S. Seferos, W.L. Daniel, M.D. Massich, P.C. Patel, and C.A. Mirkin. 2010. *Angewandte Chemie International Edition* 49 (19): 3280–3294.
40. DeLong, R.K., C.M. Reynolds, Y. Malcolm, A. Schaeffer, T. Severs, and A. Wanekaya. 2010. *Nanotechnology, Science and Applications* 3 (1): 53–63.
41. De Villiers, M.M., D.P. Otto, S.J. Strydom, and Y.M. Lvov. 2011. *Advanced Drug Delivery Reviews* 63 (9): 701–715.
42. Kim, K., S.-W. Huang, S. Ashkenazi, M. O'Donnell, A. Agarwal, N.A. Kotov, M.F. Denny, and M.J. Kaplan. 2007. *Applied Physics Letters* 90 (22): 223901.
43. Kirui, D.K., S. Krishnan, A.D. Strickland, and C.A. Batt. 2011. *Macromolecular Bioscience* 11 (6): 779–788.
44. Alkilany, A.M., L.B. Thompson, and C.J. Murphy. 2010. *ACS Applied Materials & Interfaces* 2 (12): 3417–3421.
45. Oishi, M., J. Nakaogami, T. Ishii, and Y. Nagasaki. 2006. *Chemistry Letters* 35 (9): 1046–1047.
46. Ghosh, P., G. Han, M. De, C.K. Kim, and V.M. Rotello. 2008. *Advanced Drug Delivery Reviews* 60 (11): 1307–1315.
47. Rosi, N.L., D.A. Giljohann, C.S. Thaxton, A.K. Lytton-Jean, M.S. Han, and C.A. Mirkin. 2006. *Science* 312 (5776): 1027–1030.
48. Nam, J.-M., C.S. Thaxton, and C.A. Mirkin. 2003. *Science* 301 (5641): 1884–1886.
49. Medley, C.D., J.E. Smith, Z. Tang, Y. Wu, S. Bamrungsap, and W. Tan. 2008. *Analytical Chemistry* 80 (4): 1067–1072.
50. Lee, J.S., M.S. Han, and C.A. Mirkin. 2007. *Angewandte Chemie International Edition* 46 (22): 4093–4096.
51. Lee, J.-S., P.A. Ulmann, M.S. Han, and C.A. Mirkin. 2008. *Nano Letters* 8 (2): 529–533.
52. Thaxton, C.S., D.G. Georganopoulou, and C.A. Mirkin. 2006. *Clinica Chimica Acta* 363 (1): 120–126.

53. Park, S.Y., A.K. Lytton-Jean, B. Lee, S. Weigand, G.C. Schatz, and C.A. Mirkin. 2008. *Nature* 451 (7178): 553–556.
54. Dreaden, E.C., M.A. Mackey, X. Huang, B. Kang, and M.A. El-Sayed. 2011. *Chemical Society Reviews* 40 (7): 3391–3404.
55. Bazak, R., M. Houri, S. El Achy, S. Kamel, and T. Refaat. 2015. *Journal of Cancer Research and Clinical Oncology* 141 (5): 769–784.
56. Llevot, A., and D. Astruc. 2012. *Chemical Society Reviews* 41 (1): 242–257.
57. Cho, K., X. Wang, S. Nie, and D.M. Shin. 2008. *Clinical Cancer Research* 14 (5): 1310–1316.
58. Byrne, J.D., T. Betancourt, and L. Brannon-Peppas. 2008. *Advanced Drug Delivery Reviews* 60 (15): 1615–1626.
59. Zhang, F., E. Lees, F. Amin, P. Rivera_Gil, F. Yang, P. Mulvaney, and W.J. Parak. 2011. *Small* 7 (22): 3113–3127.
60. Sperling, R.A., P.R. Gil, F. Zhang, M. Zanella, and W.J. Parak. 2008. *Chemical Society Reviews* 37 (9): 1896–1908.
61. Sokolov, K., M. Follen, J. Aaron, I. Pavlova, A. Malpica, R. Lotan, and R. Richards-Kortum. 2003. *Cancer Research* 63 (9): 1999–2004.
62. Rayavarapu, R.G., W. Petersen, C. Ungureanu, J.N. Post, T.G. van Leeuwen, and S. J. Manohar. 2007. *Biomedical Imaging* 2007 (1): 5–5.
63. Amendola, V., and M. Meneghetti. 2007. *Journal of Materials Chemistry* 17 (44): 4705–4710.
64. Kumar, A., H. Ma, X. Zhang, K. Huang, S. Jin, J. Liu, T. Wei, W. Cao, G. Zou, and X.-J. Liang. 2012. *Biomaterials* 33 (4): 1180–1189.
65. Nguyen, C.T., J.T. Nguyen, S. Rutledge, J. Zhang, C. Wang, and G.C. Walker. 2010. *Cancer Letters* 292 (1): 91–97.
66. Bickford, L.R., J. Chang, K. Fu, J. Sun, Y. Hu, A. Gobin, T.-K. Yu, and R.A. Drezek. 2008. *NanoBiotechnology* 4 (1–4): 1–8.
67. Zhang, Y., and J. Zhang. 2005. *Journal of Colloid and Interface Science* 283 (2): 352–357.
68. Popovtzer, R., A. Agrawal, N.A. Kotov, A. Popovtzer, J. Balter, T.E. Carey, and R. Kopelman. 2008. *Nano Letters* 8 (12): 4593–4596.
69. Tong, L., Q. Wei, A. Wei, and J.X. Cheng. 2009. *Photochemistry and Photobiology* 85 (1): 21–32.
70. Reuveni, T., M. Motiei, Z. Romman, A. Popovtzer, and R. Popovtzer. 2011. *International Journal of Nanomedicine* 6 (2859): e64.
71. Qian, X., X.-H. Peng, D.O. Ansari, Q. Yin-Goen, G.Z. Chen, D.M. Shin, L. Yang, A.N. Young, M.D. Wang, and S. Nie. 2008. *Nature Biotechnology* 26 (1): 83–90.
72. Huang, X., I.H. El-Sayed, W. Qian, and M.A. El-Sayed. 2006. *Journal of the American Chemical Society* 128 (6): 2115–2120.
73. El-Sayed, I.H., X. Huang, and M.A. El-Sayed. 2006. *Cancer Letters* 239 (1): 129–135.
74. Chen, J., D. Wang, J. Xi, L. Au, A. Siekkinen, A. Warsen, Z.-Y. Li, H. Zhang, Y. Xia, and X. Li. 2007. *Nano Letters* 7 (5): 1318–1322.
75. Cobley, C.M., L. Au, J. Chen, and Y. Xia. 2010. *Expert Opinion on Drug Delivery* 7 (5): 577–587.
76. Melancon, M.P., W. Lu, Z. Yang, R. Zhang, Z. Cheng, A.M. Elliot, J. Stafford, T. Olson, J. Z. Zhang, and C. Li. 2008. *Molecular Cancer Therapeutics* 7 (6): 1730–1739.
77. Loo, C., A. Lowery, N. Halas, J. West, and R. Drezek. 2005. *Nano Letters* 5 (4): 709–711.
78. Lévy, R. 2006. *ChemBioChem* 7 (8): 1141–1145.
79. Lévy, R., N.T. Thanh, R.C. Doty, I. Hussain, R.J. Nichols, D.J. Schiffrin, M. Brust, and D.G. Fernig. 2004. *Journal of the American Chemical Society* 126 (32): 10076–10084.
80. Slocik, J.M., M.O. Stone, and R.R. Naik. 2005. *Small* 1 (11): 1048–1052.
81. Tkachenko, A.G., H. Xie, D. Coleman, W. Glomm, J. Ryan, M.F. Anderson, S. Franzen, and D.L. Feldheim. 2003. *Journal of the American Chemical Society* 125 (16): 4700–4701.
82. Slocik, J.M., J.S. Zabinski, D.M. Phillips, and R.R. Naik. 2008. *Small* 4 (5): 548–551.
83. Wust, P., B. Hildebrandt, G. Sreenivasa, B. Rau, J. Gellermann, H. Riess, R. Felix, and P. Schlag. 2002. *The Lancet Oncology* 3 (8): 487–497.

84. Chua, T.C., G. Robertson, W. Liauw, R. Farrell, T.D. Yan, and D.L.J. Morris. 2009. *Journal of Cancer Research and Clinical Oncology* 135 (12): 1637–1645.
85. Chicheł, A., J. Skowronek, M. Kubaszewska, and M. Kanikowski. 2007. *Reports of Practical Oncology and Radiotherapy* 12 (5): 267–275.
86. Stauffer, P.R. 2005. *International Journal of Hyperthermia* 21 (8): 731–744.
87. I.P Soares, P., I. MM Ferreira, R. AGBN Igreja, C. MM Novo, and J. PMR Borges. 2012. *Recent Patents on Anti-Cancer Drug Discovery* 7 (1): 64–73.
88. Camerin, M., S. Rello, A. Villanueva, X. Ping, M.E. Kenney, M.A. Rodgers, and G. Jori. 2005. *European Journal of Cancer* 41 (8): 1203–1212.
89. Huang, X., P.K. Jain, I.H. El-Sayed, and M.A. El-Sayed. 2008. *Lasers in Medical Science* 23 (3): 217–228.
90. Wang, J. 2005. *Electroanalysis* 17 (1): 7–14.
91. Michalet, X., F. Pinaud, L. Bentolila, J. Tsay, S. Doose, J. Li, G. Sundaresan, A. Wu, S. Gambhir, and S. Weiss. 2005. *Science* 307 (5709): 538–544.
92. Eustis, S., and M.A. El-Sayed. 2006. *Chemical Society Reviews* 35 (3): 209–217.
93. Huang, X., and M.A. El-Sayed. 2010. *Journal of Advanced Research* 1 (1): 13–28.
94. Jain, P.K., X. Huang, I.H. El-Sayed, and M.A. El-Sayed. 2008. *Accounts of Chemical Research* 41 (12): 1578–1586.
95. Njoki, P.N., I.-I.S. Lim, D. Mott, H.-Y. Park, B. Khan, S. Mishra, R. Sujakumar, J. Luo, and C.-J. Zhong. 2007. *The Journal of Physical Chemistry C* 111 (40): 14664–14669.
96. Huang, X., S. Neretina, and M.A. El-Sayed. 2009. *Advanced Materials* 21 (48): 4880–4910.
97. Au, L., J. Chen, L.V. Wang, and Y. Xia. 2010. *Methods in Molecular Biology*, 83–99.
98. Dykman, L., and N. Khlebtsov. 2012. *Chemical Society Reviews* 41 (6): 2256–2282.
99. Kennedy, L.C., L.R. Bickford, N.A. Lewinski, A.J. Coughlin, Y. Hu, E.S. Day, J.L. West, and R.A. Drezek. 2011. *Small* 7 (2): 169–183.
100. Perrault, S.D., C. Walkey, T. Jennings, H.C. Fischer, and W.C. Chan. 2009. *Nano Letters* 9 (5): 1909–1915.
101. Pissuwan, D., T. Niidome, and M.B. Cortie. 2011. *Journal of Controlled Release* 149 (1): 65–71.
102. van der Meel, R., L.J. Vehmeijer, R.J. Kok, G. Storm, and E.V. van Gaal. 2013. *Advanced Drug Delivery Reviews* 65 (10): 1284–1298.
103. Dickerson, E.B., E.C. Dreaden, X. Huang, I.H. El-Sayed, H. Chu, S. Pushpanketh, J.F. McDonald, and M.A. El-Sayed. 2008. *Cancer Letters* 269 (1): 57–66.
104. Wang, S., W. Xi, F. Cai, X. Zhao, Z. Xu, J. Qian, and S. He. 2015. *Theranostics* 5 (3): 251.
105. Wu, H.-L., C.-H. Kuo, and M.H. Huang. 2010. *Langmuir* 26 (14): 12307–12313.
106. Barbosa, S., A. Agrawal, L. Rodríguez-Lorenzo, I. Pastoriza-Santos, R.A. Alvarez-Puebla, A. Kornowski, H. Weller, and L.M. Liz-Marzán. 2010. *Langmuir* 26 (18): 14943–14950.
107. Schwartzberg, A.M., T.Y. Olson, C.E. Talley, and J.Z. Zhang. 2006. *The Journal of Physical Chemistry B* 110 (40): 19935–19944.

Chapter 14
Application of Gold Nanorods in Cardiovascular Science

Jack G. Goldsmith, Heather L'Ecuyer, Delphine Dean and Edie C. Goldsmith

Abstract Cardiovascular disease is a significant global Cardiovascular disease is a significant global health problem. Effectively treating it and exercise, but also demands the development of novel tools for rapid diagnosis and new therapeutics for treatment. The field of nanomaterials is making significant contributions to multiple health care problems in the areas of disease detection, imaging and drug delivery. Gold nanoparticles are particularly promising due to their ease of synthesis, biocompatibility and unique optical properties. In particular, gold nanorods having received much attention for their potential in the diagnosis and treatment of cancer, are now being examined for other biomedical applications. This chapter highlights efforts using gold nanorods in cardiovascular research in such areas as detection of cardiovascular disease, understanding cardiac cell response to nanomaterials and the ability of gold nanorods to alter the mechanical properties of model tissue constructs and cardiac valves.

Keywords Gold · Nanorods · Nanomaterials · Cardiac · Heart · Fibroblast · Valve

14.1 Introduction

According to the World Health Organization, cardiovascular disease is a leading cause of death around the world [1]. More individuals die globally from cardiovascular disease than from any other noncommunicable disease (including cancers, diabetes and respiratory diseases). Cardiovascular disease is not just one disorder, but rather a family of diseases targeting either the heart itself or the blood vessels

J.G. Goldsmith · E.C. Goldsmith (✉)
Department of Cell Biology and Anatomy, University of South Carolina School of Medicine, 6439 Garners Ferry Rd, Columbia, SC 29209, USA
e-mail: edie.goldsmith@uscmed.sc.edu

H. L'Ecuyer · D. Dean
Department of Bioengineering, Clemson University, Clemson, SC 29643, USA

S.E. Hunyadi Murph et al. (eds.), *Anisotropic and Shape-Selective Nanomaterials*, Nanostructure Science and Technology, DOI 10.1007/978-3-319-59662-4_14

responsible for transporting blood throughout the body. Familiar examples of cardiovascular disease include atherosclerosis, myocardial infarction, heart failure, hypertrophic cardiomyopathy and valvular heart disease. Despite advances in diagnosing cardiovascular disease, understanding cellular/molecular changes that occur during cardiovascular disease and increased therapeutic options, new tools are needed to provide for earlier detection of individuals at risk for cardiovascular disease, to monitor disease progression, predict which patients will or will not respond to current therapies and the development of novel interventions for those living with cardiovascular disease. Nanomaterials have already made contributions in these areas [2–4] and hold great promise for impacting how cardiovascular disease is detected and managed in the future.

Nanoparticles, defined as having at least one dimension smaller than 100 nm, are finding wide use as potential biomedical agents. There are a wide variety of materials currently used to produce nanoparticles—metals, silica, biodegradable and/or biocompatible polymers, biomolecules (i.e. lipids, dendrimers, protein)—each with their own inherent advantages and disadvantages. Based on the physiochemical properties of the material itself, nanoparticles of different shapes and sizes have been developed for applications such as drug delivery [5, 6], imaging agents [7, 8], sensors [9, 10], photothermal therapy [11, 12], and diagnostic [13] tools. Gold nanoparticles are an important category of nanomaterials and are currently being investigated in many of these areas [14, 15]. While a major therapeutic focus of gold nanoparticles has been in the area of cancer [14], investigators are expanding studies to include other diseases, including those that affect the cardiovascular system. Although gold nanoparticles with varied shapes have been examined for applicability in the cardiovascular system [16–22], this chapter will focus on studies specifically examining the applicability of gold nanorods.

14.2 Application of Gold Nanorods as Agents to Detect Cardiovascular Disease

Gold nanorods (AuNRs) have been synthesized for use in improving bioimaging approaches, biosensor technology, and targeted therapies. For instance, gold nanorods make ideal probes for optical imaging techniques due to their tunable optical properties [23, 24]. In addition, because of their unique optical absorption properties, gold nanorods are currently being tested for use in hyperthermia treatments of a wide range of cancers [24, 25]. As discussed below, the application of gold nanorods in detecting damage to heart muscle has also been developed recently.

Myocardial infarction (MI), commonly referred to as a heart attack, is due to the occlusion of coronary vessels responsible for supplying the heart with oxygenated blood. Deprived of oxygenated blood, cardiac muscle cells (myocytes) die and are ultimately replaced by scar tissue. Rapid diagnosis of an MI leads to early intervention and is critical in minimizing the loss of cardiac muscle. A classical marker used to diagnose acute MI is cardiac troponin I (cTnI) [26, 27], a protein involved in regulating muscle contraction which is released from myocytes as they die. Current methods in clinical labs used to detect cTnI in patient blood include enzyme-linked immunosorbent assay (ELISA), HPLC and polymerase chain reaction (PCR); all of which take hours to perform. Several studies have shown that AuNRs can be used to detect cTnI levels, potentially offering a more rapid screen for MI [28–33].

Relying on the localized surface plasmon resonance properties of gold nanorods, which exhibit two resonance bands, specifically transverse and longitudinal plasmon bands [28, 29], several groups have worked to improve cTnI detection. In 2009, Guo et al. reported the use of surface modified AuNRs for in situ detection of human cTnI [30]. In this study, poly(styrenesulfonate) (PSS)-terminated AuNRs with an aspect ratio of 5.5 were further modified with an anti-human cTnI antibody. Incubating antibody-modified AuNRs with varying concentrations of human cTnI resulted in a dose-dependent shift in absorbance of the longitudinal band as well as a broadening of this band at high cTnI concentrations (200 ng/mL; see Fig. 6 in Ref. [30]). The incubation time of 15 min before detection by UV-Vis spectroscopy represents a significant time improvement compared to the standard ELISA method and the authors reported a detection limit of 10 ng/mL (see Fig. 6 in Ref. [30]).

While cTnI may be a "gold standard" in the diagnosis of acute MI, several other proteins including myoglobin, cardiac troponin T, creatine kinase and lactate dehydrogenase are also released from necrotic myocytes [26]. To improve diagnostic accuracy, Tang and Casas developed a multiplexed AuNR detection system which can simultaneously detect both cTnI and myoglobin in solution [31]. Carboxy-coated AuNRs with two distinct longitudinal plasmon bands (830 and 640 nm) were coated with antibodies to either cTnI or myoglobin, respectively. Simultaneous exposure of both particles to myoglobin resulted in a red-shift of the 640 nm band (AuNRs derivatized with the anti-myoglobin antibody) with no change in the 830 nm band (cTnI detecting AuNRs). Similarly, when both particles were exposed to cTnI, only the particles with the 830 nm band displayed a red-shift in their spectrum (see Fig. 4 in Ref. [31]). When myoglobin and cTnI were present in the same solution, discrete spectral shifts were observed for each antibody-coated AuNR depending upon how much of the target protein, cTnI or myoglobin, was present in the sample (see Fig. 4 in Ref. [31]). The detection of target protein (myoglobin or cTnI) was similar in both a mixture of the two proteins or when the AuNRs were incubated with each purified target protein individually, demonstrating the specificity of the approach.

An alternative to the AuNR-antibody based detection methods described above, which have limitations including antibody cost and stability, is the immobilization of biorecognition elements (BRE) on AuNRs. Using a human cTnI binding peptide immobilized on AuNRs, Tadepalli et al. compared the ability of this BRE to detect cTnI versus an anti-cTnI antibody derivatized AuNR [32]. BRE and antibody-conjugated AuNRs were adsorbed onto filter paper, creating a paper-based detection device, and homogenous distribution of the derivatized AuNRs on the paper confirmed before exposure to troponin. While both modified AuNRs specifically detected troponin and demonstrated dose-dependent shifts in longitudinal surface plasmon resonance, peptide-derivatized AuNRs had an order of magnitude improvement in detection limit compared to antibody-based particles (see Fig. 3 in Ref. [32]), even though both particles demonstrated similar affinities for cTnI.

All three studies described thus far have demonstrated the feasibility of using modified AuNRs to detect cTnI in simple solutions (purified cTnI or cTnI + myoglobin). Tadepalli et al. extended these observations, making them clinically relevant by testing detection capabilities of their paper-immobilized AuNRs using solutions of cTnI containing 10% human plasma [32]. These paper-based peptide-AuNR devices again demonstrated dose-dependent resonance shifts and were able to detect cTnI in the complex solution at a level of 353 pg/mL, a concentration well within the physiologically relevant concentrations of cTnI in human plasma (0.01–10 ng/mL) [33].

The work described above clearly demonstrates the potential for AuNRs in the detection of acute MI. Peptide and antibody-modified AuNRs demonstrated target protein specificity with short exposure times (<2 h) and limits of detection within clinically relevant target protein concentrations. Additional efforts to improve sensitivity as well as the need to insure selectivity when confronted with complex biological samples, such as human plasma or whole blood samples, are essential to advance this work into a clinical environment. Identification or examination of biomarkers associated with other cardiovascular diseases, such as atherosclerosis or heart failure, would further enhance the biomedical application of AuNRs.

Atherosclerosis, commonly referred to as hardening of the arteries, develops when cholesterol, triglycerides and macrophages/foam cells accumulate within the walls of blood vessels forming plaques, causing the diameter of the vessels to decrease. Subsequent rupture of plaques that form during atherosclerosis leads to myocardial infarction. A variety of detection methods are currently employed to examine atherosclerotic plaques but most are only used after symptoms are present in an individual. Ankri et al. devised a method for AuNR detection of active atherosclerotic plaques based on the uptake of AuNRs by macrophages, a cell population enriched in vulnerable plaques at risk for rupture [34]. Using a rat carotid artery balloon injury model, AuNRs were injected after the injury was created in one carotid artery and both injured and uninjured arteries were scanned using a diffusion reflection system. Uptake of gold nanorods in the injured artery resulted in a strong decrease in light diffusion (see Fig. 3 in Ref. [34]). High-resolution computed tomography (CT) imaging post-AuNR injection confirmed increased localization of AuNRs in the injury site was likely due to uptake by macrophages [34]. The non-injured artery demonstrated far less nanorod localization and nanorods that were detected demonstrated a homogenous distribution along the artery wall [34]. As an extension of this method, macrophage uptake of AuNRs has also been detected by either flow cytometry or by diffusion reflection using tissue-like constructs to mimic imaging conditions in vivo [35]. The in vivo detection of atherosclerotic plaques by gold nanorod uptake and non-invasive diffusion reflection imaging may provide a novel technique to monitor progression of atherosclerosis and identifying plaques in danger of rupture.

14.3 Gold Nanorods as Reporters of Material Deformation and Mechanical Environment

While many of the applications employing AuNRs as sensors have focused on their use in the detection of biomolecules [36–39], it has also been demonstrated that these particles can be used to monitor changes in mechanical environment. Taking advantage of the light scattering properties of AuNRs, Orendorff et al. described the use of nanorods to measure deformation of an elastic material [40]. Cetyltrimethyl ammonium bromide (CTAB)-capped AuNRs were suspended in solution with polyvinyl alcohol (PVA) and heat cured to generate transparent, elastic films with AuNRs distributed throughout the film. In an alternative approach, AuNRs were embedded in the surface of polydimethylsiloxane (PDMS) films by physical transfer. Nanorod dispersion in PVA films was more homogeneous than for PDMS, potentially due to the method of film preparation or chemical properties of the film; however, both samples were suitable for imaging studies. Films were stretched to 20–25% of their original length and imaged, before and after deformation, using a light microscope in dark-field mode. Based on the change in the pattern of light scattered from the AuNRs after deformation, digital image correlation software was able to calculate the displacement of nanorods within the films and allowed the authors to map the strain field produced by the deformation. This "proof of concept" study demonstrated the potential for AuNRs as reporters that can respond to changes in mechanical environment.

Mechanical signaling is an important cue mediating cell behavior and cardiac fibroblasts have been shown to respond to changes in their mechanical environment in a number of ways [41–44]. While the afore cited studies focus on the response of cardiac fibroblasts to externally applied mechanical loads, very little is known about the local changes in the mechanical environment produced by cardiac fibroblasts as they interact with, and modify, the extracellular matrix (ECM). Extending the imaging method developed by Orendorff et al. [40] into a biological system, it has been possible to quantify the strain that cardiac fibroblasts exert on a collagen matrix [45–47]. Two-dimensional collagen films were prepared in a two step process: an initial layer of type I collagen was adsorbed onto a slide followed by a second layer to which CTAB-coated AuNRs were added. CTAB-terminated AuNRs could be used in this application due to the short time during which cells would be in contact with the nanorods. Light scattered by the nanorods disbursed in the collagen film provided a pattern which could be imaged by dark-field microscopy over time as the fibroblasts deformed the collagen matrix. Analysis of matrix deformation was carried out using digital image correlation software by tracking changes in position of the scattered light from the nanorods and local cell-induced strain field maps constructed. Using this method, it was observed that cardiac fibroblasts exerting traction on the collagen matrix could produce matrix deformations on the order of 18 m. The authors noted the displacements measure with the AuNRs were in line with previous reports and an advantage of using gold nanorods compared to methods employing fluorescent markers is the absence of photobleaching.

This imaging technique has also been used to examine age-related differences in the ability of cardiac fibroblasts to mechanically manipulate their local ECM environment. In this study [47], primary neonatal and adult cardiac fibroblasts were plated on 2-dimensional (2D) films of type I collagen containing CTAB-terminated AuNRs. Collagen-AuNR constructs were created by initial adsorption of a neutral type I collagen solution onto a slide followed by the addition of a second layer of collagen to which AuNRs were added. After polymerization of the AuNR-collagen layer, fibroblasts were allowed to attach to the collagen prior to imaging. To enable visualization of the fibroblasts in addition to imaging the AuNRs, cells were briefly labeled with a membrane-permeable, vital fluorescent dye. Dark-field and epifluorescence microscopy images were collected over a one hour time period and cell-induced strain fields were calculated using digital image correlation analysis. Neonatal fibroblasts exhibited maximal peak principal strains of 6.3% (tensile) and −5.5% (compressive) while adult fibroblasts achieved maximal values of 3.7% (tensile) and −3.3% (compressive). Although the data the authors reported for maximal principal strains, mean principal strains and the distribution of strain values were not significantly different between the two cell types, this study along with Stone et al. [45] demonstrated the utility of AuNRs as a tool to examine the local mechanical environment surrounding cells.

14.4 Using Gold Nanorods to Direct Cell Behavior

As gold nanomaterials become tools for biomedical research and potential therapeutics, it is important to understand how these materials can impact normal cellular behavior. The studies described above using AuNRs as imaging agents used relatively small quantities of nanorods and short exposure times to cells. What happens to cellular function, specifically cardiac fibroblasts, when the concentration of nanorods and the exposure time is increased? An in vitro study using a biologically relevant 3D model of cardiac fibroblast-mediated collagen remodeling addressed this question.

Three-dimensional collagen gels have long been used to examine tissue remodeling [48] and several studies have employed this model system to examine factors that impact ECM remodeling by cardiac fibroblasts [49–52]. To ascertain if AuNRs could impact the ability of cardiac fibroblast to remodel 3D collagen gels, Sisco et al. prepared AuNRs capped with poly(styrene sulfonate) (PSS) and seeded them into 3D collagen gels along with cardiac fibroblasts [53]. Nanorods were coated with PSS to mask the CTAB surfactant used during AuNR synthesis, reducing any potential cytotoxicity. PSS-coated AuNRs inhibited fibroblast-mediated contraction of collagen gels in a dose-dependent manner and qualitative observations suggested PSS-AuNRs also altered gel stiffness. Alterations in cellular contractile function were not associated with changes in cell viability or activation of matrix metalloproteinases, enzymes involved in collagen remodeling [54], but were associated with significant decreases in type I collagen and α-smooth muscle actin expression. This suggests that the presence of the nanorods in the collagen gels impaired the ability of cardiac fibroblasts to transform into myofibroblasts and thus impedes the normally expected remodeling process. Therapeutic approaches to modulate myofibroblast activity during cardiovascular disease are severely limited; however, polyelectrolyte coated AuNRs may provide a novel tool for regulating these cells.

While the observation that nanorods can alter cardiac fibroblast-mediated remodeling was exciting, to be biomedically useful it is also important to understand the mechanism(s) through which these particles exert their biological effects. It has been well documented that proteins can adsorb onto the surface of nanomaterials creating a "protein corona" [55–57], but how these immobilized proteins can impact cell function is less clear [58]. To determine if protein adsorption onto polyelectrolyte-coated AuNRs could contribute to the observed effect these particles had on cardiac fibroblast function, the protein corona of AuNRs exposed to fibroblast conditioned culture media was characterized [59]. This study compared AuNRs coated with negatively (PSS) or positively (PDADMAC) charged polyelectrolytes or polyethylene glycol (PEG) as a control which should exhibit limited protein binding. PEGylated-AuNRs had minimal effects on cardiac fibroblast mediated gel contraction whereas both PSS and PDADMAC terminated nanorod containing gels had large decreases in contraction, suggesting that particle charge plays a role in blocking contraction. Identification of proteins bound to the surface of coated nanorods by mass spectrometry revealed that the charged particles adsorbed a more diverse range of proteins compared to the PEGylated nanorods and while some proteins bound to all three types of nanorods used in the study others demonstrated a charge preference. One protein identified in this study, biglycan, has been shown to modulate cell behavior and its sequestration on the nanorods was postulated as a mechanism for reduced fibroblast remodeling activity. Depletion or inactivation of biomolecules through interaction with nanoparticles could have a significant impact on the biomedical applications of these materials. Characterization of the protein corona may need to be factored into evaluation of these materials for therapeutic usage, particularly for non-biodegradable nanoparticles such as AuNRs.

An alternative to the protein corona hypothesis is that AuNRs could alter cellular engagement of the collagen matrix within the gel thereby preventing collagen gel contraction. Wilson et al. explored this idea by examining the potential interaction between type I collagen and polyelectrolyte capped AuNRs [60]. In previous studies using the collagen gel contraction model [53, 59–61], 3D collagen gels were prepared by combining neutral type I collagen solution, cardiac fibroblasts and polyelectrolyte-coated AuNRs with gel formation achieved by incubating the solution at 37 °C. Using a turbidity assay to monitor collagen polymerization in a cell-free system, Wilson et al. discovered that AuNR surface charge altered polymerization kinetics, with negatively charged (PSS, poly(acrylic acid) PAA) AuNRs accelerating collagen fibril formation while positively charged (poly(allylamine) hydrochloride PAH, PDADMAC) nanorods delayed fibril assembly. Similar effects on collagen fibrillogenesis were observed using AuNRs coated with the sulfated glycosaminoglycans heparan sulfate and chondroitin sulfate [61]. Examination of collagen fibril organization within 3D gels indicated relatively small changes in fibril organization but that AuNRs tended to associate with the collagen, an observation that was supported by TEM, and were not located in the fluid-filled pores of the gels [53, 60]. In AuNR-collagen gel constructs seeded with cardiac fibroblasts, no noticeable changes in cell morphology were detected, suggesting the inclusion of nanorods in the gel constructs does not impair fibroblast-collagen interactions. Previous work suggested a qualitative change in collagen gel stiffness in the presence of PSS-coated AuNRs [53] which was examined quantitatively by dynamic torsional shear testing [60]. Rheological evaluation of cell-free AuNR containing collagen gels indicated that gels doped with negatively charged AuNRs (PSS or PAA) were stiffer than controls lacking AuNRs while gels containing positively charged AuNRs (PAH or PDADMAC) had mechanical properties similar to control gels. These results offer the exciting possibility that negatively charged AuNRs introduced into an actively organizing collagen matrix, such as scar formation post-MI, would influence the assembly of collagen within the scar and ultimately the mechanical properties of the resulting scar tissue.

A significant obstacle for tissue engineering, including development of engineering cardiac tissue, is the ability to direct cell patterns and vascularization within the engineered construct. Development of methods that would allow for pattern integration during construct formation or directed patterning in situ could advance this field. Taking advantage of the near-infrared (NIR) absorption properties of gold nanorods Hribar et al. were able to direct endothelial cell migration and alignment into hollow, vessel like tubes [62]. PEGylated AuNRs were combined with collagen and endothelial cells and following gelation exposed to 800 nm light using a NIR laser. Taking advantage of the AuNRs conversion of the energy absorbed from NIR excitation into heat (the photothermal effect), irradiation resulted in local heat denaturation of collagen within the gel and formation of channels whose features could be controlled by laser power which was optimized to maximize endothelial cell viability. Over a two week period of time, endothelial cells migrated towards the channels created by AuNRs and aligned themselves within the channels in such a way so as to produce hollow tubes lined with a single layer of endothelial cells that are morphologically similar to simple blood vessels. This method provides a novel tool for directing cell organization within a biomaterial and if combined with AuNRs of different NIR absorption maxima could allow for the stepwise organization of multiple cell types, such as myocytes and fibroblasts, within an engineered tissue construct.

Another important consideration in the development of replacement cardiac muscle tissue is that myocyte organization and alignment, along with electrical coupling, is essential to achieve synchronous contraction. Ganji et al. examined the importance of electrical stimulation in myocyte organization with a bioengineered construct using gold nanotubes/nanowires embedded in biodegradable polyurethane (PU-GNT/NW) [63]. H9C2 cardiomyocytes were seeded onto PU-GNT/NW constructs, subjected to external electrical stimulation and evaluated for cell growth properties. Electrical stimulation in the PU-GNT/NW constructs enhanced cell spreading and resulted in a higher degree of cellular confluence and alignment with morphological characteristics similar to those observed in cardiac myocytes in vivo. The incorporation of gold nanotubes/wires into a biodegradable scaffold provides a promising new platform for cardiac tissue engineering.

14.5 Using Gold Nanorods to Alter the Material Properties of Cardiac Valves

Atrioventricular valves separate the heart into four chambers and when functioning properly ensure that blood does not flow back into the atria from the ventricles during ventricular contraction. Multiple pathological conditions, including post-MI ventricular remodeling, rheumatic fever, or degenerative disorders can result in incomplete closure of the mitral valve causing regurgitation of blood back into the left atrium [64]. In congestive heart failure patients, activation of fibroblasts within the mitral valve can trigger increased collagen deposition yielding stiffer mitral valves [65, 66] while in myxomatous valves increased production of matrix metalloproteinases leads to excessive degradation of collagen and “floppy” valves [67]. In both cases, mitral valve regurgitation occurs. Given studies indicating that AuNRs have the capacity to alter cardiac fibroblast function as well as the mechanical properties of collagen matrices, the effects of AuNRs on the mechanical properties of mitral valves was examined by our group.

To determine if polyelectrolyte coated AuNRs can alter the mechanical properties of heart valves, mitral valves were isolated from freshly harvested porcine hearts obtained from a local abattoir. Biopsy punches (5 mm in diameter) were taken from the valves and divided into the following experimental groups—no injection, saline injection, PSS–or PDADMAC-terminated nanorod (aspect ratio ~ 15) injections of either low or high nanorod concentrations. Injections were made in the center of the punch and subjected to mechanical testing using atomic force microscopy (AFM) with a 5 m spherical indenter tip over a period of 96 h post-injection. Nanoindentation and stress-relaxation data was collected from 5 points on each valve, beginning near the injection site and moving radially outward to 2.5 mm away from the injection site (Fig. 14.1). As shown in Fig. 14.1,

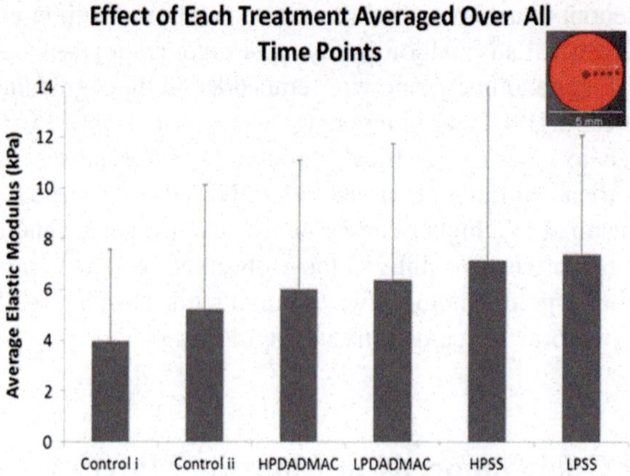

Fig. 14.1 Polyelectrolyte-coated gold nanorods alter valve stiffness. Punch biopsies from porcine mitral valves were untreated (*Control i*), injected with saline (*Control ii*), or injected with high (H) or low (L) concentrations of positively (PDADMAC) or negatively (PSS) charged AuNRs. The elastic moduli were calculated from nanoindentation curves using the Hertz Linear Elastic analytical model at each time point (24, 48, 72, 96 h post injection) for each data acquisition point and then averaged over the entire experimental time period. *Inset* shows AuNR injection site in *center* of valve biopsy and data acquisition points relative to injection site. n = 75 force curves for each experimental condition

injection of PDADMAC–and PSS-coated AuNRs tended to increase valve stiffness (increased elastic modulus) independent of nanorod concentration when examined over the entire 96 h experimental time period and including all five measurement locations on each valve biopsy. When compared against control samples at each time point, PDADMAC-coated nanorods resulted in a large increase in elastic modulus initially after injection which decreased over time (Fig. 14.2A) while PSS coated AuNRs resulted in more sustained changes in valve stiffness (Fig. 14.2B). Examining the impact distance from the injection had on stiffness revealed significant increases in stiffness at the position closest to the injection site (Fig. 14.3) for both PDADMAC and PSS terminated nanorods, indicating that the observed differences represent localized changed in mechanical properties. While these data showed significant changes in elastic modulus due to the nanorods, stress-relaxation tests suggested that neither PSS nor PDADMAC nanorods had an effect on the viscous properties of valves. Taken together these data suggest that AuNRs can be used to modulate the mechanical properties of native valve tissue but that these effects have both temporal and distance to injection site components. Additional work is needed to optimize particle characteristics such that sustained changes in mechanical environment can be achieved.

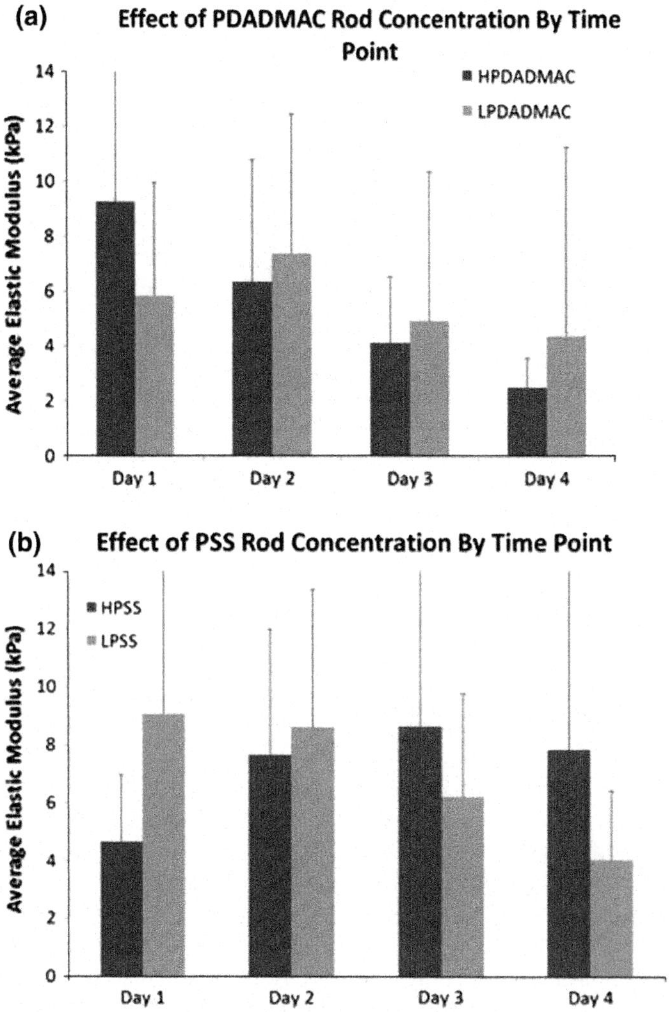

Fig. 14.2 Temporal changes in mitral valve stiffness with AuNR injection. Measurement of the elastic modulus of samples injected with PDADMAC-coated AuNRs demonstrated a statistically significant ($p < 0.05$) increase in stiffness 24 and 48 h post-injection which decreased gradually over time (**a**) whereas PSS-AuNR injections resulted in an elevated, but more sustained change in stiffness (**b**). $n = 75$ force curves for each experimental condition

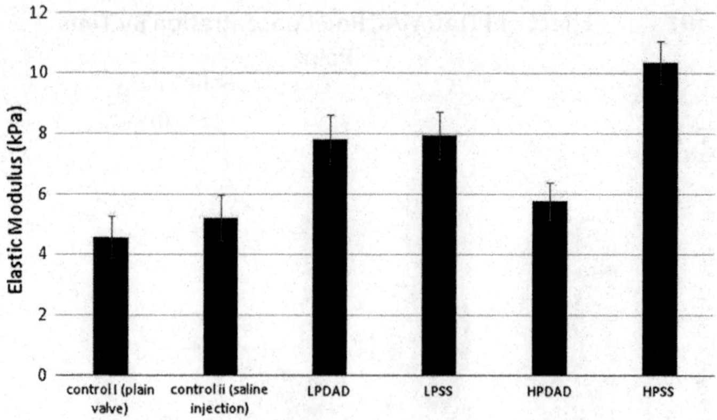

Fig. 14.3 Gold nanorods stimulate local changes in valve mechanical environment. Data from indentation points at or nearest to the injection site demonstrate statistically significant increases in elastic moduli for all AuNR injection conditions compared to both control samples. L—low AuNR concentration; H—high AuNR concentration; PDAD—PDADMAC. $p < 0.05$ for all samples compared to both controls

14.6 Conclusions and Future Directions

Gold nanorods have tremendous potential for use in diagnosing and treating cardiovascular disease. These nanomaterials have already been shown to be able to detect changes in clinically significant biomolecules involved in the diagnosis of MI. Future work needs to confirm that the spectroscopic detection methods used are compatible with blood/serum samples. Expansion to include biomarkers of other cardiovascular diseases, such as atherosclerosis, cardiac hypertrophy and heart failure, could have a significant impact on disease detection and monitoring. The incorporation of these AuNRs into a portable, rapid detection devices would be of great clinical value. The ability of polyelectrolyte-coated gold nanorods to alter not only fibroblast phenotype but the stiffness of collagen matrices has important clinical implications for the treatment of heart disease as well other conditions in which fibrosis, the accumulation of excess collagen, impairs normal tissue function. Future studies need to address the mechanism of nanorod delivery to ensure targeted delivery to the heart and confirm that the ability of AuNRs to modulate cell phenotype and material properties in vitro are maintained in vivo. Someday in the future it may really be possible to say that an individual has a heart of gold, thanks to advancing therapeutic applications of gold nanorods.

Acknowledgements The authors would like to thank the National Institutes of Health (HL097214) and the Mid-Atlantic Affiliate of the American Heart Association (13GRNT17070086) for funding.

References

1. World Health Organization. 2014. *Global status report on noncommunicable diseases*, 9–20. Geneva: WHO.
2. Jain, A., P. Kesharwani, N.K. Garg, A. Jain, P. Nirbhavane, N. Dwivedi, S. Banerjee, A.K. Iyer, and M.C.I.M. Amin. 2015. Nano-constructed carriers loaded with antioxidant: Boon for cardiovascular system. *Current Pharmaceutical Design* 21: 4456–4464.
3. Sharma, P.A., R. Maheshwari, M. Tekade, and R.K. Tekade. 2015. Nanomaterial based approaches for the diagnosis and therapy of cardiovascular diseases. *Current Pharmaceutical Design* 21: 4465–4478.
4. Behera, S.S., K. Pramanik, and M.K. Nayak. 2015. Recent advancement in the treatment of cardiovascular diseases: Conventional therapy to nanotechnology. *Current Pharmaceutical Design* 21: 4479–4497.
5. Sundar, D.S., M.G. Antoniraj., C.S. Kumar., S.S. Mohapatra., N.N. Houreld., K. Ruckmani. 2016. Recent trends of biocompatible and biodegradable nanoparticles in drug delivery: A review. *Current Medicinal Chemistry* 23 (32): 3730-3751 (doi:10.2174/0929867323666160607103854).
6. Nitta, S.K., and K. Numata. 2013. Biopolymer-based nanoparticles for drug/gene delivery and tissue engineering. *International Journal of Molecular Sciences* 14: 1629–1654.
7. Padmanabhan, P., A. Kumar, S. Kumar, R. Chaudhary, and B. Gulyas. 2016. Nanoparticles in practice for molecular-imaging applications: An overview. *Acta Biomaterialia* 41: 1–16.
8. Zhu, D., F. Liu, L. Ma, D. Liu, and Z. Wang. 2013. Nanoparticle-based systems for T_1-weighted magnetic resonance imaging contrast agents. *International Journal of Molecular Sciences* 14: 10591–10607.
9. Uusitalo, L.M., and N. Hempel. 2012. Recent advances in intracellular and in vivo ROS sensing: Focus on nanoparticle and nanotube applications. *International Journal of Molecular Sciences* 13: 10660–10679.
10. Wang, J. 2005. Nanomaterial-based amplified transduction of biomolecular interactions. *Small* 1: 1036–1043.
11. Wang, J., and J. Qiu. 2016. A review of organic nanomaterials in photothermal cancer therapy. *Cancer Research Frontiers* 2: 67–84.
12. Jabeen, F., M. Najam-ul-Haq, R. Javeed, C.W. Huck, and G.K. Bonn. 2014. Au-nanomaterials as a superior choice for near-infrared photothermal therapy. *Molecules* 19: 20580–20593.
13. Chen, G., I. Roy, C. Yang, and P. Prasad. 2016. Nanochemistry and nanomedicine for nanoparticle-based diagnostics and therapy. *Chemical Reviews* 116: 2826–2885.
14. Dykman, L., and N. Khlebtsov. 2012. Gold nanoparticles in biomedical applications: Recent advances and perspectives. *Chemical Society Reviews* 41: 2256–2282.
15. Dreaden, E.C., A.M. Alkilany, X. Huang, C.J. Murphy, and M.A. El-Sayed. 2012. The golden age: Gold nanoparticles for biomedicine. *Chemical Society Reviews* 41: 2740–2779.
16. Ostdiek, A.M., J.R. Ivey, S.A. Hansen, R. Gopaldas, and S.A. Grant. 2016. Feasibility of a nanomaterial-tissue patch for vascular and cardiac reconstruction. *Journal of Biomedical Materials Research B* 104: 449–457.
17. Baei, P., S. Jalili-Firoozinezhad, S. Rajabi-Zeleti, M. Tafazzoli-Shadpour, H. Baharvand, and N. Aghdami. 2016. Electrically conductive gold nanoparticle-chitosan thermosensitive hydrogels for cardiac tissue engineering. *Materials Science and Engineering C* 63: 131–141.
18. Yang, C., A. Tian, and Z. Li. 2016. Reversible cardiac hypertrophy induced by PEG-coated gold nanoparticles in mice. *Scientific Reports* 6: 20203.
19. Liu, G., M. Qi, Y. Zhang, C. Cao, and E.M. Goldys. 2016. Nanocomposites of gold nanoparticles and graphene oxide towards a stable label-free electrochemical immunosensor for detection of cardiac marker troponin-I. *Analytica Chimica Acta* 909: 1–8.

20. Sridhar, S., J.R. Venugopal, R. Sridhar, and S. Ramakrishna. 2015. Cardiogenic differentiation of mesenchymal stem cells with gold nanoparticle loaded functionalized nanofibers. *Colloids and Surfaces B* 134: 346–354.
21. Payam, B., S. Jalili-Firoozinezhad, S. Rajabi-Zeleti, M. Tafazzoli-Shadpour, H. Baharvand, and N. Aghdami. 2016. Electrically conductive gold nanoparticle-chitosan thermosensitive hydrogels for cardiac tissue engineering. *Materials Sciences and Engineering C* 63: 131–141.
22. Fleischer, S., M. Shevach, R. Feiner, and T. Dvir. 2014. Coiled fiber scaffolds embedded with gold nanoparticles improve the performance of engineered cardiac tissues. *Nanoscale* 6: 9410–9414.
23. Huang, X., I.H. El-Sayed, W. Qian, and M.A. El-Sayed. 2006. Cancer cell imaging and photothermal therapy in the near-infrared region using gold nanorod. *Journal of the American Chemical Society* 128 (6): 2115–2120.
24. Wackenhut, F., A.V. Failla, and A.J. Meixner. 2015. Single gold nanorods as optical probes for spectral imaging. *Analytical and Bioanalytical Chemistry* 407: 4029–4034.
25. Krishnan, S., Z. DiagaQin, Y. Wang, J. Randrianalisoa, V. Raeesi, W.C. Chan, W. Linski, and J.C. Bischof. 2016. Quantitative comparison of photothermal heat generation between gold nanospheres and nanorods. *Scientific Reports* 6: 29836.
26. Antman, E., J.P. Bassand, W. Klein, M. Ohman, J.L.L. Sendon, L. Ryden, M. Simoons, and M. Tendera. 2000. Myocardial infarction redefined—A consensus document of the joint European Society of Cardiology/American College of Cardiology Committee for the redefinition of myocardial infarction. *Journal of American College of Cardiology* 36: 959–969.
27. Apple, F.S., A.H.B. Wu, and A.S. Jaffe. 2002. European Society of Cardiology and American College of Cardiology guidelines for redefinition of myocardial infarction: How to use existing assays clinically and for clinical trials. *American Heart Journal* 144: 981–986.
28. Murphy, C.J., T.K. Sau, A. Gole, and C.J. Orendorff. 2005. Surfactant-directed synthesis and optical properties of one-dimensional plasmonic metallic nanostructures. *MRS Bulletin* 30: 349–355.
29. El-Sayed, M.A. 2001. Some interesting properties of metals confined in time and nanometer space of different shapes. *Accounts of Chemical Research* 34: 257–264.
30. Guo, Z.R., C.R. Gu, X. Fan, Z.P. Bian, H.F. Wu, D. Yang, N. Gu, and J.N. Zhang. 2009. Fabrication of anti-human cardiac troponin I immunogold nanorods for sensing acute myocardial damage. *Nanoscale Research Letters* 4: 1428–1433.
31. Tang, L., and J. Casas. 2014. Quantification of cardiac biomarkers using label-free and multiplexed gold nanorod bioprobes for myocardial infarction diagnosis. *Biosensors and Bioelectronics* 61: 70–75.
32. Tadepalli, S., Z. Kuang, Q. Jiang, K.K. Liu, M.A. Fisher, J.J. Morrissey, E.D. Kharasch, J.M. Slocik, R.R. Naik, and S. Singamaneni. 2015. Peptide functionalized gold nanorods for the sensitive detection of a cardiac biomarker using plasmonic paper devices. *Scientific Reports* 5: 16206.
33. Apple, F.S., R. Ler, and M.M. Murakami. 2012. Determination of 19 cardiac troponin I and T assay 99th percentile values from a common presumably healthy population. *Clinical Chemistry* 58: 1574–1581.
34. Ankri, R., D. Leshem-Lev, D. Fixler, R. Popovtzer, M. Motiei, R. Kornowski, E. Hochhauser, and E.I. Lev. 2014. Gold Nanorods as absorption contrast agents for the noninvasive detection of arterial vascular disorders based on diffusion reflection measurements. *Nano Letters* 14: 2681–2687.
35. Ankri, R., S. Melzer, A. Tarnok, and D. Fixler. 2015. Detection of gold nanorods uptake by macrophages using scattering analyses combined with diffusion reflection measurements as a potential took for in vivo atherosclerosis tracking. *International Journal of Nanomedicine* 10: 4437–4446.
36. Huang, H., F. Liu, S. Huang, S. Yuan, B. Liao, S. Yi, Y. Zeng, and P.K. Chu. 2012. Sensitive and simultaneous detection of different disease markers using multiplexed gold nanorods. *Analytica Chimica Acta* 755: 108–114.

37. Truong, P.L., B.W. Kiim, and S.J. Sim. 2012. Rational aspect ratio and suitable antibody coverage of gold nanorod for ultra-sensitive detection of a cancer biomarker. *Lab on a Chip* 12: 1102–1109.
38. Huang, H., S. Huang, X. Liu, Y. Zeng, X. Yu, B. Liao, and Y. Chen. 2009. Label-free optical biosensors based on Au2S-coated gold Nanorods. *Biosensors and Bioelectronics* 24: 3025–3029.
39. Mayer, K.M., S. Lee, H. Liao, B.C. Rostro, A. Fuentes, P.T. Scully, C.L. Nehl, and J.H. Hafner. 2008. A label-free immunoassay based upon localized surface plasmon resonance of gold Nanorods. *ACS Nano* 2: 687–692.
40. Orendorff, C.J., S.C. Baxter, E.C. Goldsmith, and C.J. Murphy. 2005. Light scattering from gold nanorods: Tracking material deformation. *Nanotechnology* 16: 2601–2605.
41. Fuseler, J.W., C.F. Millette, J.M. Davis, and W. Carver. 2007. Fractal and image analysis of morphological changes in the actin cytoskeleton of neonatal cardiac fibroblasts in response to mechanical stretch. *Microscopy and Microanalysis* 13: 133–143.
42. Kamkin, A., I. Liseleva, I. Lozinsky, K.D. Wagner, G. Isenberg, and H. Scholz. 2005. The role of mechanosensitive fibroblasts in the heart. In *Mechanosensitivity in cells and tissues*, ed. Kamkin, A., I. Kiseleva. Moscow: Academia.
43. Atance, J., M.J. Yost, and W. Carver. 2004. Influence of the extracellular matrix on the regulation of cardiac fibroblast behavior by mechanical stretch. *Journal of Cellular Physiology* 200: 377–386.
44. Carver, W., M.L. Nagpal, M. Nachtigal, T.K. Borg, and L. Terracio. 1991. Collagen expression in mechanically stimulated cardiac fibroblasts. *Circulation Research* 69: 116–122.
45. Stone, J.W., P.N. Sisco, E.C. Goldsmith, S.C. Baxter, and C.J. Murphy. 2007. Using gold nanorods to probe cell-induced collagen deformation. *NanoLetters* 7: 116–119.
46. Chernak, D.J., P.N. Sisco, E.C. Goldsmith, S.C. Baxter, and C.J. Murphy. 2013. High aspect ratio gold Nanorods: Their synthesis and application to image cell-induced strain fields in collagen films. *Methods of Molecular Biology* 1026: 1–20.
47. Wilson, C.G., J.W. Stone, V. Fowlkes, M.O. Morales, C.J. Murphy, S.C. Baxter, and E.C. Goldsmith. 2011. Age-dependent expression of collagen receptors and deformation of type I collagen substrates by rat cardiac fibroblasts. *Microscopy and Microanalysis* 17: 555–562.
48. Grinnell, F. 2003. Fibroblast biology in three-dimensional collagen matrices. *Trends in Cell Biology* 13: 264–269.
49. Carver, W., I. Molano, T.A. Reaves, T.K. Borg, and L. Terracio. 1995. Role of alpha 1 beta 1 integrin complex in collagen gel contraction in vitro by fibroblasts. *Journal of Cellular Physiology* 165: 425–437.
50. Baxter, S.C., M.O. Morales, and E.C. Goldsmith. 2008. Adaptive changed in cardiac fibroblast morphology and collagen organization as a result of mechanical environment. *Cell Biochemistry and Biophysics* 51: 33–44.
51. Law, B.A., and W.E. Carver. 2013. Activation of cardiac fibroblast by ethanol is blocked by TGF– inhibition. *Alcoholism, Clinical and Experimental Research* 37: 1286–1294.
52. Svystonyuk, D.A., J.M. Nqu, H.E. Mewhort, B.D. Lipon, G. Teng, D.G. Guzzardi, G. Malik, D.D. Belke, and P.W. Fedak. 2015. Fibroblast growth factor-2 regulates human cardiac myofibroblast-mediated extracellular matrix remodeling. *Journal of Translational Medicine* 13: 147–157.
53. Sisco, P.N., C.G. Wilson, E. Mironova, S.C. Baxter, C.J. Murphy, and E.C. Goldsmith. 2008. The effect of gold Nanorods on cell-mediated collagen remodeling. *NanoLetters* 8: 3409–3412.
54. Borg, K.T., W. Burgess, L. Terracio, and T.K. Borg. 1997. Expression of metalloproteases by cardiac myocytes and fibroblasts in vitro. *Cardiovascular Pathology* 6: 261–269.
55. Lundqvist, M., J. Stigler, G. Elia, I. Lynch, T. Cedervall, et al. 2008. Nanoparticle size and surface properties determine the protein corona with possible implications for biological impacts. *Proceedings of the National Academy of Sciences* 105: 14265–14270.
56. Walczyk, D., F.B. Bombelli, M.P. Monopoli, I. Lynch, and K.A. Dawson. 2010. What the cell sees in bionanoscience. *Journal of the American Chemical Society* 132: 5761–5768.

57. Monopoli, M.P., D. Walczyk, A. Campbell, G. Elia, I. Lynch, et al. 2011. Physical-chemical aspects of protein corona: relevance to in vitro and in vivo biological impacts of nanoparticles. *Journal of the American Chemical Society* 133: 2525–2534.
58. Maiorano, G., S. Sabella, B. Sorce, V. Brunetti, M.A. Malvindi, et al. 2010. Effects of cell culture media on the dynamic formation of protein-nanoparticle complexes and influence on the cellular response. *ACS Nano* 4: 7481–7491.
59. Sisco, P.N., C.G. Wilson, D. Chernak, J.C. Clark, E.M. Grzincic, K. Ako-Asare, E.C. Goldsmith, and C.J. Murphy. 2014. Adsorption of cellular proteins to polyelectrolyte-functionalized gold nanorods: A mechanism for nanoparticle regulation of cell phenotype. *PLoS ONE* 9: e86670.
60. Wilson, C.G., P.N. Sisco, F.A. Gadala-Maria, C.J. Murphy, and E.C. Goldsmith. 2009. Polyelectrolyte-coated gold nanorods and their interaction with type I collagen. *Biomaterials* 30: 5639–5648.
61. Wilson, C.G., P.N. Sisco, E.C. Goldsmith, and C.J. Murphy. 2009. Glycosaminoglycan-functionalized gold nanorods: Interactions with cardiac cells and type I collagen. *Journal of Materials Chemistry* 19: 6332–6340.
62. Hribar, K.C., K. Meggs, J. Lui, W. Zhu, X. Qu, and S. Chen. 2015. Three-dimensional direct cell patterning in collagen hydrogels with near-infrared femtosecond laser. *Scientific Reports* 5: 17203.
63. Ganji, Y., Q. Li, E.S. Quabius, M. Bottner, C. Selhuber-Unkel, and M. Kasra. 2016. Cardiomyocyte behavior on biodegradable polyurethane/gold nanocomposite scaffolds under electrical stimulation. *Materials Science and Engineering C* 59: 10–18.
64. Fedak, P.W.M., F.M. McCarthy, and R.O. Bonow. 2008. Evolving concepts and technologies in mitral valve repair. *Circulation* 117: 963–974.
65. Grande-Allen, K.J., J.E. Barber, K.M. Klatka, P.L. Houghtaling, I. Vesely, C.S. Moravec, and P.M. McCarthy. 2005. *Journal of Thoracic and Cardiovascular Surgery* 130: 783–790.
66. Grande-Allen, K.J., A.G. Borowski, R.W. Troughton, P.L. Houghtaling, N.R. DiPaola, C.S. Moravec, I. Vesely, and B.P. Griffin. 2005. Apparently normal mitral valves in patients with heart failure demonstrate biochemical and structural derangements. *Journal of the American College of Cardiology* 45: 54–61.
67. Rabkin, E., M. Aikawa, J.R. Stone, Y. Fukumoto, P. Libby, and F.J. Schoen. 2001. Activated interstitial myofibroblasts express catabolic enzymes and mediate matrix remodeling in myxomatous heart valves. *Circulation* 104: 2525–2532.

Chapter 15
Architectured Nanomembranes

Michael Z. Hu and Matthew R. Sturgeon

Abstract This paper has reviewed the frontier field of "architectured membranes" that contain anisotropic oriented porous nanostructures of inorganic materials. Three example types of architectured membranes were discussed with some relevant results from our own research: (1) anodized thin-layer titania membranes on porous anodized aluminum oxide (AAO) substrates of different pore sizes, (2) porous glass membranes on alumina substrate, and (3) guest-host membranes based on infiltration of yttrium-stabilized zirconia inside the pore channels of AAO matrices.

Keywords Membranes · Nanostructures · Mesoporous · Nanoporous · Separations

15.1 Introduction

Membranes are utilized in a variety of industrial processes from water treatment to gas purification to energy production. The main feature that defines a membrane is permeability. Membranes can separate out solids, liquids, and gases [1] from reaction mixtures based on different permeabilities by allowing transportation through the membrane of only the desired species. Utilizing membranes to purify a gas streams is a useful tool with the growing demands of the hydrogen economy [2] and concern of rising CO_2 levels [3]. Ions can also be conducted through membranes, allowing for utilization in energy production (i.e. fuel cells [4] and Li ion batteries [5]). With a specific transportation in mind, a membrane can be engineered to meet the demands of that process by changing its physical or chemical properties [6].

In these times of clean energy policies, new technologies must be explored to maintain efficiencies, keep operating costs at a minimum, all while removing

M.Z. Hu (✉) · M.R. Sturgeon
Oak Ridge National Laboratory, Oak Ridge, TN 37831-6181, USA
e-mail: hum1@ornl.gov

S.E. Hunyadi Murph et al. (eds.), *Anisotropic and Shape-Selective Nanomaterials*,
Nanostructure Science and Technology, DOI 10.1007/978-3-319-59662-4_15

contaminants and reducing pollutants. The nanomembranes are designed for use in energy intensive industrial applications to improve overall performance while reducing operating costs and minimizing environmental impacts. A DOE study speculated that nanomaterials used in membrane and catalyst technologies are expected to have a financial impact of over $10 billion in process savings, provide an energy savings of up to 1.1 quadrillion BTU/year, as well as reducing wastewater and toxic emissions [7]. It is overwhelming clear that there is great potential for economic and environmental benefits to utilizing nanomaterials.

There have been several studies on how nanostructured materials can improve performance in electrical devices. Guo et al. [8] report that ion diffusion distances are dramatically decreased in nanostructured materials applied in energy storage/conversion devices thus improving devices like photovoltaic cells where faster electron travel will improve efficiency by reducing electron-hole recombination. The large aspect ratios of nanomaterials used in batteries can provide large electrode/electrolyte contact surface areas. The nature of the nanostructured membranes should prevent some of the down falls of working with nanoparticles such as agglomeration and low thermodynamic stability; allowing for utilization in higher temperature/energy applications. In their review on oriented nanostructures for energy conversion and storage Liu et al. also emphasize the importance controlled pore channels and alignment of nanocrystalline phases play in optimizing electron and ion transport in photovoltaics, batteries, supercapacitors, and thermoelectrics [9].

As energy prices continue to increase and the impending EPA's clean air act [10, 11] the need for clean energy and reducing energy requirements in industrial processes is paramount. There exist several potential benefits in utilizing architectured nanomembranes in industrial applications/processes such as electrical applications, [12–14] conductive ion membranes for use in batteries and fuel cells, [5, 15–17] solar applications (dye sensitized solar cells, [18–40] hydrogen production via photocatalytic water splitting, [13, 28, 41, 42] and photocatalysts CO_2 [36, 37, 43, 44]), catalytic membranes, [45–53] and energy intensive separations [54–57].

Architectured nanomembranes are produced by intentional manipulation of the membrane's pore size. The architectured nanomembranes are more than nanoporous inorganic structures/thin films. They employ intentionally ordered and oriented pore arrays with engineered alignment. These unique structures are fabricated through molecular engineering methodologies. These methodologies include, electrochemical anodization, [23, 28, 32–34, 37, 58–62] glass drawing, [63–67] and molecular self-assembly [68, 69]. The exploration of producing large dimensions and large quantities of nanomembranes via electrochemical anodization and manipulation of sputtered glass films based on glass drawing methodologies is presented here-in.

Anodization is a well-known electrochemical process in which a metal surface is electrochemically oxidized and etched creating unique nanostructures (nanopores/nanotubes). It is extremely useful in creating straight aligned Al_2O_3 nanopores [58, 59, 70–72] or metal oxide (TiO_2) nanotubes [28, 32, 37, 43, 60, 61, 73, 74]. The anodization process employs a metal substrate in an electrolyte solution. Several studies have shown that metal deposited on different substrates can be anodized [35, 75–81]. Employing current deposition technologies, large dimension metallic films can be coated onto several different kinds of substrates. This allows for great versatility in the application of the resultant materials and viable industrial implementation. The variety of electrolyte solutions allows for use of multiple metal/substrate combinations without fear of incompatibilities.

Glass drawing also produces very well aligned nanopores [63–67, 82–84]. However due to the nature of the draw—bundle—draw process it is rather difficult to obtain membranes of large area dimensions. The etch rate of the core also dictates how thick the sample can be [64]. Only specific glasses can be used; however the glass can be modified before, during, or after the drawing process to fabricate a membrane again suitable for a variety of applications. The draw—bundle—draw process can be a painstakingly slow and unable to produce large quantities of large dimension membranes. A new method was adopted where in a glass surface is sputter coated onto a substrate. Heat treatment of the glass will cause phase separation [85, 86]. Once separated, the phases have different etch rates and the more etch able glass is removed leaving behind a nanoporous film (i.e. membrane) with controllable pore sizes.

15.2 Synthesis Methodologies

The overall methodology to synthesize the architectured nanomembrane is illustrated in Fig. 15.1. The engineering approach to synthesis was designed in a way that would produce large area dimension membranes in large industrial scale quantities.

Two specific methods were tested according to Fig. 15.1. One was the deposition and anodization of titanium to create supported titanium oxide nanotubes. The other was sputtering a borosilicate glass composite on a substrate followed by heat treatment and preferential etching. The deposition process is a well-known industrial technique and would serve well to produce these materials on an industrial scale.

The deposition and anodization of Ti films on porous alumina substrate is illustrated in Fig. 15.2. The membrane thickness can be controlled by Ti film thickness as well as anodization processing. Complete anodization of Ti film would leave tube bottoms connected to porous alumina support. The tube bottoms are very thin and a low concentration etchant would etch off the tube bottom without etching the support.

Fig. 15.1 Engineering methodology to produce large quantities of large dimension nanomembranes

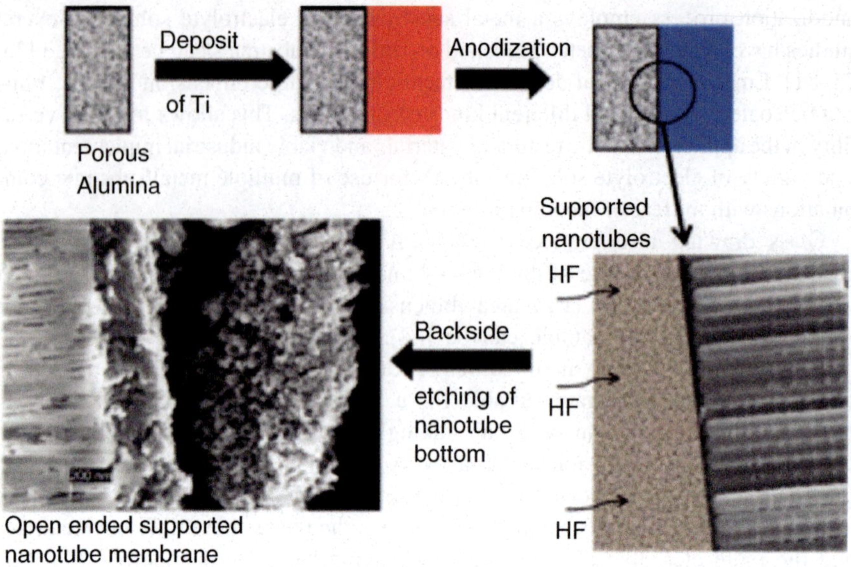

Fig. 15.2 Deposition and anodization of Ti film and processing of film to produce supported architectured nanomembrane

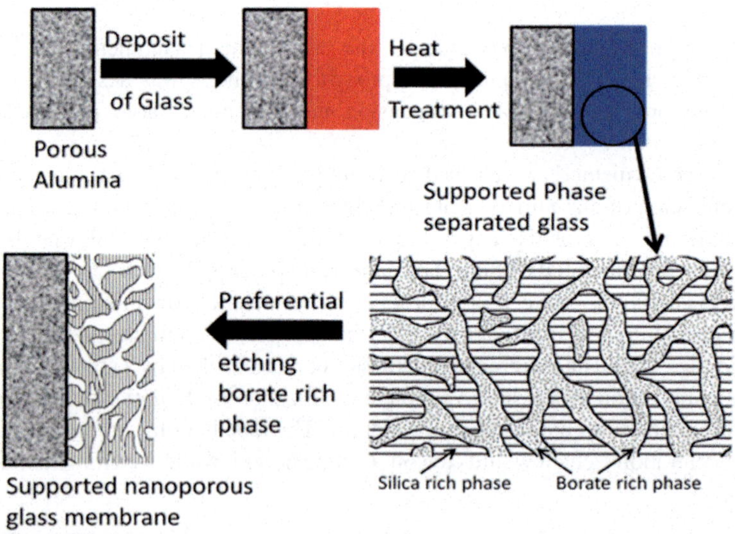

Fig. 15.3 Method for producing supported nanoporous glass membrane via deposition, phase separation, and preferential etching

Creating nanoporous membranes from glass is conducted in a similar deposition, which processing method and is illustrated in Fig. 15.3.

As stated before, the membrane thickness can be controlled by deposition of glass on substrate surface. The pore size can be controlled by heat treatment, longer heating time will create more phase separated samples; there by increasing pore size upon etching away the borate rich phase.

15.3 Experimental

In the production of titania nanotube membranes, anodized alumina (AAO) disks of varying pore sizes, 13 ± 2, 18 ± 3, 35 ± 3, 55 ± 6, and 80 ± 8 nm were used as received by Synkera, AAO disks of 200 nm were used as received from Alltech/Whatman. VWR micro slides glass slides were cleaned with acetone, isopropanol, and DI water prior to deposition. AAO disks were glued to glass slides using Ladd conducting graphite glue. Ti films were deposited onto supports via electron beam deposition using a Varian e-beam evaporator equipped with an Airco Temescal model CV-8 electron beam. *Wet 'n' Wild* brand nail polish was used as a polymer in sample preparation. Ethylene glycol (Fisher Scientific) and ammonium fluoride (J.T. Baker 99.0%) were used as received as in the electrolyte solution. A Hewlett Packard HP 6653A 0-35 V/0-15A DC power supply used in the anodization experiments. A 99.9% Pt plate (Alfa Aesar) was used as a cathode in anodization.

Glass was sputtered coated onto a porous alumina disks (42 mm diameter) used as received from AdValue Technology using a RF sputtering system equipped with a Kurt J. Lesker sputter source. Samples were etched in HF solutions. SEM images were obtained using a Zeiss Merlin field emission SEM.

15.3.1 Production of Titania Nanotube Membranes

Titanium was deposited on both glass slides and AAO disks of varying pore size using two different deposition techniques; DC—magnetron sputtering at a rate of 4.2 Å/s and electron beam deposition at a rate of 7 Å/s. Both techniques require the sample to be placed in the sample chamber upside down. Therefore the AAO disks were glued onto a glass slide using a very small dot (2–3 mm in diameter) of either graphite glue or polymer (*Wet 'n' Wild* brand nail polish). Three or Four AAO disks could be glued to a single slide for Ti deposition. Once deposition was complete the coated AAO were carefully removed from the glass slide prior to anodization.

To prepare the Ti coated samples for anodization the disks or glass samples were glued to a glass slide by using a dot (4–5 mm) of graphite glue. An insulated Cu wire was then attached to the top of the disk with a small dot (2–3 mm) of graphite glue. In order to protect the electrical connection during anodization (while the sample is immersed in electrolyte) the graphite glue was coated with polymer (Fig. 15.4).

Samples were anodized in an ethylene glycol solution containing 0.3 wt% ammonium fluoride and 2% DI water. The prepared Ti coated AAO disks served as the anode with a 1 × 1 in. Pt plate as the cathode. The Pt electrode was cleaned ultrasonically for 1 min each in chloroform, acetone, isopropanol, and DI water, sequentially. The electrodes were connected to the power supply at an interspacing

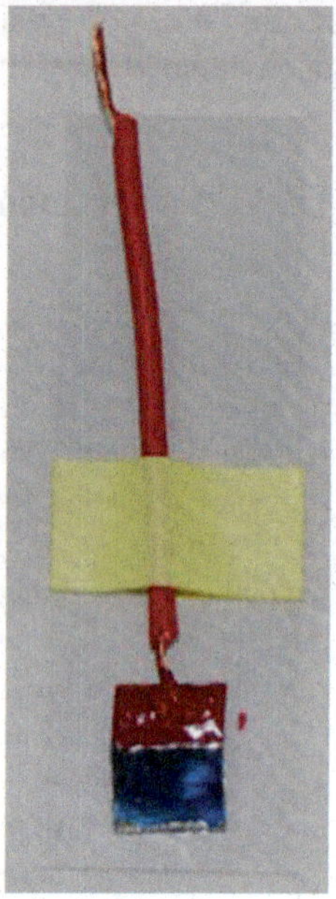

Fig. 15.4 Ti coated glass sample prepared for anodization

of either 2.0 or 1.0 cm. A Teflon® container was then filled with 50 mL of the electrolyte solution. The entire Ti coated AAO disk was immersed in electrolyte while only half of the Pt electrode was immersed. Samples were anodized at a voltage of 20 V and current of 0.001 A. Anodization times of greater than one hour produced nanotubes. After anodization the disk samples were rinsed with DI water and isopropanol. To remove the disk from the glass slide they were sonicated in acetone for 1 min. Anodized samples were back side etched with a 5% HF solution. Samples were then left to dry in air.

15.3.2 *Production of Nanoporous Glass Membranes*

An RF sputtering system was used to deposit a layer of glass onto a porous alumina. The chamber was pumped down to 10^{-2} Torr by a mechanical pump, followed by a turbo pump that further reduced the base-pressure down to 10^{-5} Torr. The sputter target was made up of borosilicate glass. Depositions were conducted in Ar and Ar-O_2 gas mixture and at room temperature. The glass film was then phase separated by heating in a box furnace under ambient atmosphere. The temperature was increased at a rate of 2 °C/min, to a final temperature of 700 °C with a hold time of 20 min. After heating the temperature was decreased at the same rate. The samples were then etched in a 5% HF solution for 1, 3, and 5 min.

15.4 Results and Discussion

(1) *Supported Titania Nanotube Membranes*

AAO was used as a model porous substrate, as the pore sizes are highly regular and the relatively smooth surface (compared to composite alumina) is ideal for coating. DC sputtering was first used to deposit Ti films of thickness 200 and 600 nm on AAO with a pore size of 200 nm (Fig. 15.5).

In the 200 nm film, there are clearly visible voids in the film with agglomerates of larger particles (100–300 nm). There are no voids in the 600 nm Ti film however the particle agglomerates are much larger, >1 μm (Fig. 15.6).

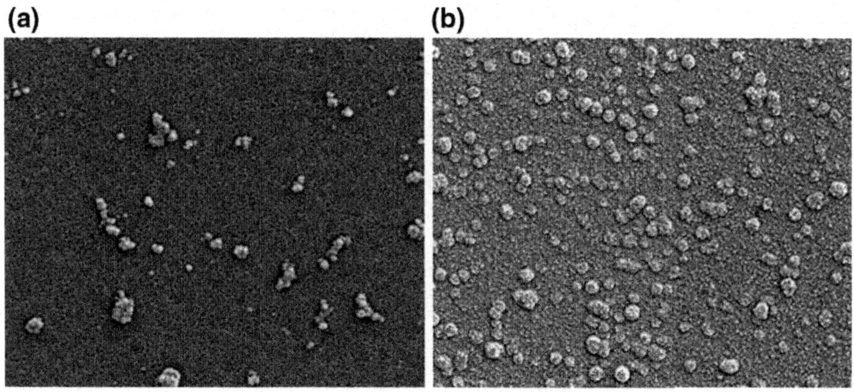

Fig. 15.5 The surface morphologies of the Ti Sputtered films: 200 nm (**a**), 600 nm (**b**) image width is 8–9 μm

Fig. 15.6 SEM image of a ~3 μm particle from the DC sputter deposition of 600 nm Ti on AAO

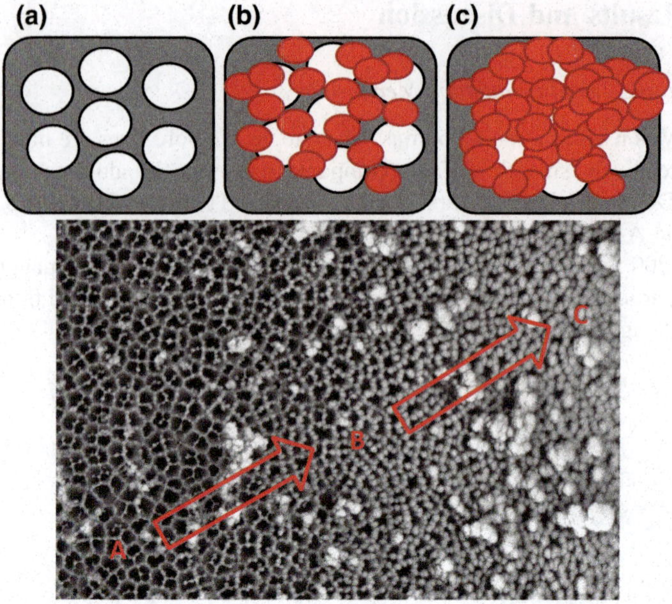

Fig. 15.7 Method of Ti sputter deposition on AAO substrate showing bare AAO support grid (**a**), beginning of coating (**b**) and incomplete coverage (**c**) in the same sample. Image width is 7.193 μm

We use the schematic in Fig. 15.7 to illustrate the agglomeration formation process on the surface of AAO substrate. The A, B, and C schematic correspond to the letter label spot in SEM image.

Both the 200 and 600 nm films were unsuccessfully anodized, with the Ti film dissolving off of the substrate during anodization. The pore size of the AAO seemed to be affecting the deposition of the Ti film and a smooth film was unattainable. AAO of three smaller pore sizes, 13, 18, and 55 nm were then coated with 200 nm Ti using DC sputtering. The sputter deposition gave a nice uniform Ti film on all AAO disks and the glass substrate (Fig. 15.8).

SEM analysis shows that these films are uniform and continuous. However other sputtered samples had non-uniform Ti coatings. DC sputter coating of Ti did not reliably reproduce uniform films with smooth surfaces. Due to these inconsistencies in DC sputtering, electron beam deposition was then employed in the deposition of Ti films. Initial e-beam depositions produced uniform 200 nm thick Ti films on 200 nm pore size AAO (Fig. 15.9) as well as a glass slide substrate. The Ti films consisted of uniform particles. Particle sizes on glass substrate were all less than 50 nm. On AAO the particles were less than 100 nm in size and gave a continuous film at 200 nm film thickness, previously unattainable using DC sputter deposition on 200 nm pore size AAO.

Films deposited via e-beam deposition had the advantage on not only being uniform and highly reproducible the sample chamber is much larger than that in the

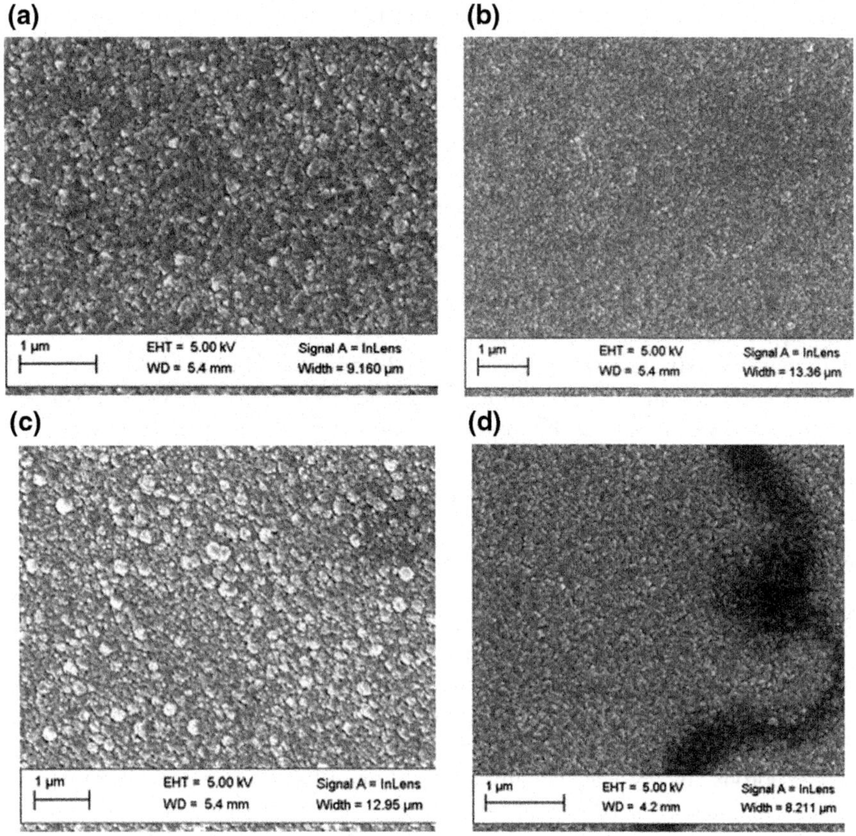

Fig. 15.8 SEM images of 200 nm Ti sputter coated on 13 nm pore size AAO (**a**), 18 nm pore size AAO (**b**), 55 nm pore size AAO (**c**), and glass substrate (**d**)

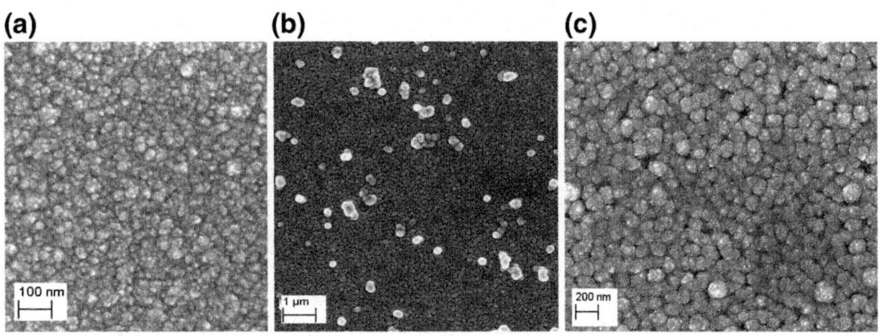

Fig. 15.9 SEM images of e-beam deposited 200 nm Ti film on glass substrate (**a**) and AAO (**b**, **c**)

sputtering system. This allowed for the deposition of Ti on multiple substrates in a single run. As many as fifteen substrates (of ~1-cm diameter disks) could be coated in a single e-beam deposition run, as opposed to only three substrates in a sputter deposition run. Being able to coat five times as many substrates in a run is a significant increase on an industrial scale.

The surface of the Ti films was smoother on glass substrates than AAO substrates. In order to obtain a working procedure for the anodization of deposited films, early anodization experiments were performed on Ti coated glass substrates. These early anodizations of the 200 nm Ti films were short (less than 30 min) and produced very uniform porous surface features (Fig. 15.10). However no nanotubes were observed.

In longer anodization times (greater than 30 min) the Ti film would dissolve off of the substrate, with the dissolution rate greatest at the air to electrolyte solution surface interface. It was at this point that the samples were prepared to be fully immersed in the electrolyte solution. Further anodizations of 200 nm thick Ti films, samples fully immersed in electrolyte, still exhibited Ti films dissolving off the substrate surface during anodization. For this reason thicker Ti films (500 nm) were deposited. The thicker films could with stand longer anodization times. Anodization times of 2 h successfully produced titania nanotubes on glass substrates (Fig. 15.11).

The glass surface had a non-uniform color over the anodized area, with colors ranging from purple to blue to sea green to light yellow. It was determined from profile studies of the supported nanotubes (Fig. 15.12) that the tubes were *c.a.* 717 nm in length covered by a porous coating *c.a.* 82.5 nm thick.

The porous coating was partially removed upon further sonication, area of 0.15 by 0.16 mm were virtually clear of the porous coating. The nanotubes were not damaged during this extra sonication.

E-beam deposition of 200 nm thick Ti films on 200 nm pore size AAO gave continuous films (unlike DC sputtering deposition). The films were made up of aggregates of particles, as seen before. However these particles were all of a

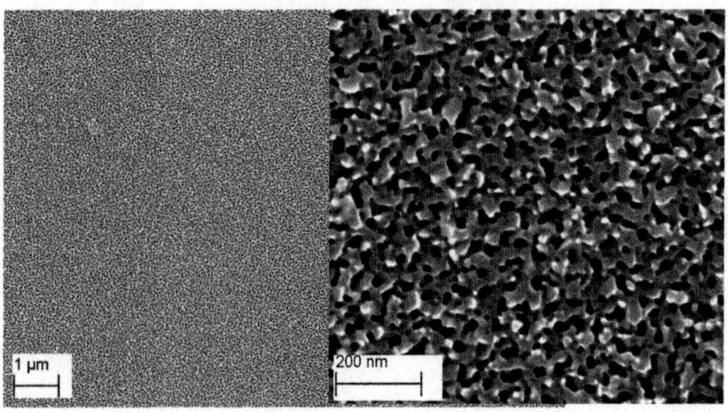

Fig. 15.10 SEM images of anodized Ti films on glass substrate

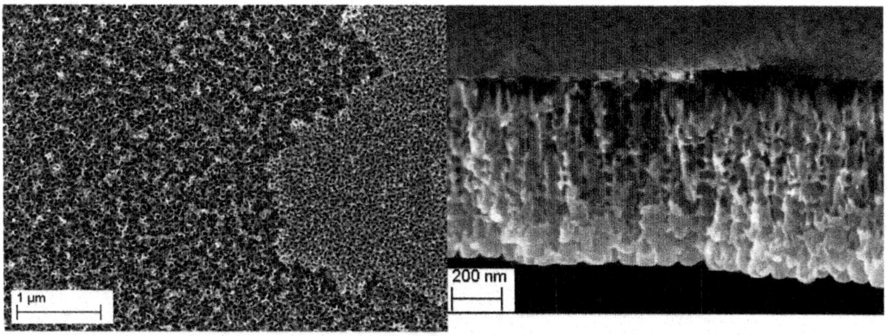

Fig. 15.11 SEM of titania nanotubes on glass substrate

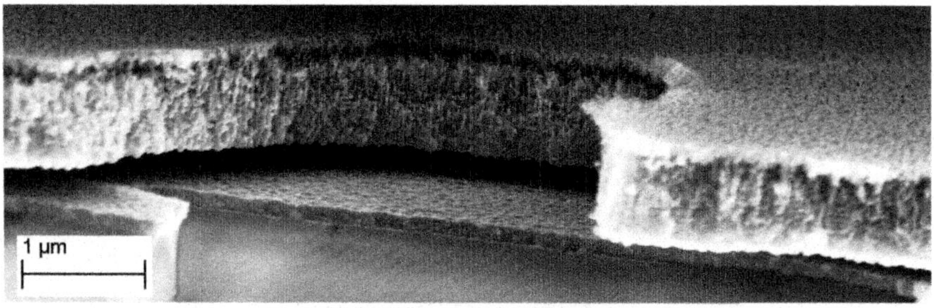

Fig. 15.12 SEM profile view of titania nanotubes on glass

Fig. 15.13 SEM image of anodized 200 nm Ti film on 200 nm pore size AAO. *A*, *B*, and *C* correspond to EDX spectra in Fig. 15.14

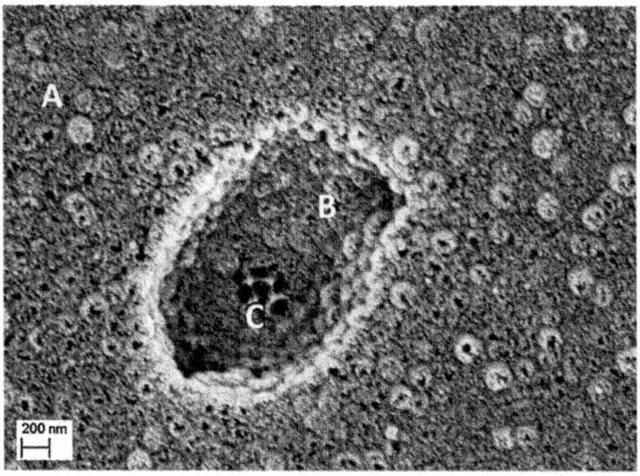

uniform size. Anodization of this 200 nm Ti film on 200 nm pore size AAO was successful in creating nanotube. Figure 15.13 shows an SEM image of the anodized surface with corresponding EDX spectra (Fig. 15.14). However there were visible voids in the anodized surface.

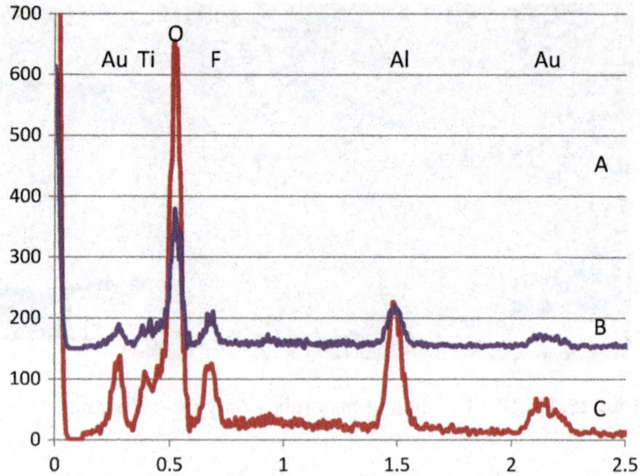

Fig. 15.14 EDX spectra of spots in Fig. 15.13, snowing the elemental composition as indicated by peak positions. X-coordinate shows the range of energy (keV) and Y-coordinates shows the intensity measured by the detector

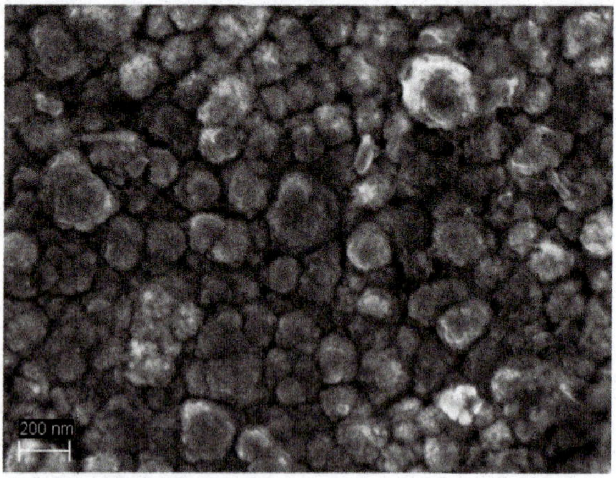

Fig. 15.15 SEM image of 200 nm Ti film on 80 nm pore size AAO

It is evident from the SEM images and the EDX spectra that the anodized film is not uniform and as stated above there were visible voids in the coating. Therefore 200 nm Ti films were deposited on AAO with smaller pore sizes. Films were deposited on AAO with pore sizes of 80 and 18 nm. The films on the 80 nm samples were continuous and uniform with average Ti particle size less than 200 nm (Fig. 15.15). Anodization of this film produced some very interesting features (Fig. 15.16).

Upon SEM analysis it was determined that the particles were actually being individually anodized, as opposed to the entire film. In order to obtain continuous layer of titania nanotube that would be converted into a supported nanomembrane a smoother film would be required. Therefore AAO with 18 nm pore size were coated with Ti, producing very smooth, continuous, uniform Ti film. Short anodization (10 min) of this film successfully produced nanotubes (Fig. 15.17).

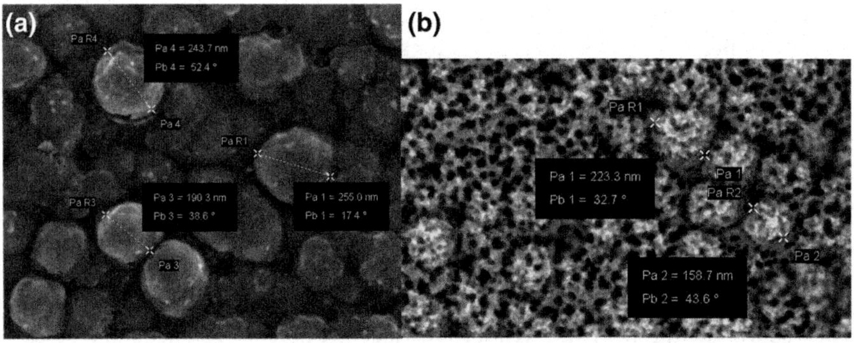

Fig. 15.16 SEM image of 200 nm film on 80 nm pore size AAO before (**a**) and after (**b**) anodization

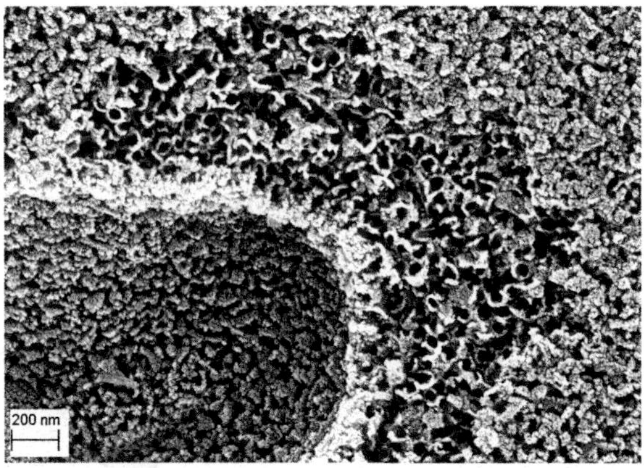

Fig. 15.17 SEM image of anodized 200 nm Ti on 18 nm pore size AAO

The tubes were covered with a similar porous coating as seen in the Ti/glass samples. The anodized surface was very non-uniform with visible pits. To the naked eye the anodized surface had visible non-uniformities in color, again ranging from purple to tan. At this point it was determined that thicker Ti films with longer anodization times would produce the desired supported nanotube arrays. Increased conductivity of the electrolyte have also shown to produce nanotube arrays with no porous coating [32]. According to the equation for conductance, G [87]:

$$G = \frac{kA}{l}$$

where G is the conductance, A is area of electrodes, l is the distance between the electrodes, and k is the conductivity; simply decreasing l, the distance between the electrodes would increase conductivity. In the following anodization experiments the working distance of the electrodes (the prepared Ti/AAO and Pt plate) was reduced from 2 cm (used in all previous experiments) to 1 cm.

AAO with a pore size of 13 nm was coated (via e-beam deposition) with a 500 nm Ti film. Not only was the film very smooth, continuous, and uniform; it was dense as well. The film could withstand anodization times up to 3 h even though anodization times of 1 h were sufficient to produce nanotubes (Fig. 15.18).

The nanotube array produced from 1 h anodization was mostly clear of debris, however there were still some spots covered (Fig. 15.19). The array produced from a 3 h anodization was clear of debris.

Complete anodization of the Ti film was apparent from transparent samples as well as SEM profile images (Fig. 15.20).

Once the films were anodized the bottoms of nanotubes would need to be removed in order produce the architectured nanomembrane. The nanotube bottoms were successfully backside etched in 5% HF (Fig. 15.21).

The successful etching of the titania nanotube array resulted in a supported architectured nanomembrane of *c.a.* 500 nm thick with *c.a.* 40 nm pore size.

(2) *Supported Nanoporous Etched Glass Membranes*

This part of work is heavily influenced by the work of Simpson and D'Urso [86]. They showed that a porous glass material could be produced from etching phase separated glass. However they were never successful in creating a supported glass material. To create a supported nanoporous glass membrane, glass from a borosilicate sample was RF sputtered onto a porous alumina substrate (Fig. 15.22).

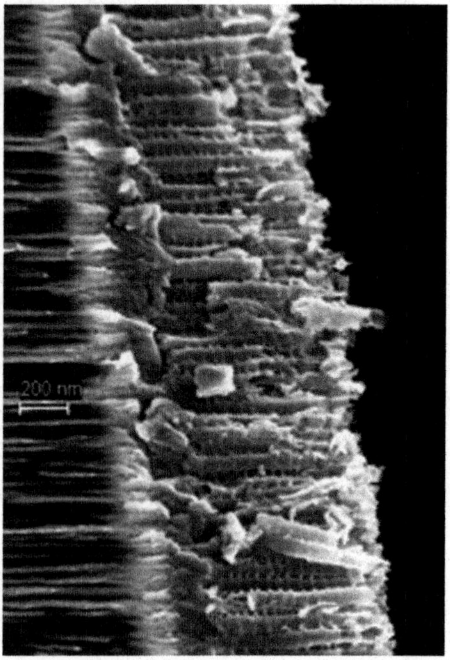

Fig. 15.18 SEM image of TiO_2 nanotubes produced from a 1 h anodization of 500 nm Ti film on AAO (13 nm pore size)

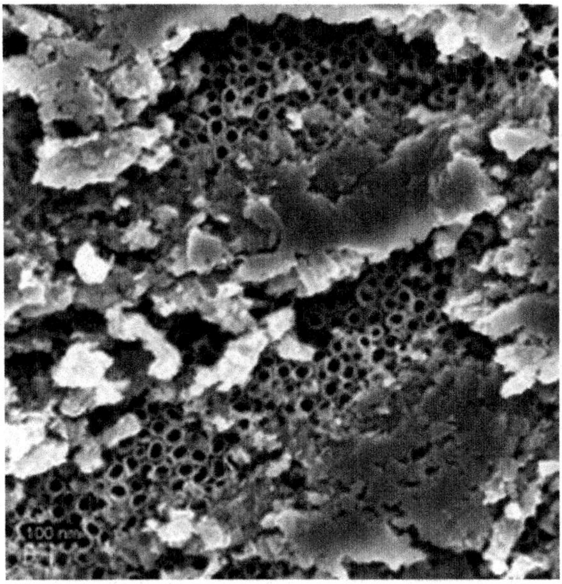

Fig. 15.19 SEM image showing nanotubes from a 1 h anodization of 500 nm Ti film on 13 nm pore size AAO

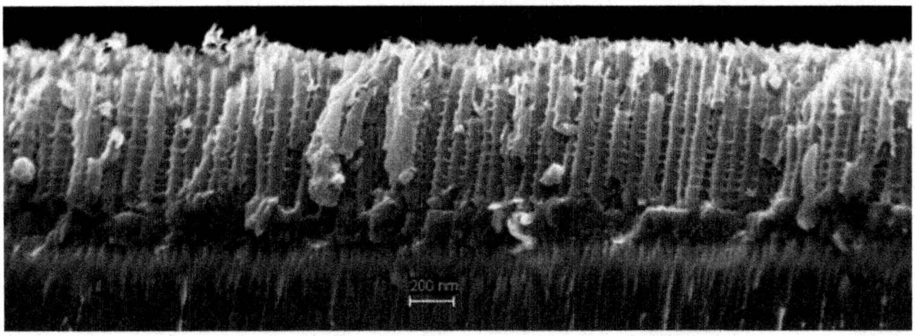

Fig. 15.20 SEM profile image showing complete anodization (~3 h) of Ti film on AAO, showing parallel arrays of nanotubes connected with "bridges"

Due to the rough surface morphology and characteristics of the alumina substrate, the thickness of the glass film was not determined. The borate and silicate phases in the glass were then separated by heating. Only limited studies were performed on the degree of phase separation due to heating as this was a proof of principle conceptual study. The coated substrate was heated to 700 °C for 20 min. The sample was then etched in a 5% HF solution for 1, 3, and 5 min. The 1 min etch time was sufficient in creating nanopores (pore size less than 100 nm) in the glass coating (Fig. 15.23).

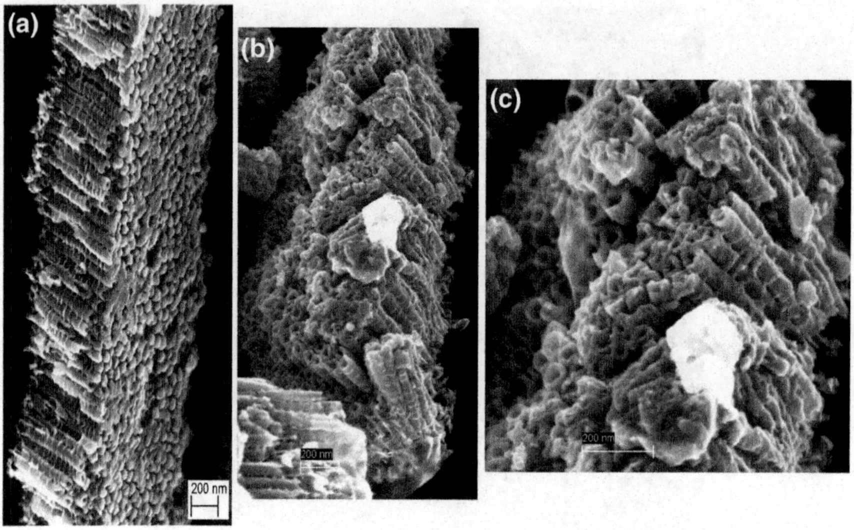

Fig. 15.21 SEM image showing the successful etching of nanotube bottoms; before etching (**a**) and after etching (**b**, **c**)

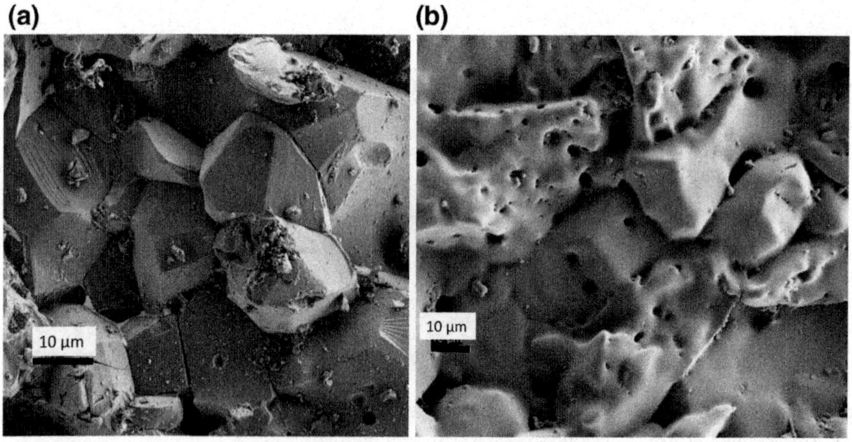

Fig. 15.22 Alumina substrate before (**a**) and after (**b**) glass deposition

Due to the complexity of the alumina support it was not determined if the nanopores penetrated through the entire glass coating or not. However, from the SEM image shown in Fig. 15.23, it does appear that there is a supported porous glass nanomembrane.

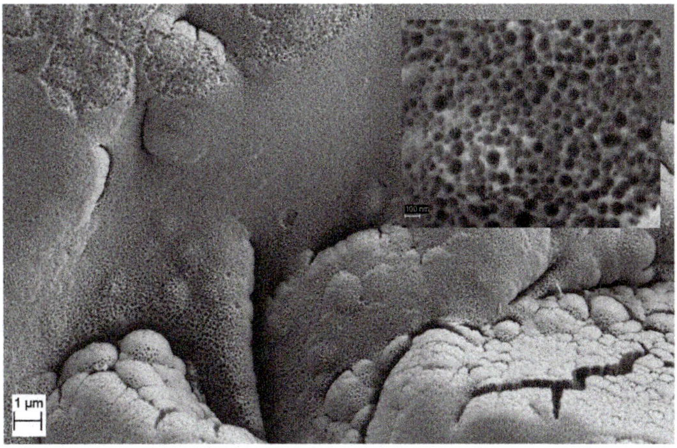

Fig. 15.23 SEM image of nanopores etched into glass coated alumina

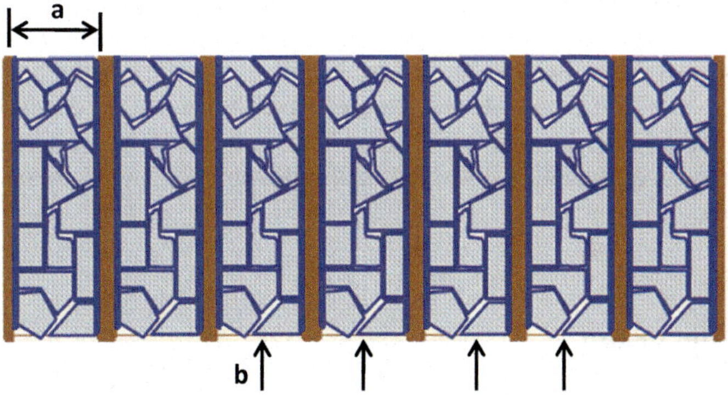

Fig. 15.24 A schematic of Hu's guest—host composite nanomembrane. Channel pore size (*a*) can be as low as 10 nm. Current flow is indicated by the *arrows* (*b*) [48]

(3) *YSZ in AAO—A Guest Nanomaterial in a Host Architectured Nanopore Channels*

Another type of architectured membranes consist of "guest" materials inside the pores of a host matrix like AAO. An example of guest material is the electrolyte used in SOFCs, i.e., yttria stabilized zirconia (YSZ). Hu and Kosacki synthesized a guest host nanocomposite membrane consisting YSZ (guest) in mesoporous AAO (host) for use as an oxide conducting nanomembrane (Fig. 15.24) [48].

They obtained the highest ionic conductivity ever reported for YSZ. The nanocomposite maximizes both the grain boundary effect (YSZ grain size—grains <100 nm show ten times the conductivity of larger grains) and the film thickness

Fig. 15.25 Highest ionic conductivity achieved in nanocrystalline YSZ enclosed inside the parallel array of nanopore channles of AAO

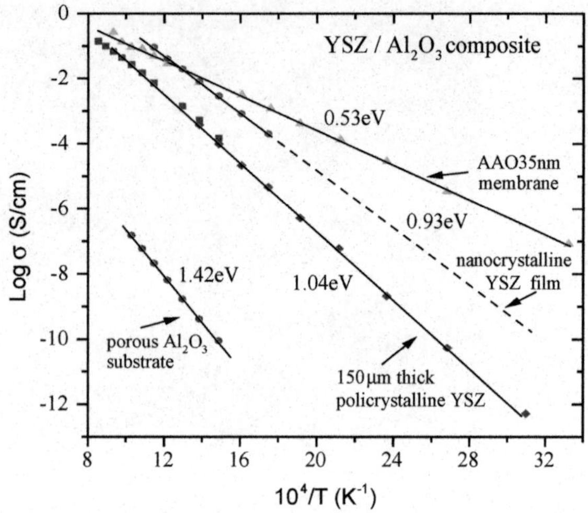

effect (thin films <60 nm of YSZ induce 10–10^3 times increase in conductivity, Fig. 15.25). Detailed conductivity enhancement mechanisms were discussed in a previously filed patent [48]. With increased conductivity the operating temperature for fuel cells can be reduced. Hu and Kosacki have shown that the conductivity data can be measured even at room temperature. This new class of guest-host electrolyte membrane reduces operating costs and makes possible for low-temperature SOFC applications.

15.5 Conclusion

Applications of architectured nanomembranes include (but are not limited to): electrical applications, conductive ion membranes for use in batteries and fuel cells, solar applications (dye sensitized solar cells, hydrogen production via photocatalytic water splitting, and photocatalysts), catalytic membranes, and energy intensive separations. There exist many engineering R&D needs required to manufacture these nanomembranes. Two such methods have been studied and reported herein. The anodization method has the ability of producing titania nanotube on a substrate. E-beam deposition of Ti films on any substrate (with varying topography i.e. tubular) can create a film that once anodized will leave behind nanotubes. Backside etching through the porous substrate will produce a supported architectured nanomembrane. The thickness and pore size of the membrane can be controlled through anodization techniques. The glass membranes have benefits unique to them as well. The pore size is controllable through heating. The longer the sample is heated the greater the degree of phase separation of the highly etchable borate phase from the more stable silicate phase. Once etched the membranes can be further manipulated through the vast amount of available glass chemistry.

Acknowledgements We thank partial sponsorship from ORNL-Seed Money fund, DOE/Industrial Technologies Office fund, and DOE/BETO fund. Part of the materials deposition and characterization (including e-beam deposition and SEM) was conducted at the Center for Nanophase Materials Sciences, which is sponsored by the ORNL Scientific User Facilities Division and DOE Office of Basic Research Sciences.

References

1. Baker, R.W. 2002. Future directions of membrane gas separation technology. *Industrial and Engineering Chemistry Research* 41 (6): 1393–1411.
2. Ritter, J.A., and A.D. Ebner. 2007. State-of-the-art adsorption and membrane separation processes for hydrogen production in the chemical and petrochemical industries. *Separation Science and Technology* 42 (6): 1123–1193.
3. Ritter, J.A., and A.D. Ebner. 2007. *Carbon Dioxide Separation Technology: R&D Needs for the Chemical and Petrochemical Industries.* DOE report: http://www.chemicalvision2020.org/pdfs/CO2_Separation_Report_V2020_final.pdf.
4. DOE/EERE. 2004. *Fuel Cells Power for the 21st Century.* DOE Report: http://www1.eere.energy.gov/hydrogenandfuelcells/.
5. Zhang, S.S. 2007. A review on the separators of liquid electrolyte Li-ion batteries. *Journal of Power Sources* 164 (1): 351–364.
6. BCS. 2005. *Materials for Separation Technologies: Energy and Emission Reduction Opportunities.* Industrial Technologies Program DOE.
7. Thayer, G.R., F. Roach, and L. Dauelsberg. 2006. *DOE Report: Estimated Energy Savings and Financial Impacts of Nanomaterials by Design on Selected Applications in the Chemical Industry.* Los Alomos Chemical Industry Vision 2020. http://www.chemicalvision2020.org/pdfs/LANL_Nano_Rept_Final_4-17-06.pdf.
8. Guo, Y.G., J.S. Hu, and L.J. Wan. 2008. Nanostructured materials for electrochemical energy conversion and storage devices. *Advanced Materials* 20 (15): 2878–2887.
9. Liu, J., G.Z. Cao, Z.G. Yang, D.H. Wang, D. Dubois, X.D. Zhou, G.L. Graff, L.R. Pederson, and J.G. Zhang. 2008. Oriented nanostructures for energy conversion and storage. *ChemSusChem* 1 (8–9): 676–697.
10. Johnson, J. 2010. EPA's move on CO_2 reaps 24 lawsuits. In *Chemical and Engineering News*, 6.
11. EPA. 2010. *Clean Air Act.* http://www.epa.gov/air/caa/.
12. Banerjee, P., I. Perez, L. Henn-Lecordier, S.B. Lee, and G.W. Rubloff. 2009. Nanotubular metal-insulator-metal capacitor arrays for energy storage. *Nature Nanotechnology* 4 (5): 292–296.
13. http://x-journals.com/2009/nanotech-batteries-for-a-new-energy-future/, 2009.
14. Simon, U., and M.E. Franke. 2000. Electrical properties of nanoscaled host/guest compounds. *Microporous and Mesoporous Materials* 41 (1–3): 1–36.

15. Kim, M., Y.C. Nho, and J.H. Park. 2010. Electrochemical performances of inorganic membrane coated electrodes for li-ion batteries. *Journal of Solid State Electrochemistry* 14 (5): 769–773.
16. http://www.hydrogen.energy.gov/. U.S. Department of Energy Hydrogen Program, 2011.
17. Oleksandrov, S., J.W. Lee, J.H. Jang, S. Haam, and C.H. Chung. 2009. Proton-conductive nanochannel membrane for fuel-cell applications. *Journal of Nanoscience and Nanotechnology* 9 (2): 1551–1554.
18. Chen, X., and S.S. Mao. 2007. Titanium dioxide nanomaterials: Synthesis, properties, modifications, and applications. *Chemical Reviews* 107 (7): 2891–2959.
19. Directors, D.N.L. 1997. *Technology Opportunities to Reduce U.S. Greenhous Gas Emissions.* www.ornl.gov/~webworks/cppr/y2003/rpt/110512.pdf.
20. Gong, D., C.A. Grimes, O.K. Varghese, W.C. Hu, R.S. Singh, Z. Chen, and E.C. Dickey. 2001. Titanium oxide nanotube arrays prepared by anodic oxidation. *Journal of Materials Research* 16 (12): 3331–3334.
21. Mor, G.K., M.A. Carvalho, O.K. Varghese, M.V. Pishko, and C.A. Grimes. 2004. A room-temperature TiO_2-nanotube hydrogen sensor able to self-clean photoactively from environmental contamination. *Journal of Materials Research* 19 (2): 628–634.
22. Varghese, O.K., G.K. Mor, C.A. Grimes, M. Paulose, and N. Mukherjee. 2004. A titania nanotube-array room-temperature sensor for selective detection of hydrogen at low concentrations. *Journal of Nanoscience and Nanotechnology* 4 (7): 733–737.
23. Wang, W.Z., O.K. Varghese, M. Paulose, C.A. Grimes, Q.L. Wang, and E.C. Dickey. 2004. A study on the growth and structure of titania nanotubes. *Journal of Materials Research* 19 (2): 417–422.
24. Cai, Q.Y., M. Paulose, O.K. Varghese, and C.A. Grimes. 2005. The effect of electrolyte composition on the fabrication of self-organized titanium oxide nanotube arrays by anodic oxidation. *Journal of Materials Research* 20 (1): 230–236.
25. Shankar, K., M. Paulose, G.K. Mor, O.K. Varghese, and C.A. Grimes. 2005. A study on the spectral photoresponse and photoelectrochemical properties of flame-annealed titania nanotube-arrays. *Journal of Physics D-Applied Physics* 38 (18): 3543–3549.
26. Varghese, O.K., M. Paulose, K. Shankar, G.K. Mor, and C.A. Grimes. 2005. Water-photolysis properties of micron-length highly-ordered titania nanotube-arrays. *Journal of Nanoscience and Nanotechnology* 5 (7): 1158–1165.
27. Mor, G.K., K. Shankar, M. Paulose, O.K. Varghese, and C.A. Grimes. 2006. Use of highly-ordered TiO_2 nanotube arrays in dye-sensitized solar cells. *Nano Letters* 6 (2): 215–218.
28. Mor, G.K., O.K. Varghese, M. Paulose, K. Shankar, and C.A. Grimes. 2006. A review on highly ordered, vertically oriented TiO_2 nanotube arrays: Fabrication, material properties, and solar energy applications. *Solar Energy Materials and Solar Cells* 90 (14): 2011–2075.
29. Paulose, M., G.K. Mor, O.K. Varghese, K. Shankar, and C.A. Grimes. 2006. Visible light photoelectrochemical and water-photoelectrolysis properties of titania nanotube arrays. *Journal of Photochemistry and Photobiology a-Chemistry* 178 (1): 8–15.
30. Paulose, M., K. Shankar, O.K. Varghese, G.K. Mor, and C.A. Grimes. 2006. Application of highly-ordered TiO_2 nanotube-arrays in heterojunction dye-sensitized solar cells. *Journal of Physics D-Applied Physics* 39 (12): 2498–2503.
31. Ruan, C.M., M. Paulose, O.K. Varghese, and C.A. Grimes. 2006. Enhanced photo electrochemical-response in highly ordered TiO_2 nanotube-arrays anodized in boric acid containing electrolyte. *Solar Energy Materials and Solar Cells* 90 (9): 1283–1295.
32. Grimes, C.A. 2007. Synthesis and application of highly ordered arrays of TiO_2 nanotubes. *Journal of Materials Chemistry* 17 (15): 1451–1457.
33. Shankar, K., G.K. Mor, H.E. Prakasam, S. Yoriya, M. Paulose, O.K. Varghese, and C.A. Grimes. 2007. Highly-ordered TiO_2 nanotube arrays up to 220 mu m in length: use in water photoelectrolysis and dye-sensitized solar cells. *Nanotechnology* 18 (6).
34. Shankar, K., J.I. Basham, N.K. Allam, O.K. Varghese, G.K. Mor, X.J. Feng, M. Paulose, J.A. Seabold, K.S. Choi, and C.A. Grimes. 2009. Recent advances in the use of TiO_2 nanotube and

nanowire arrays for oxidative photoelectrochemistry. *Journal of Physical Chemistry C* 113 (16): 6327–6359.
35. Varghese, O.K., M. Paulose, and C.A. Grimes. 2009. Long vertically aligned titania nanotubes on transparent conducting oxide for highly efficient solar cells. *Nature Nanotechnology* 4 (9): 592–597.
36. Varghese, O.K., M. Paulose, T.J. LaTempa, and C.A. Grimes. 2009. High-rate solar photocatalytic conversion of CO_2 and water vapor to hydrocarbon fuels. *Nano Letters* 9 (2): 731–737.
37. Rani, S., S.C. Roy, M. Paulose, O.K. Varghese, G.K. Mor, S. Kim, S. Yoriya, T.J. LaTempa, and C.A. Grimes. 2010. Synthesis and applications of electrochemically self-assembled titania nanotube arrays. *Physical Chemistry Chemical Physics* 12 (12): 2780–2800.
38. Mor, G.K., K. Shankar, M. Paulose, O.K. Varghese, and C.A. Grimes. 2004. Enhanced photocleavage of water using titania nanotube arrays. *Nano Letters* 5 (1): 191–195.
39. Jennings, J.R., A. Ghicov, L.M. Peter, P. Schmuki, and A.B. Walker. 2008. Dye-sensitized solar cells based on oriented TiO_2 nanotube arrays: Transport, trapping, and transfer of electrons. *Journal of the American Chemical Society* 130 (40): 13364–13372.
40. Banerjee, S., M. Misra, S.K. Mohapatra, C. Howard, and S.K. Kamilla. 2010. Formation of chelating agent driven anodized TiO_2 nanotubular membrane and its photovoltaic application. *Nanotechnology* 21 (14).
41. Gong, J.J., Y.K. Lai, and C.J. Lin. 2010. Electrochemically multi-anodized TiO_2 nanotube arrays for enhancing hydrogen generation by photoelectrocatalytic water splitting. *Electrochimica Acta* 55 (16): 4776–4782.
42. Richter, C., and L. Menon. 2010. Impact of adsorbed alkali ions on photoelectrochemical hydrogen production by titania nanotubes. *Energy & Environmental Science* 3 (4): 427–433.
43. Albu, S.P., A. Ghicov, J.M. Macak, R. Hahn, and P. Schmuki. 2007. Self-organized, free-standing TiO_2 nanotube membrane for flow-through photocatalytic applications. *Nano Letters* 7 (5): 1286–1289.
44. Tan, L.K., M.K. Kumar, W.W. An, and H. Gao. 2010. Transparent, well-aligned TiO_2 nanotube arrays with controllable dimensions on glass substrates for photocatalytic applications. *ACS Applied Materials & Interfaces* 2 (2): 498–503.
45. Kormann, H.P., G. Schmid, K. Pelzer, K. Philippot, and B. Chaudret. 2004. Gas phase catalysis by metal nanoparticles in nanoporous alumina membranes. *Zeitschrift fur Anorganische und Allgemeine Chemie* 630 (12): 1913–1918.
46. Stair, P., C. Marshall, G. Xiong, H. Feng, M. Pellin, J. Elam, L. Curtiss, L. Iton, H. Kung, M. Kung, and H.H. Wang. 2006. Novel, uniform nanostructured catalytic membranes. *Topics in Catalysis* 39 (3): 181–186.
47. Dotzauer, D.M., J.H. Dai, L. Sun, and M.L. Bruening. 2006. Catalytic membranes prepared using layer-by-layer adsorption of polyelectrolyte/metal nanoparticle films in porous supports. *Nano Letters* 6 (10): 2268–2272.
48. Hu, M.Z., and I. Kosacki. 2010. *Design and Synthesis of Guest-Host Nanostructures to Enhance Ionic Conductivity Across Nanocomposite Membranes.* US Patent US 7,641,997 B2.
49. Pellin, M.J., J.N. Hryn, and J.W. Elam. 2005. *Catalytic Nanoporous Membranes US 2005/0065028 A1*. United States Patent Application.
50. Pellin, M.J., J.N. Hryn, and J.W. Elam. 2009. *Catalytic Nanoporous Membranes US 7,625,840 B2*. United States Patent Application.
51. Pellin, M.J., J.N. Hryn, and J.W. Elam. 2010. *Catalytic Nanoporous Membranes US 2010/0075827 A1*. U.S.P.A. Publication, Editor.
52. Routkevitch, D., and O.G. Polyakov. 2007. *Composite Membranes and Methods for Making Same US 2007/0256562 A1*. U.S.P.A. Publication, Editor. Synkera Technologies INC.
53. Guo, Y.G., J.S. Hu, H.M. Zhang, H.P. Liang, L.J. Wan, and C.L. Bai. 2005. Tin/platinum bimetallic nanotube array and its electrocatalytic activity for methanol oxidation. *Advanced Materials* 17 (6): 746–+.

54. Pokropivnyi, V.V. 2002. Two-dimensional nanocomposites: Photonic crystals and nanomembranes (review). II: Properties and applications. *Powder Metallurgy and Metal Ceramics* 41 (7–8): 369–381.
55. Paulose, M., H.E. Prakasam, O.K. Varghese, L. Peng, K.C. Popat, G.K. Mor, T.A. Desai, and C.A. Grimes. 2007. TiO_2 nanotube arrays of 1000 μm length by anodization of titanium foil: Phenol red diffusion. *Journal of Physical Chemistry C* 111 (41): 14992–14997.
56. Albu, S.P., A. Ghicov, S. Berger, H. Jha, and P. Schmuki. 2010. TiO_2 nanotube layers: Flexible and electrically active flow-through membranes. *Electrochemistry Communications* 12 (10): 1352–1355.
57. Roy, P., T. Dey, K. Lee, D. Kim, B. Fabry, and P. Schmuki. 2010. Size-selective separation of macromolecules by nanochannel titania membrane with self-cleaning (declogging) ability. *Journal of the American Chemical Society* 132 (23): 7893–7895.
58. Masuda, H., F. Hasegwa, and S. Ono. 1997. Self-ordering of cell arrangement of anodic porous alumina formed in sulfuric acid solution. *Journal of the Electrochemical Society* 144 (5): L127–L130.
59. Masuda, H., H. Yamada, M. Satoh, H. Asoh, M. Nakao, and T. Tamamura. 1997. Highly ordered nanochannel-array architecture in anodic alumina. *Applied Physics Letters* 71 (19): 2770–2772.
60. Prakasam, H.E., K. Shankar, M. Paulose, O.K. Varghese, and C.A. Grimes. 2007. A new benchmark for TiO_2 nanotube array growth by anodization. *Journal of Physical Chemistry C* 111 (20): 7235–7241.
61. Ghicov, A., and P. Schmuki. 2009. Self-ordering electrochemistry: A review on growth and functionality of TiO_2 nanotubes and other self-aligned MO_x structures. *Chemical Communications* 20: 2791–2808.
62. Paulose, M., L. Peng, K.C. Popat, O.K. Varghese, T.J. LaTempa, N.Z. Bao, T.A. Desai, and C.A. Grimes. 2008. Fabrication of mechanically robust, large area, polycrystalline nanotubular/porous TiO_2 membranes. *Journal of Membrane Science* 319 (1–2): 199–205.
63. Tonucci, R.J., B.L. Justus, A.J. Campillo, and C.E. Ford. 1992. Nanochannel array glass. *Science* 258 (5083): 783–785.
64. D'Urso, B., J.T. Simpson, and M. Kalyanaraman. 2007. Nanocone array glass. *Journal of Micromechanics and Microengineering* 17 (4): 717–721.
65. Tonucci, R.J., and G.K. Hubler. 2007. Materials characterization and nanofabrication methods—Nanochannel glass materials. In *Advances in Nanophotonics Ii*, ed. Sibilia, C., and D.S. Wiersma, 59–71. Melville: Amer Inst Physics.
66. Ma, Z.Y., L.Y. Ma, and M. Su. 2008. *Engineering* Three-dimensional micromirror arrays by fiber-drawing nanomanufacturing for solar energy conversion. *Advanced Materials* 20 (19): 3734–+.
67. Hendricks, T.R., I.N. Ivanov, D.A. Schaeffer, P.A. Menchhofer, and J.T. Simpson. 2010. Processing of loose carbon nanotubes into isolated, high density submicron channels. *Nanotechnology* 21 (11).
68. Guliants, V.V., M.A. Carreon, and Y.S. Lin. 2004. Ordered mesoporous and macroporous inorganic films and membranes. *Journal of Membrane Science* 235 (1–2): 53–72.
69. Wang, Y.H., and W.D. Zhou. 2010. A review on inorganic nanostructure self-assembly. *Journal of Nanoscience and Nanotechnology* 10 (3): 1563–1583.
70. Tsuya, N., T. Tokushima, M. Shiraki, Y. Wakui, Y. Saito, H. Nakamura, S. Hayano, A. Furugori, and M. Tanaka. 1986. Alumite disk using anordic oxidation. *IEEE Transactions on Magnetics* 22 (5): 1140–1145.
71. Jessensky, O., F. Muller, and U. Gosele. 1998. Self-organized formation of hexagonal pore structures in anodic alumina. *Journal of the Electrochemical Society* 145 (11): 3735–3740.
72. Jessensky, O., F. Muller, and U. Gosele. 1998. Self-organized formation of hexagonal pore arrays in anodic alumina. *Applied Physics Letters* 72 (10): 1173–1175.
73. Paulose, M., K. Shankar, S. Yoriya, H.E. Prakasam, O.K. Varghese, G.K. Mor, T.A. Latempa, A. Fitzgerald, and C.A. Grimes. 2006. Anodic growth of highly ordered TiO_2

nanotube arrays to 134 μm in length. *Journal of Physical Chemistry B* 110 (33): 16179–16184.
74. Macak, J.M., H. Tsuchiya, A. Ghicov, K. Yasuda, R. Hahn, S. Bauer, and P. Schmuki. 2007. TiO_2 nanotubes: Self-organized electrochemical formation, properties and applications. *Current Opinion in Solid State and Materials Science* 11 (1–2): 3–18.
75. Chen, C.C., J.H. Chen, C.G. Chao, and W.C. Say. 2005. Electrochemical characteristics of surface of titanium formed by electrolytic polishing and anodizing. *Journal of Materials Science* 40 (15): 4053–4059.
76. Chu, S.Z., S. Inoue, K. Wada, S. Hishita, and K. Kurashima. 2005. Self-organized nanoporous anodic titania films and ordered titania nanodots/nanorods on glass. *Advanced Functional Materials* 15 (8): 1343–1349.
77. Sadek, A.Z., H.D. Zheng, K. Latham, W. Wlodarski, and K. Kalantar-Zadeh. 2009. Anodization of Ti thin film deposited on ITO. *Langmuir* 25 (1): 509–514.
78. Wang, J., and Z.Q. Lin. 2009. Anodic formation of ordered TiO_2 nanotube arrays: Effects of electrolyte temperature and anodization potential. *Journal of Physical Chemistry C* 113 (10): 4026–4030.
79. Kar, A., R. Pando, and V. Subramanian. 2010. Photoelectrochemical responses of anodized titanium oxide films. *Journal of Materials Research* 25 (1): 82–88.
80. Lin, J., J.F. Chen, and X.F. Chen. 2010. Facile fabrication of free-standing TiO_2 nanotube membranes with both ends open via self-detaching anodization. *Electrochemistry Communications* 12 (8): 1062–1065.
81. Biswas, S., M. Shahjahan, M.F. Hossain, and T. Takahashi. 2010. Synthesis of thick TiO_2 nanotube arrays on transparent substrate by anodization technique. *Electrochemistry Communications* 12 (5): 668–671.
82. Park, Y.I., M. Nagai, J.D. Kim, and K. Kobayashi. 2004. Inorganic proton-conducting gel glass/porous alumina nanocomposite. *Journal of Power Sources* 137 (2): 175–182.
83. Yang, J., and Y. Jaluria. 2009. Transport processes governing the drawing of a hollow optical fiber. *Journal of Heat Transfer-Transactions of the ASME* 131 (7).
84. Yang, J., and Y. Jaluria. 2009. Feasibility and optimization of the hollow optical fiber drawing process. *International Journal of Heat and Mass Transfer* 52 (17–18): 4108–4116.
85. Mazurin, O.V.1984. *Phase Separation in Glass*. The American Ceramic Society.
86. Simpson, J.T., and B. D'Urso. 2007. *Transparent, Super-Hydrophobic, Disordered Composite Material.* US Patent number: US 2007/0184247 A1.
87. Atkins, P., and J. de Paula. 2006. *Physical Chemistry*, 8th ed. New York: W.H. Freeman and Company.

Summary and Final Thoughts

The goal for this book was to provide an overview of anisotropic nanoparticles, with particular attention paid to how properties and applications can be tuned by selecting an appropriately shaped nanoparticle. The journey began by considering many different methods to create nanoparticles with different shapes. By employing elegant self-assembly techniques or precise top-down electron beam writing, anisotropic nanoparticles of almost any conceivable shape and composition can be formed. These new nanoparticle shapes bring new challenges, and greater care must be taken when characterizing anisotropic nanoparticles. As discussed in Chap. 4, laboratory analyses very often assume the sampling of isotropic materials, and therefore, it is important to consider the orientation and built-in assumptions of both the technique and sample during characterization. Finally, the most important driving force in nanoscale research is the intended application of nanomaterials. An extensive portion of the book has been devoted to detailed studies and reports describing the advantages and implications of using anisotropic and shape-selective nanoparticles for various applications, and the topics covered in this text ranged from plasmonics to catalysis to energy generation. These different chapters illustrate how nanoparticle shape influences behavior. For example, Goldsmith et al. described how Au nanorods can enhance cardiovascular disease detection and treatment. While on the other hand, Aich et al. discussed how nanorods and other shapes of nanoparticles can have different effects on bio-toxicity that depend very strongly on their shape, size, and aggregation state. The information presented in these and other chapters show that anisotropic nanoparticles may behave in unexpected ways when placed in different conditions, and furthermore, these unique behaviors can induce profound effects on their surrounding environment, both good and bad.

The overall view of the syntheses, characterizations, and applications of anisotropic and shape-selective nanoparticles is that, while plenty of work remains, these fields are quite well-developed. Moving forward, we expect to see the persistent

S.E. Hunyadi Murph et al. (eds.), *Anisotropic and Shape-Selective Nanomaterials*,
Nanostructure Science and Technology, DOI 10.1007/978-3-319-59662-4

increase in the precision of nanomaterial production and experimental measurements that have propelled research in this field so far. Scientists and engineers will continue to push the outer boundary of the ability to confine and control matter at the nanoscale, but it is as important, if not more so, to bring the trailing edges into mainstream society. Indeed, a significant challenge for nanotechnology for the foreseeable future is the transition of the remarkable results seen in laboratories into everyday applications. The main hurdles to the widespread adoption of nanotechnologies are the difficulties of scaling up processes and reducing costs associated with nanomaterial production. Health and environmental implications of widespread use of nanomaterials will also require careful consideration and could pose another obstruction to commercial applications. It is important that the quest for the industrial adoption of nanotechnologies, shape-selective or otherwise, should not be judged solely as a profitmaking scheme, though commercial viability is paramount for any technology. Over the next century, rapidly growing urban population centers and a changing climate will place tremendous demands on a limited supply of resources and our ability to provide security and healthcare. Broad implementation of innovative technologies is our best hope to meet these challenges. Throughout this book, we have seen demonstrations in the laboratory setting that anisotropic nanoparticles can help solve these impending problems. Now, clever thinking and engineering is required to transmit what is found in the lab out into the real world.

Editors

Index

S.E. Hunyadi Murph et al. (eds.), *Anisotropic and Shape-Selective Nanomaterials*,
Nanostructure Science and Technology, DOI 10.1007/978-3-319-59662-4

Printed by Printforce, the Netherlands